Regenerative Sanitation:
A New Paradigm For Sanitation 4.0

AF729562

Regenerative Sanitation: A New Paradigm For Sanitation 4.0

First Edition

Thammarat Koottatep
Asian Institute of Technology, Bangkok, Thailand

Peter Emmanuel Cookey
Rivers State College of Health Science and Technology, Port Harcourt, Nigeria

Chongrak Polprasert
Thammasat University, Bangkok, Thailand

IWA PUBLISHING

Published by **IWA Publishing**
Alliance House
12 Caxton Street
London SW1H 0QS, UK
Telephone: +44 (0)20 7654 5500
Fax: +44 (0)20 7654 5555
Email: publications@iwap.co.uk
Web: www.iwapublishing.com

First published 2019
© 2019 IWA Publishing

Apart from any fair dealing for the purposes of research or private study, or criticism or review, as permitted under the UK Copyright, Designs and Patents Act (1998), no part of this publication may be reproduced, stored or transmitted in any form or by any means, without the prior permission in writing of the publisher, or, in the case of photographic reproduction, in accordance with the terms of licenses issued by the Copyright Licensing Agency in the UK, or in accordance with the terms of licenses issued by the appropriate reproduction rights organization outside the UK. Enquiries concerning reproduction outside the terms stated here should be sent to IWA Publishing at the address printed above.

The publisher makes no representation, express or implied, with regard to the accuracy of the information contained in this book and cannot accept any legal responsibility or liability for errors or omissions that may be made.

Disclaimer
The information provided and the opinions given in this publication are not necessarily those of IWA and should not be acted upon without independent consideration and professional advice. IWA and the Editors and Authors will not accept responsibility for any loss or damage suffered by any person acting or refraining from acting upon any material contained in this publication.

British Library Cataloguing in Publication Data
A CIP catalogue record for this book is available from the British Library

ISBN: 9781780409672 (print)
ISBN: 9781780409689 (eBook)
ISBN: 9781780409696 (ePub)

Contents

About the authors

Thammarat Koottatep is a Professor of the Environmental Engineering Management of the Asian Institute of Technology (AIT), Thailand. He is an internationally recognized authority on wastewater treatment technology and management. His major scholarly contributions include publications of more than 50 refereed international journal papers, three books and nine book chapters. He has invented sanitation technologies, one of which is patented to his credit and several that are filing. He has jointly developed a professional master's degree program in Water Technology and Management and mentored 16 doctoral students. He has secured significant funded projects including research and training grants and, most notably, Bill & Melinda Gates Foundation grants for an AIT project, 'Decentralized Wastewater Management in Developing Countries: Design, Operation and Monitoring'. He has contributed significantly to capacity building in wastewater management in Thailand and abroad, including capacity strengthening for policymakers. Professor Koottatep won the AIT 2016 Research Leader Award on the basis of his innovative and practical approach to research, focus and ability to carry out large-scale research projects.

Peter Emmanuel Cookey is a Chief Lecturer and sanitation professional with the School of Environmental Health, Rivers State College of Health Science and Technology, Port Harcourt, Nigeria. He has an interdisciplinary academic background in environmental health, physiology and pharmacology, environmental microbiology, environment science and engineering management, governance and environmental management. His working career traverses occupational domains such as governance, regulations, inspection, monitoring, research, academics, consulting and training for private and public entities, advising governments on policies, legislation and reforms in sanitation and water as well as serving on global sanitation bodies. Peter was an ardent and active member of the Water Supply and Sanitation Collaborative Council (WSSCC), Geneva (representing middle, northern and western African countries) as well as a pioneer member of the International Standard Development Committee of the Alliance for Water Stewardship (AWS) team that developed and delivered the global water stewardship standards. As an environmental health professional primarily focused on sanitation and water, he taught and mentored future sanitation professionals for seventeen years and conducted related research in the

Niger Delta during that period. Peter facilitated the Rivers State Water and Sanitation Sector reforms, which drew down significant funding from the World Bank and African Development Bank as well as full government support. He was also co-chair for the drafting of proposed policy statements and legislative documents with the Rivers State Ministry of Water Resources and Rural Development. As a member of the Greater Port Harcourt Development Committee, he was directly involved in the design and development of the Greater Port Harcourt Mega-city water and sanitation projects among others. All of these experiences have given this erudite scholar and seeker in-depth insights into the core issues of sanitation and a rich understanding of most thorny issues.

Chongrak Polprasert is currently Professor of the Civil/Environmental Engineering at the Faculty of Engineering, Thammasat University and Professor Emeritus at the Asian Institute of Technology (AIT), Thailand. He received his PhD in civil/environmental engineering under the Fulbright scholarship from the University of Washington, Seattle, U.S.A. Professor Polprasert was a faculty member at the Asian Institute of Technology (AIT) from 1997 to 2009, held the AEON Group Chair of Environmental Engineering from 1991 to 1995 and was Dean of the School of Environment, Resources and Development from 1996 to 2005. He was Director and Professor at the Sirindhorn International Institute of Technology (SIIT) of Thammasat University, Thailand, from 2009 to 2012. His research over the past 35 years has been in the areas of sanitation, waste reuse and recycling and hazardous waste engineering and management.

Professor Polprasert has served in the U.S. National Academy of Sciences panel on productive utilization of wastes in developing countries and the Water Environment Federation task force on natural systems for wastewater treatment. He is a recipient of the Biwako Prize for Ecology (Japan), Outstanding Scientist award (Thailand), Thailand Research Fund publication award, Outstanding Researcher award in engineering and industry (Thailand), an elected fellow of the Graduate School of Engineering, the University of Tokyo, Japan, and a fellow of the Royal Society of Thailand. He has been invited to serve as Visiting Professor/Scholar at the UNESCO-IHE Institute for Water Education, Delft, the Netherlands; Lulea University, Sweden; Tohoku University and Kyoto University, Japan, and occasionally acts as a consultant to both private and public organizations internationally.

Preface

The concept of regenerative sanitation (ReGenSan) was conceived from a debate on the need to expand the practicalities of sustainable sanitation (SuSan) in order to avoid sanitation falling behind again at the close of the SDG no. 6 deadline in 2030. The various challenges that sanitation faced, from the first UN international drinking water and sanitation decade, the 1970s, to the MDGs (Millennium Development Goals) era, were explored. One major recurring highlight was that sanitation has received much hype, but has never been given its own core focus. It gets lost in water, health, hygiene, poverty reduction, gender issues and technology and in the end concerns addressed are often not sanitation-specific; so much so that the solutions sought and provided bypass the existing sanitation challenges. To make things even more complicated, such solutions were often basically restricted to linear technologies, which created more ecological and social problems. Even when sanitation is given due consideration, it is regarded as a culprit that needs to be eliminated or removed out of sight. One thing is certain nowadays: sanitation is so much more complex and dynamic than was originally conceived and deserves a more holistic and integrated approach.

Questions such as: what then is the way forward for sanitation in this SDG era that will lead to universal access to adequate and equitable sanitation? How could we present sanitation without losing its essence? How can we guide research and development towards holistic, integrated and systemic solutions? are explored and resolved in this book. With so many years in practice, teaching and research, the authors have acquired insights into the challenges, shortfalls and gains of sanitation practice and research. Professor Koottatep spent over 18 years searching for solutions to sanitation's myriad challenges, particularly wastewater and faecal sludge management, and he ended up with more questions than answers as the situation globally seemed to defy all efforts. The poor conclusion at the end of the MDGs stirred up concerns and raised red flags about the way sanitation was considered and practised. As a teacher, trainer and researcher in so many regions and topics, he began to wonder if there was not really more to sanitation than the dynamic complexities that are inherent in generic and contextual cases. Several partnerships with Professor Polprasert (one of the first researchers to directly investigate the components of excreta) established his concerns as being concrete and shared by others.

Professor Polpraser had committed 40 years of his working life trying to gain insights into the secrets of the by-products of human digestion (urine and excreta). He researched extensively the recycling potentials and possible reuse of sanitation matter, as well as the best way to collect and treat organic waste. As a pioneer of sanitation, particularly in the tropics, Asia and the developing world, he, too, was concerned with the persistent global dilemma. With studies and reports raising more questions on how to reduce the impact of sanitation on public health and environmental quality, he investigated along with Professor Koottatep the use of constructed wetlands for septage and faecal sludge treatment and recycling options . In 1989, he gathered his findings in a book titled *Organic Waste Recycling* (published by IWA Publishing, UK) and three revised editions followed within a span of 18 years, all of which considered technology and management in organic waste recycling. It was during a brainstorming session while working on the fourth edition (*Organic Waste Recycling: Technology, Management and Sustainability*) that Professor Koottatep posed the question to Dr Cookey of the possibility of using sanitation to regenerate communities recovering from disaster or facing food crisis.

Dr Cookey, with 24 years directly in the sanitation sector (as chief environmental health officer in sanitation at state level, lecturer/trainer, consultant, researcher and global player), was quick to latch on to the idea with zeal and passion. He spent seven years on the steering committee of the Water Supply and Sanitation Collaborative Council (WSSCC), Geneva (representing middle, northern and western Africa countries), trying to reconcile the challenges in dealings with policymakers, regulators, users and intervention/aid agencies with the stories from other regions across the world. He realized that his country's situation was only peculiar as it concerned place and context. He also questioned the fact that the main workforce, especially in Africa and former British colonies, that are responsible for monitoring and regulating sanitation practices (environmental health officers) are mostly left out of sanitation debates and interventions. He concluded that there was something wrong with the way sanitation was perceived, practised and taught. And so, an innocent question created the tripod union that started a revolution to explore and review the arguments; thus Regenerative Sanitation was born.

This book proposes Regenerative Sanitation as the next era of sanitation (Sanitation 4.0); indicating a paradigm shift from techno/toilet focus in solutions development to more focus on benefits, safety, functionality, replicability, reliability and moving from pilots to marketplace. And if there is anything to be emphasized in this book, it is that sanitation is so much more than the disposal and treatment of human waste. It is too big not to be given its own individual focus in the global development agenda because it deals with unavoidable human physiological activities that occur several times a day. It is time to rethink sanitation and change the narrative towards more progressive trajectories. Sanitation is as huge an issue as can be imagined (seven billion people passing faeces more than twice a day and urine at least four times a day, plus all the other aspects); so redirecting the ship towards resource recovery and reuse rather than just amelioration is needed. Sanitation's impact on food security, livelihood support, urban regeneration, rural development and even local economies, directly or indirectly, is enormous. Resources are also generated for fertilizer, animal feed, construction, horticulture etc.

Unfortunately, over the years sanitation has been mistreated as unfashionable and thought of as increasing disease burden, mortality rate, environmental risks and economic losses – there is much more talk of the burden of sanitation than of its gains. And even though there has been a recent rise in investigations into recovery and reuse of human waste, the SDGs still propagate public health and environmental protection while keeping sanitation in the shadows of water and health/hygiene, rather than harnessing its rich resources. Any progressive movement targeting 2030 will need to shift and change the way we consider sanitation. It is not just a transition from technology and toilet focus, but a complete overhaul of sanitation concepts to include users, contexts and locations and marketplace and to target those at the base-of-the-pyramid (BoP) in society. Therefore, this book aims to stir up a new way of thinking about sanitation for practitioners, teachers/trainers, policymakers, regulators, researchers and students (and maybe to make the profession attractive to the younger and incoming workforce).

The structure of the book is inter- and trans- disciplinary, to show that sanitation management cannot be restricted to one field or discipline alone. It is a collaborative assignment that involves integrated processes across boundary lines and stakeholders. Therefore, engineers, scientists and technologists will have to share the stage with psychologists, sociologists, philosophers, anthropologists, political scientists, economists and others to provide effective and practical solutions; with the users/customers at the centre. The concept developed and proposed moves away from a compendium of technologies so as to provide a foundation for the study of sanitation. It is written as both a textbook for students and a practitioner's guide. A conceptual framework and a regenerative sanitation theory with principles are presented and expanded upon, while each chapter provides exercises and recommendations for future research. In addition, there is a simple historical review of how Sanitation 4.0 was conceived in Chapter 1, further expanded in Chapter 7.

This work was written within a year and relied on the direct and indirect support and assistance of many people. We are grateful to the researchers and staff at the Naturally Acceptable and Technological Sustainable (NATS) centre, post-doctoral researchers and graduate students at the Asian Institute of Technology, Bangkok, Thailand for the constructive suggestions on the concept and contents of the book. Significant contributions by Mayowa Abiodun Peter-Cookey, EarthWatch Research Institute, Port Harcourt, Nigeria and Tooba Masood, Asian Institute of Technology, Bangkok, Thailand, are greatly acknowledged in the overall improvement of the quality of the book.

We would like to express our sincere appreciation to Mark Hammond of IWA Publishing, London, and the entire IWA staff for their professional support in the publication of this first edition.

Thammarat Koottatep
Peter Emmanuel Cookey
Chongrak Polprasert
Bangkok, Thailand
February 2019

Abbreviations and symbols

ABS	American Bureau of Shipping
AD	Anaerobic Digestion
ADB	Asian Development Bank
AGCA	Associated General Contractors of America
AIT	Asian Institute of Technology
ANSI	American National Standards Institute
ApT	Appropriate Technology
ARSO	African Organization for Standardization
AS	Standards Australia
ASAL	Arid and Semi-arid Lands
ASCE	American Society of Civil Engineers
ASHRAE	American Society of Heating, Refrigerating and Air Conditioning Engineers
ASTM	American Society for Testing Materials
AWE	Aquaculture-Wetland Ecosystem
AWF	Alliance for Water Efficiency
BaB	Biomimicry and Biophilia
BGC	Bio-Geo-Chemical
BMGF	Bill & Melinda Gates Foundation
BOD	Biochemical Oxygen Demand
BoP	Base-of-the-Pyramid
BPDST	Biogas Project For Domestic Sewage Treatment
BPL	Below Poverty Line
BSI	British Standards Institution

CalTech	California Institute of Technology, USA
CAWST	Centre for Affordable Water and Sanitation Technology
CBOs	Community Based Organizations
CCTV	Closed Circuit Television
CEAMA	Morelos State Water and Environmental Commission
CEN	Comité Européen de Normalisation (European Committee for Standardization)
CHP	Combined Heat and Power Generation
CiM	Continuous improvement
CLTS	Community Led-Total Sanitation
CoCT	City of Cape Town
ConSan	Conventional Sanitation
COPANT	Pan American Standards Commission
CPDA	Clay Pipe Development Association
CRSP	Central Rural Sanitation Programme
CSF	Centralized Sanitation Facilities
CSIR	Council for Scientific and Industrial Research
CSO	Combined Sewer Overflow
CSTEP	Center for Study of Science, Technology and Policy
CWTRC	California Wastewater Training & Research Center
DAF	Dissolved Air Flotation
DALY	Disability Adjusted Life Year
DBD	Demand-Based Designs
DCS	Distributed Centralized Systems
DeBoP	Design for the Base-of-the-Pyramid
DeN	Design for Nature
DeRaR	Design For Recovery And Reuse
DeSaP	Design for Service and Products
DESAR	Decentralization Sanitation And Reuse
DFID	Department for International Development
DGTSs	Domestic Greywater Treatment Systems
DOHWA	Department of Health Western Australia
DSF	Decentralized Sanitation Facilities
DSF	Design For Service Framework
DPR	Direct Potable Reuse
DWA	German Association for Water, Wastewater and Waste
DWAF	Department of Water Affairs and Forestry, Pretoria, South Africa
DWWTS	Domestic Wastewater Treatment Systems
EcoSan	Ecological Sanitation
EDIE	Environmental Data Interactive Exchange

EFP	Excreta-Fertilizer Products
EIDT	Existing Improved Design Technology
ELI	Environmental Law Institute
EMF	Ellen MacArthur Foundation
EU	European Union
FaST	Functionalism and System Theory
FLT	Fresh Life Toilets
FNSS	Formal National Sanitation Standards
FPG	Fit-for-Purpose Governance
FS	Faecal Sludge
FSM	Faecal Sludge Management
FSS	Formal Sanitation Standards
FWFD	Faecal Waste Flow Diagram
FWS	Free Water Surface
GDP	Gross Domestic Products
GIS	Geographic Information System
GLASS	Global Assessment and Analysis of Sanitation and Water
GNP	Gross National Product
GoF	Governance Function
GoI	Government of India
GPS	Geographic Positioning System
GTZ	Deutsche Gesellschaft für Technische Zusammenarbeit
HES	Health Education Service
HET	High Efficiency Toilets
HS	Hybridized solutions
HTC	Hydrothermal Carbonization
IAs	Institutional Arrangements
IDWSSD	International Drinking Water Supply and Sanitation Decade
IFSVC	Integrated Functional Sanitation Service Chain
IMA	Institute of Management Accountants
IPR	Indirect Potable Reuse
ISEAL	International Social and Environmental Accreditation and Labelling
ISO	International Organization for Standardization
ISS	Informal Sanitation Standards
IWA	International Water Association
JECES	Japan Education Center of Environmental Sanitation
JMP	Joint Monitoring Programme
JSWA	Japan Sewage Works Association
KaS	Knowledge and Skills

LCA	Lifecycle Assessment
LCCA	LifeCycle Costs Approach
LIXIL	Link to Good Living
LoSWaS	Local Scale Wastewater Services
MDGs	Millennium Development Goals
MFA	Material Flow Analysis
MINALOC	Ministry of Local Government, Rwanda
MINECOFIN	Ministry of Finance and Economic Planning, Rwanda
MINEDUC	Ministry of Education, Rwanda
MININFRA	Ministry of Infrastructure, Rwanda
MINISANTE	Ministry of Health, Rwanda
MLIT	Ministry of Land, Infrastructure, Transport and Tourism, Japan
MMA	Modernized Mixtures Approach
MOA	Ministry of Agriculture, China
MOE	Ministry of the Environment, Japan
NDHS	Nigeria Demographic and Health Survey
NEWEL	Nutrients-Energy-Water-Environment-Land
NGO	Non-Governmental Organization
NISER	Nigerian Institute of Social and Economic Research
NoDT	Nouveau Design Technology
NOWRA	National Onsite Wastewater Recycling Association
NSBs	National Standard Bodies
NSF	National Sanitation Foundation International
NSF	National Science Foundation
NToB	No-Transfer-of-Burden
NZS	New Zealand Standards
OD	Open Defecation
ODF	Open Defecation Free
OECD	Organisation of Economic Cooperation and Development
OFWAT	Office of Water Services
OHS	Occupational Health And Safety
ORs	Operational Responsibilities
OSS	On-site Sanitation Systems
OWTS	On-site Wastewater Treatment Systems
PaP	Participation and Partnership
PaS	Place and Scale
PASC	Pacific Area Standards Congress
PDSS	President Deputy of Strategic Supervision
PLUM	Passive Latrine Use Monitor

PSC	Psycho-Socio-Cultural
PSEP	Psycho-Socio-Ecophilia
PSS	Public Sanitation Service
RA	Rural Agriculture
RaSR	Recycling and Safe Reuse
RBP	Resource-based Paradigm
RD	Rural Development
RDDT	Regenerative Development and Design Theory
ReDT	Restorative Design Tech
ReGenSan	Regenerative Sanitation
ReGenSanC	Regenerative Sanitation Concept
ReGenSanP	Regenerative Sanitation Paradigm
ReGenSanT	Regenerative Sanitation Theory
ReS	Resource SubSystem
RFID	Radio Frequency Identification Technology (RFID)
RoDF	Rehabilitation of Dysfunctional Facilities
ROS	Resource-Oriented Sanitation
RRR	Resource Recovery and Reuse
SABS	South African Bureau of Standards
SaG	Sanitation Governance
SaPs	Sanitation Paradigms
SARSO	South Asian Regional Standards Organization
SC	Service Chain
SCF	Sanitation Conceptual Framework
SCR	Struvite Crystallizer Reactor
SD	Sustainable Development
SDC	Swiss Agency for Development and Cooperation
SDE	Sanitation-Derived Energy
SDO	Standard Development Organizations
SDFs	Sanitation-Derived Fertilizers
SDG	Sustainable Development Goals
SDNs	Sanitation-Derived Nutrients
SDPs	Sanitation-Derived Products
SE	Socio-Economic
SEI	Stockholm Environment Institute
SER	Society for Ecological Restoration
SES	Social-Ecological SubSystem
SF	Subsurface Flow
SFD	Shit Flow Diagram

SHI	Systemic Holistic Integration
SIs	Sanitation Infrastructure Standards
SIS	Sanitation Infrastructure Standardization
SLTS	School-Led Total Sanitation
SMEs	Sanitation Management Entities
SMS	Short Message Service
SON	Standards Organisation of Nigeria
SS	Sewage Sludge
SSA	Sub-Saharan Africa
SSC	Sanitation Service Chain
SSC-RaMs	Sanitation Service Chain Raw Materials
SSD	Source Separation And Decentralization
SSDC	Sanitation Service Delivery Chain
SSF	Satellite Sanitation Facilities
SSL	Sanitation Service Ladder
SOM	Soil Organic Matter
STE	Socio-Techno-Ecological
STED	Septic Tank Effluent Drainage
STR	Social-Technical Regimes
SuSan	Sustainable Sanitation
SuSanA	Sustainable Sanitation Alliance
SVC	Sanitation Value Chain
SVCA	Sanitation Value Chain Analysis
SWA	Sanitation and Water for All
TBC	Toilet Board Coalition
TeS	Technology SubSystem
TISI	Thai Industrial Standards Institute
TOP	Top-of-the-Pyramid
TSC	Total Sanitation Campaign
UD	Urine Diversion
UDCT	Urine-Diversion-Composting-Toilet
UDDT	Urine-Diversion Dry Toilets
UDT	Urine-Diverting Toilets
UF/RO	Ultrafiltration/Reverse Osmosis
UN	United Nations
UNCED	United Nations Conference on Environment and Development
UNDP	United Nations Development Programme
UN-Water	United Nations Water
UNESC	United Nations Economic and Social Council

UNESCO	United Nations Educational, Scientific and Cultural Organization
UNGA	United Nations General Assembly
UNHRC	United Nations Human Rights Council
UNICEF	United Nations Children Fund
UR	Urban Regeneration
USAID	United States Agency for International Development
USDE	United States Department of Energy
USDOE	U.S. Department of Energy
USEPA	United States Environmental Protection Agency
USLG	Urban Sustainability Learning Group
UWWTD	Urban Wastewater Treatment Directives
VC	Value Chain
VC	Vacuum Closet
VF	Vertical-flow
VIPs	Ventilated Improved Pits
WASAC	Water and Sanitation Corporation, Rwanda
WCED	World Commission on Environment and Development
WCT	Waterless Composting Toilet
WECD	World Commission on Environment and Development
WEDC	Water, Engineering and Development Centre
WEF	Water and Environment Federation
WERF	Water Environment Research Foundation
WE&RF	Water Environment & Reuse Foundation
WHO	World Health Organization
WSDHWMP	Washington State Department of Health Wastewater Management Program
WSP	Water and Sanitation Program
WSSD	World Summit on Sustainable Development
WWAP	United Nations World Water Assessment Programme
WWPT	Wastewater Treatment Plants

Atomic weight and number of elements

Element	Symbol	Atomic Number	Atomic Weight
Aluminium	Al	13	26.97
Antimony	Sb	51	121.76
Argon	A	18	39.94
Arsenic	As	33	74.91
Barium	Ba	56	137.36
Beryllium	Be	4	9.02
Bismuth	Bi	83	209.00
Boron	B	5	10.82
Bromine	Br	35	79.92
Cadmium	Cd	48	112.41
Calcium	Ca	20	40.08
Carbon	C	6	12.01
Cerium	Ce	58	140.13
Caesium	Cs	55	132.91
Chlorine	Cl	17	35.46
Chromium	Cr	24	52.01
Cobalt	Co	27	58.94
Copper	Cu	29	63.57
Dysprosium	Dy	66	162.46
Erbium	Er	68	167.20

(*Continued*)

Element	Symbol	Atomic Number	Atomic Weight
Europium	Eu	63	152.00
Fluorine	F	9	19.00
Gadolinium	Cd	64	156.90
Gallium	Ga	31	69.72
Germanium	Ce	32	72.60
Gold	Au	79	197.20
Hafnium	Hf	72	178.60
Helium	He	2	4.00
Holmium	Ho	67	164.94
Hydrogen	H	1	1.01
Indium	In	49	114.76
Iodine	I	53	126.92
Iridium	Ir	77	193.1
Iron	Fe	26	55.84
Krypton	Kr	36	83.70
Lanthanum	La	57	138.92
Lead	Pb	82	207.21
Lithium	Li	3	6.94
Lutecium	Lu	71	175.00
Magnesium	Mg	12	24.32
Manganese	Mn	25	54.93
Mercury	Hg	80	200.61
Molybdenum	Mo	42	95.95
Neodymium	Nd	60	144.27
Neon	Ne	10	20.18
Nickel	Ni	28	58.69
Niobium	Nb	41	92.91
Nitrogen	N	7	14.01
Osmium	Os	76	190.20
Oxygen	O	8	16.00
Palladium	Pd	46	106.70
Phosphorus	P	15	30.98
Platinum	Pt	78	195.23
Potassium	K	19	39.10
Praseodymium	Pr	59	140.92
Protactinium	Pa	91	231.00

(*Continued*)

Element	Symbol	Atomic Number	Atomic Weight
Radium	Ra	88	226.05
Radon	Rn	86	222.00
Rhenium	Re	75	186.31
Rhodium	Rh	45	102.91
Rubidium	Rb	37	85.48
Ruthenium	Ru	44	101.70
Samarium	Sm	62	150.43
Scandium	Sc	21	45.10
Selenium	Se	34	78.96
Silicon	Si	14	28.06
Silver	Ag	47	107.88
Sodium	Na	11	22.10
Strontium	Sr	38	87.63
Sulphur	S	16	32.06
Tantalum	Ta	73	180.88
Tellurium	Te	52	127.61
Terbium	Tb	65	159.20
Thallium	Tl	81	204.39
Thorium	Th	90	232.10
Thulium	Tm	69	169.40
Tin	Sn	50	118.70
Titanium	Ti	22	47.90
Tungsten	W	74	183.92
Uranium	U	92	238.07
Vanadium	V	23	50.95
Xenon	Xe	54	131.30
Ytterbium	Yb	70	173.04
Yttrium	Y	39	88.92
Zinc	Zn	30	65.38
Zirconium	Zr	40	91.22

Source: www.petrik.com.

Conversion factors for SI units

Measurement	Unit	Multiplier	SI unit
Area	in^2	6.45	cm^2
	ft^2	0.09	m^2
	yd^2	0.84	m^2
	acre	0.40	ha
	acre	4047	m^2
Flow	ft^3/s	0.03	m^3/s
	gal/min	0.06	L/s
	gal/d	4.4×10^{-5}	L/s
		0.04	m^3/s
	mgd	43.81	L/s
	ft^3/m	0.47	L/s
Heat	Btu	1.06	kJ
	Btu	2.928×10^{-4}	kWh
	Btu	0.25	kcal
	hp	745.70	watt
	hp	10.70	kcal/min
Length	in	2.54	cm
	in	25.40	mm
	ft	30.48	cm
	ft	0.30	m
	yd	0.91	m

(*Continued*)

Measurement	Unit	Multiplier	SI unit
	mile	1.61	km
Load/Pressure	lb/ft^2	4.88	kg/m^2
	atmosphere	10,333	kg/m^2
	lb/in^2	703	kg/m^2
Mass	lb	0.45	kg
Temperature	°**F**	**(°F–32)/1.8**	°**C**

Chapter 1
Introduction

'the significant problems we face cannot be solved at the same level of thinking we were at when we created them'

Albert Einstein

1.1 BACKGROUND

A quick run through history will reveal that sanitation management evolved from the Industrial Revolution and mimicked the linear economic model of production and consumption of the era to develop linear conventional sanitation technological solutions. These linear models of sanitation management have achieved a great deal of success over the years, but they seem to be increasingly challenged by indications that deeper changes in the operating systems are necessary (Ellen MacArthur Foundation, 2012, 2013). The efficiency and effectiveness of these resource-intensive linear sanitation systems (whether sewer or non-sewer) have been questioned and thus created rising concerns about sustainability as well as the ability to actually expand access and improve service. These are huge concerns as the current global statistics for sanitation indicate that a wide margin still exists between desired results and reality (Fam & Mitchell, 2012; Hutton & Varughese, 2016; Lopes *et al.*, 2012). The MDGs aimed to halve the 2.4 billion people without adequate sanitation, but today we still have an estimated 2.3 billion people without improved sanitation facilities and about 892 million practising open defecation (Giné-Garriga *et al.*, 2017; Joint Monitoring Programme (JMP), 2015; O'Reilly *et al.*, 2017; WHO & UNICEF, 2017). In fifteen years of the MDGs, however, all efforts and resources did not even come near halving the 2.4 billion and now the Sustainable Development Goals (SDGs) aspire for universal access and improved services across the globe for adequate and equitable sanitation (WHO & UNICEF, 2017). But then, these could be just lofty ideals as the SDGs do not actually give particular attention to sanitation, which seems to be considered an afterthought to water and wastewater.

© IWA Publishing 2019. Regenerative Sanitation: A New Paradigm for Sanitation 4.0
Editor(s): Thammarat Koottatep, Peter Emmanuel Cookey and Chongrak Polprasert
doi: 10.2166/9781780409689_0001

The risk here is that sanitation could again be overlooked by the water targets, as was the case for the MDGs (Andersson & Minoia, 2017). It is expedient that sanitation be addressed on its own core platform and extracted from the clutches of water and hygiene (health) because managing sanitation and providing required services are truly not the same as managing water and hygiene. Doing things the same way will always yield the same results as 15 years of the MDG has revealed, with over 2.3 billion people still without basic sanitation. The risks are far-reaching too with grave impacts on water quality, health and the environment (note: controlling these risks is not the same as providing sanitation solutions, but rather protection for water, health and the environment from sanitation as well as other practices). It is, therefore, essential to push forward in the direction of sanitation solutions that yield increased efficiency and effectiveness, reduced costs and risks, livelihood support and ecological sustainability as well as accelerated drive towards expanded access and improved services by 2030.

Clearly, sanitation needs to head away from the conventional approach of techno-focus that has been predominant over the years and during the MDGs era without much-needed success. It will not be easy to transition from these rigid, locked-in conventional sanitation technologies characterized by incremental progression along existing trajectories rather than radical innovations, but transformation to holistic, integrated and systemic solutions that address the whole spectrum of sanitation is essential (Fam & Mitchell, 2012; Lopes *et al.*, 2012). Technological innovations are not enough to bring the progressive results desired in the sanitation arena, but mutually reinforcing institutional, psycho-social, ecological, cultural and resource-oriented transformations is required (Geels, 2005). There is overwhelming evidence that focusing on technology or infrastructure (small and large) alone will not provide needed solutions as many 'failed technologies' in sanitation did not fail due to technical deficiencies only, but also, and even more so, on system misfit in terms of scale, or the social, geographical, cultural or even economic contexts in which they were implemented (van Vliet *et al.*, 2011). Studies have shown how psycho-social-economic, cultural conditions and lock-in effects prevent uptake of new technologies due to their lack of integration into the existing technical and institutional structure. This is because new technology might negatively impact the existing sociotechnical systems if it contradicts other interests and could lead to a dead end (Hiessl *et al.*, 2003; Panebianco & Pahl-Wostl, 2006; van Vliet *et al.*, 2011). Some examples of challenges in the absence of systemic-holistic-integrated perspectives in sanitation programming are illustrated in Boxes 1.1 to 1.4.

BOX 1.1 UGANDA

An Agenda 21 project in Jinja involving community latrines with a digester producing biogas as an energy source for domestic cooking and lighting failed soon after its start in 2001. The households complained about bad odours and were reported to be reluctant to use their own human waste for cooking, and so eventually someone removed the pipes. The project failure can hardly be attributed to the hardware itself; rather the way it was implemented as well as prevailing social and cultural conditions (UN-Habitat, 2002; van Vliet *et al.*, 2011).

BOX 1.2 SOUTH AFRICA

Conventional gravity sewerage installed in Cape Town's informal settlements proved difficult due to various social, built and natural environment challenges (Sonnenberg, 2013; Taing *et al.*, 2013). In order to resolve this problem, the Officials of the City of Cape Town (CoCT) decided to explore vacuum sewer technology as an alternative solution. However, the vacuum sewer failed immediately after its commissioning in early 2009 and so residents demanded its removal and that its flush toilets be replaced with previously phased-out bucket toilets (Taing, 2017). The Vacuum Sewer System (VSS) was hampered from inception by users' and service providers' (CoCT's) poor management and governance (DWAF, 2008a, b; Taing *et al.*, 2013).

BOX 1.3 THE NETHERLANDS

The city of Wageningen in 2002 suffered the same fate as Cape Town, South Africa. It became apparent to both the municipality and project proponents, even before allocation of the accommodation to residents, that all was not well with the vacuum sewerage and anaerobic treatment system in a new residential site. At the start of the project, many aspects of the experiment with vacuum toilets and anaerobic treatment systems were still unresolved, from seemingly trivial issues like the shape and colour of toilets to more crucial aspects of management and transfer of technology. Both the municipality and project developers decided in 2003 to withdraw, causing the project to fail even before the technology could be installed (van Vliet, 2006; van Vliet *et al.*, 2011).

BOX 1.4 SWEDEN

The Urine Diversion (UD) toilet systems in Sweden revealed the significance of human-centred issues relating to pyschosociocultural acceptance and governance support as sine qua non for the process of sanitation access expansion and service improvement. The Swedish experience, which installed about 135,000 UD units over the last two decades (Kvarnström *et al.*, 2006), was the primary source of influence and inspiration for UD projects globally. The engagement of end users as critical stakeholders was a key determinant in the successful adoption of the UD systems, as the most enduring UD systems in Sweden are those that were collaboratively organized and/or managed by end users (e.g. cooperative housing estates, eco-villages and private summer houses) as well as the strong endorsement of the coastal municipality of Tanum, the first to mandate UD systems for all new housing development (Fam & Mitchell, 2012). However, while the concept of UD is gaining global momentum, the advancement of UD within Sweden (the birthplace of the modern UD toilet) is slowing down since its peak in the mid-1990s with a number of high-profile pilot projects dismantled and two manufacturers of UD systems ceasing production. This decline is due to dwindling support from Tanum Municipality due to planning and development of large scale wastewater treatment plants connecting households to centralized sewers, which reduced the appeal for small-scale systems such as UD for nutrient recovery (Fam & Mitchell, 2012).

Current sanitation infrastructural designs are at odds with today's environmental, economic and social sustainability paradigms (Apul, 2010) as well as meeting the aspiration of the goals and targets of SDG no. 6. Developed countries are just as affected as the conventional centralized linear systems that were substantially revamped after World War Two are close to or past their useful design lifespan (usually considered to be 50–60 years) and need to undergo major rehabilitation/refurbishment (Capodaglio, 2017; van Vliet *et al.*, 2011). Thus, developed countries, due to resource constraint, are confronted with the twin challenges of retrofitting and upgrading ageing and decaying sanitation infrastructures (Ashley *et al.*, 2011; Bracken *et al.*, 2006; Brands, 2014; Butler & Parkinson, 1997; Delleur, 2003; Drangert, 1998; Wilsenach *et al.*, 2003; WWAP, 2017). To address the contemporary sanitation dilemma, systemic-holistic-integrated strategies must be considered across the sanitation spectrum and interlinkages to agriculture, carbon cycles, environment and health, resource reuse and recovery, psycho-social, economic and cultural factors, ecological sustainability, institutions and governance, and the myriad roles of stakeholders in the process of sanitation (Brands, 2014; Marshall & Farahbakhsh, 2013; van Vliet *et al.*, 2011).

In the face of all these, a new approach to sanitation management and solutions design named Regenerative Sanitation (ReGenSan) is proposed. ReGenSan aims to provide the foundation for a new way of thinking by adapting the principles of regeneration and functional-living-system theory to the specific context of sanitation. To this effect, *sanitation is regenerative when it integrates the psycho-socio-ecological elements (place, scale, culture/religion, status, economy and governance), resource recovery (reuse, recycling), ecosystem (geographical and ecological) and technological elements (storage, collection, transport, treatment, reuse) perspectives into one systemic whole for the rejuvenation and revitalization of human excreta/urine management in a manner that mimics and contributes to nature's system.* Therefore, ReGenSan could provide the opportunity to systematically address complex sanitation challenges from multi- and transdisciplinary perspectives, as it approaches solution provision from a holistic-integrated perspective with an ecological systemic worldview. ReGenSan is designed to be place-/scale-based where comprehensive sanitation solutions and management are provided within what is termed 'sani-sheds' and with the intent to mimic nature and be psycho-socially acceptable and affordable, livelihood-supportive and technically appropriate as well as enabled by specific governance mechanisms. It aims to offer strategies that support the development of novel and place-peculiar sanitation facilities, restoration and upgrade of old, dysfunctional and unimproved sanitation infrastructure, ensuring continuous maintenance of existing improved sanitation facilities as well as recovery and reuse of valuable resources from excreta, urine and wastewater management in order to expand access and improve services. This is in consonance with the idea of the circular economy, which is redefining products and services to eliminate waste, while minimizing negative environmental impacts (Capodaglio, 2017; EMF, 2012, 2013; IWA, 2016; TBC, 2016) and contributing in a greater measure to resource conservation (Lia Buarque, 2012). Thus, technology will no longer just be about storage, collection, transport, treatment and reuse; focus will be on the most reliable, sensible, acceptable, appropriate and resourceful ways to implement technology.

ReGenSan aspires to ensure that sanitation systems do not transfer ecological and public health burdens and do not mix human waste with other wastes, but align sanitation facilities with ecological processes to ensure 'zero' or 'near minimal' discharge, and ensure that sanitation systems provide net benefits to the people and environment. It is common knowledge that current sanitation practices have large negative effects on ecosystem services, especially in the urban centres of developing and developed countries. But sanitation systems can be designed to benefit particular ecosystems in order to mitigate or reduce such negative impacts (Graham, 2003; Rees, 1999; Zari, 2012). Moreover, ReGenSan could contribute positively to ecosystem services by ensuring benefit-sharing and focused reduction on environmental and public health impacts. In addition, regenerative goals demand that processed-waste volumes do not exceed the capacity of the environment, thus ensuring that there is no-transfer-of-burden (NToB) (Lyle, 1994).

1.2 PERSPECTIVES OF THE REGENERATIVE PHENOMENA

The design of ReGenSan was drawn from interdisciplinary and multidisciplinary analysis of literature on regenerative science, biology, medicine, development, designs, built environment and architecture, industrial, ecology, agriculture, economics and sustainability. The regenerative concept has been used extensively in the fields of biological and medical sciences, urban development, agriculture, economics, architectural design and the built environment, but not yet in sanitation management. ReGenSan particularly draws insights from the following.

1.2.1 Regenerative science and medicine

Medical, biological and ecological sciences have been exploring the regenerative capacities of living systems like plants, animals and humans, resulting in major advances in medical treatment such as cancer, ecological preservation and restoration and endangered species protection. Regenerative science investigates the interaction of living tissues with other materials and the ability of animal tissues to rebuild and recover internally and without external effort (Aida & Carrel, 2014; Illingworth, 1974; Kragl *et al.*, 2009; Sampogna *et al.*, 2015) and supports the idea that life is always in a constant state of renewal, restoration, replacement or repair (Graham, 2003; Guterstam & Todd, 1990; Kibert *et al.*, 2002; Lenhoff & Lenhoff, 1991; Maienschein, 2011; McHarg, 1969; Mitsch, 1993; Mitsch & Jørgensen, 1989, 2003; Porcellini, 2009; Sunderland, 2008, 2010, Van der Ryn & Cowan, 2007; Whitman *et al.*, 1997). Findings from such investigations contributed to the use of the ecosystem approach in waste treatment as encapsulated in the principles of ecological engineering and biomimicry, which stand as core principles of ReGenSan (Todd & Josephson, 1996; Todd & Todd, 1980, 1984, 1994). Relatively, ecological and nature-based sanitation technologies have attributes that separate them from conventional technologies and are unique in their application to a wide range of sewage treatment technologies (Lyle, 1994; Mitsch, 1993; Todd & Josephson, 1996) in that the restorative capacity of the ecological system is reinforced (du Plessis & Brandon, 2015) to improve system integrity and performance for the purpose of impeding the rate of depletion and degradation (Mang & Reed, 2012a).

Therefore, ReGenSan utilizes the 'listen and learn' principle of regenerative science and medicine to learn from nature and align sanitation solutions with nature's potential to create new (nouveau) solutions, restore and retrofit previously damaged sanitation systems and recover valuable resources from human waste. In essence, restorative, renewal and replace strategies could be valuable tools for innovative repair and rehabilitation of sites and locations damaged by previous and current sanitation practices (SER, 2004).

1.2.2 Regenerative development and design

Regenerative development and design aims to create systems that are capable of restoring health to both human communities and ecosystems, especially in the fields of architecture and urban development (Zari, 2012). The application of regenerative design in the built environment engages the natural world as the medium and generator of human settlement. It focuses on conservation and performance through reduction in environmental impacts of buildings as well as treatment of the environment as an equal shareholder in the built environment (Littman, 2009; Reed, 2007). Figure 1.1 depicts regenerative design in the built environment and shows the movement from the reductionist-mechanistic design thinking on the lower left to natural systems design thinking on the upper right, which is all part of the journey to a regenerative system (Fullerton, 2015; Mang & Reed, 2012b).

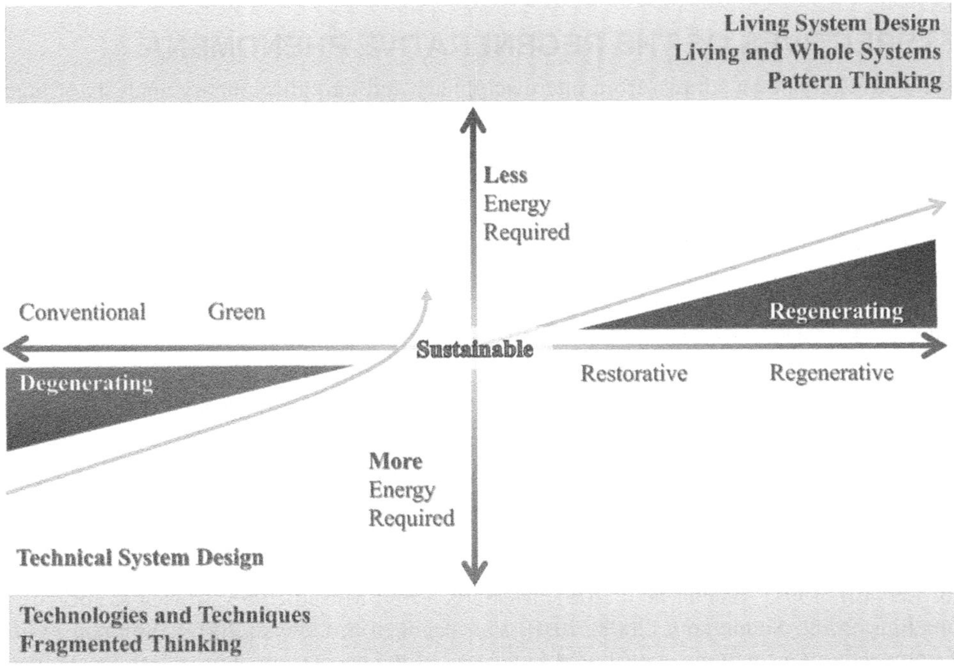

Figure 1.1 Contrast of technical system design and living system design (*Source*: Mang & Reed, 2012b).

Regenerative design and development pushes beyond sustainable development principles by seeking to redevelop systems with absolute effectiveness while allowing for the co-evolution of human species with other species (Lyle, 1994) with a net positive approach to sustainability that departs from the dominant sustainability narratives (Cole *et al.*, 2012; Robinson & Cole, 2015). It seeks to create systems that integrate the needs of society with the integrity of nature through processes that restore, renew and revitalize their own sources of energy and materials. Regenerative design acknowledges that humans are more entangled with the complex systems of environment and the biosphere than the conventional mode of linear cause and effect thinking, which argues that the best design guides for regenerating our environments are ecological principles i.e. the fundamental laws inherent to the natural world (Vester, 2004). Furthermore, regenerative design is a system of technologies and strategies based on an understanding of the inner workings of ecosystems to generate designs that regenerate rather than deplete underlying life support systems and resources within socioecological wholes, while regenerative development refers to technologies and strategies for generating a patterned whole-system understanding of the self-evolving and self-organizing capacities of a 'place' (Mang & Reed, 2012a, b; Reed, 2007).

Thus, regenerative design and development investigates how humans can participate in ecosystems through development to create optimum health for both human communities (physically, psychologically, socially, culturally and economically) and other living organisms and systems (Jenkin & Pedersen, 2009). It can be aptly said that regeneration is more about a process of engagement than a set of outcomes and this process of engagement has significant environmental, economic, social and cultural benefits (Jenkin & Pedersen, 2009). By ensuring that human and natural systems are adequately coordinated to produce positive impacts (Akturk, 2016), it emphasizes the importance of the unique and diverse human and non-human elements of each 'place' (socioecological systems) (Cole *et al.*, 2006). The idea of 'place' is a way for people to envision the unity of humans and natural systems (Mang *et al.*, 2016; Zari & Jenkin, 2010). In each 'place' on earth, natural and cultural systems express themselves uniquely and differently and this implies that sanitation systems should be tailored to the unique characteristics of the 'place' by exploring the opportunities and solutions that are indigenous to the specific 'place'. Regenerative design and development asserts that development can and should contribute to the capacity of all natural, cultural and economic systems that affect a 'place' (to grow and evolve their health and ongoing viability), as shown in Figure 1.2.

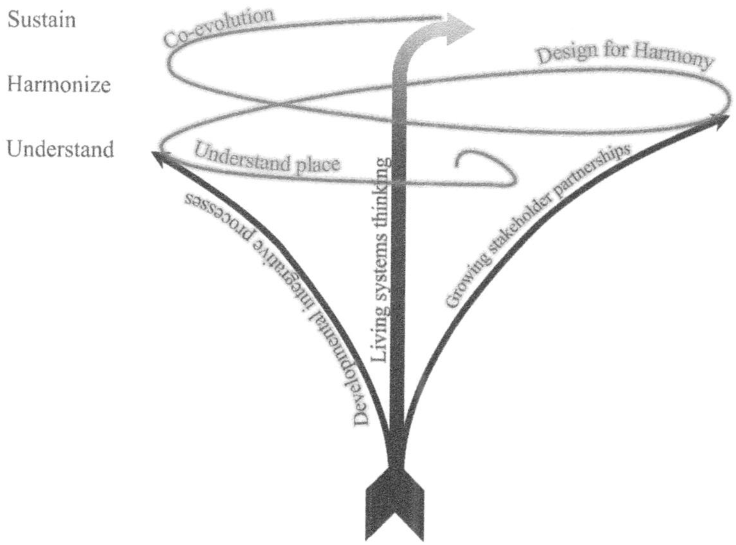

Figure 1.2 Overall framework of regenerative development and design (*Source*: Mang & Reed, 2012a).

Figure 1.2 depicts the three regenerative design and development phases – understanding/ conceptualizing right relationship to 'place', design for harmony and co-evolution and three development processes of growing stakeholder partnerships, living systems thinking and integrative developmental processes – that are ke to creating and sustaining the holism required to actualize the concept (Mang & Reed, 2012a, b). According to Lyle (1994), the regenerative concept replaces the present linear system with cyclical flows and provides for continuous replacement through self-functional processes of the energy and materials used in its operations. In the functional order of natural ecosystems, materials are always reused through the process of conversion, distribution, assimilation, filtration, storage and production to continue their roles in nature's cycle. There are strong links between the concept of regenerative design, development and sustainability with current calls for systemic, holistic and integrative sanitation solutions and management.

A shift from the reductionist-mechanistic worldview (Renger *et al.*, 2015) of previous and current approaches to sanitation to an ecological-systemic worldview that embraces the complex and dynamic nature of sanitation is suggested, whereby interacting units are integrated and inter-synergistically linked for improved performance. This will shift the focus of users and service providers from toilets and treatment systems to a true integration of human dimension that will provide more adaptive and comprehensive, improved sanitation solutions. It is interesting to note that sanitation systems are complex processes and interactions between physical, social, economic, technological and ecological components, which interlink and influence each other (Mele *et al.*, 2010; Seadon, 2006, 2010; Tilley *et al.*, 2014a, b). In addition, it is not the sanitation infrastructure itself that is regenerated in the literal sense of self-healing and self-organizing attributes, but the acts of providing sanitation solutions such as technology, psycho-social-cultural contexts, economic and institutional issues as well as resource recovery that are regenerative and provide the needed catalyst for positive change (Robinson & Cole, 2015).

1.2.3 Regenerative agriculture

Regenerative agriculture is a holistic systems approach to agriculture that aims to improve the resources it uses and avoid resource depletion and destruction. It is designed to improve soil health and restore degraded soil biodiversity through organic practices such as, among others, residual mulching and composting (Rodale Institute, 2017). It is assumed that regenerative agriculture, also called regenerative organic agriculture practices, could positively impact carbon and water cycles, integrate crop farming with animal farming and increase biodiversity and ecosystem health and resilience as well as improve soil fertility and total health (Rodale Institute, 2017). Core principles of regenerative agriculture resonate with the call for change in sanitation perspectives, especially in working with wholes and not parts, context-specific designs and holistic decisions; and focus on the specificity of the 'place' and integrating crop and animal farming to mutual benefits. Regenerative agriculture is a relatively new paradigm in the field of agriculture that borrows largely from the general living system theory, agro-ecology, organic agriculture, permaculture and holistic management (Rodale Institute, 2017).

1.2.4 Regenerative economy and capitalism

Proponents of regenerative economy are focused on designing an economic and investment model that serves humanity while also preserving the ecosystem. It is a shift towards an ecological systemic, holistic and integrative worldview and paradigm with principles that focus on the peculiarities of place, cooperation between man and the earth's system, empowered participation and wholeness (Brown & Garver, 2008; Lietaer *et al.*, 2012). According to Fullerton (2015) regenerative economy should promote and sustain human prosperity and well-being in an economy of permanence because economic sustainability cannot simply depend on measuring designed outputs; economy and financial systems should be based on principles that ensure enduring system health. In addition, reliable inputs and healthy outputs are to be maintained by preserving and protecting all key parts of the broader societal and environmental systems of the economy as well as strengthening human networks to preserve vitality and resilience. One argument that connects with the new thinking on sanitation is that regenerative economy does not argue for or against capitalism or socialism, just as there are no arguments for decentralized or centralized sanitation or even existing frameworks. The aim is to achieve an effective integration of the best of both paradigms with an infusion of an understanding of how natural living systems' principles can be adopted to produce shared vitality and prosperity. In a regenerative economy, the network of human interactions with each other and the natural world is the 'system'.

Therefore, economic and financial growth goes beyond gross domestic product (GDP) or gross national product (GNP) to the presumption that economic vigour is rather dependent on human and societal vitality, rooted in ecological health and the inclusive development of human capabilities and potentials (Fullerton, 2015). It thrives on willingness to learn, adapt to change and continually evolve as a whole system. Regenerative capitalism is the product of a regenerative economy that works with the symbiotically related principles of regenerative economy to determine how to disburse and pursue capital investment.

Five out of these eight principles as enumerated by Fullerton (2015) relate closely to the regenerative concept adapted in architectural design, urban development and agriculture, which were adopted for this work. They include:

(I) Recognizing that humanity is also a part of an interconnected whole inclusive of the ecosystem and all global localities,
(II) Wealth is viewed holistically as it concerns all parts of the whole with all forms of capital,
(III) Innovation, adaptiveness and responsiveness are crucial,
(IV) All parts of the whole are empowered to participate in systems' growth and how they benefit from it, and
(V) Honours community and place by respecting specific peculiarities in ecology, geography, society, culture, history etc.

1.3 REGENERATIVE SANITATION WORLDVIEW

The above regenerative perspectives are derived from a shift in worldviews that has led conceptualization in the same direction. New knowledge, new thinking and new technology are built by developing successful ideas based on a set of premises and values (i.e. paradigms drawn from a worldview), which remain true until they are challenged. Sanitation technological solutions and management need a shift in worldview and paradigmatic conceptions (Du Plessis & Brandon, 2015). To improve the overall sanitation system performance and achieve strong sustainability, a worldview shift is essential from the reductionist-mechanistic linear conventional sanitation worldview to more cyclical-integrated-systemic-ecological perspectives (Marshall & Farahbakhsh, 2013; Thibodeau *et al.*, 2014; Tilley *et al.*, 2014a). A worldview can be described as a collection of concepts, theorems and assumptions that provides a coherent way of looking at and thinking about the world (Aerts *et al.*, 2007; Kearney, 1984) as well as interconnected systems of belief (DeWitt, 2010) that act as a filter through which phenomena are perceived and comprehended (Miller & West, 1993). According to Aerts *et al.* (2007), worldview shapes how individuals interpret and interact with the world around them, defining what can be known or done and how and what goals should be, or whether they should even be pursued (Mang & Reed, 2012a; Koltko-Rivera, 2004).

The reductionist-mechanistic worldview is essentially the Newtonian-Cartesian modern civilization philosophy, operating on the assumption that the properties of the whole can be reduced to and deduced from the sum of the properties of the parts (as is possible with mechanical and other complicated cybernetic systems) and it places a high emphasis on prediction and centralized control to ensure system performance (Du Plessis & Cole, 2011; Heylighen, 2006; Robertson & Choi, 2010). At the core of this worldview is the assumption that the universe as a whole and all of its material components, including living organisms, function like mechanical systems that are governed by universal laws (Benne & Mang, 2015).

The ecological-systemic worldview, on the other hand, draws on an understanding of nature and its processes and relationships (Capra, 1997). It is a much broader concept than that encapsulated in classical ecology or even ecological economics. 'Ecological' implies an understanding that we are dealing with living systems and all that comes with such systems, including connections, flows, relationships, interdependence, evolution and consciousness (Du Plessis & Brandon, 2015). The ecological-systemic worldview emphasizes interconnectedness and its orientation is more holistic in nature, recognizing that a thorough understanding of any system requires knowledge of the nature of the interactions among its parts as well as the nature of its interdependencies with other parts of the larger system being embedded in the whole (Checkland, 1997; Robertson & Choi, 2010). It recognizes the self-organizing capacity inherent in all natural systems (Jantsch, 1980), which are self-managing and self-regulating in that their distinct and diverse parts engage in patterns of interaction that maintain a dynamic equilibrium (Kauffman, 1995) of the whole while emphasizing the co-evolutionary and mutually reinforcing dynamics with their environments (Capra, 2002). The ecological-systemic worldview, which incorporates the general living system theory (von Bertalanffy, 1968), considers systems as integrated wholes and focuses on interactions among the parts and the larger environment in which they are embedded (Robertson & Choi, 2010; Unsal, 2016).

Sanitation systems designed from the reductionist-mechanistic worldview normally optimize the efficiency of the individual constituent elements by identifying discrete performance requirements, setting specific measurable goals and targets and following designated formulas, rules and criteria. The sustainable performance of the whole is optimized by aggregating the solutions for the different parts and generic technocentric design, and top-down solutions and approaches are prescribed to similar circumstances with accommodations for regional distinctions and cultural differences. Humans are often seen as outsiders from nature and are responsible for being good stewards of natural resources and protecting them from harmful human activities through conservation activities. These approaches do not recognize or leverage the interconnectivity and dynamic nature of socio-ecological systems, and thus have limited impacts on the sustainability, vitality and resilience of the systems (Benne & Mang, 2015). Meanwhile, designing sanitation solutions from the ecological-systemic worldview is based on concepts, designs and processes that draw on an understanding of the unique dynamics and potential of how life works in a 'place' and of the distinctive role and value-contributing potential of the sanitation systems (Benne & Mang, 2015). Thus, the ecological-systemic worldview seeks to provide better understanding of the knowledge of interconnections among the sanitation subsystems, dimensions and components as well as holistically integrating methodologies that address the interlinkages of the sociocultural, environmental, economic and technical spheres (Marshall & Farahbakhsh, 2013). It adopts an integrated perspective that addresses improved sanitation planning, takes advantage of economic opportunities, incorporates specialized and appropriate technology and follows up with behavioural change, which could ensure access and sustainable use as well as operations and maintenance of sanitation intervention (Tilley *et al.*, 2014a).

However, the ecological-systemic worldview does not necessarily negate the reductionist-mechanistic worldview, but adds to the knowledge base by providing a different perspective that reveals different types and kinds of knowledge to improve sanitation services. Both worldviews could provide valuable insights when applied within the appropriate context of analysis and its realm of validity (Du Plessis & Brandon, 2015; Wilber, 2000). The knowledge and laws of sanitation systems revealed by the reductionist-mechanistic worldview are still immensely useful when it comes to engineering and technology, while the ecological-systemic worldview works towards ensuring systemic and holistic integration of all sanitation components without separating technology from the economic and psycho-social contexts (Altaf, 2011; Du Plessis & Brandon, 2015; Nilsson & Olsson, 2014). This book proposes the ecological-systemic worldview as an overall worldview that incorporates the necessary principles from the reductionist-mechanistic worldview so as to come up with regenerative sanitation technology and management solutions.

1.4 PARADIGMS OF SANITATION

Consequently, a shift in worldview will also cause a shift in paradigm and recent years have seen an increase in discussion and debates about a paradigm shift in sanitation management. From the conventional traditional linear sanitation technological approaches of 'flush and discharge' or 'drop and store'

(Andersson *et al.*, 2016a, b; Esrey, 2001) to more systemic-integrated approaches between physical, pyschosocial, economic, ecological and technological aspects (IWA 24:2016; Marshall & Farahbakhsh, 2013; Schübeler, 1996; Seadon, 2006; Spaargaren *et al.*, 2007, van Vliet *et al.*, 2011). A paradigm is a subset of a worldview that provides accepted models or patterns of ideas or basic assumptions (within the context of that worldview) about how something should be perceived, thought about, valued, done or made (Benne & Mang, 2015; Harman, 1970). It is also an agreed way of thinking about the world or an agreed set of valid approaches to investigating a scientific and technological phenomenon (Pahl-Wostl *et al.*, 2011). It provides insights into what is to be observed and scrutinized, kinds of research questions to ask and answer, how questions are to be structured and how results of scientific investigations are to be interpreted (Halbe *et al.*, 2013; Kuhn, 1962). Paradigms also provide an intellectual and operational environment for scientific and technological research to take place and shape the nature of the problems to be addressed (Leclerc, 2005; Pahl-Wostl, 2011).

In this case, sanitation paradigms (SaPs) are the different accepted forms and approaches of conceptualizing and handling sanitation management and technology over the years. A SaP trails processes and systems development for research, technology and management design and implementation of sanitation solutions and services. Most existing SaPs are technocentric, resource-oriented, service-oriented, hygiene-focused, water-quality-focused, participatory-focused and behavioural-change focused. There have, however, been calls for a paradigm shift, especially with the dismal results of the sanitation targets in the MDGs. The rationales for this call for a paradigm shift include:

(I) Sanitation systems are increasingly recognized as complex and integrated in processes between physical, social, economic, ecological and technological aspects (IWA 24:2016; Marshall & Farahbakhsh, 2013; Schübeler, 1996; Seadon, 2006; Spaaragaren *et al.*, 2007; van Vliet *et al.*, 2011);

(II) Contextual factors such as culture, economy, institutional control and climate are considered as critical elements that can affect public acceptance of technical and infrastructural feasibility for innovation (Domènech & Saurí, 2010; Larsen *et al.*, 2010; McConville *et al.*, 2014; Whittington *et al.*, 1993);

(III) Technological development processes and approaches are now highly contextual and dependent on a number of sociocultural-economic and ecological factors (Balkema *et al.*, 2002; Hegger *et al.*, 2007; McConville *et al.*, 2014; Raven, 2007);

(IV) Nature-based solutions such as ecological engineering and eco-technologies are known to represent plausible sustainable approaches for solving global sanitation problems, especially in the use of constructed wetlands and other ecosystem solutions (Barros *et al.*, 2008; Elzein *et al.*, 2016; European Commission, 2015; Guterstam & Todd, 1990; Lens *et al.*, 2001; Mahmood *et al.*, 2013; Nesshöver *et al.*, 2017; Polprasert & Koottatep, 2017; Vymazal, 2005; Wu *et al.*, 2016);

(V) Governance is considered a critical success and supporting mechanism that provides the enabling environment for effective and efficient service delivery, and financing sustainability (Hallerod *et al.*, 2013; Joshi *et al.*, 2015; Rodic, 2015; Schertenleib, 2005; Speer, 2012);

(VI) Resource recovery and reuse are acknowledged as key components of comprehensive integrated sanitation solutions with the potential for natural resource conservation, and have the capacity to leverage the long-term efficacy of treatment schemes and livelihood support (Benetto *et al.*, 2009; Cofie *et al.*, 2016; Hu *et al.*, 2016; Keraita *et al.*, 2014; Murray & Buckley, 2009; Nikiema *et al.*, 2014; Otoo *et al.*, 2016; Rao *et al.*, 2016; USEPA, 2012; Werner *et al.*, 2009; WWAP, 2017);

(VII) Technology-based indicators for determining 'improved' and 'unimproved' sanitation instead of functional-service-based indicators constrain innovation (Banerjee & Morella, 2011; Domènech & Saurí, 2010; Kuznyetsov, 2007; Kvarnström *et al.*, 2011; Larsen *et al.*, 2010; McConville *et al.*, 2014; Potter *et al.*, 2011; Sutton, 2008; van Vliet *et al.*, 2011; WSP, 2008); and

(VIII) Inadequate and dysfunctional sanitation infrastructures are due to structural failures, material degradation, poor construction and workmanship as well as ageing facilities and lack of appropriate operations and maintenance (Brikke, 2000; Brepolsa *et al.*, 2008; Cookey *et al.*, 2016). These dysfunctionalities and inadequacies are costly for society and its economic impacts are estimated to cost 18 African countries a colossal sum of USD 5.5 billion per year (Bartram & Cairncross, 2010; Bos & Gijzen, 2005; Cheng *et al.*, 2012; Ekane *et al.*, 2014; Fewtrell *et al.*, 2005; Pruss *et al.*, 2002; Prüss-Üstün *et al.*, 2008; UNDP, 2006; WSP, 2014; Yardley, 2010). Strategic and innovative plans and programmes to retrofit these systems could accelerate the targets of SDG no. 6 on sanitation.

The major paradigms that have influenced sanitation research, science, technology, management, policy and practice have graduated from one level to another and produced several kinds of technological solution (Cooper, 2001; O'Reilly *et al.*, 2017; Persson, 2017; Zeldovich, 2017). These sanitation paradigms include: conventional sanitation paradigm (sanitation 1.0); ecological sanitation paradigm (sanitation 2.0); and sustainable sanitation paradigm (sanitation 3.0).

1.4.1 Conventional sanitation paradigm (sanitation 1.0)

The conventional paradigm approach to sanitation falls under the category of either waterborne 'flush and discharge' or dry 'drop and store' (Werner *et al.*, 2009) systems technology based on the linear flow premise that excreta is waste suitable only for disposal, which makes recycling impossible (Andersson *et al.*, 2016a, b; Esrey, 2001) (Figure 1.3).

Although there are innovative approaches within this system incorporating resource recovery and reuse as incremental improvement strategies, paradigm shifts are being advocated for overcoming these challenges, especially through the adoption of integrated sanitation solutions that go above and beyond the focus of technology (Geels, 2006; Larsen, 2011; Marshall & Farahbakhsh, 2013; Otterpohl, 2002; Thibodeau *et al.*, 2014; Tilley *et al.*, 2014a; Zeeman, 2009).

1.4.2 Ecological sanitation paradigm (sanitation 2.0)

The ecological paradigm approach to sanitation is based on an overall view of material flows of the 'closed-loop system' designed for incremental progression over the conventional systems (Mang & Reed, 2012a; Werner *et al.*, 2009) (Figure 1.4). It can also be described as a resource-oriented, ecological and source separation-based sanitation approach. It is a reconceptualization of conventional sanitation from 'flush and discharge' to 'drop and reuse' models (Haq & Cambridge, 2012; Hu *et al.*, 2016; Langergrabera & Muelleggera, 2005) and is designed to close the nutrient loop between sanitation and agriculture even though the technologies are not ecological per se (Esrey, 2001; Langergrabera & Muelleggera, 2005; Werner *et al.*, 2004, 2009). The major drawback is basically the cost and cultural constraints on the usage of treated excreta and sewage (Hu *et al.*, 2016). In any case, it is still based on the reductionist-mechanistic worldview of the conventional sanitation paradigm and so focuses on technology and the back end of the loop.

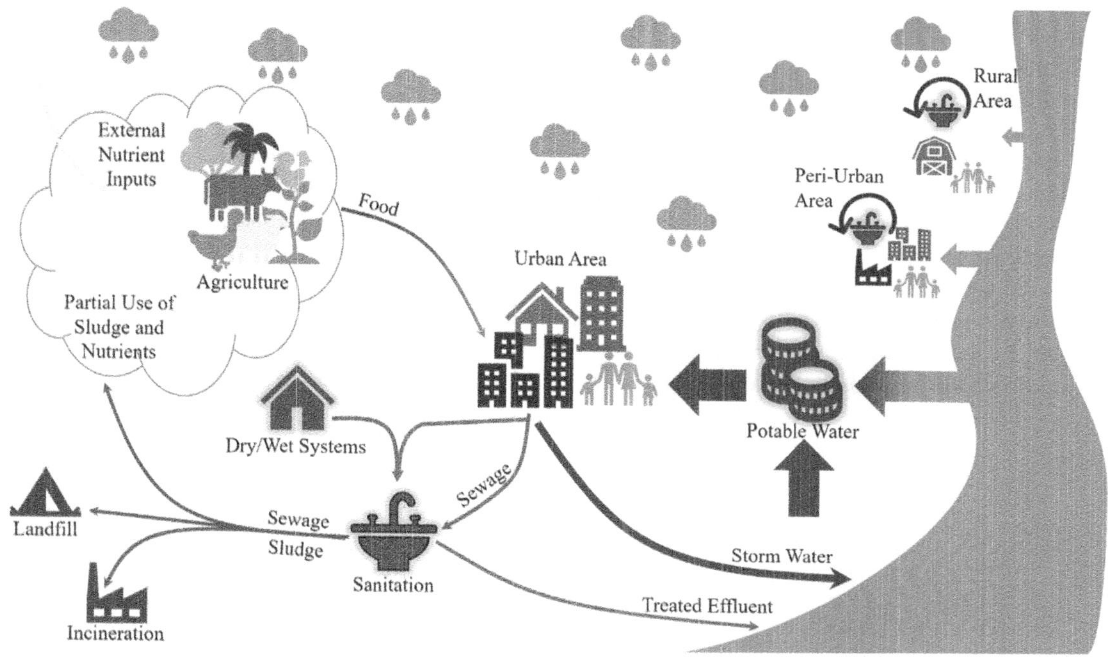

Figure 1.3 Conceptual diagram of conventional sanitation. Arrows are non-quantitative representations of resource (water, nutrients, organic matter) flows associated with conventional sanitation and connected processes (*Source*: GTZ, 2002).

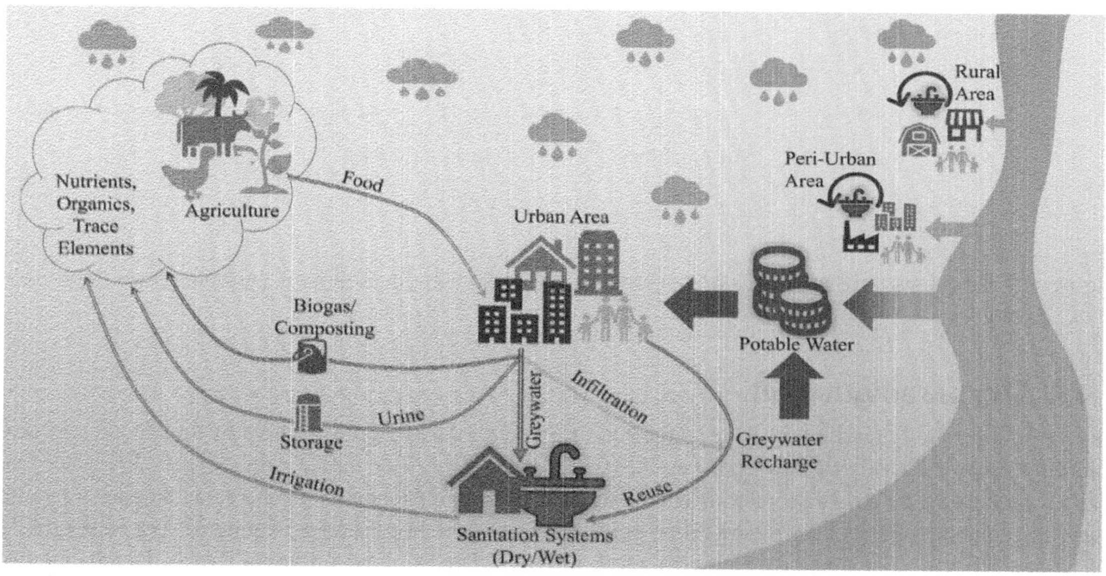

Figure 1.4 Conceptual diagram of ecological sanitation/source separation and decentralization. Arrows are non-quantitative representations of resource (water, nutrients, organic matter) flows associated with sanitation and connected processes (*Source*: GTZ, 2002).

1.4.3 Sustainable sanitation paradigm (sanitation 3.0)

The sustainable sanitation paradigm recognizes not only technology, but also social, environmental and economic aspects of sanitation (SUSANA, 2008). It includes means of recycling human excrement that have no negative impact – or even a positive impact – on local and global resources (Brands, 2014). Its core principles include: human health protection, affordability, environmental sustainability and institutional appropriateness (Chinyama *et al.*, 2012; Mara *et al.*, 2007). Sustainable sanitation aims to address the challenges of conventional and ecological sanitation by adopting the generic principles of sustainable development (SD), but it is constrained by oversimplification of complex dynamic problems such as sanitation technology and management (Marshall & Farahbakhsh, 2013). Sanitation sustainability cannot be achieved in the absence of the whole-system thinking and fit-for-purpose solutions that can only be found by addressing challenges at a system level to reveal mutually inter-synergistic advantageous interactions as well as undesirable ones (Anarow *et al.*, 2003; Rocky Mountain Institute, 2006) (Figure 1.5).

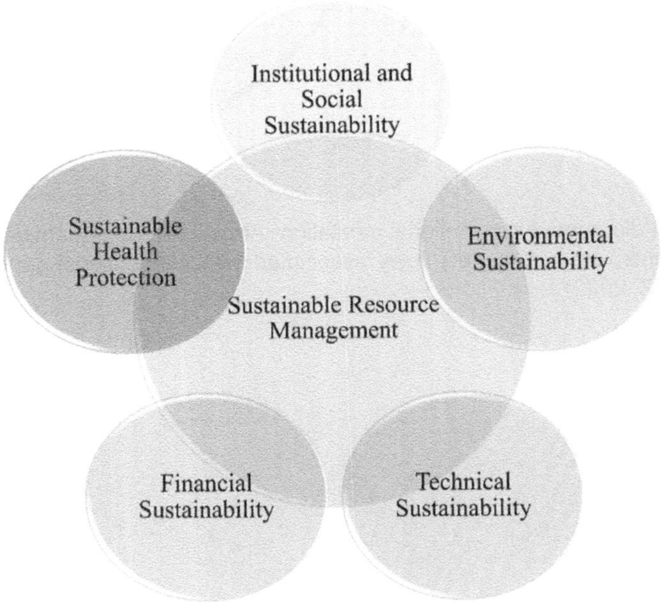

Figure 1.5 Key sustainability dimensions in sustainable sanitation management. (Adopted from Andersson *et al.*, 2016a, b).

1.4.4 The paradigm shift

Several scholars and practitioners have suggested the need for a paradigm shift towards holistic, integrated, multidimensional-transdisciplinary perspectives (Galli *et al.*, 2014; Marshall & Farahbakhsh, 2013; Seadon, 2010; van Vliet *et al.*, 2011; Zeeman, 2009). Some proposed nomenclatures for this new paradigm, but did not really come up with a clear framework to guide this transition. For instance, Agudelo-Vera *et al.* (2012), Swart and Palsma (2013) and Wielemaker *et al.* (2016) called it 'new sanitation' (or in Dutch 'nieuwe sanitatie'); Fan *et al.* (2017) proposed 'sanitation 2.0'; Heggar *et al.* (2007) and van Vliet *et al.* (2010, 2011) named it 'hybrid or modernized mixture'; Verstraete *et al.* (2009) and Nansubuga *et al.* (2016) suggested 'cluster decentralized systems' and Marshall and Farahbakhsh (2013) and Tilley *et al.* (2014a) proposed 'integrated approach'; while Galli *et al.* (2014) suggested 'whole-system approach'. Evidently, the need for a paradigm shift in the field of sanitation technology and management is long overdue. To this effect, a paradigm shift towards regenerative sanitation is proposed to capture the essence of the change advocated beyond technology. The word 'regenerative' here implies complete revitalization and rejuvenation of the entire sanitation system to improve service and expand access as well as a deliberate partnership with nature (Figure 1.6).

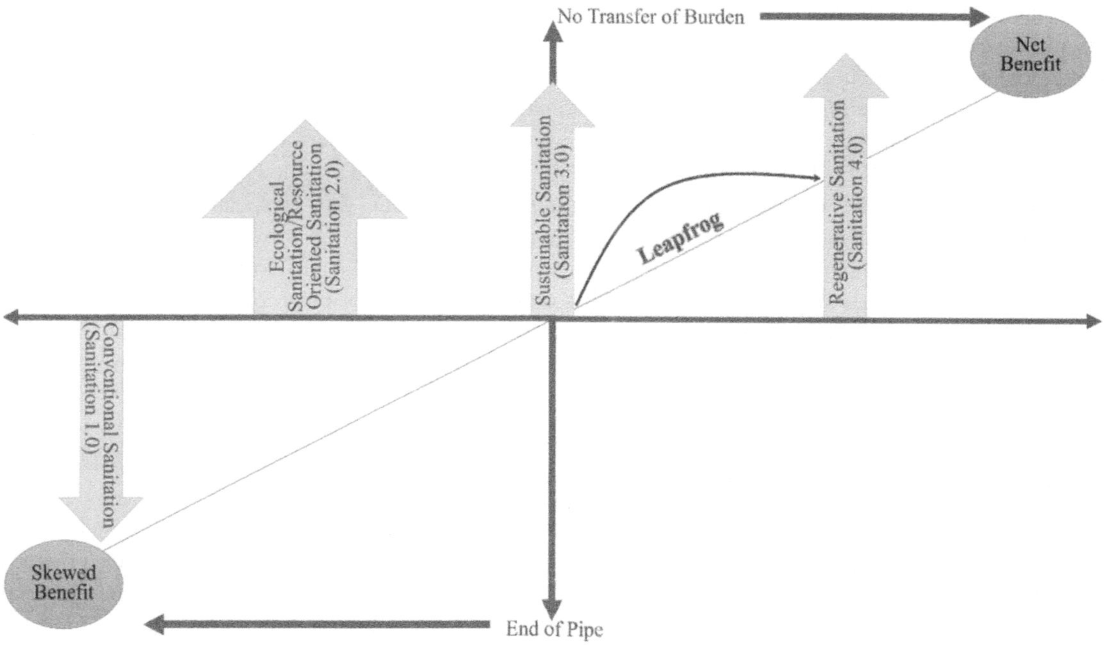

Figure 1.6 Trajectory of sanitation paradigms.

Figure 1.6 shows the progression of paradigms that influenced design and infrastructure solutions for sanitation management over the years and presents a new paradigmatic influence for the future of sanitation. The paradigms are grouped into four levels (Sanitation 1.0–4.0) and fall between two divides i.e. 'end-of-pipe to skewed-benefits' and 'no-transfer-of-burden to net benefits'. On the lower left, the conventional sanitation paradigm (Sanitation 1.0), which was mainly based on the reductionist-mechanistic worldview, focused on end-of-pipe technologies that produced benefits skewed towards environmental and public health only. Subsequently, incremental progress led to the concepts built from the ecological sanitation paradigm (Sanitation 2.0), which found it difficult to move beyond pilot-scale levels due to psycho-social challenges, and then the sustainable sanitation paradigm (Sanitation 3.0) based on the concept of maintaining the status quo (i.e. no harm-no gain) whereby focus is on ensuring that sanitation practice does not cause harm without seeking any gains from it. However, in spite of the ecological and sustainable paradigms, the reductionist-mechanistic worldview persisted and focus remained on end-of-pipe technologies with skewed benefits. Consequently, the regenerative sanitation paradigm (Sanitation 4.0) is proposed as a holistic, integrated and systemic management of all sanitation components (and not just technology) to ensure no-transfer-of-burden on environmental and public health and deliver net benefits to areas including agriculture, aquaculture, economies and livelihoods.

1.4.5 Regenerative sanitation paradigm (sanitation 4.0)

The paradigm of ReGenSan, called sanitation 4.0, does not aim to present an undebatable truth (Meadows, 1999), but addresses the need for a change in the way sanitation is viewed considering the growing and seemingly giant-like challenges of 2030 and beyond. It is about learning new ways, seeking fresh answers from old ways, adapting to change with resilience and flexibility, thinking before acting and working to integrate and synergize the whole process of sanitation management into a whole system (and not just working from individual parts). The ReGenSan paradigm (sanitation 4.0), therefore, is conceptualized based on the principles of systemic internal processes for restoration and sustainability and addresses sanitation as a comprehensive and holistic integrated system incorporating all the unique characteristics of sanitation management from the starting point of a specific and peculiar 'place' (location). It sees sanitation as resource/service management rather than problem management. At its core is a process trail that works from within the system and the peculiarities of the 'place' to ensure resilience and adaptability. ReGenSan is at the same time a paradigm, framework and perspective for sanitation solutions. The overall aim of ReGenSan is to refocus the perspectives of researchers, professionals, practitioners, advocacy agents, policymakers and solution-providers from attacking a problem (of health, hygiene and water quality) to providing services and solutions; and ultimately, expand access as well as improve service towards achieving SDG no. 6 by 2030. The corresponding conceptual framework presented in this book is designed to reflect the process for the kind of sanitation practice that would fulfil this ambitious goal. It is hoped that this conceptualization will contribute to the success of sanitation management globally.

1.5 SANITATION CONCEPTUAL FRAMEWORKS

Concepts towards providing solutions or understanding phenomena are derived from existing paradigms. A conceptual framework is a domain-specific language tailored to a specific knowledge area or study discipline (Abelson & Sussman, 1987; Bentley, 1986; Hinkel *et al.*, 2014; Hudak, 1998). The elements of a framework are concepts, which are terms associated with a meaning and the conceptual relationship between them (Hinkel *et al.*, 2014). Therefore, a conceptual framework facilitates the exchange of arguments about a knowledge domain through having clear and unambiguous accounts of the concepts at stake and their relationships (Hinkel, 2008; Ionescu *et al.*, 2009; Wolf *et al.*, 2013). Sanitation conceptual frameworks (SCF) provide the premises and the parameters for research, investigations, interpretation, programme development and implementation of action plans for improving service and expanding access to meet SDG no. 6 by 2030.

A SCF will lay out key factors, constructs or variables, and presume the relationships among them (Campbell *et al.*, 2015; Miles & Huberman, 1994), as well as provide a template used for acknowledging and organizing the elements of related systems of interest to support further investigations. They will also illustrate the connected parts and potential interdependencies between broad system variables (Binder *et al.*, 2013; Repella, 2014) and support the formulations and comparisons of hypothesis, models and theories as well as provide general sets of concepts assumed to be applicable to the whole (Ostrom, 2005). The goal of such framework development is to stay as neutral as possible to allow the representation of different theories within the framework (Hinkel *et al.*, 2014). It is, therefore, expedient to investigate and understand some existing conceptual frameworks in the field of sanitation, such as:

(I) Sanitation ladder – the most popular and one of the oldest, which has been used in many forms and cases ranging from toilets/latrines provision, full service chain delivery, hygiene behaviour, human right issues, financing sanitation, latrine ownership and MDG/SDG Joint Monitoring

Programmes as well as in the community-led total sanitation for the eradication of open defecation (Exley *et al.*, 2014; Gine *et al.*, 2011; JMP, 2010; Kar & Chambers, 2008; Kvarnström *et al.*, 2011; Obeng *et al.*, 2015; O'Reilly *et al.*, 2017; Potter *et al.*, 2011; Verhagen & Carrasco, 2013; WHO/UNICEF, 2017);

(II) Sanitation service chain – used in the analysis of physical flow of faecal waste through sewered and non-sewered systems, development of faecal waste flow matrix and diagrams to summarize city- or community-level outcomes and highlight bottleneck in faecal waste management. It is also used for business analysis of faecal sludge management and in the development of the global sanitation service technology filter tool (Blackett *et al.*, 2014; Chowdhry & Kone, 2012; Rao *et al.*, 2016; Strande *et al.*, 2014; Trémolet & Rama, 2012; USAID, 2016; WSP, 2014).

(III) Decision support – used in designs, testing, selections, sustainability and performance modelling of technologies from centralized to decentralized sanitation systems (Kalbermatten *et al.*, 1980, 1982; Katukiza *et al.*, 2010, 2012; Libralato *et al.*, 2012; Loetscher & Keller, 2002; Magid *et al.*, 2006; Maurer *et al.*, 2012; Massoud *et al.*, 2009; McConville *et al.*, 2014; Ramoa *et al.*, 2015; Roefs *et al.*, 2017; Simha *et al.*, 2017; Stenström *et al.*, 2011; Tilley *et al.*, 2014b; Tobias *et al.*, 2017; Zurbrügg & Tilley, 2007, 2009);

(IV) Designing for service – focused on resource utilization options for sanitation systems. These include concepts that deal with resource-oriented sanitation, recovery and reuse, integrated resource management, cradle-to-cradle, ecological sanitation and circular flow as well as recirculation of nutrients and organic matter (McDonough & Braungart, 2001a, b; Brands, 2014; Hodges, 2006; Langergrabera & Muelleggera, 2005; Larsen *et al.*, 2009, 2011; Murray & Buckley, 2009; Wang *et al.*, 2006; Werner *et al.*, 2009; Wilsenach *et al.*, 2003; Wielemaker *et al.*, 2016);

(V) Spatial planning diagnostics – provides approaches and methodologies for detailed assessment of sanitation situations of regions, cities, urban, formal and informal settlement as well as rural areas to determine appropriate levels of intervention activities for programme implementation (ASHWAS, 2011; Bright-Davies *et al.*, 2016; GHK Research & Training, 2000; ISF-UTS & SNV, 2016; IWA, 2014; Kerstens *et al.*, 2016; Lüthi *et al.*, 2011a, b; Parkinson *et al.*, 2014; Schmitt *et al.*, 2017);

(VI) Equity and inclusion – designed to improve access to sanitation services for the society's vulnerable groups and people with disabilities (Desai *et al.*, 2016; Gosling, 2010; Patkar & Gosling, 2011);

(VII) Behavioural change – this software tool is used in various sanitation, hygiene and handwashing programmes, community-led total sanitation (CLTS) as well as sanitation solutions adoption and uptake (Dreibelbis *et al.*, 2013; Dwipayanti *et al.*, 2017; Kar & Chambers, 2008; Kvarnström *et al.*, 2004; O'Reilly & Louis, 2014; Peal *et al.*, 2010; Whittington *et al.*, 1993); and

(VIII) Capacity development assessment – designed to assess effectiveness and efficiency of sanitation capacity development programmes (Barat *et al.*, 2014; Crocker *et al.*, 2016; Gunawardana *et al.*, 2013; Mvulirwenande *et al.*, 2013; Ngai *et al.*, 2014; Pascual Sanz *et al.*, 2013).

These conceptual frameworks, however, are primarily technocentric and basically designed for implementation of conventional sanitation and to some extent ecological sanitation paradigms. Furthermore, there are no specific frameworks for governance and participation in sanitation management. Therefore, a holistic, integrated and systemic sanitation conceptual framework that embraces the complexities inherent in providing sanitation solutions rather than simplistic recipes is demanded to enhance access expansion and improved service delivery (Meinzen-Dick, 2007; Pahl-Wostl, 2007; Pahl-Wostl *et al.*, 2010). Hence, the main contribution of this book is to provide such a framework (or model) known as the regenerative sanitation framework, which is not technocentric and toilet-latrine-focused but where all key critical aspects of sanitation subsystems, dimensions and components are interlinked for total rejuvenation and revitalization of the whole spectrum of sanitation management systems as well as for the development and implementation of innovative solutions. Providing comprehensive sanitation solutions for the current global challenges depends not only on availability of technology, but also on many other factors like management of sanitation delivery, quality of institutions that manage them, prevailing psycho-socio-cultural, economic and ecological conditions that influence sanitation planning, technology design and development, management processes and practice, as well as financial and governance mechanisms (Starkl *et al.*, 2015; Tilley *et al.*, 2014a; Zurbrugg & Tilley, 2009).

Sanitation in this book is used to mean the systemic integration of measures for managing human waste in a safe and resourceful pattern adapted to fit specific and peculiar contexts (Box 1.5). It is a systemic integration in that it flows from the 'place' of human interactions (pyschosocio-cultural-economic) with the ecosystem (ecological, bio-geo-physical) to interconnect with the governance (institutional, policy and regulatory mechanisms), resource recovery (recovery and reuse of excreta/urine-flushed water i.e. black and greywater) and technology (infrastructure-utilities) to improve service delivery and access expansion.

BOX 1.5 IN SUMMARY

Sanitation is the treatment and management of by-products of human digestion.

The issue at present is not whether such an approach is desirable, but rather how can this be achieved in the real world? The central idea is the need for broader integration with all the system components of sanitation, minimized transport, local orientation, hybridization of centralized and decentralized systems, source separation and resource recovery, management and governance (Agudelo-Vera *et al.*, 2012; Larsen & Maurer, 2011; Lens *et al.*, 2001; Kujawa-Roeleveld & Zeeman, 2006; Larsen *et al.*, 2009; Maurer *et al.*, 2012; Nansubuga *et al.*, 2016; Tchobanoglous & Leverenz, 2013; van Vliet *et al.*, 2010, 2011; Zeeman, 2012).

1.6 CHANGE IS IMPERATIVE

For the SDG to do better than the MDG in sanitation, new strategies and concepts must be developed. ReGenSan offers a new perspective in debates at various levels on sanitation and its management towards the need for a paradigm shift. Nevertheless it is not quite new per se, as shall be seen in subsequent chapters, but builds on the strengths of historical successes in combination with an understanding of current trends to offer a new kind of prerogative outlook. It attempts to avoid the argument of what is wrong or right, but weaves around existing concepts and ideas to create an integrative system to guide thoughts towards greater and enduring successes.

In addition, instead of seeing sanitation as a problem or rather a storehouse of resource, it views sanitation as an entity in itself that is resourceful, with psycho-social, economic, cultural and ecological angles that could support livelihood and societal interactions. In fact, a sanitation economy could be envisaged in that there will be producers, consumers, service providers, policymakers and regulators. ReGenSan works as a complex and dynamic phenomenon that is influenced by society, livelihood, status, economy, culture, tradition, geography and ecology, all of which can change gradually or suddenly. This implies the need for continuous and constant change to match the changes in society, geography and ecology, especially with climate change and livelihoods concerns.

1.7 EXERCISES

(I) What is the regenerative sanitation paradigm and how does it differ from the other pre-existing paradigms?

(II) What is the relationship between the ecological-systemic worldview and the regenerative paradigm and how does it support the regenerative sanitation concept?

(III) How does regenerative sanitation relate to:

- (a) Regenerative science and medicine
- (b) Regenerative development and design
- (c) Regenerative agriculture
- (d) Regenerative economy and capitalism

(IV) What are the common features in the five fields above and how are they uniquely applied?

(V) Why do you think the conventional sanitation paradigm has been predominant through the years?

(VI) Why do you think sanitation solutions have been more technology-focused and less systemic-holistic-integrative in all pre-existing paradigms?

(VII) What worldview and paradigms are the highlighted conceptual frameworks (and any others you can identify) based on, in your opinion, and why?

(VIII) In your opinion, why do we need a paradigm shift in sanitation? How will this affect the provision of sanitation services and infrastructure? What are the possibilities of such a shift and its impact on the way sanitation is perceived and addressed? And how will this affect sanitation technologies?

(IX) Where do you see technology design and development in a concept like regenerative sanitation? What do you think is involved in the design process of sanitation technologies in a regenerative future?

(X) Do you agree that change in the way of sanitation is handled is imperative if we are to meet SDG no. 6 goals and targets for sanitation by 2030?

(XI) How do you think regenerative sanitation can impact the sanitation targets of SDG no. 6?

(XII) What determines the dynamics of a transition from the reductionist-mechanistic conventional sanitation paradigm to a more systemic, holistic and integrative technological and management solution? What are the barriers and the drivers needed for this transition?

1.8 RECOMMENDATIONS FOR FUTURE RESEARCH

There are a number of knowledge gaps around the application of the concept of regenerative design and development in sanitation research; therefore further research is needed for better understanding of the ReGenSan concept:

(I) Investigate the ways worldviews and existing paradigms affect sanitation management and technology design.

(II) Empirical investigation of existing sanitation conceptual frameworks to determine their efficacy, effectiveness, efficiency and fitness for purpose as well as optimization.

(III) Explore the possibilities of the interlinkages between the existing frameworks for synergistic results.
(IV) Explore further research on pre-existing paradigms as well as research to deepen understanding and create a mix for conceptualization.
(V) Critical review to determine the meaning of sanitation, what it entails, how it works and its impacts on humanity and ecosystem.

REFERENCES

Abelson H. and Sussman G. J. (1987). Lisp: a language for stratified design. *Byte*, **13**(2), 207–218.

Aerts D., Apostel L., De Moor B., Hellemans S., Maex E., van Belle H. and van der Veken J. (2007). World Views. From Fragmentation to Integration, Internet edn. VUB Press, Brussels. http://www.vub.ac.be/CLEA/pub/books/worldviews.pdf (accessed 2 February 2008).

Agudelo-Vera C. M., Mels A., Keesman K. and Rijnaarts H. (2012). The urban harvest approach as an aid for sustainable urban resource planning. *Journal of Industrial Ecology*, **16**, 839–850.

Aida L. and Carrel A. (2014). Visionary vascular surgeon and pioneer in organ transplantation. *Journal of Medical Biography*, (3), 172–175.

Akturk A. (2016). Regenerative Design and Development for a Sustainable Future: Definitions and Tool Evaluation. M. S. thesis, University of Minnesota, USA. Retrieved from the University of Minnesota Digital Conservancy. http://hdl.handle.net/11299/182113

Altaf S. W. (2011). So Much Aid, So Little Development: Stories from Pakistan. Johns Hopkins University Press, Baltimore, USA.

Anarow B., Greener C., Gupta V., Kinsley M., Henderson J., Page C. and Parrot K. (2003). Whole-Systems framework for sustainable consumption and production. *Denmark: Danish Ministry of the Environment*, **807**, 1–51.

Andersson K., Dickin S. and Rosemarin A. (2016a). Towards 'sustainable' sanitation: challenges and opportunities in urban areas. *Sustainability*, **8**, 1289.

Andersson K., Rosemarin A., Lamizana B., Kvarnström E., McConville J., Seidu R., Dickin S. and Trimmer C. (2016b). Sanitation, Wastewater Management and Sustainability: From Waste Disposal to Resource Recovery. United Nations Environment Programme and Stockholm Environment Institute, Nairobi and Stockholm. https://www.sei-international.org/mediamanager/documents/Publications/SEI-UNEP-2016-SanWWM&Sustainability.pdf (accessed 9 November 2017).

Andersson M. and Minoia P. (2017). Ecological sanitation: a sustainable goal with local choices. A case study from Taita Hills, Kenya. *African Geographical Review*, **36**(2), 183–199, doi: 10.1080/19376812.2015.1134336

Apul D. (2010). Ecological design principles and their implications on water infrastructure engineering. *Journal of Green Building*, **5**(3), 147e164, doi: 10.3992/jgb.5.3.147

Ashley K., Cordell D. and Mavinic D. (2011). A brief history of phosphorus: from the philosopher's stone to nutrient recovery and reuse. *Chemosphere*, **84**(6), 737–746, doi: 10.1016/j.chemosphere.2011.03.001. PMID: 21481914

ASHWAS (2011). ASHWAS Process Handbook: Planning and Execution Guide for Participatory Surveys on Household Water and Sanitation. Arghyam, Bengaluru 560 008, Karnataka, India.

Balkema A. J., Preisig H. A., Otterpohl R. and Lambert F. J. D. (2002). Indicators for the sustainability assessment of wastewater treatment systems. *Urban Water*, **4**, 153–161.

Banerjee S. G. and Morella E. (2011). Africa's Water and Sanitation Infrastructure: Access, Affordability, and Alternatives. Directions in Development; Infrastructure. World Bank, Washington D.C., USA. http://documents.worldbank.org/curated/en/712211468202191672/Africas-water-and-sanitation-infrastructure-access-affordability-and-alternatives (accessed 27 May 2017).

Barat M. A., Bhandari B. and Wali S. (2014). Capacity building and training services in the WASH sector in Afghanistan. 37th WEDC International Conference, Hanoi, Vietnam, pp. 1–5.

Barros P., Ruiz I. and Soto M. (2008). Performance of an anaerobic digester-constructed wetland system for a small community. *Ecological Engineering*, **33**(2), 142–149.

Bartram J. and Cairncross S. (2010). Hygiene, Sanitation, and Water: forgotten Foundations of Health. *PLoS Medicine*, **7**(11), 1–9, doi: 10.1371/journal.pmed.1000367

Benetto E., Nguyen D., Lohmann T., Schmitt B. and Schosseler P. (2009). Life cycle assessment of ecological sanitation system for small-scale wastewater treatment. *Science of the Total Environment*, 1506–1516, doi: 10.1016/j.scitotenv.2008.11.016

Benne B. and Mang P. (2015). Working regeneratively across scales-insights from nature applied to the built environment. *Journal of Cleaner Production*, **109**, 42–52. http://dx.doi.org/10.1016/j.jclepro.2015.02.037

Bentley J. (1986). Programming pearls: little languages. *Communications of the ACM*, **29**(8), 711–721. http://dx.doi.org/10.1145/6424.315691

Binder C. R., Hinkel J., Bots P. W. G. and Pahl-Wostl C. (2013). Comparison of frameworks for analyzing social-ecological systems. *Ecology and Society*, **18**(4), 26, 1–19.

Blackett I., Hawkins P. and Heymans C. (2014). Targeting the Urban Poor and Improving Services in Small Towns. The Missing Link in Sanitation Service Delivery: A Review of Fecal Sludge Management in 12 Cities. Water and Sanitation Program (WSP) Research Brief. April 2014. WSP, World Bank, Washington DC, USA.

Bos A. and Gijzen H. (October 2005). Health benefits versus costs of water supply and sanitation. *Water*, **21**(7.5), 1–72.

Bracken P., Wachtier A., Panesar A. R. and Lange J. (2006). The road not taken: how traditional excreta and greywater management may point the way to a sustainable future. *Water Science & Technology: Water Supply*, **7**(1), 219–227, doi: 10. 2166/ws.2007.025

Brands E. (2014). Prospects and challenges for sustainable sanitation in developed nations: a critical review. *Environmental Reviews*, **22**, 1–18, doi: 10.1139/er-2013-0082

Brepolsa C. H., Dorgeloh E., Frechen F. B., Fuchs W., Haider S., Joss A., de Korte K., Ruiken C. H., Schier W., van der Roest H., Wett M. and Wozniak T. H. (2008). Upgrading and retrofitting of municipal wastewater treatment plants by means of membrane bioreactor (MBR) technology. *Desalination*, **231**, 20–26.

Bright-Davies L., Schmidt A., Duma L. and Mbuduka F. (2016). 5 step planning guide: which solutions go where? Case Study: Dar es Salaam, Tanzania Solutions for residential & commercial areas. Produced by BORDA.

Brikke F. (2000). Operation and Maintenance of Rural Water Supply and Sanitation Systems: A Training Package for Managers and Planners. World Health Organization Geneva, Switzerland. http://apps.who.int/iris/bitstream/10665/66716/1/WHO_SDE_WSH_00.2_p1-124.pdf (accessed 30 April 2017).

Brown P. and Garver G. (2008). In Right Relationship, Building a Whole Earth Economy. Berrett-Koehler Publishers, San Francisco, USA.

Butler D. and Parkinson J. (1997). Towards sustainable urban drainage. *Water Science and Technology*, **35**(9), 53–63, doi: 10.1016/S0273-1223(97)00184-4

Campbell O. M. R., Benova L., Gon G., Afsana K. and Cumming O. (2015). Getting the basic rights – the role of water, sanitation and hygiene in maternal and reproductive health: a conceptual framework. *Tropical Medicine and International Health*, **20**(3), 252–267, doi: 10.1111/tmi.12439

Capodaglio A. G. (2017). Integrated, decentralized wastewater management for resource recovery in rural and peri-urban areas. *Resources*, **6**, 22, 1–20, doi: 10.3390/resources6020022

Capra F. (1997). The Web of Life. Doubleday-Anchor Books, New York, USA.

Capra F. (2002). The Hidden Connections: Integrating the Biological, Cognitive, and Social Dimensions of Life into a Science of Sustainability. Doubleday, New York, USA.

Checkland P. (1997). Systems Thinking, Systems Practice. John Wiley & Sons Ltd, Chichester, UK.

Cheng J. J., Schuster-Wallace C. J., Watt S., Newbold B. K. and Mente A. (2012). An ecological quantification of the relationships between water, sanitation and infant, child, and maternal mortality. *Archives of Environmental Health*, **11**(1), 1–8, doi: 10.1186/1476-069X-11-4

Chinyama A., Chipato P. T. and Mangore E. (2012). Sustainable sanitation systems for low income urban areas – a case of the city of Bulawayo, Zimbabwe. *Physics and Chemistry of the Earth*, **50–52**, 233–238, doi: 10.1016/j.pce.2012.08.010

Chowdhry S. and Kone D. (2012). Business Analysis of Fecal Sludge Management: Emptying and Transportation Services in Africa and Asia. Draft Final Report. Sponsored by the Bill & Melinda Gates Foundation. http://www.susana.org/_resources/documents/default/2-1662-chowdhury-2012-business.pdf (accessed 27 May 2017).

Cofie O., Nikiema J., Impraim R., Adamtey N., Paul J. and Kone D. (2016). Co-Composting of Solid Waste and Fecal Sludge for Nutrient and Organic Matter Recovery. International Water Management Institute (IWMI), Colombo, Sri Lanka, CGIAR Research Program on Water, Land and Ecosystems (WLE). 47p. (Resource Recovery and Reuse Series 3). http://dx.doi.org/10.5337/2016.204

Cole R. J., Charest S. and Schroeder S. (2006). Beyond Green: Drawing on Nature (For the Royal Architectural Institute of Canada's 'Beyond Green: Adaptive, Restorative and Regenerative Design' course – SDCB 305). The University of British Columbia.

Cole R. J., Busby P., Guenther R., Briney L., Blaviesciunaite A. and Alencar T. (2012). A regenerative design framework: setting new aspirations and initiating new discussions. *Building Research & Information*, **40**(1), 95–111, doi: 10.1080/09613218.2011.616098

Cookey P. E., Koottatep T., van der Steen P. and Lens P. N. L. (2016). Public health risk assessment tool: strategy to improve public policy framework for onsite wastewater treatment systems (OWTS). *Journal of Water, Sanitation and Hygiene for Development*, **06.1**, 74–88, doi: 10.2166/washdev.2016.081

Cooper P. F. (2001). Historical aspects of wastewater treatment: Chapter 2. In: Decentralised Sanitation and Reuse – Concepts, Systems and Implementation, P. Lens, G. Zeeman and G. Lettinga (eds), IWA Publishing, the Hague, Netherlands.

Crocker J., Shields K. F., Venkataramanan V., Saywell D. and Bartram J. (2016). Building capacity for water, sanitation, and hygiene programming: training evaluation theory applied to CLTS management training in Kenya. *Social Science & Medicine*, **166**, 66–76.

Delleur J. W. (2003). The evolution of urban hydrology: past, present, and future. *Journal of Hydraulic Engineering*, **129**(8), 563–573, doi: 10.1061/(ASCE)0733-9429(2003)129:8(563)

Desai G., Willbur J., Ram P., Jensen J. N., Lenker J. and Smith K. (2016). Sanitation for all: a framework for research and practice to improve equity for people with disabilities. 39th WEDC International Conference, Kumasi, Ghana.

DeWitt R. (2010). Worldviews: An Introduction to the History and Philosophy of Science. Wiley-Blackwell, Oxford, UK.

Domènech L. and Saurí D. (2010). Socio-technical transitions in water scarcity contexts: public acceptance of greywater reuse technologies in the Metropolitan Area of Barcelona. *Resources Conservation and Recycling*, **55**, 53–62.

Drangert J. O. (1998). Fighting the urine blindness to provide more sanitation options. *Water SA*, **24**(2), 157–164. http://www.wrc.org.za/Knowledge%20Hub%20Documents/Water%20SA%20Journals/Manuscripts/1998/02/WaterSA_1998_02_apr98_p157.pdf (accessed 6 November 2013).

Dreibelbis R., Winch P. J., Leontsini E., Hulland K. R. S., Ram P. K., Unicomb L. and Luby S. P. (2013). The Integrated Behavioural Model for Water, Sanitation, and Hygiene: a systematic review of behavioural models and a framework for designing and evaluating behavioural change interventions in infrastructure-restricted settings. *BMC Public Health*, **13**, 10–15. http://www.biomedcentral.com/1471-2458/13/1015 (accessed 29 May 2017).

Du Plessis C. and Brandon P. (2015). An ecological worldview as basis for a regenerative sustainability paradigm for the built environment. *Journal of Cleaner Production*, **109**, 53–61. https://doi.org/10.1016/j.jclepro.2014.09.098

Du Plessis C. and Cole R. J. (2011). Motivating change: shifting the paradigm. *Building Research & Information*, **39**(5), 436–449.

DWAF (Department of Water Affairs and Forestry) (2008a). the National Sanitation Bucket Replacement Programme – Lessons Learnt. http://www.ewisa.co.za/misc/WIN_SA_Facts/WIN%20SA/TheNSRPLessonLearnt.pdf (accessed 12 July 2017).

DWAF (Department of Water Affairs and Forestry) (2008b). Free Basic Sanitation Implementation Strategy. Department of Water Affairs and Forestry, Pretoria. https://www.dwaf.gov.za/dir_ws/wspd/policyinfo.aspx?filen=556 (accessed 1 February 2010).

Dwipayanti N. M. U., Phung T. D., Shannon R. and Chu C. (2017). Towards sustained sanitation services: a review of existing frameworks and an alternative framework combining ecological and sanitation life stage approaches. *Journal of Water, Sanitation and Hygiene for Development*, (1), 25–42, doi: 10.2166/washdev.2017.086

Ekane N., Nykvist B., Kjellén M., Noel S. and Weitz N. (2014). Multi-level sanitation governance: understanding and overcoming the challenges in the sanitation sector in Sub-Saharan Africa. Working paper no. 2014–4. Stockholm

Environment Institute. https://www.sei-international.org/mediamanager/documents/Publications/Water-sanitation/sei-workingpaper-ekane-multi-level-sanitation.pdf (accessed 8 May 2017).

Elzein Z., Abdou A. and Abd Elgawad I. (2016). Constructed wetlands as a sustainable wastewater treatment method in communities. *Procedia Environmental Sciences*, **34**, 605–617.

EMF (2012). Towards the Circular Economy: Opportunities for the consumer goods sector, vol. 1. https://www.ellenmacarthurfoundation.org/assets/downloads/publications/Ellen-MacArthur-Foundation-Towards-the-Circular-Economy-vol.1.pdf (accessed 7 September 2017).

EMF (2013). Towards the Circular Economy: Opportunities for the consumer goods sector, vol. 2. https://www.ellenmacarthurfoundation.org/assets/downloads/publications/TCE_Report-2013.pdf (accessed 7 September 2017).

Esrey S. A. (2001). Towards a recycling society ecological sanitation – closing the loop to food security. *Water Science & Technology*, **43**(4), 177–187.

European Commission (2015). In: Innovation, D.-G.F.R.A. (ed.), Towards an EU Research and Innovation Policy Agenda for Nature-based Solutions & Re-naturing Cities – Final Report of the Horizon 2020 Expert Group. European Commission, DirectorateGeneral for Research and Innovation, Brussels, Belgium, p. 74.

Exley J. L. R., Liseka B., Cumming O. and Ensink J. H. J. (2014). The sanitation ladder, what constitutes an improved form of sanitation? *Advances in Environmental Science and Technology*, **49**, 1086–1094, doi: 10.1021/es503945x

Fam D. M. and Mitchell C. A. (2012). Sustainable innovation in wastewater management: lessons for nutrient recovery and reuse. *Local Environment: The International Journal of Justice and Sustainability*, **18**(7), 769–780, doi: 10.1080/13549839.2012.716408

Fan B., Hu M., Wang H., Xu M., Qu B. and Zhu S. (2017). Get in sanitation 2.0 by opportunity of rural China: scheme, simulating application and life cycle assessment. *Journal of Cleaner Production*, **147**, 86–95.

Fewtrell L., Kaufmann R. B., Kay D., Enanoria W., Haller L. and Colford J. M. (2005). Water, sanitation, and hygiene interventions to reduce diarrhoea in less developed countries: a systematic review and meta-analysis. *The Lancet Infectious Diseases*, **5**(1), 42–52, doi: 10.1016/S1473-3099(04)01253-8

Fullerton J. (2015). Regenerative Capitalism – How Universal Principles and Patterns Will Shape Our New Economy. The Regenerative Framework & White Paper. http://capitalinstitute.org/wp-content/uploads/2015/04/2015-Regenerative-Capitalism-4-20-15-final.pdf (accessed 6 October 2017).

Galli G., Nothomb C. and Baetings E. (2014). Towards Systemic Change in Urban Sanitation. IRC Working Paper. The Hague, Netherlands.

Geels F. (2005). Co-evolution of technology and society: the transition in water supply and personal hygiene in the Netherlands (1850–1930) – a case study in multilevel perspective. *Technology in Society*, **27**, 363–397.

Geels F. W. (2006). The hygienic transition from cesspools to sewer systems (1840–1930): the dynamics of regime transformation. *Research Policy*, **35**(7), 1069–1082.

GHK Research and Training (2000). Strategic Planning for Municipal Sanitation: A Guide. Published in association with Water, Engineering and Development Centre (WEDC) and Water and Sanitation Program for South Asia (WSP-SA), London. https://www.ircwash.org/sites/default/files/302.5-00ST-16187.pdf (accessed 13 May 2017).

Gine R., Jimenez A. and Perez-Foguet A. (2011). The Future of Water, Sanitation and Hygiene: Innovation, Adaptation and Engagement in a Changing World – A Closer Look at the Sanitation Ladder: Issues of Monitoring the Sector. Water, Engineering and Development Centre International Conference. 35th WEDC International Conference: The Future of Water, Sanitation and Hygiene in Low-Income Countries: Innovation, Adaptation and Engagement in a Changing World, Loughborough, UK. http://hdl.handle.net/2117/14199

Giné-Garriga R., Flores-Baquero O., de Palencia A. J. and Pérez-Foguet A. (2017). Monitoring sanitation and hygiene in the 2030 Agenda for Sustainable Development: a review through the lens of human rights. *Science of the Total Environment*, **580**, 1108–1119.

Gosling L. (2010). Equity and inclusion: a rights-based approach. A WaterAid report, London, UK. file:///C:/Users/NATS/Downloads/equity%20and%20inclusion%20framework.pdf (accessed 20 June 2017).

Graham P. M. (2003). Building Ecology. First Principles for a Sustainable Built Environment. Blackwell Science Ltd, Oxford, UK.

GTZ (Deutsche Gesellschaft für Technische Zusammenarbeit) (2002). Ecosan – recycling beats disposal. Eschborn (Germany) 7 GTZ. http://www.waterfund.go.ke/watersource/Downloads/002.%20ecosan-projectinfo-2002.pdf (accessed 9 September 2018).

Gunawardana I., Leendertse K. and Handoko W. (2013). Monitoring outcomes and impacts of capacity development in the water sector: a Cap-Net UNDP experience. *Water Policy*, **15**, 226–241. http://dx.doi.org/10.2166/wp.2013.121

Guterstam B. and Todd J. (1990). Ecological engineering for wastewater treatment and its application in New England and Sweden. *Ambio*, **19**, 173–175.

Halbe J., Pahl-Wostl C., Sendzimir J. and Adamowski J. (2013). Towards adaptive and integrated management paradigms to meet the challenges of water governance. *Water Science & Technology*, **67**(11), 2651–2660, doi: 10.2166/wst.2013.146

Hallerod B., Rothstein B., Daoud A. and Nandy S. (2013). Bad governance and poor children: a comparative analysis of government efficiency and severe child deprivation in 68 low-and middle-income countries. *World Development*, **48**(1), 19–31.

Haq G. and Cambridge H. (2012). Exploiting the co-benefits of ecological sanitation. *Current Opinion in Environmental Sustainability*, **4**(4), 431–435.

Harman W. (1970). An Incomplete Guide to the Future. W.W. Norton, San Francisco, USA.

Hegger D. L. T., van Vliet J. and van Vliet B. J. M. (2007). Niche management and its contribution to regime change: the case of innovation in sanitation. *Technology Analysis and Strategic Management*, **19**, 729–746.

Heylighen F. (2006). The Newtonian World View. Principia Cybernetica Web. Online. http://pespmc1.vub.ac.be/NEWTONWV.html (accessed 15 May 2014).

Hiessl H., Toussaint D., Becker M., Dyrbusch A., Geisler S., Herbst H. and Prager J. U. (2003). Alternativen der kommunalen Wasserversorgung und Abwasserentsorgung (Alternatives of Municipal Water Supply and Sanitation), AKWA 2100. Physica-Verlag, Heidelberg, Germany.

Hinkel J. (2008). Transdisciplinary Knowledge Integration: Cases from Integrated Assessment and Vulnerability Assessment. PhD thesis, Wageningen University, Wageningen, Netherlands. http://ciret-transdisciplinarity.org/biblio/biblio_pdf/Hinkel.pdf

Hinkel J., Bots P. W. G. and Schluter M. (2014). Enhancing the Ostrom social–ecological system framework through formalization. *Ecology and Society*, **19**(3), 51, 1–19, doi: 10.5751/ES-06475-190351.

Hodges N. (2006). Regenerative Design Theory and Practice: A Demonstration of the Integrated Framework in a Resort Development at Mountain Lake, VA. Master's thesis, Virginia Polytechnic and State University. https://theses.lib.vt.edu/theses/available/etd-05082006-101306/

Hu M., Fan B., Wang H., Qu B. and Zhu S. (2016). Constructing the ecological sanitation: a review on technology and methods. *Journal of Cleaner Production*, **125**, 1–21. http://dx.doi.org/10.1016/j.jclepro.2016.03.012

Hudak P. (1998). Modular domain specific languages and tools. Proceedings of the 5th International Conference on Software Reuse, IEEE, Bangalore, India, pp. 134–142. http://dx.doi.org/10.1109/ICSR.1998.685738

Hutton G. and Varughese M. (2016). The Costs of Meeting the 2030 Sustainable Development Goal Targets on Drinking Water, Sanitation, and Hygiene. Water and Sanitation Program (WSP): Technical Paper 103171. http://documents.worldbank.org/curated/en/415441467988938343/pdf/103171-PUB-Box394556B-PUBLIC-EPI-K8543-ADD-SERIES.pdf (accessed 5 October 2017).

Illingworth C. M. (1974). Trapped fingers and amputated finger tips in children. *European Journal of Pediatric Surgery*, **9**(6), 853–858.

Ionescu C., Klein R. J. T., Hinkel J., Kumar K. S. K. and Klein R. (2009). Towards a formal framework of vulnerability to climate change. *Environmental Modelling and Assessment*, **14**(1), 1–16. http://dx.doi.org/10.1007/s10666-008-9179-x

ISF-UTS & SNV (2016). 'Are we doing the right thing? Critical questioning for city sanitation planning'. Institute for Sustainable Futures, University of Technology Sydney and SNV Netherlands Development Organization. https://

www.uts.edu.au/sites/default/files/Arewedoingtherightthing-criticalquestioningforcitysanitationplanning.pdf (accessed 2 May 2017).

IWA (2014). Sanitation 21 – a planning framework for improving city-wide sanitation services. Publication supported by Sandec-EAWAG and Switzerland, GIZ, Germany.

IWA (2016a). Water Utility Pathways in a Circular Economy. http://www.iwa-network.org/wp-content/uploads/2016/07/IWA_Circular_Economy_screen-1.pdf (accessed 7 September 2016).

IWA (24:2016b). Non-sewered sanitation systems – General safety and performance requirements for design and testing. https://www.iso.org/standard/70604.html (accessed 9 May 2017).

Jantsch E. (1980). The Self-Organizing Universe: Scientific and Human Applications of the Emerging Paradigm of Evolution. Pergamon, New York, USA.

Jenkin S. and Pedersen Z. M. (2009). Rethinking Our Built Environments: Towards a Sustainable Future. A Research Document. Ministry for the Environment, Manatu Mo Te Taiao, Wellington, New Zealand. https://www.researchgate.net/publication/261476724_Rethinking_our_built_environments_Towards_a_sustainable_future (accessed 10 February 2018).

Joint Monitoring Programme (2015). Progress on Sanitation and Drinking Water: 2015 Update and MDG Assessment. Report of the Joint Monitoring Programme of the World Health Organization, Geneva and UNICEF, New York, USA. https://data.unicef.org/wp-content/uploads/2015/12/Progress-on-Sanitation-and-Drinking-Water_234.pdf (accessed 10 February 2018).

Joint Monitoring Programme (JMP) (2010). The drinking-water and sanitation ladders. Report of the Joint Monitoring Programme of the World Health Organization, Geneva and UNICEF, New York, USA. www.wssinfo.org/definitions-methods/watsan-ladder/ (accessed 13 March 2012).

Joshi D. K., Hughes B. B. and Sisk T. D. (2015). Improving governance for the post-2015 sustainable development goals: scenario forecasting the next 50 years. *World Development*, **70**, 286–302, 2015 0305–750X. http://dx.doi.org/10.1016/j.worlddev.2015.01.013

Kalbermatten J. M., Julius D. S., Mara D. D. and Gunnerson C. G. (1980). Appropriate Technology for Water Supply and Sanitation A Planner's Guide. The International Bank for Reconstruction and Development/the World Bank. https://www.ircwash.org/sites/default/files/201-80AP-19010.pdf (accessed 27 May 2017).

Kalbermatten J. M., Julius D. S. and Cunnerson C. G. (1982). Appropriate sanitation alternatives: a technical and economic appraisal. World Bank Studies in Water Supply and Sanitation no. 1. The World Bank. http://documents.worldbank.org/curated/en/637021468740368977/Appropriate-sanitation-alternatives-a-technical-and-economic-appraisal (accessed 27 May 2017).

Kar K. and Chambers R. (2008). Handbook on Community-Led Total Sanitation, Plan International (UK). London, UK.

Katukiza A. Y., Ronteltap M., Oleja A., Niwagaba C. B., Kansiime F. and Lens P. N. L. (2010). Selection of sustainable sanitation technologies for urban slums – a case of Bwaise III in Kampala, Uganda. *Science of the Total Environment*, **409**, 52–62.

Katukiza A. Y., Ronteltap M., Niwagaba C. B., Foppen J. W. A., Kansiime F. and Lens P. N. L. (2012). Sustainable sanitation technology options for urban slums. *Biotechnology Advances*, **30**, 964–978.

Kauffman S. (1995). At Home in the Universe: The Search for Laws of Self-Organization and Complexity. Oxford University Press, New York, USA.

Kearney M. (1984). Worldviews. Chandler & Sharp, Novato, USA.

Keraita B., Drechsel P., Klutse A. and Cofie O. (2014). On-farm treatment options for wastewater, greywater and fecal sludge with special reference to West Africa. CGIAR Research Program on Water, Land and Ecosystems (WLE). 32p. (Resource Recovery and Reuse Series 1). Colombo, Sri Lanka: International Water Management Institute (IWMI).

Kerstens S. M., Spiller M., Leusbrock I. and Zeeman G. (2016). A new approach to nationwide sanitation planning for developing countries: case study of Indonesia. *Science of the Total Environment*, **550**, 676–689.

Kibert C. J., Sendzimir J. and Guy G. B. (2002). Construction Ecology: Nature as the Basis for Green Buildings. Spon Press, London, UK and New York, USA.

Koltko-Rivera M. E. (2004). The psychology of worldviews. *Review of General Psychology*, **8**, 3–58.

Kragl M., Knapp D., Nacu E., Khattak S., Maden M. and Epperlein H. H. (2009). Cells keep a memory of their tissue origin during axolotl limb regeneration. *Bionature*, **460**, 60–65.

Kuhn T. S. (1962). The Structure of Scientific Revolutions. University of Chicago Press, Chicago, USA.

Kujawa-Roeleveld K. and Zeeman G. (2006). Anaerobic treatment in decentralized and source-separation-based sanitation concepts. *Reviews in Environmental Science and Bio/Technology*, **5**, 115–139.

Kuznyetsov V. (2007). Millennium Development Goals – Problems of Turning Commitment into Reality. Stockholm International Water Institute: Abstract Volume World Water Week, Stockholm, August 12–18, 2007.

Kvarnström E., Bracken P., Karrman E., Finnson A. and Saywell D. (2004). People-Centred Approaches to Water and Environmental Sanitation. Proceedings of the 30th WEDC International Conference, Vientiane, Lao, PDR: WEDC.

Kvarnström E., Emilsson K., Stintzing A. R., Johansson M., Jönsson H., af Petersens E., Schönning C., Christensen J., Hellström D., Qvarnström L., Ridderstolpe P. and Drangert J. (2006). Urine Diversion: One Step Towards Sustainable Sanitation. Stockholm Environmental Institute, Stockholm, Sweden.

Kvarnström E., McConville J., Bracken P., Johansson M. and Fogde M. (2011). The sanitation ladder – a need for a revamp? *Journal of Water Sanitation and Hygiene for Development*, **1**(1), 3–12.

Langergrabera G. and Muelleggera E. (2005). Ecological Sanitation – a way to solve global sanitation problems? *Environment International*, **31**(2005), 433–444.

Larsen T. A. (2011). Redesigning wastewater infrastructure to improve resource efficiency. *Water Science and Technology*, **63**(11), 2535–2541.

Larsen T. A. and Maurer M. (2011). Source separation and decentralization. In: Treatise on Water Science, P. Wilderer (ed.), Elsevier, Oxford, UK, pp. 203–229.

Larsen T. A., Alder A. C., Eggen R. I., Maurer A. and Lienert J. (2009). Source separation: will we see a paradigm shift in wastewater handling. *Environmental Science & Technology*, **43**, 6121–6125.

Larsen T. A., Maurer M., Eggen R. I. L., Pronk W. and Lienert J. (2010). Decision support in urban water management based on generic scenarios: the example of NoMix technology. *Journal of Environmental Management*, **91**, 2676–2687.

Leclerc M. (2005). The ecohydraulics paradigm shift: IAHR enters a New Era. *Journal of Hydraulic Research*, **43**(4 suppl.), 63–64.

Lenhoff H. M. and Lenhoff S. G. (1991). Abraham Trembley and the origins of research on regeneration in animals. In: A History of Regeneration Research. Milestones in the Evolution of a Science, C. E. Dinsmore (ed.), Cambridge University Press, Cambridge, UK, pp. 47–66.

Lens P., Zeeman G. and Lettinga G. (2001). Decentralized Sanitation and Reuse: Concepts, Systems and Implementation. IWA Publishing, London, UK.

Lia Buarque D. M. G. (2012). Sociotechnical design for a sustainable world. *Theoretical Issues in Ergonomics Science*, **13**(2), 240–269, doi: 10.1080/1463922X.2011.641230

Libralato G., Ghirardini A. V. and Avezzu F. (2012). To centralize or to decentralize: an overview of the most recent trends in wastewater treatment management. *Journal of Environmental Management*, **94**(1), 61–68.

Lietaer B., Arnsperger C., Goerner S. and Brunnhuber S. (2012). Money and Sustainability, The Missing Link. Triarchy Press, Charmouth, UK.

Littman J. A. (2009). Regenerative Architecture: A Pathway Beyond Sustainability. Master's thesis, University of Massachusetts, Amherst, USA. http://scholarworks.umass.edu/theses/303 (accessed 5 April, 2017).

Loetscher T. and Keller J. (2002). A decision support system for selecting sanitation systems in developing countries. *Socio-Economic Planning Sciences*, **36**, 267–290.

Lopes A. M., Fam D. and Williams J. (2012). Designing sustainable sanitation: involving design in innovative, transdisciplinary research. *Design Studies*, **33**, 298–317, doi: 10.1016/j.destud.2011.08.005

Lüthi C., Morel A., Tilley E. and Ulrich L. (2011a). Community-Led Urban Environmental Sanitation Planning (CLUES); Swiss Federal Institute of Aquatic Science and Technology (EAWAG) Dübendorf, Switzerland. http://www.susana.org/_resources/documents/default/2-1300-cluesguid.pdf (accessed 14 May 2017).

Lüthi C., Panesar A., Schütze T., Norström A., McConville J., Parkinson J., Saywell D. and Ingle R. (2011b). Sustainable Sanitation in Cities: A Framework for Action. Papiroz Publishing House, Rijswijk, Netherlands.

Lyle J. T. (1994). Regenerative Design for Sustainable Development. John Wiley & Sons, Hoboken.

Magid J., Eilersen A. M., Wrisberg S. and Henze M. (2006). Possibilities and barriers for recirculation of nutrients and organic matter from urban to rural areas: A technical theoretical framework applied to the medium-sized town Hillerød, Denmark. *Ecological Engineering*, **28**, 44–54.

Mahmood Q., Pervez A., Zeb B. S., Zaffar H., Yaqoob H., Waseem M., Zahidullah and Afsheen S. (2013). Natural treatment systems as sustainable ecotechnologies for the developing countries. *BioMed Research International*, **79**, 63–73. http://dx.doi.org/10.1155/2013/796373

Maienschein J. (2011). Regenerative medicine's historical roots in regeneration, transplantation, and translation. *Journal of Developmental Biology*, **358**, 278–284, doi:10.1016/j.ydbio.2010.06.014

Mang P. and Reed B. (2012a). Designing from place: a regenerative framework and methodology. *Building Research & Information*, **40**(1), 23–38.

Mang P. and Reed B. (2012b). Regenerative Development and Design. Regenesis Group and Story of Place, Encyclo. Sustainability Science & Technology.

Mang P., Haggard B. and Regenesis (2016). Regenerative Development and Design: A Framework for Evolving Sustainability. John Wiley & Sons, Inc., New York, USA.

Mara D., Drangert J., Viet Anhc N., Tonderski A., Gulyas H. and Tonderski K. (2007). Selection of sustainable sanitation arrangements. *Water Policy*, **9**, 305–318.

Marshall R. E. and Farahbakhsh K. (2013). Systems approaches to integrated solid waste management in developing countries. *Waste Management*, **33**, 988–1003. http://dx.doi.org/10.1016/j.wasman.2012.12.023

Massoud M. A., Tarhini A. and Nasr J. A. (2009). Decentralized approaches to wastewater treatment and management: Applicability in developing countries. *Journal of Environmental Management*, **90**, 652–659, doi: 10.1016/j.jenvman.2008.07.001

Maurer M., Bufardi A., Tilley E., Zurbrügg C. and Truffer B. (2012). A compatibility-based procedure designed to generate potential sanitation system alternatives. *Journal of Environmental Management*, **104**, 51–61.

McConville J., Kunzle R., Messmer U., Udert K. M. and Larsen T. A. (2014). Decision support for redesigning wastewater treatment technologies. *Environmental Science & Technology*, **48**(20), 12238–12246.

McDonough W. and Braungart B. (2001a). The Next Industrial Revolution, William McDonough, Michael Braungart and the Birth of the Sustainable Economy. Earthome Productions, Stevenson, USA.

McDonough W. and Braungart B. (2001b). Cradle to Cradle: Remaking the Way We Make Things. Northpoint Press, New York, USA.

McHarg I. L. (1969). Design with Nature. Doubleday, New York, USA.

Meadows D. H. (1999). Leverage Points: Places to intervene in a system. The Sustainability Institute, Hartland VT, USA. https://www.sustainabilityinstitute.org/pubs/Leverage-Points.pdf (accessed 27 February 2017).

Meinzen-Dick R. (2007). Beyond panaceas in water institutions. *PNAS* **104**(39), 15200–15205. https://doi.org/10.1073/pnas.070229

Mele C., Pels J. and Polese F. (2010). A brief review of systems theories and their managerial applications. *Service Science*, **2**(1–2), 126–135. http://dx.doi.org/10.1287/serv.2.1_2.126

Miles M. B. and Huberman A. M. (1994). Qualitative Data Analysis: An Expanded Source Book. Sage, Newbury Park, USA, Cited in Jabareen Y. (2009) Building a conceptual framework: philosophy, definitions and procedures. International Journal for Qualitative Methods, 2008.

Miller M. E. and West A. N. (1993). Influences of world view on personality, epistemology, and choice of profession. In: Development in the Work-Place, J. Demick and P. M. Miller (eds), Erlbaum, Hillsdale, USA, pp. 3–19.

Mitsch W. J. (1993). Ecological engineering a cooperative role with the planetary life-support system. *Advances in Environmental Science and Technology*, **27**(3), 438–445.

Mitsch W. J. and Jørgensen S. E. (1989). Ecological Engineering: An Introduction to Ecotechnology. Wiley, New York, USA.

Mitsch W. J. and Jørgensen S. E. (2003). Ecological engineering: a field whose time has come. *Ecological Engineering*, **20**, 363–377.

Murray A. and Buckley C. (2009). Designing reuse-oriented sanitation infrastructure: the design for service planning approach. In: Wastewater Irrigation and Health Assessing and Mitigating Risk in Low-income Countries, P. Drechsel, C. A. Scott, L. Raschid-Sally, M. Redwood and A. Bahri (eds), Earthscan, London, UK.

Mvulirwenande S., Alaerts G. and de Montalvo U. W. (2013). From knowledge and capacity development to performance improvement in water supply: the importance of competence integration and use. *Water Policy*, **15**, 267–281. http://dx.doi.org/10.2166/wp.2013.023

Nansubuga I., Banadda N., Verstraete W. and Rabaey K. (2016). A review of sustainable sanitation systems in Africa. *Reviews in Environmental Science and Bio/Technology*, **15**, 465–478, doi: 10.1007/s11157-016-9400-3

Nesshöver C., Assmuth T., Irvine K. N., Rusch G. M., Waylen K. A., Delbaere B., Haase D., Jones-Walters L., Keune H., Kovacs E., Krauze K., Külvik M., Rey F., van Dijk J., Vistad O. I., Wilkinson M. E. and Wittmer H. (2017). The science, policy and practice of nature-based solutions: an interdisciplinary perspective. *Science of the Total Environment*, **579**, 1215–1227.

Ngai T. K. K., Coff B. E., Manzano E., Seel K. and Elson P. (2014). Evaluation of Education and Training in Water and Sanitation Technology: Case Studies in Nepal and Peru. 37th WEDC International Conference, Hanoi, Vietnam, pp. 1–7.

Nikiema J., Cofie O. and Impraim R. (2014). Technological options for safe resource recovery from fecal sludge. Resource Recovery and Reuse Series 2. International Water Management Institute (IWMI), Sri Lanka. ISSN 2478-0510; e-ISSN 2478-0529; ISBN 978-92-9090-804-3.

Nilsson M. and Olsson L. (2014). Sanitation in an Informal Settlement – A User-Oriented Study on Sanitation Solutions and Factors Influencing Decisions for Implementation in Cochabamba, Bolivia. MS thesis, Chalmers University of Technology, Gothenburg, Sweden. http://publications.lib.chalmers.se/records/fulltext/207030/207030.pdf (accessed 16 February 2017).

O'Reilly K. and Louis E. (2014). The toilet tripod: understanding successful sanitation in rural India. *Health & Place*, **29**, 43–51.

O'Reilly K., Dhanju R. and Goel A. (2017). Exploring 'The Remote' and 'The Rural': open defecation and latrine use in Uttarakhand, India. *World Development*, **93**, 193–205, 2017 0305-750X. https://doi.org/10.1016/j.worlddev.2016.12.022

Obeng P. A., Keraita B., Oduro-Kwarteng S., Bregnhøj H., Abaidoo R. C. and Konradsen F. (2015). 'The latrine ownership ladder: a conceptual framework for enhancing sanitation uptake in low-income peri-urban settings'. *Management of Environmental Quality: An International Journal*, **26**(5), 752–763. https://doi.org/10.1108/MEQ-05-2014-0079

Ostrom E. (2005). Understanding Institutional Diversity. Princeton University Press, Princeton, USA.

Otoo M., Drechsel P., Danso G., Gebrezgabher S., Rao K. and Madurangi G. (2016) Testing the Implementation Potential of Resource Recovery and Reuse Business Models: From Baseline Surveys to Feasibility Studies and Business Plans. International Water Management Institute (IWMI). CGIAR Research Program on Water, Land and Ecosystems (WLE), Colombo, Sri Lanka. 59p. (Resource Recovery and Reuse Series 10), doi: 10.5337/2016.206, https://cgspace.cgiar.org/handle/10568/78431 (accessed 27 May 2017).

Otterpohl R. (2002). Options for alternative types of sewerage and treatment systems directed to improvement of the overall performance. *Water Science and Technology*, **45**(3), 149–158.

Pahl-Wostl C. (2007). Transitions towards adaptive management of water facing climate and global change. *Water Resources Management*, **21**, 49–62, doi: 10.1007/s11269-006-9040-4

Pahl-Wostl C., Holtz G., Kastens B. and Knieper C. (2010). Analyzing complex water governance regimes: the management and transition framework. *Environmental Science & Policy*, **13**, 571–581.

Pahl-Wostl C., Jeffrey P., Isendahl N. and Brugnach M. (2011). Maturing the new water management paradigm: progressing from aspiration to practice. *Water Resources Management*, **25**, 837–856, doi: 10.1007/s11269-010-9729-2

Panebianco S. and Pahl-Wostl C. (2006). Modelling socio-technical transformations in wastewater treatment – a methodological proposal. *Technovation*, **26**, 1090–1100, doi: 10.1016/j.technovation.2005.09.017

Parkinson J., Lüthi C. and Walther D. (GIZ) (2014). Sanitation 21 – A Planning Framework for Improving City-wide Sanitation Services. IWA, EAWAG-Sandec, GIZ. http://www.iwa-network.org/filemanager-uploads/IWA-Sanitation-21_22_09_14-LR.pdf (accessed 2 May 2017).

Pascual Sanz M., Veenstra S., de Montalvo U. W., van Tulder R. and Alaerts G. (2013). What counts as 'results' in capacity development partnerships between water operators? A multi-path approach toward accountability, adaptation and learning. *Water Policy*, **15**, 242–266. http://dx.doi.org/10.2166/wp.2013.022

Patkar A. and Gosling L. (2011). Equity and Inclusion in Sanitation and Hygiene in Africa A Regional Synthesis Paper. Water Supply and Sanitation Collaborative Council and WaterAid. https://www.sanitationmonitoringtoolkit.com/images/SMTdocuments/52_equity_and_inclusion_synthesis_africa_working_paper_for_africasan_final.pdf (accessed 20 June 2017).

Peal A., Evans B. and van der Voorden C. (2010). Hygiene and Sanitation Software: An Overview of Approaches. Water Supply & Sanitation Collaborative Council, Geneva, Switzerland. http://wsscc.org/wp-content/uploads/2016/06/Hygiene-and-Sanitation-Software-An-overview-of-approaches-WSSCC.pdf (accessed 1 June 2017).

Persson C. P. (2017). 1,000-year-old Viking toilet uncovered in Denmark. *Science Nordic*. http://sciencenordic.com/1000-year-old-viking-toilet-uncovered-denmark (accessed 3 July 2017).

Polprasert C. and Koottatep T. (2017). Organic Waste Recycling – Technology, Management and Sustainability, 4th edn. IWA Publishing, the Hague, Netherlands.

Porcellini A. (2009). Regenerative medicine: a review. *Revista Brasileira de Hematologia e Hemoterapia*, **31**(supl.2), São Paulo, Brazil. Online version. http://dx.doi.org/10.1590/S1516-84842009000800017

Potter A., Klutse A., Snehalatha M., Batchelor C., Uandela A., Naafs A., Fonseca C. and Moriarty P. (2011). Assessing sanitation service levels. IRC International Water and Sanitation Centre, Working paper 3, 2nd edn. https://www.ircwash.org/sites/default/files/Potter-2011-Assessing.pdf (accessed 18 May 2017).

Pruss A., Kay D., Fewtrell L. and Bartram J. (2002). Estimating the burden of disease from water, sanitation, and hygiene at a global level. *Environmental Health Perspectives*, **110**(5), 537–542.

Prüss-Üstün A., Bos R., Gore F. and Bartram J. (2008). Safer Water, Better Health: Costs, Benefits and Sustainability of Interventions to Protect and Promote Health. Geneva, Switzerland. http://www.who.int/water_sanitation_health/publications/safer_water/en/ (accessed 16 February 2017).

Ramoa A., Matos J. and Luthi C. (2015). System-Based Decision Trees for the Selection of Sanitation Technologies. 38th WEDC International Conference, Loughborough University, Loughborough, UK. http://wedc.lboro.ac.uk/resources/conference/38/Ramoa-2250.pdf (accessed 26 June 2017).

Rao K. C., Kvarnström E., Di Mario L. and Drechsel P. (2016). Business models for fecal sludge management. CGIAR Research Program on Water, Land and Ecosystems (WLE). (Resource Recovery and Reuse Series 6). International Water Management Institute (IWMI), Colombo, Sri Lanka, doi: 10.5337/2016.21

Raven R. P. J. M. (2007). Co-evolution of waste and electricity regimes: multi-regime dynamics in the Netherlands (1969–2003). *Energy Policy*, **35**(4), 2197–2208. doi.10.1016/j.enpol.2006.07.005

Reed B. (2007). Shifting from 'sustainability' to regeneration. *Building Research & Information*, **35**(6), 674–680, doi: 10.1080/09613210701475753

Rees W. (1999). The built environment and the ecosphere: a global perspective. *Building Research & Information*, **27** (4–5), 206–220.

Renger B. C., Birkeland J. L. and Midmore D. J. (2015). Net-positive building carbon sequestration. *Building Research & Information*, **43**(1), 11–24.

Repella A. (2014). Framework Perspectives for Water: a small sample of the range of frameworks for addressing questions about water. Water Diplomacy Network Blog. http://blog.waterdiplomacy.org/2014/01/framework-perspectives-for-water/ (accessed 6 June 2017).

Robertson P. J. and Choi T. (2010). Ecological governance: organizing principles for an emerging era. *Public Administration Review*, **70**(1), 89–99, doi: 10.1111/j.1540-6210.2010.02250.x

Robinson J. and Cole R. J. (2015). Theoretical underpinnings of regenerative sustainability. *Building Research & Information*, **43**(2), 133–143, doi: 10.1080/09613218.2014.979082

Rocky Mountain Institute (2006). Hypercar design and technology. https://old.rmi.org/sitepages/pid390.php (accessed on 7 February 2018).

Rodale Institute (2017). Regenerative Organic Agriculture and Climate Change: A Down-to-Earth Solution to Global Warming. https://rodaleinstitute.org/assets/WhitePaper.pdf (accessed 6 October 2017).

Rodic L. (2015). Waste governance. In: Global Waste Outlook, D. C. Wilson (ed.), ISWA/UNEP, Nairobi, Kenya. http://www.unep.org/ietc/Portals/136/Publications/Waste%20Management/GWMO%20report/GWMO%20full%20report.pdf (accessed 12 September 2016).

Roefs I., Meulman B., Vreeburg J. H. G. and Spiller M. (2017). Centralised, decentralised or hybrid sanitation systems? Economic evaluation under urban development uncertainty and phased expansion. *Australian Journal of Marine and Freshwater Research*, **109**, 274–286.

Sampogna G., Guraya S. Y. and Forgione A. (2015). Regenerative medicine: historical roots and potential strategies in modern medicine. *Journal of Microscopy and Ultrastructure*, **3**, 101–107.

Schertenleib R. (2005). From conventional to advanced environmental sanitation. *Water Science and Technology*, **51**(10), 7–14.

Schmitt R. J. P., Morgenroth E. and Larsen T. A. (2017). Robust planning of sanitation services in urban informal settlements: an analytical framework. *Australian Journal of Marine and Freshwater Research*, **110**, 297–312.

Schübeler P. (1996). Conceptual framework for municipal solid waste management in low-income countries. In: UNDP/UNCHS/World Bank/SDC Collaborative Programme on Municipal Solid Waste Management in Low-Income Countries, K. Wehrle and J. Christen (eds), St. Gallen, Switzerland.

Seadon J. K. (2006). Integrated waste management – looking beyond the solid waste horizon. *Waste Management*, **26** (12), 1327–1336.

Seadon J. K. (2010). Sustainable waste management systems. *Journal of Cleaner Production*, **18**(16–17), 1639–1651.

SER (Society for Ecological Restoration, Science and Policy Working Group) (2004). The SER Primer on Ecological Restoration. http://www.ser.org/ (accessed 15 May 2017).

Simha P., Lalander C., Vinnerås B. and Ganesapillai M. (2017). Farmer attitudes and perceptions to the re–use of fertiliser products from resource–oriented sanitation systems – the case of Vellore, South India. *Science of the Total Environment*, **581–582**, 885–896.

Sonnenberg E. (2013). Toilet provision. *Cape Times*, 1 Oct 2013.

Spaargaren G., Mommaas H., Van Den Burg S., Maas L., Drissen E., Dagevos H., Bargeman B., Putman L., Nijhuis J., Verbeek D. and Sargant E. (2007). More sustainable lifestyles and consumption patterns: a theoretical perspective for the analysis of transition processes within consumption domains. Contrast Research Report, TMP project. Wageningen: Environmental Policy Group, Wageningen University; Telos: Tilburg University; RIVM; LEI, Netherlands.

Speer J. (2012). Participatory governance reform: a good strategy for increasing government responsiveness and improving public services? *World Development*, **40**(12), 2379–2398.

Starkl M., Brunner N., Feil M. and Hauser A. (2015). Ensuring sustainability of non-networked sanitation technologies: an approach to standardization. *Advances in Environmental Science and Technology*, **49**, 6411–6418, doi: 10.1021/acs.est.5b00887

Stenström T. A., Seidu R., Ekane N. and Zurbrügg C. (2011). Microbial Exposure and Health Assessments in Sanitation Technologies and Systems. SEI report, EcoSanRes Series 2011-1.

L. Strande, M. Ronteltap and D. Brdjanovic (2014). Faecal Sludge Management: Systems Approach for Implementation and Operation. IWA Publishing, London, UK.

Sunderland M. (2008). Studying Development: The Value of Diversity, Theory, and Synthesis. PhD thesis, Arizona State University, USA.

Sunderland M. (2010). Regeneration: Thomas Hunt Morgan's window into development. *Journal of the History of Biology*, **43**, 325–361.

Sutton S. (2008). The risks of a technology-based MDG indicator for rural water supply. Access to Sanitation and Safe Water: Global Partnerships and Local Actions. Conference Proceedings of the 33rd WEDC International Conference Accra, Ghana, pp. 500–505.

SuSanA (ed.) (2008). Towards More Sustainable Sanitation Solutions (Version 1). Sustainable Sanitation Alliance (SuSanA), Eschborn. https://www.susana.org/_resources/documents/default/3-267-7-1452594644.pdf (accessed 21 June 2017).

Swart D. B. and Palsma A. J. B. (2013). The Netherlands: 'Nieuwe Sanitatie'. In: Source Separation and Decentralization for Wastewater Management, T. A. Larsen, K. M. Udert and J. Lienert (eds), IWA Publishing, London, UK, pp. 431–438.

Taing L. (2017). How sociopolitical factors affected the implementation of Cape Town's vacuum sewer. *Environmental Science: Water Research & Technology*, **3**, 513, doi: 10.1039/c6ew00089d

Taing L., Armitage N., Ashipala N. and Spiegel A. (2013). TIPS for sewering informal settlements: Technology, Institutions, People and Services, WRC Report No. TT 557/13, Water Research Commission, Pretoria, South Africa.

TBC (Toilet Board Coalition) (2016). Sanitation in the Circular Economy: Transformation to a commercially valuable, self-sustaining, biological system. http://www.toiletboard.org/media/17-Sanitation_in_the_Circular_Economy.pdf (accessed 7 September 2017).

Tchobanoglous G. and Leverenz H. (2013). The rationale for decentralization of wastewater infrastructure. In: Source Separation and Decentralization for Wastewater Management, T. A. Larsen, K. M. Udert and J. Lienert (eds), IWA Publishing, London, UK, pp. 101–115.

Thibodeau C., Monette F., Bulle C. and Glaus M. (2014). Comparison of blackwater source-separation and conventional sanitation systems using life cycle assessment. *Journal of Cleaner Production*, **67**, 45–57. https://doi.org/10.1016/j.jclepro.2013.12.012

Tilley E., Strande L., Luthi C., Mosler H., Udert K. M., Gebauer H. and Hering J. G. (2014a). Looking beyond technology: an integrated approach to water, sanitation and hygiene in low income countries. *Advances in Environmental Science and Technology*, **48**(17), 9965–9970, doi: 10.1021/es501645d

Tilley E., Ulrich L., Luthi C., Reymond P. H. and Zurbrugg C. (2014b). Compendium of Sanitation Systems and Technologies, 2nd rev. edn. Swiss Federal Institute of Aquatic Science and Technology (EAWAG), Dübendorf, Switzerland.

Tobias R., O'Keefe M., Künzle R., Gebauer H., Gründl H., Morgenroth E., Pronk W. and Larsen T. A. (2017). Early testing of new sanitation technology for urban slums: the case of the Blue Diversion Toilet. *Science of the Total Environment*, **576**, 264–272.

Todd J. and Josephson B. (1996). The design of living technologies for waste treatment. *Ecological Engineering*, **6**, 109–136. https://doi.org/10.1016/0925-8574(95)00054-2

Todd J. and Todd N. J. (1980). Tomorrow is Our Permanent Address: The Search for an Ecological Science of Design. Harper and Row, New York, USA.

Todd N. J. and Todd J. (1984). Bioshelters, Ocean Arks, City Farming: Ecology as the Basis of Design. Sierra Club Books, San Francisco, USA.

Todd N. J. and Todd J. (1994). From Eco-Cities to Living Machines: Principles of Ecological Design. North Atlantic Books, Berkeley, USA.

Trémolet S. and Rama M. (2012). Tracking National Financial Flows into Sanitation, Hygiene and Drinking Water. World Health Organization. http://apps.who.int/iris/handle/10665/75225 (accessed 10 February 2018).

UNDP (2006). Human Development Report 2006: Beyond Scarcity: Power, Poverty and the Global Water Crisis. Palgrave Macmillan, Basingstoke, New York. http://www.undp.org/content/dam/undp/library/corporate/HDR/2006%20Global%20HDR/HDR-2006-Beyond%20scarcity-Power-poverty-and-the-global-water-crisis.pdf (accessed 10 February 2018).

UN-Habitat (2002). Innovations in Local Governance and Decentralization in East Africa: An inventory. UN Habitat, Nairobi, Kenya.

Unsal F. (2016). Sustainability, Systems Thinking and Ecological Worldview. 2016 International Conference on Power, Energy Engineering and Management (PEEM 2016). http://dpi-proceedings.com/index.php/dteees/article/view/5050/4676 (accessed 31 August 2017).

USAID (2016). Global Sanitation Service Chain Technology Manual: An analysis of sanitation technology innovations. http://www.path.org/publications/files/PATH_dt_ssd_tech_manual.pdf (accessed 20 June 2017).

USEPA (U.S. Environmental Protection Agency) (2012). Guidelines for Water Reuse. EPA/600/R-12/618. Produced by CDM Smith Inc., under a Cooperative Research and Development Agreement (CRADA) with the U.S. Environmental Protection Agency (EPA).

van der Ryn S. and Cowan S. (2007). Ecological Design, 10th anniversary edn. Island Press, Washington D.C., USA. https://islandpress.org/books/ecological-design-tenth-anniversary-edition (accessed 10 February 2018).

van Vliet B. (2006). The sustainable transformation of sanitation. In: Reflexive Governance for Sustainable Development, J. P. Voss, D. Bauknecht and R. Kemp (eds), Edward Elgar, Cheltenham, UK, pp. 337–354.

van Vliet B. J. M., Spaargaren G. and Oosterveer P. (2011). Sanitation under challenge: contributions from the social sciences. *Water Policy*, **13**, 797–809.

Verhagen J. and Carrasco M. (2013). Full-Chain Sanitation Services that Last: Non-sewered Sanitation Services. July 2013, IRC International Water and Sanitation Centre, the Hague, Netherlands. https://www.ircwash.org/sites/default/files/201306_sanitation_framework_final.pdf (accessed 17 June 2017).

Verstraete W., Van de Caveye P. and Diamantis V. (2009). Maximum use of resources present in domestic 'used water'. *Bioresource Technology*, **100**, 5537–5545.

Vester F. (2004). A Report to the Club of Rome. Deutscher Tashenbuch Verlag, Munich, Germany.

von Bertalanffy L. (1956). General system theory. In: General System, Yearbook of the Society for the Advancement of General System Theory, F. E. Emery (ed.). https://monoskop.org/images/7/77/Von_Bertalanffy_Ludwig_General_System_Theory_1968.pdf (accessed 10 February 2018).

von Bertalanffy L. (1968). General System theory: Foundations, Development, Applications. George Braziller, New York, USA.

Vymazal J. (2005). Constructed wetlands for wastewater treatment. *Ecological Engineering*, **25**(5), 475–477.

Wang L., Peng J., Wang B. and Yang L. (2006). Design and operation of an eco-system for municipal wastewater treatment and utilization. *Water Science and Technology*, **54**(11–12), 429–436, doi: 10.2166/wst.2006.923. PMID: 17302348

Werner C., Avendanõ V., Demsat S., Eicher I., Hernandez L., Jung C., Kraus S., Lacayo I., Neupane K., Rabiega A. and Wafler M. (ed.) (2004). Ecosan – Closing the Loop. Proceedings of the 2nd International Symposium on Ecological Sanitation, 7–11 April 2003, Lübeck, Germany. Eschborn (Germany) GTZ; 19893-00-012791-7 p. 1004. http://www.gtz.de/ecosan/english/symposium2-proceedings-eng.htm (accessed 10 February 2018).

Werner C., Panesar A., Rud S. B. and Olt C. U. (2009). Ecological sanitation: principles, technologies and project for examples sustainable wastewater and excreta management. *Desalination*, **248**, 392–401.

Whitman D. H., Berry R. L. and Green D. M. (1997). Platelet gel: an autologous alternative to fibrin glue with applications in oral and maxillofacial surgery. *Journal of Oral and Maxillofacial Surgery*, **55**(11), 1294–1299.

Whittington D., Lauria D. and Choe K. (1993). Household sanitation in Kumasi, Ghana: a description of current practices, attitudes, and perceptions. *World Development*, **21**(5), 733–748. https://doi.org/10.1016/0305-750X(93)90030-D

WHO and UNICEF (2017). Progress on Drinking Water, Sanitation and Hygiene: Update and SDG Baselines. World Health Organization (WHO) and the United Nations Children's Fund (UNICEF), Geneva. http://www.who.int/mediacentre/news/releases/2017/launch-version-report-jmp-water-sanitation-hygiene.pdf?ua=1 (accessed 5 October 2017).

WHO/UNICEF (2017). Progress on Drinking Water, Sanitation and Hygiene: 2017 Update and SDG Baselines. Geneva: World Health Organization (WHO) and the United Nations Children's Fund (UNICEF). Licence: CC BY-NC-SA 3.0 IGO.

Wielemaker R. C., Weijma J. and Zeeman G. (2016). Harvest to harvest: recovering nutrients with New Sanitation systems for reuse in Urban Agriculture. *Resources, Conservation and Recycling*, **128**, 438–450. http://dx.doi.org/10.1016/j.resconrec.2016.09.015

Wilber K. (2000). Sex, Ecology, Spirituality, 2nd edn. Shambala, Boston.

Wilsenach J. A., Maurer M., Larsen T. A. and van Loosdrecht M. C. M. (2003). From waste treatment to integrated resource management. *Water Science and Technology*, **48**(1), 1–9. © IWA Publishing.

Wolf S., Hinkel J., Hallier M., Bisaro A., Linke D., Ionescu C. and Klein R. J. T. (2013). Clarifying vulnerability definitions and assessments using formalisation. *International Journal of Climate Change Strategies and Management*, **5**(1), 54–70. http://dx.doi.org/10.1108/17568691311299363

WSP (2014). The Missing Link in Sanitation Service Delivery: A Review of Fecal Sludge Management in 12 Cities. Water and Sanitation Program: Research Brief. https://www.wsp.org/sites/wsp.org/files/publications/WSP-Fecal-Sludge-12-City-Review-Research-Brief.pdf (accessed 20 June 2017).

Wu S., Carvalho P. N., Müller J. A., Manoj V. R. and Dong R. (2016). Sanitation in constructed wetlands: a review on the removal of human pathogens and fecal indicators. *Science of the Total Environment*, **541**, 8–22.

WWAP (United Nations World Water Assessment Programme) (2017). The United Nations World Water Development Report 2017. Wastewater: The Untapped Resource. UNESCO, Paris.

Yardley S. (2010). Joining the Dots: Why Better Water, Sanitation and Hygiene are Necessary for Progress on Maternal, Newborn and Child Health. Publication of Tearfund, Teddington, UK. http://tilz.tearfund.org/~/media/files/tilz/research/j20066%20wash%20report%20web.pdf (accessed 10 February 2018).

Zari M. P. and Jenkins S. (2010). Rethinking cutting edge sustainable design: from eco-efficiency to regenerative development. Paper presented at the Sustainable Building Conference 2010 (SB10, Wellington, New Zealand. https://www.branz.co.nz/cms_show_download.php?id=2ab15295f01d3078189c96855a6fa055226d58b4 (accessed 05 April 2017).

Zari M. P. (2012). Ecosystem services analysis for the design of regenerative built environments. *Building Research & Information*, **40**(1), 54–64, doi: 10.1080/09613218.2011.628547

Zeeman G. (2009). Centralised or decentralised sanitation chains? In: Nieuwe Uitdagingen; 61e Vakantiecursus in Drinkwatervoorziening, 28e Vakantiecursus in Riolering en Afvalwaterbehandeling. (New challenges; 61st Holiday Course in Drinking Water Supply, 28th Holiday Course in Sewerage and Wastewater Treatment), J. C. Dijk, J. Q. J. C. Verberk and A. A. van Woerden (eds), Water Management Academic Press, Delft, Netherlands, pp. 101–109.

Zeeman G. (2012). New Sanitation: Bridging Cities and Agriculture. Wageningen University, Wageningen, Netherlands. http://edepot.wur.nl/246120 (accessed 10 February 2018).

Zeldovich L. (2017). Reinventing the toilet. Mosaic Science Online. https://mosaicscience.com/story/poo-toilet-waste-energy-madagascar-loowatt-future (accessed 3 July 2017).

Zurbrügg C. and Tilley E. (2007). Evaluation of Existing Low Cost Conventional as Well as Innovative Sanitation System and Technologies – NETSSAF Deliverable D22&23. Swiss Federal Institute of Aquatic Science and Technology (EAWAG), Dübendorf, Switzerland.

Zurbrugg C. and Tilley E. (2009). A system perspective in sanitation – Human waste from cradle to grave and reincarnation. *Desalination*, **248**, 410–417.

Chapter 2

Regenerative sanitation foundations

'we need to replace the industrial mind with the ecological mind'

Wes Jackson

2.1 INTRODUCTION

The previous chapter highlighted existing worldviews, paradigms and concepts of sanitation spanning many decades. The sanitation journey (from the period of arbitrary disposal to today's prevailing sanitation technologies) has all been based on the conventional sanitation (ConSan) paradigm and its corresponding concepts, as seen in Sections 1.4.1 and 1.4.2 in Chapter 1 (Cooper, 2001; O'Reilly *et al.*, 2017; Persson, 2017; Zeldovich, 2017). Paradigms are shaped by controlling worldviews and so regardless of the direction of a shift in paradigms, if it remains under the same worldview, not much would change. The mechanistic-reductionist worldview of the ConSan paradigm was also the foundation of the ecological (EcoSan) and sustainable sanitation (SuSan) paradigms, which were also technocentric, and inevitably, in spite of its identified shortcomings ConSan remained dominant over them for many years.

Nevertheless, the need for a paradigm shift in sanitation has been reiterated in multiple studies and debates (see Chapter 1), but since the worldview remains the same, change has not been as effective as desired. The challenge of EcoSan and SuSan was that they tended to lean more to the mechanistic-reductionist worldview, which focuses mainly on technological solutions and ignores other subsystems, dimensions and components of sanitation systems. It seems more justifiable that such paradigms as EcoSan and SuSan should lean more towards an ecological systemic worldview rather than focusing on just 'parts' and 'technology' (which is the main challenge of all the three previous paradigms). Most conceptualizations conceived and borne out of the these paradigms (e.g. sanitation ladder, service chain, behavioural change, decision support – Chapter 1, Section 1.5) do not address sanitation as a holistic, complex and dynamic system, but focus on singular aspects in a Newtonian-Cartesian reductionist manner through the mechanistic thinking of the Industrial Revolution that technology can fix everything.

© IWA Publishing 2019. Regenerative Sanitation: A New Paradigm for Sanitation 4.0
Editor(s): Thammarat Koottatep, Peter Emmanuel Cookey and Chongrak Polprasert
doi: 10.2166/9781780409689_0035

In the end, minimal successes in sanitation service improvement and access expansion are recorded with overwhelming disappointment. Sanitation, today, has become a global challenge, with much funding directed towards it for many years and pilot cases of various approaches, but almost no uptakes or lasting solutions. Sanitation targets for the MDGs are still way off-track and progress towards the SDGs could be on the same path. Sanitation programmes cannot continue with business as usual and expect different results. It is time for a new generation of thought in sanitation and this book aims to act as a catalyst to conceive a new paradigm as well as theories and principles for management and technology solutions in the race towards 2030 and beyond.

In essence, a new generation of thought in sanitation will obviously require a shift in paradigm – not necessarily nullifying the old, but tweaking it to strengthen its weaknesses and integrate all parts. As indicated in Chapter 1, this new paradigm will need an initial shift from the mechanistic-reductionist worldview to the ecological-systemic worldview. Meadows (1999) strongly suggested that paradigms are the key and most important place to intervene for solutions in any system – especially complex systems such as sanitation management. Change in parameters (number of people with or without), size (capacity of existing facilities), structure of facilities (types of toilets/treatment facilities) and established goals (halving vulnerable population/ensuring universal access), as promising as they might be, may not yield the desired results without a paradigm shift. Paradigms are the sources of systems that lead to goals, information, structure, feedbacks, rules and everything else about the system (Meadows, 1999). It is increasingly evident that in order to move in the direction of progress, the sanitation management system needs to be restructured and most scholars and practitioners agree that this restructuring will stem from holistic, integrated, systemic and ecological foundations.

After much contemplation and exploration of studies and practical cases across the globe, the *regenerative sanitation paradigm was articulated, which brought about the regenerative sanitation concept* outlined in Chapter 1. This shift sees sanitation as resource/service management rather than problem management and aims to reduce transfer of burden, be resource-oriented, conserve (recover) energy/water and incorporate restorative and continuous improvement principles. At its core is a process trail that works from within the system and the peculiarities of the 'place' to ensure resilience and adaptability. The overall aim of ReGenSan is to refocus the perspectives of researchers, professionals, practitioners, advocacy agents, policymakers, donors and funders as well as solution-providers from attacking a problem to providing a service and solutions; and ultimately, expanding access as well as improving service towards achieving the SDG no. 6 sanitation targets by 2030 (Boxes 2.1 and 2.2).

BOX 2.1 REGENERATIVE SANITATION PARADIGM (ReGenSanP)

The regenerative sanitation paradigm (ReGenSanP) demands a systemic and holistic approach to sanitation management that mimics the principles of nature, respects the peculiarities of place, considers the ecology and geography and integrates all aspects of sanitation while taking into account interactions between the psycho-social, management, resource utilization and technology aspects to take advantage of the multidimensional and multidisciplinary characteristics to provide sanitation solutions.

This chapter introduces a new generation of thought in sanitation from these points by proposing a regenerative sanitation theory and some underlying principles. Chapter 3 then presents a conceptual framework designed to reflect the process for the kind of sanitation practice that would fulfil this ambitious goal.

BOX 2.2 REGENERATIVE SANITATION CONCEPT (ReGenSanC)

The regenerative sanitation concept (ReGenSanC) focuses on revitalization and rejuvenation of the entire sanitation management system to improve service and expand access through deliberate partnership with nature in a holistic, systemic and integrative manner where sanitation is addressed as a whole system with functional parts that are symbiotically connected. It is based on the principles of systemic internal processes for restoration and sustainability and addresses sanitation comprehensively while incorporating all the unique characteristics of sani-management from the starting point of a specific and peculiar 'place' (location).

2.2 REGENERATIVE SANITATION ASSUMPTIONS

This ReGenSan concept is not an attempt to right certain wrongs and offer better options, but to fill an opportunity for progressive change towards a future where the potential of sanitation is harnessed and its negative impacts minimally reduced. It aims to tackle the increasing demand for a shift in the way sanitation is addressed by providing a framework for planning, research and development in sanitation management – while at the same time redirecting thought on sanitation matters from strictly techno-mechanistic to systemic, holistic and ecological explorations of a complex and dynamic phenomenon.

The pursuit of an understanding of this kind of systemic-holistic thinking that views sanitation as complex and dynamic with integrative characteristics led to a series of assumptions and a theoretical foundation upon which a conceptual framework for sanitation management was built. The following assumptions were made with hypothetical statements to heighten theoretical clarity, for better understanding of the ReGenSan, and the key related theories were adopted in order to propose a regenerative sanitation theory (Table 2.1).

2.3 REGENERATIVE SANITATION THEORETICAL FOUNDATIONS

The theoretical foundations for ReGenSan, therefore, consist of the following theories: functionalism and system theory, regenerative development and design theory and centralization theory as well as decentralization theory.

2.3.1 Functionalism and system theory (FaST)

Functionalism and system theory (FaST) facilitates the understanding of complex systems that consider whole systems as a structure of interrelated parts working together in harmony to ensure stability, resilience, continuity, balance and healthy flow of the whole (Mooney *et al.*, 2013; Sato, 2011). Just as the various organs of the body work together to keep the entire system functioning and regulated, the various parts of the system work together to keep the whole system functioning and stable (Griffiths *et al.*, 2015; Seiffert & Loch, 2005; Spencer, 1998; Vester, 2004). The most important property of functional systems is that they are made up of several dynamic parts that are not isolated, but closely interlinked, forming a complex structure (Seiffert & Loch, 2005) and function as a recurrent activity of the parts in the whole as well as the contribution it makes to structural continuity (Griffiths *et al.*, 2015; Radcliffe-Brown, 1952). Every part in a system is influenced by all the other parts and when there is a dysfunction in one part, it affects the stability and balance of the whole system, causing a disequilibrium that affects the complete system (Griffiths *et al.*, 2015; Sato, 2011) and this requires a change in

Table 2.1 ReGenSan Assumptions and Corresponding Hypothetical Statements.

S/No	Assumptions	Hypothetical Statements
1	Current technocentric sanitation approaches address issues from the mechanistic-reductionist perspective of parts rather than from the systemic-holistic perspectives. They solve problems in one part and create problems for other parts.	Sanitation should be looked at from integrated, holistic and systemic perspectives, not just technocentric, and from how the parts contribute towards the healthy flow of the whole system. There is no need to compartmentalize and fragmentize sanitation service delivery into centralization and/or decentralization; rather it should be looked at from the perspective of the whole, which will require hybridization approaches to create the needed synergy for improved service delivery and access expansion.
2	Attempts to design one-cap-fits-all solutions have failed to recognize the peculiarities of the 'place' and 'scale', which impacted negatively on sanitation services because of its inability to address the pyschosocio-cultural, economic and ecological challenges.	The peculiarities of the 'place' are of utmost importance in the implementation of sociotechnical innovations for sanitation systems, especially in the areas of technological design, selection and management solutions.
3	Current sanitation approaches do not seek to align with nature or support the health of natural systems.	Sanitation management and technological approaches should strive to support and work with nature as well as attempt to mimic the systemic flow of natural and living systems.
4	The major challenges of sanitation militating against expansion of access and service improvement in developing countries are lack of 'place' and 'scale' management, standardization, regulatory requirements and system specifications, as well as inadequate rehabilitation strategies, especially in the implementation of decentralized systems.	Sanitation management needs to be organized within the 'place' and 'scale' of 'sani-sheds' as in 'sewer sheds' and 'watershed', which is the natural hydrogeological and geographical pattern for ensuring adequate systems and product specifications, management and governance strategies, participation and partnership and adoption of appropriate technology as well as benefit-sharing among others.
5	Current on-site sanitation technologies (particularly in developing countries) usually result in the generation of large quantities of faecal sludge and wastewater transferred from one point of generation to another outside location for treatment and/or disposal, thereby creating social, environmental, public health and ecological burdens in that new location.	Sanitation technologies and management strategies should be designed to minimize the transfer of by-products (faecal sludge and wastewater) from one location to another as well as ensure resource recovery, utilization and reuse within the 'place' and 'scale' to minimize transfer of burdens.
6	Most sanitation technologies are susceptible to design, operations and acceptability failure	Sanitation technology solutions should be designed with internal sustaining power,

(*Continued*)

Table 2.1 ReGenSan Assumptions and Corresponding Hypothetical Statements (*Continued*).

S/No	Assumptions	Hypothetical Statements
	at critical points, especially at implementation. There are many sani-innovations that never made it beyond pilot phase. Furthermore, most innovations in sanitation often pursue idealistic and utopian outcomes that are not built on sound theoretical and conceptual foundations and so hardly survive the marketplace and selection uptake.	flexibility, resilience, ease of operations and maintenance and consideration for users' acceptability. Sanitation technological solutions and management strategies should be based on strong theoretical and conceptual foundations upon which innovations and novelty could be designed and developed.
7	There are so many dysfunctional and failed sanitation infrastructures, and considering that 2.4 billion people globally are making use of unimproved sanitation facilities, it would make more business sense to work to restore and upgrade dysfunctional, failing and unimproved facilities than to build new ones that might well not be acceptable, especially in developing countries.	Expanding access and improving service in most developing countries will require more targeted efforts towards restoration and retrofitting of dysfunctional, failing and existing unimproved sanitation infrastructures.
8	**The biggest sanitation technology challenges over the past decades are connected to the neglect of psycho-social and cultural issues in delivering sanitation services.**	**Like the self-flourishing and self-regulating characteristics of living and natural systems, sanitation systems should be able to feed the psycho-social, economic, cultural and ecological needs as well as stability of the 'place' and 'scale' of its location.**
9	Sanitation infrastructures in developed countries are close to their end of life and many are beginning to fail and break down due to age and changing conditions. Climate change, new trends and levels of urbanization and the drive towards regenerative cities will call for new sanitation management strategies even in developed countries.	Sanitation management strategies cannot be restricted to developing and poor countries alone, and so systemic, holistic and integrated strategies should be explored for design and development of new solutions as well as retrofitting existing infrastructure for all to work within a system based on ecological principles.
10	Most failed sanitation technologies and management solutions in developing countries were transferred without consideration for differences in sociocultural, economic, geographical and ecological characteristics.	Based on many design factors, sanitation solutions that work in one location may not work in another development and planning should come from the starting point of the 'place' and 'scale'.

operations. This implies that a system's functionality is dependent on the symbiotic relationship between its parts and how they function together; that is, any system is dependent on its own internal interdependencies (Marshall & Farahbakhsh, 2013).

The fundamental notion of the functionalism and system theory is its focus on interactions (Von Bertalanffy, 1956, 1968) and the sole shift of attention from the 'part' to the 'whole' (Benne & Mang, 2015; Capra, 1997; Checkland, 1997; DeWitt, 2010; Du Plessis, 2009; Hes & du Pleasis, 2014; Heylighen, 2006; Jackson, 2003; Koltko-Rivera, 2004; Mele *et al.*, 2010; Miller & West, 1993;

Weinberg, 2001; Von Baertalaffy, 1956). FaST is designed to escape mechanistic-reductionism and embrace the functionalism dynamic complexities of systemic, holistic and integrated perspectives (Patton, 2002). It ensures that in any given situation at least three levels of inquiries are considered: the system (what?), the subsystem (how?) and the wider system (why?) (Checkland, 1981; Unsal, 2016). Its application promotes holistic integration and extends beyond technology-focus; it exploits opportunities of inter-synergistic interactions between all the sanitation subsystems, dimensions and components (Anarow *et al.*, 2003; Rocky Mountain Institute, 2006). A sanitation system that considers the functions of all subsystems, dimensions and components will provide adequate technological and management solutions by considering that sanitation problems are rarely solved at the individual household level alone. In that case, it is essential that sanitation systems' boundaries be extended to accommodate systemic, holistic and integrated sanitation management processes. This gives support to the concept of sani-sheds, which systematically, holistically and integratively combine technological processes such as storage, collection, transport, treatment, recycling and reuse and minimal final discharge within the 'place' (Tilley *et al.*, 2014a, b; Zurbrügg & Tilley, 2009). Functional sanitation systems should:

(I) Incorporate planning,
(II) Take advantage of the inherent resources in faecal matter,
(III) Benefit from appropriate and best applicable technology for 'place' and 'scale',
(IV) Ensure effective and efficient operations and maintenance,
(V) Address relevant psycho-socio-cultural issues that help integrate users' perspectives, and
(VI) Incorporate rehabilitation of dysfunctional infrastructure.

ReGenSan embraces functional systemic thinking for the complex and dynamic nature of sanitation where the interacting units function systematically, holistically and integratively to improve performance. Significantly, sanitation systems are complex processes and interactions between physical, social, economic, technological and ecological components interlink and influence each other; where there is a dysfunction in one, there is a dysfunction in all and the whole system is then weakened. This, inevitably, affects access expansion and service improvement (Mele *et al.*, 2010; Seadon, 2006, 2010; Tilley *et al.*, 2014a, b).

2.3.2 Regenerative development and design theory (RDDT)

The regenerative development and design theory (RDDT) springs from the need to focus on reducing the negative impacts of human development activities on the planet to creating net-positive impacts that align cultural and natural patterns, to provide opportunities for positive outcomes rather than just doing less damage (Hoxie *et al.*, 2012; Lyle, 1994). It seeks to create systems that integrate the needs of society with the integrity of nature through processes that restore, renew and revitalize their own sources of energy and materials (Mang & Reed, 2012a, b). The theory aims to restore or create the capacity of ecosystems and bio-geo-chemical cycles (carbon, hydrological, nitrogen, etc.) to function optimally without human intervention. It further creates new potential, as humans are able to evolve with the ecosystems they are part of (Jenkin & Pedersen, 2009). This theory could replace the present one-way linear material flow that devours its own sources of sustenance often described as 'degeneration' to cyclical flow systems of 'regeneration' that work for optimization and resilience (Lyle, 1994). Everything does not rest on technology alone; the core is the value of human relationship with nature (Akturk, 2016). Patterns are enmeshed in natural and social processes to make their purposes far more complex and function as integral parts of the communities they serve and also involve the communities in design and operations. RDDT seeks to replace throughput flows with cyclical flows at source, consumption centres and sinks to change behaviour pattern (Lyle, 1994; Figure 2.1).

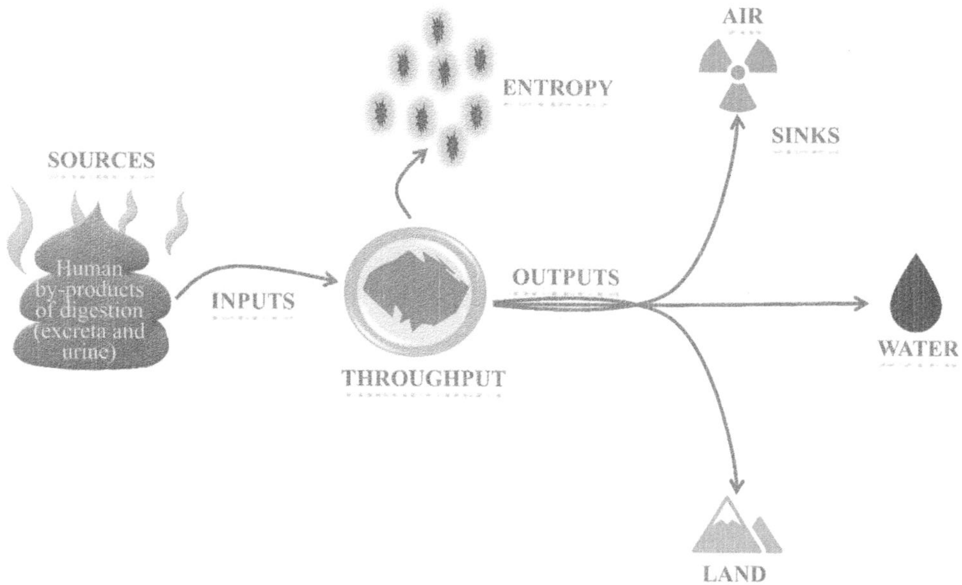

Figure 2.1 Degenerative linear systems of throughput flows (adapted from Lyle, 1994).

Lyle (1994) proposes six basic processes required to shift from linear to cyclical processes and 12 strategies for the achievement of regenerative design and development. The six basic processes are:

(I) Conversion,
(II) Distribution,
(III) Assimilation,
(IV) Filtration,
(V) Storage, and
(VI) Thought.

Conversion is the mechanism by which energy is turned into usable energy, while distribution is the means by which the energy reaches community members. Assimilation is the collection and return of materials to the landscape, while filtration is the process that maintains a clean and healthy ecosystem and storage supports stabilization of products, such as water and food. Finally, regenerative design is guided by thoughts. Lyle's 12 regenerative strategies are:

(I) Letting nature do the work,
(II) Considering nature as both model and context,
(III) Aggregating not isolating,
(IV) Seeking optimum levels for multiple functions, not the maximum or minimum level for any one,
(V) Matching technology to need,
(VI) Using information and knowledge,
(VII) Providing multiple pathways,
(VIII) Seeking a common solution to disparate problems,

(IX) Managing storage as a key to sustainability,
(X) Shaping form to guide flow,
(XI) Shaping form to manifest process, and
(XII) Prioritizing for sustainability (Lyle, 1994).

These processes and strategies support three important natural sanitation technological approaches: composting, natural sewage treatment and bioremediation. They also illustrate the very functions and capacities that both plants and microorganisms can play in sanitation solutions, especially in sewage treatment, without using the elaborate, energy-intensive, often unreliable mechanical devices from conventional linear sewage treatment plants (that is, letting nature do the work of sewage treatment) (Lyle, 1994).

Thayer (1994) uses RDDT to create positive associations between the natural and technical world and shows how they impact culture and behavioural change. Thayer's five characteristics of RDDT are:

(I) Generate renewal energy without ecological destabilization,
(II) Maximize recycling resources, nutrients and by-products with minimum water production,
(III) Maintain local structure and function without reducing the diversity or stability of the surrounding ecosystem,
(IV) Preserve and serve local human communities rather than change or destroy them, and
(V) Incorporate technologies that support these goals.

McDonough and Braungart's (2001) theory of 'cradle-to-cradle' brings some of the primary principles underpinning RDDT to the forefront. McDonough-Braungart's principles include:

(I) Recognize that waste equals food,
(II) Respect diversity, and
(III) Reconcile energy consumption to current solar income.

The 'waste equals food' is very relevant to the ReGenSan concept, which embodies the concept that in nature nothing produced is useless. RDDT provides the platform for integrated approaches to planning, design and development for effective and efficient use of resources needed for expansion of access and service improvement in sanitation delivery (Jenkin & Pedersen, 2009). The cradle-to-cradle concept eliminates the idea of regarding faecal matter (human waste) as waste, and regards it rather as a resource system that revolutionizes the paradigm of 'cradle-to-grave' – a linear conventional sanitation system (Jenkin & Pedersen, 2009; McDonough & Braungart, 2002). The goal is to enhance resource recycling and safe reuse of sanitation products that enhance benefit-sharing and no-transfer-of-burden while using appropriate solutions. Therefore, the regenerative goal makes it clear that the volume of waste to be processed should be limited to the capacity of the environment to assimilate it as well as ensure that these materials are returned to the processes of natural and human ecosystems (Lyle, 1994). This closely connects ReGenSan to regenerative agriculture as the subject of food sufficiency as well as carbon emission reduction. ReGenSan commits to systemic, holistic and integrated processes that provide better understanding of the knowledge of functions and interconnections among sanitation subsystems, dimensions and components.

In the context of sanitation technological and management solutions, RDDT provides an extensive range of possible combinations of social-ecological-systems and technical management strategies, concepts and designs. The storage component could act as the stabilization (aerobic and anaerobic digestion) process for by-products of human digestion from users' interface and the filtration could represent the application of various polishing and treatment processes for faecal matter and other associated waste materials to render them safe for reuse and disposal into the ecosystem. The next stage could then be conversion comprising the various processes of biomass recovery and recycling from the by-products collected, while assimilation could involve the process of mineralization of the recycled unusable portion or remnants of the biomass that are returned to the ecosystem for sustenance through emissions. The next process could be distribution, the process of ensuring that the recovered and recycled resources reach community members in the sani-sheds (possibility for regenerative agriculture), and then finally advancing the ReGenSan concept through the power of information and knowledge. Adopting RDDT into sanitation practice could richly change previous sanitation systems' axiomatic mode of operations of front end and back end (IWA 24:2016). Box 2.3 presents ReGenSan systems components and processes.

BOX 2.3 ReGenSan SYSTEMS COMPONENTS AND PROCESSES

ReGenSan systems components and processes comprise:

(I) Front end, which refers to users' interface,
(II) Mid end, which makes up storage and treatment processes, and
(III) Back end, which includes resource recovery/utilization and disposal.

Therefore, ReGenSan technological and management solutions are designed and structured to let nature do the work, by drawing on natural processes of decomposition to reintegrate materials into the ecosystem (Figure 2.2).

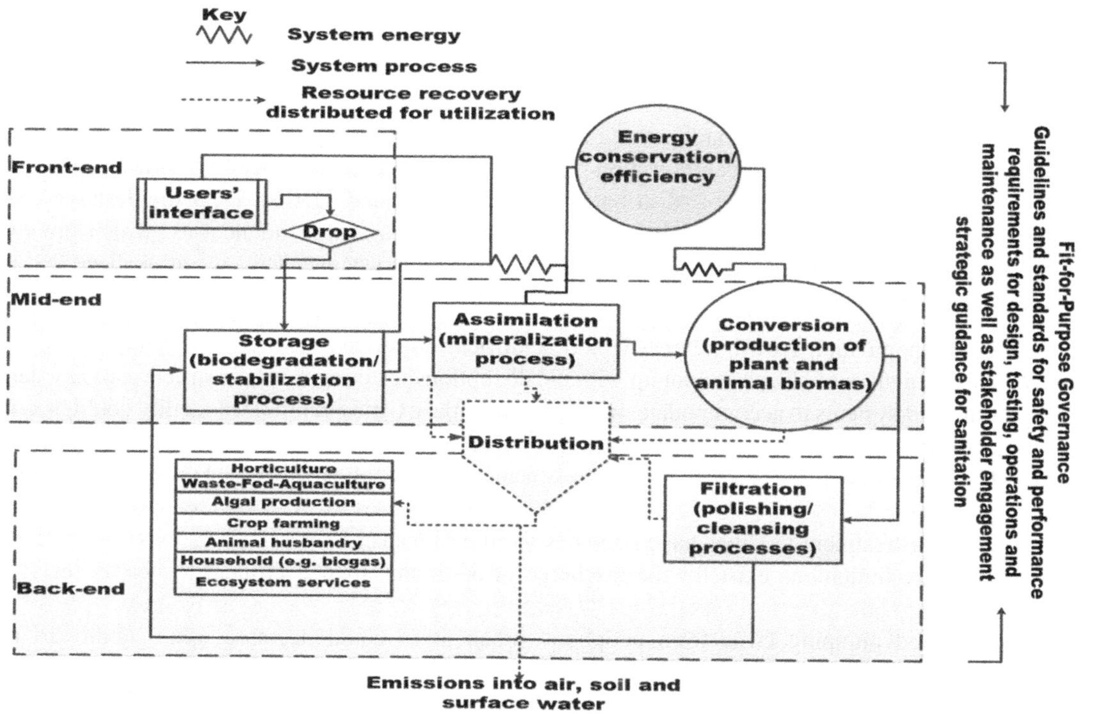

Figure 2.2 Regenerative sanitation cyclical process flow (adapted from Lyle, 1994).

2.3.3 Centralization theory

The centralization theory originated from municipal governance systems and was driven by the benefit of economics of scale, the knowledge of traditional conventional water treatment methods and limited ability for remote monitoring and control (Ahluwalia, 2012). Centralization involves management processes that collect, treat and discharge large quantities of wastewater by the use of pumps, piping materials and energy, and hinges on the notion that often 'out of sight' requires less public participation and awareness (Giri *et al.*, 2006; Go & Demir, 2006; Massoud *et al.*, 2009; USEPA, 2004; West, 2001; Wilderer & Schreff, 2000). It technically supports the collection and drainage of wastewater generated from households, industries and sometimes stormwater from large urban and peri-urban areas to central locations for treatment, reclamation and surface discharge away from human settlements (Tchobanoglous & Leverenz, 2013; Tchobanoglous *et al.*, 2009; Wilderer & Schreff, 2000). This concept has long been implemented and utilized as a way to treat wastewater in highly populated and urbanized areas and is often publicly owned to manage large volumes of wastewater for large communities, which carry huge investments and associated costs that generally make it difficult and time-consuming to establish (Bakir, 2001; Suriyachan *et al.*, 2012)

This centralization theory has given birth to combined or separate sewage systems for bulk wastewater disposal provided in large wastewater catchment areas, which could also be split up into separate sewage systems and subsequently connected to various treatment plants. In addition, it supports centralized treatment plants that offer a multiplicity of systems and system variations for wastewater treatment. These include various wastewater treatment levels such as mechanical and mechanical-biological treatment, nutrient removal, disinfection and use of membrane removal of minute particles (Orth, 2007). Centralized systems have provided considerable benefits to modern society, especially in the improvement of public health through the removal and treatment of wastewater (Brown *et al.*, 2009; Sharma *et al.*, 2013). However, demographic changes including the ageing population, socio-economic factors, climate change, biodiversity, energy use, water supply and consumption and ageing infrastructure have put increasing pressure on centralized sanitation systems (Brown *et al.*, 2011; Marlow *et al.*, 2013, Sapkota *et al.*, 2015; Sharma *et al.*, 2010) (Figure 2.3).

Consequently, proponents of centralization have come to accept the need to break up central systems, which led to the concept of 'distributed centralized systems' (DCS). DCS divide large cities into discrete sections, typically according to political and or geographic boundaries, which are then served by different but linked centralized sanitation treatment systems. This is rationalized by the recognition that:

(I) Future population growth in areas would be difficult to service,
(II) Unwillingness of citizens to put up with the disruptions resulting from construction of new large transport systems to accommodate future growth at the existing centralized facility could be quite challenging,
(III) Capacity limitations for collection and treatment could be problematic based on anticipated future growth,
(IV) Existing treatment facilities have gone beyond their lifespan,
(V) Capacity limitations exist for the discharge of additional treated effluent to nearby receiving waters,
(VI) Increased pumping costs from peripheral urban areas could decrease users' ability to pay (affordability), and
(VII) Opportunities for resource recovery, recycling and safe reuse in the upper reaches of the collection systems are high in DCS (Tchobanoglous & Leverenz, 2013).

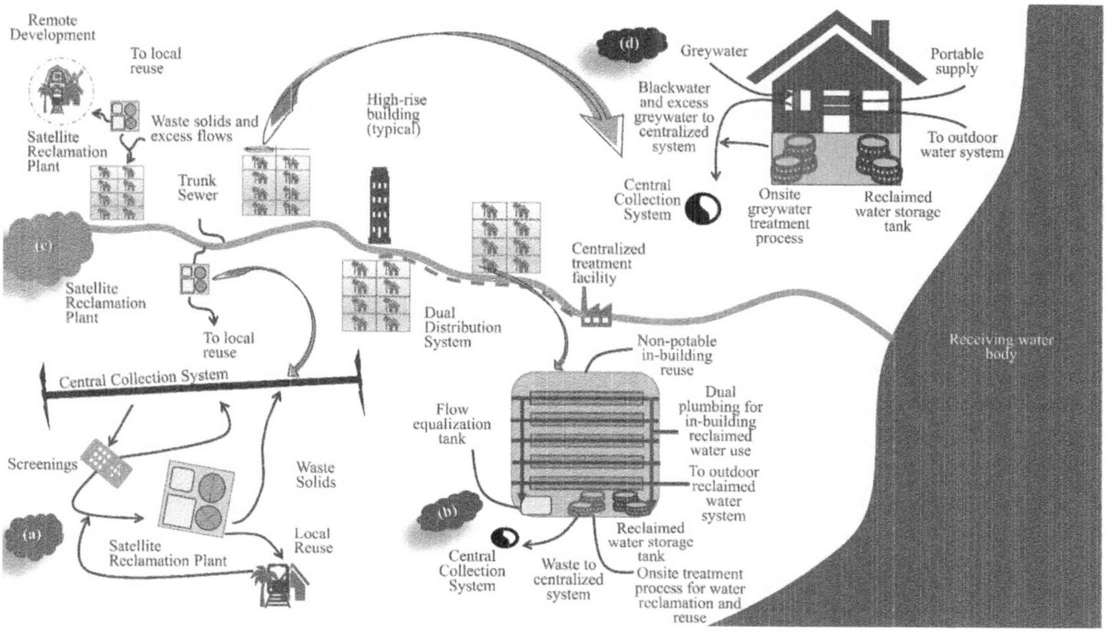

Figure 2.3 Centralized wastewater collection system with satellite systems (adapted from Tchobanoglous & Leverenz, 2013). *Note:* (a) large extraction type, (b) commercial building interception type, (c) upstream interception type and (d) individual home type.

Centralization in sanitation faces overarching challenges, particularly in protecting and maintaining public health, environmental protection, resource recovery and sustainability (Brands, 2014; Butler & Parkinson, 1997; Delleur, 2003). Although so much has been achieved with community and public health protection, large volumes of untreated or inadequately treated wastewater still continue to be released into the environment, resulting in deterioration of water quality and thereby public health (Hutton *et al.*, 2007; Nhapi *et al.*, 2006; World Bank, 2003). The most daunting challenges, however, for centralized systems remain environmental protection, resource recovery and sustainability (Bracken *et al.*, 2006). Mixing of different wastewater streams (as is practised in most centralized systems) makes recovering different resources such as water, energy and nutrients difficult. In addition, dilution of wastewater streams containing pathogens and toxic compounds, such as heavy metals and organic micropollutants, makes treatment more complex and demands more resources such as energy, money, space and expertise, while still putting pressure on the environment through emissions (Balkema *et al.*, 2002). In spite of these challenges, the theory contributes to the ReGenSan concept of sani-sheds and scales-based technology and management (as in sewer sheds).

2.3.4 Decentralization theory

The decentralization theory is often associated with local governance and is increasingly recognized as a potentially suitable way to contribute towards reducing the large population lacking access to improved sanitation (Bieker *et al.*, 2010; IDRC, 2010; Larsen & Maurer, 2011) as well as increasing the effectiveness and efficiency of sanitation technology and resource recovery (Libralato *et al.*, 2012). Decentralization collects, treats and disposes/reuses treated wastewater at or near the source of generation and aims to operate at small scale (Suriyachan *et al.*, 2012; USEPA, 2004). It not only reduces the effects on the environment and on public health, but also increases resource recovery and reuse depending on the type of technical options and local setting (Massoud *et al.*, 2009). Historically, this theory is usually applied to on-site sanitation systems adopted for properties in remote and small communities where the implementation of centralized sanitation systems may not be viable due to technical, economic or environmental constraints or not chosen by homeowners (Sharma *et al.*, 2013; USEPA, 2002) (Figure 2.4). Currently though, it is not only implemented in peri-urban, rural and remote areas, but also in conjunction with centralized systems to reduce the stress on existing ageing centralized urban sanitation systems (Bieker *et al.*, 2010). The emphasis is on localized systems that lend themselves to recycling because it is expected that sustainability can only be achieved through closing the nutrient cycles (Libralato *et al.*, 2012). Decentralization processes seem to satisfy all the traditional centralized treatment requirements, with some added values relating mainly to the ability to minimize residual effluent contamination as well as ecosystem disruption by removing emerging micropollutants such as metals, pharmaceuticals and personal care products (Borsuk *et al.*, 2008).

Small treatment plants, however, are not entirely reflective of the decentralized theory. The fact that small sanitation treatment technologies act locally in a decentralized manner does not imply that they follow the principles of decentralization (as these plants may not even be small) such as:

(I) On-site scale that covers treatment technologies and/or management systems that provide sanitation services at the scale of individual properties,

(II) Cluster or development scale that operates under some form of common ownership model and services two or more dwellings or a whole development, with wastewater treated in proximity to dwellings and in most cases includes resource recovery and stormwater harvesting (Geisinger & Chartier, 2005),

(III) Distributed or community-type systems providing services to large developments (i.e. $n > 100$ properties) with the service owned and operated by utilities.

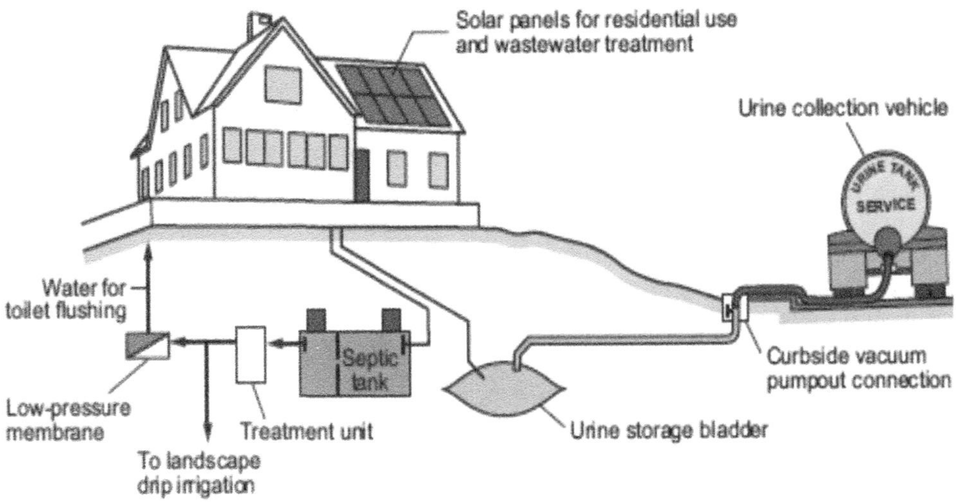

Figure 2.4 Decentralized system of home reuse system with urine separation and urine storage (*Source*: Tchobanoglous & Leverenz, 2013).

Decentralized sanitation systems have been projected to continue to grow based on the expansion of development in urban fringes, natural areas and urban areas (Sharma *et al.*, 2013; Tchobanoglous & Leverenz, 2008, 2013). The decentralization theory contributes to ReGenSan to further improve the concept of place- and scale-based sanitation strategies and source separation for recovery of valuable resources from human excreta.

The purpose of ReGenSanT (Box 2.4) is to promote the total rejuvenation and revitalization of all aspects of sanitation systems based on quality and functionality of infrastructures and not merely counting numbers (Kvarnström *et al.*, 2011). It aims to guide the new generational thought on sanitation management and this theory is still open to strengthening as further explorations and investigations are considered.

BOX 2.4 REGENERATIVE SANITATION THEORY (ReGenSanT)

Based on the preceding section, the theory of regenerative sanitation theory (ReGenSanT) proposes that sanitation technology and management approaches must consider the system functions and parts in a holistic, systemic and integrated manner, from the starting point of place and scale and work with nature to produce more locally fit-for-purpose, acceptable and practical solutions, which will in turn lead to expanded access and improved service delivery.

2.4 PRINCIPLES OF REGENERATIVE SANITATION

Critical to the pursuit of ReGenSan is identifying the core principles that will drive its operations and implementation to ensure its sustainability, resilience, adaptability and effectiveness. Based on the ReGenSanT posited above and perusals of other stated theories, the following 10 principles, though not exhaustive, are conceived to guide planning, research and development in regenerative sanitation. These synergistic and interconnected principles also serve as directives for the corresponding framework (to be presented in Chapter 3). Table 2.2 shows the summary of the 10 key principles of ReGenSan, while detailed explanations are presented below, in alphabetical order and not necessarily in their order of importance.

These principles, described below, are designed to work together as an integrated whole and not as individual parts – and it is only when they are inputted together that they give the desired results.

2.4.1 Appropriate technology (ApT)

ApT suggests that technology is neither neutral nor does it evolve along a single path, and recognizes that different cultural and geographical groups should have specific sanitation technologies responsive to their varying conditions (Illich, 1973; Schumacher, 1973; Willoughby, 1990). ApT provides context-specific series of infrastructure, methods, approaches, strategies and services designed for the management of resources derived from human by-products of digestion (excreta and urine) (Tilley *et al.*, 2014a, b). ApT is a marked departure from the original concept of 'appropriate technologies' – of low-cost alternatives to conventional sewerage systems designed to keep the cost of infrastructure low (Mohammed Shahalam, 1985; Kalbermatten *et al.*, 1980; Shuval *et al.*, 1981). ApT encompasses every level of technology, from the very simple to the highly complex, provided that it is tailored specifically to people's needs based on the requirements of the 'place' and/or 'scale' (Lyle, 1994). The key element of appropriateness is encouraging the right choice and technology that fit the 'place', and goes beyond toilets to the final end use of excreta-derived end-products resources and disposal of residue as well as

Table 2.2 Summary of ReGenSan Principles and Interpretation.

S/No	Regenerative Sanitation Principles	Summary Interpretation
1	Appropriate technology (ApT)	Technology that is appropriate for particular 'place' and 'scale'.
2	Biomimicry and Biophilia (BaB)	Sanitation solutions/strategies that mimic and support natural systems.
3	Fit-for-purpose governance (FPG)	Governance mechanisms that fit 'place', 'scale' and purpose for functional, efficient and effective sanitation requirement.
4	Hybridized solutions (HS)	Sanitation solutions that could synergize centralized and decentralized options to deliver effective and efficient services to particular 'place' and 'scale'.
5	Notransfer of-burden (NToB)	Sanitation solutions that do not transfer health, environment, social, economic and ecological burdens to society.
6	Systemic holistic integration (SHI)	Sanitation solutions that interconnect and synergize subsystems, dimensions and components to achieve functional, efficient, and effective whole systems.
7	Place and scale (PaS)	Sanitation solutions that indigenously align to location (place) and sites (scale) for optimization.
8	Participation and partnership (PaP)	Sanitation solutions that involve stakeholders at all spheres and levels of localized contexts.
9	Recycling and safe reuse (RaSR)	Sanitation solutions that recover and safely reuse valuable resources from by-products of human digestion.
10	Rehabilitation of dysfunctional facilities (RoDF)	Sanitation solutions that rehabilitate and retrofit dysfunctional, unimproved and ageing infrastructure

engaging in extensive interaction with the would-be users (Hollick, 1982; Kalbermatten *et al.*, 1982; Tilley *et al.*, 2014b). ApT seeks to provide sanitation systems that are of high standards, technically (especially for children and people with disabilities) and hygienically safe to use (Giné-Garriga *et al.*, 2017).

ApT should strive to be ecologically adaptable and resilient, use local resources, ensure energy conservation and feasibility of resource recovery and build institutional and governance capacities to create the enabling environment (Katukiza *et al.*, 2010, 2012; Mara *et al.*, 2007; Murphy *et al.*, 2009; Nansubuga *et al.*, 2016; Saubolle, 1980). It incorporates both 'hard' and 'soft' aspects of technology and addresses issues of knowledge-transfer mechanisms, capacity-building and communication methods as well as social, cultural, religious and gender implications of technology implementation (Murphy *et al.*, 2009). The key considerations for ApT development and implementation are presented in Box 2.5.

BOX 2.5 KEY CONSIDERATIONS FOR ApT DEVELOPMENT AND IMPLEMENTATION

Key considerations for ApT development and implementation include but are not limited to:

(I) Meeting basic needs of users,
(II) Sound and quality technology,

(III) Flexible and restorative technology,
(IV) Meeting local capabilities by utilizing local materials and resources,
(V) Affordability and sustainability,
(VI) Encouraging local participation and partnership,
(VII) Psycho-social, cultural, religious and ecological appropriateness,
(VIII) Encouraging resource recovery and reuse,
(IX) Design consistent with ecological principles, and
(X) Gender and vulnerabilities considerations (Murphy *et al.*, 2009).

A condominial sanitation model in Brazil could effectively illustrate the ApT principle. This model differs from conventional sewer models that service individual housing units in that it delivers services to each housing block or any group of dwellings that could be termed a neighbourhood unit or 'condominial'. Thus, public networks no longer need to run through every plot of land present on every street, as a single connection point is provided for each city block. In addition, much closer relationships between service providers and users encourages the two parties to come to an agreement that facilitates service expansion and adaptation to local needs and constraints (Melo, 2005). This community-based approach to urban sanitation illustrates a shift from conventional sewage technology to a technically equivalent and lower-cost alternative system. The lower costs arose from the use of sound technical standards based on current scientific and technical research as well as current experience and innovation to fit the 'place' and 'scale' rather than the use of low-quality and substandard materials. This approach has been used to provide sanitation services to 2 million residents of Brasilia State for more than 10 years. Within this period, 121,000 homes were linked to the condominial system, using 1300 kilometres of condominial branches and more than 600 kilometres of public networks at average costs per person of USD 27 and per metre of sewer network of USD 16 (UN Millennium Project, 2005). Standard and quality is not compromised, but users' pay is minimized through the shared costs that spread the cost over every household covered.

2.4.2 Biomimic and biophilic (BaB)

The BaB principle focuses on human relationships that seek to utilize knowledge, solutions, mechanisms, strategies and principles found in nature to advance and foster better sanitation technological solutions and management (Badarnah & Kadri, 2015). Biomimicry designs innovation through the emulation of biological forms, processes, patterns and systems aimed at providing sanitation-technology solutions and management (Kennedy & Marting, 2016) and uses analogies to biological systems to develop solutions to tackle sanitation challenges (Gamage & Hyde, 2012; Helms *et al.*, 2009). In addition, it emulates natural processes to create efficient and innovative sanitation solutions using nature as a model, mentor and measure (Benyus, 2002). BaB advocates for a paradigm shift from sanitation systems that focus on end-off-pipe to that which replicates and intensifies the processes of nature in organic recycling (Lyle, 1994). Biophilia is an innate human inclination to affiliate with the natural world and is instrumental in promoting people's health, fitness and well-being. It recognizes that nature contributes to humanity's well-being and helps productivity and creativity, as well as seeking to incorporate an understanding of the processes of human psychological connections into sanitation-technology solutions and management. It is not just about mimicking nature, but is also about copying or learning from principles of nature such as self-organization, self-regulation and self-evolution. It also aims to adopt the characteristics of adaptability while supporting nature to do its work through sanitation-technology and management solutions, recognizing that sanitation can impact positively on communities when it is designed to work in partnership with nature (Dias, 2015; Kellert, 2012, 2016; Kellert & Wilson, 1993; Wilson, 1984). Several strategies have been developed for design concepts inspired by nature, but the two most common approaches are:

(I) Problem-based approaches, which is a top-down strategy that seeks solutions from nature for a particular engineering problem (Badarnah & Kadri, 2015; Baumeister, 2012; Gebeshuber & Drack, 2008; Speck *et al.*, 2006; Vattam *et al.*, 2007), and

(II) Solution-based approaches; that is, bottom-up strategies inspired by observations of nature, which lead to a technological design (Badarnah & Kadri, 2015; Baumeister, 2012; Gebeshuber & Drack, 2008; Speck *et al.*, 2006; Vattam *et al.*, 2007).

The BaB principle incorporates understanding of the natural systems into the design and development of sanitation technologies and management solutions (Kellert, 2016) and reflects respect and affinity for the ingenuity and intelligence of solutions hammered into a species' genes through countless and cyclical adaptations.

This calls for a basic reform in the prevailing paradigm that governs linear sanitation systems, to a more cyclical form of sanitation. The principle of BaB is designed to ensure that sanitation technologies are consistent with ecological engineering principles (Bergen *et al.*, 2001) and enhances the capability of natural treatment systems such as the adoption of wetland treatment systems for sanitation solutions (Mitsch, 1993; Mitsch & Jørgensen, 1989, 2003; Todd & Josephson, 1996; Todd & Todd, 1980, 1984, 1994). It embraces the need for advanced science and technology that can point the way towards achieving mutually beneficial sanitation infrastructure and management solutions (Kellert, 2016). BaB application is aimed at providing innovative sanitation solutions by adopting the natural principles of the ecosystems by following some primary steps, presented in Box 2.6.

BOX 2.6 BaB PRIMARY APPLICATION SOLUTIONS STEPS

BaB primary application solutions steps comprise:

(I) Identifying the troubled and needful areas,
(II) Exploring natural systems solutions,
(III) Extracting and investigating the most relatable options – often referred to as pinnacles,
(IV) Elaborating and analyzing the pinnacles for strategies and designs,
(V) Determining the practicalities and efficiencies of these strategies and designs,
(VI) Combining related strategies and designs to form synergistic linkages for fit-for-purpose solutions,
(VII) Evaluating and validating the solutions, and
(VIII) Starting over and exploring more pinnacles of further improvement is needed (Badarnah & Kadri, 2015).

A close illustration of the BaB principle is the 'flower and the butterfly' constructed wastewater wetland treatment system used in the rehabilitation of the badly damaged wastewater treatment system in the Phi Phi Island tourist centre of Krabi Province of southern Thailand) after the tsunami disaster of December 2004. Wastewater is fed to the flower vertical-flow (VF) and horizontal subsurface flow (SF) wetland units before feeding to the butterfly Free Water Surface (FWS) wetland units (Figure 2.5, Koottatep *et al.*, 2007; Polprasert & Koottatep, 2017).

Figure 2.5 Application of the principle of biomimicry and biophilia in flower and butterfly constructed wastewater treatment wetland system design (*Source*: Koottatep *et al.*, 2007; Polprasert & Koottatep, 2017).

2.4.3 Fit-for-purpose governance (FPG)

The FPG principle ensures the adequacy and functionality of management and governance structures and processes to deliver effective and efficient sanitation services and solutions (Rijke *et al.*, 2012). 'Management' refers to activities of analyzing, monitoring, developing and implementing measures aimed at enhancing the functionality of sanitation systems, while 'governance' takes into account the different actors and networks that help formulate and implement sanitation policies and/or institutional instruments as well as embracing the full complexity of regulatory processes and their interactions (Pahl-Wostl, 2009). 'Institutions' refers to the body of rules, regulations and processes that guides sanitation management actions and procedures as well as enforcement (Folke *et al.*, 2007; North, 1990; Ostrom, 1990; Pahl-Wostl, 2009; Scott, 2001). A management regime is the whole complex of technologies, institutions (formal and informal rules), environmental factors and paradigms that work together to form a base for the functioning of the management system targeted to fulfil societal functions such as sanitation services (Pahl-Wostl *et al.*, 2008). The failure of several sanitation solutions is a misfit between the institutional arrangement, management regime and technological approaches as well as with the psycho-social, cultural and ecological systems (Cookey *et al.*, 2017; Galaz *et al.*, 2008; Lebel *et al.*, 2013; Ostrom, 1990, 2010; Young, 2002a, b; Young *et al.*, 1999).

Young (2002a, b) identified functional misfit as a substantial contributory factor to the deterioration of quality of the ecosystem due to poor management. The major drawbacks of existing sanitation institutions are the challenges of fragmentation that manifest in duplication of responsibilities, cross-implementing agencies, overlaps or gaps in authority, as well as poorly designed or total absence of adequate and proper sanitation legislations and policies. Other drawbacks are the inability of laws to support more adaptive and flexible frameworks for improved sanitation service delivery (Clarvis *et al.*, 2014;

Cookey *et al.*, 2015, 2016a, b, c). The key factors for the achievement of FPG for the 'place' and 'scale' are presented in Box 2.7.

BOX 2.7 KEY FACTORS FOR FPG

The key factors for FPG comprise:

(I) Understanding the peculiarities and unique features of the 'place'
(II) Socio-economic and cultural considerations of the 'place'
(III) Delineation of sani-sheds
(IV) Specific and clear standards and regulatory framework for 'place' and 'scale' and sani-shed levels
(V) Compliance and enforcement framework
(VI) Specific and unambiguous policy instruments
(VII) Coordinating and harmonizing agency
(VIII) Stakeholder engagement, business and industry partnership
(IX) Leadership and management
(X) Research in partnership with policy and practice
(XI) Capacity development and expansion
(XII) Funding and financing mechanism
(XIII) Monitoring and evaluation
(XIV) Information and knowledge sharing

FPG should be able to capture the synergistic linkages between all management and governance components as well as determine the local contexts, especially the prevailing regulatory environment, including national and local policies and legal and institutional frameworks (Cookey *et al.*, 2016b, c, d; Ross *et al.*, 2014). The main idea behind FPG is that sanitation technology and management solutions should properly address the local circumstances in question in order to enhance efficiency and effectiveness of sanitation solutions and service delivery (Cookey, 2016a, b; Ostrom, 2005, 2010). Sanitation implementation will require effective and efficient governance, management and institutional arrangements that are tailored to fit the psycho-social, cultural, economic, geographical and ecological systems of the 'place' and 'scale' (Moss, 2007, 2012). In this regard, adaptive and flexible policy and regulatory instruments should be provided for sani-sheds to support the dynamics of solution management.

2.4.4 Hybridized solutions (HS)

The HS principle seeks to systematically integrate and combine decentralized sanitation facilities (DSF), satellite sanitation facilities (SSF) and centralized sanitation facilities (CSF) into one operating system in sani-sheds and/or sewer sheds (Daigger & Crawford, 2007). The inter-synergistic combination of centralized systems designed for the treatment of by-products of human digestion (excreta and urine), through large-scale sewer networks with low end-user involvement and the decentralized 'standalone' systems (on-site sanitation systems) that do not need connection to larger infrastructure with high end-user involvement, will result in optimized technical, social and economic benefits, increased system security and sustainability (van Vliet *et al.*, 2011). Decentralization utilizes the management of small flow through processes that operate closer to natural and ecological systems, while centralized processes utilize high-rate flows that accomplish bulk treatment on a short timescale and small footprints. The combination of these two strategies creates opportunities in the sani-sheds that provide needed cost savings through the reduction of sewer-network expansions (Gikas & Tchobanoglous, 2009; Tchobanoglous & Leverenz, 2013).

HS supports rehabilitation of dysfunctional sanitation systems whether centralized, decentralized, clusters and/or individual treatment technologies to fulfil the minimum requirements of ReGenSan. It is characterized by the comprehensive rejuvenation and revitalization of the entire spectrum of sanitation infrastructural and management solutions at place/scale through inter-synergistic linkages of all sanitation ancillary facilities to optimize system performance and enhance benefit-sharing. It also seeks to provide an adequate environment for a combination of solutions, technologies, strategies and decision-making structures to create a better fit with the pyscho-socio-cultural, economic, geographical and ecological systems. This principle lowers entry cost for sanitation solutions, increases flexibility and adaptability to demographic changes and opens up possibilities in resource management, nutrient and energy recovery and greenhouse-gas emission reduction. It also defers augmentation of existing infrastructure as well as staggering infrastructure investments (Burkhard *et al.*, 2000; Carragher *et al.*, 2012; Christian-Smith & Gleick, 2012; Daigger, 2009; Daigger & Crawford, 2007; Kluge & Moser-Nørgaard, 2008; Wolf & Störmer, 2010). HS decreases resource consumption through reduction of conveyance requirement and offers the same high standards and flexible solutions wherever certain thresholds of population density are exceeded (Bieker *et al.*, 2010; Sapkota *et al.*, 2015; Weber *et al.*, 2007). The future application of HS is anticipated to occur at a more rapid pace compared to the recent past for a variety of reasons, presented in Box 2.8.

BOX 2.8 REASONS FOR FUTURE GROWTH OF HS

Reasons for future growth of HS include but are not limited to:

(I) Population growth and distribution patterns,
(II) Climate change,
(III) Long-term water shortages, and
(IV) Need to recover energy and resources (Tchobanoglous & Leverenz, 2013)

Recent sanitation management research has explored the potential of the HS principle in the developed (European) and developing (east Africa) countries through applying the 'modernized mixture' approach. The modernized mixture approach (MMA) is a sociotechnical configuration of sanitation infrastructures of various scale and types of technologies (centralized and decentralized), level of consumer involvement and multilevel governance and management, as well as separation or combination of wastewater flows. These main clusters of variables could categorize sanitation systems (Figure 2.6, Hegger, 2007; van Vliet *et al.*, 2010, 2011). MMA aims to bridge the gap between proponents of the views 'large is robust and efficient' and 'small is beautiful' that have dominated the debates on infrastructural renewal for a long time. The approach postulates that there is no single sustainable solution, but various solutions, whose applicability depends on the 'place' and 'scale' (Hegger, 2007).

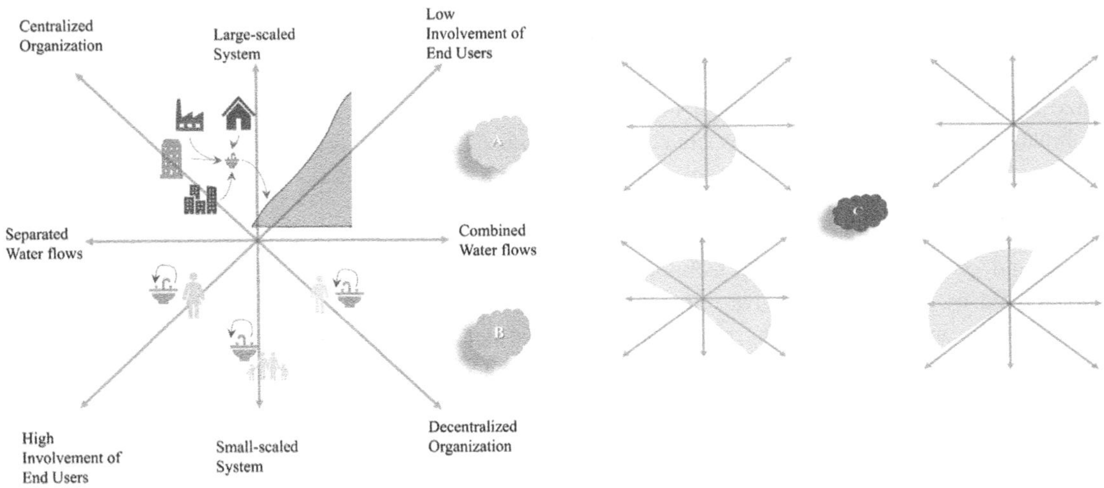

Figure 2.6 Modernized mixtures sanitation approaches (adapted from van Vliet *et al.*, 2011). *Note:* (A) conventional systems are found in the clustering of values at the top of the diagram: central organization, large-scaled systems and low user involvement; (B) alternative systems are to be found at the opposite end of the diagram: small-scale systems, responsible users, decentralized organization; and (C) various combinations of social and technical variables make up 'modernized mixtures'.

2.4.5 No-Transfer-of-burden (NToB)

The principle of no-transfer-of-burden aims at promoting sanitation systems that do not impact negatively on the ecosystem, sociocultural factors, economy and public health of the place, while reducing wastewater, faecal sludge and energy burdens for human waste treatment and transportation. This principle requires that sanitation management should be confined to the sani-sheds by ensuring that issues such as wastewater, faecal sludge, resource recovery etc. are contained and managed within the sani-shed, thus reducing materials transfer for treatment at another location. It also advocates separating human waste from other wastes and aligns sanitation facilities with ecological processes to ensure 'zero' or 'near minimal' discharge so as to ensure that sanitation systems provide net benefits to the people and environment. NToB seeks to prevent and stop further damage and harm of ecological systems and human society from sanitation technologies and management solutions. It works to balance societal, economic, health and environmental interests while at the same time meeting core sanitation values of providing ambient infrastructural solutions and management to fulfil human physiological needs (defecation and urination).

The disease burden from poor sanitation and hygiene as well as unsafe water has been estimated to be one-tenth of the global burden of diseases (Fewtrell *et al.*, 2005, 2007; WHO, 2007). This impacts negatively on the society and the economy by causing around 1.6–2.5 million deaths annually (Kosek *et al.*, 2003; LIXIL *et al.*, 2016; Mara *et al.*, 2010; Mathers *et al.*, 2006; WEDC & DFID, 1998). The burden of sanitation due to lack of access to improved sanitation cost the world about USD 222.9 billion in 2015, up from USD 182.5 billion in 2010, a rise of over USD 40 billion in just five years. This figure constitutes an average 0.9% of gross domestic product (GDP) of the Latin America, Caribbean, Africa and Asia Pacific countries impacted by poor sanitation, which is a modest decline from 1% GDP five years ago (LIXIL *et al.*, 2016).

Studies have shown that several sewerage systems globally discharge sewage without treatment and as a result more than 50% of the world's rivers, oceans and lakes are polluted with untreated wastewater; and on-site sanitation systems characterized by high rates of failure and absence of technological improvement rates, especially in the developing countries, add to this burden (Baum *et al.*, 2013; Brown & Root, 2001; Cookey *et al.*, 2016d; Jones *et al.*, 2004; Siegrist *et al.*, 2000; Starkl *et al.*, 2013, 2015). To eliminate the above challenges, sanitation technology and management solutions should be approached with the NToB principle and from localized pyscho-socio-cultural-economic, ecological, geographical, resource recovery and technology perspectives; and within the technology (Box 2.9)

BOX 2.9 REQUIREMENTS FOR IMPLEMENTATION OF NToB

Requirements for implementation of NToB shall comprise:

(I) Continuous maintenance and improvement of the currently adjudged existing-improved systems,
(II) Retrofitting and rehabilitation of unimproved and dysfunctional infrastructure, and
(III) Emerging novel solutions that reduce burden transfer and ensure net-benefit-sharing.

These approaches shall ensure that sanitation facilities isolate and/or destroy the pathogens to break the transmission pathway and ensure that valued resources are recovered and reused. Also, NToB shall seek to ensure that sanitation infrastructure is designed, developed, constructed, operated and maintained based on well-established guidelines and standards. The key success factors for NToB are presented in Box 2.10.

BOX 2.10 KEY SUCCESS FACTORS FOR NToB

The key success factors for NToB comprise:

(I) Zero failing and dysfunctional systems
(II) Drastically reduce transportation of faecal sludge from one location to another
(III) Encourage sanitation management within sani-sheds
(IV) Ensure that only unusable parts of recycled materials are released into the ecosystem
(V) Encourage reuse within sani-sheds to reduce transportation of faecal sludge from one sani-shed to another
(VI) Ensure periodic operations and maintenance of sanitation infrastructure, especially decentralized facilities, to avoid failure and dysfunction
(VII) Provide specific risk-based guidelines, standards, regulations and certification schemes for design, construction, installation, operations and maintenance of on-site systems to ensure facility integrity and proper asset management

2.4.6 Systemic holistic integration (SHI)

The SHI principle aims at integrating technology, social, economic, ecological, geographical and environmental perspectives into comprehensive sanitation design solutions. This approach encourages the participation of and development of partnerships between actors from a variety of backgrounds, disciplines and sectors to develop innovative sanitation solutions (Charnley & Lemon, 2011; Coley & Lemon, 2008; Stasinopoulos *et al.*, 2009). Dynamic-complex sanitation problems that span technological, environmental, geographical, social and economic phenomena will necessarily require the views of experts from multiple disciplines and the use of transdisciplinary skills (Charnley & Lemon, 2011). The complexity of facilitating sanitation innovations cannot be brought about through technological, health and hygiene innovations alone, but requires mutually reinforcing institutional and socio-cultural-technical transformations (Geels, 2005; Lopes *et al.*, 2012). Technological systems, of course, also need to be understood, addressed and measured; they are still important but just not sufficient enough to deliver the needed access expansion and service improvement for the sector. Sanitation technologies, then, should be systemic, holistic and integrative, with the capacity to align with nature's principles, and should be psycho-socio-culturally acceptable, economically affordable and environmentally friendly (Illich, 1973; Mara, 2003; Pattnaik & Dhal, 2015; Schumacher, 1973; Semiyaga *et al.*, 2015; Simiyu, 2017; Willoughby, 1990).

SHI discourages the current dichotomy of centralized and decentralized technologies among sanitation practitioners; rather, it encourages harmonious integration by tapping into the adoption of hybridized solutions. When sanitation practitioners understand that the acts of defecation and urination are more about the physiological need to remove by-products of digestion (excreta and urine) than anything else and that these by-product are a resource, then the scope of sanitation work will expand to include broader systemic, holistic and integrative approaches (Reed, 2007). This will transform the ways sanitation is considered, practised, designed and developed (Dubrovsky, 2004; Reed, 2007; Seiffert & Loch, 2005). In conventional sanitation management approaches, excreta and urine technologies for generation, storage, conveyance, treatment, recovery, reuse and final disposal are planned as independent operations. However, all seven are very closely interlinked and each component can influence the others. The planning for these operations requires a balance between the subsystems of population growth, land use patterns, urban growth and development, agricultural production, water resources and environmental and public health considerations (Clark, 1978). This shows the complex interactions between the physical components of the system, and conceptual components of the social and environmental spheres (Seadon, 2010). When sanitation is viewed from systemic-holistic-integrative angles, the relationship between system parts is revealed and the potential for greater sustainability is enhanced (McDougall *et al.*, 2001). It incorporates perspectives that promote enabling environments that support frameworks for the uptake of sanitation technology and management solutions, including political, legal, institutions, financial, educational, technical and social conditions (Tilley *et al.*, 2014a).

SHI strives for integration of the many interrelated processes and entities that make up sanitation technology and management solutions (Marshall & Farahbakhsh, 2013; McDougall *et al.*, 2001; Morrissey & Browne, 2004; Petts, 2000; Thomas & McDougall, 2005; van de Klundert & Anschutz, 2001), which ensures that sanitation solutions are tailored to specific 'place' and 'scale' goals by incorporating peculiar and contextual perspectives and needs (Kollikkathara *et al.*, 2009; McDougall *et al.*, 2001; van de Klundert & Anschutz, 2001). Sanitation systems that ignore psycho-socio-cultural components, users' acceptance, public participation and partnership, technical and economic aspects of sanitation are bound to fail (Carabias *et al.*, 1999; Dijkema *et al.*, 2000; Henry *et al.*, 2006; Marshall & Farahbakhsh, 2013; Morrissey & Browne, 2004; Petts, 2000) (Box 2.11).

BOX 2.11 APPLYING WHOLE-SYSTEM DESIGN IN SHI

The SHI principle can be illustrated by the application of the 'whole-system design' approach, which seeks to integrate social, economic and environmental phenomena into comprehensive sanitation technological and management design solutions. The approach encourages the development of partnerships between experts and professionals from a variety of different backgrounds, disciplines and sectors to develop and implement innovative ReGenSan solutions (Charnley & Lemon, 2011; Coley & Lemon, 2008; Stasinopoulos *et al.*, 2009).

This principle was used in the 'Funny Dunny' project involving collaboration with a broad range of interdependent stakeholders involved to operationalize the urine-diversion (UD) sanitation system in Sweden. The project modelled complex institutional arrangements at the meso-level (population size that falls between the micro- and macro-levels, such as a community or an organization) of the sanitation system. The project was able to identify the enabling conditions for new infrastructural planning processes, sympathetic regulatory and legal frameworks; altered user practices; and recast cultural meaning in the sanitation, water, agricultural and horticultural sectors and beyond (Lopes *et al.*, 2012; Sankaran *et al.*, 2010). SHI shows that integration of stakeholders across disciplines as well as at governmental level is central to well-functioning sanitation technological design and management solutions (Charnley & Lemon, 2011).

2.4.7 Place and scale (PaS)

The idea of 'place' is the envisioning of the alignment of humans to psycho-socio-cultural, economic, ecological, social and natural systems (Mang *et al.*, 2016; Zari & Jenkin, 2010). In each 'place' and 'scale' on earth, natural and cultural systems express themselves uniquely and differently, which implies that sanitation systems should be tailored to the unique characteristics of the 'place' by exploring the opportunities and solutions that are indigenous to that specific 'place' (location) and 'scale' (site), which ReGenSan refers to as 'sani-sheds'. 'Place' is a multilayer network of living systems within a geographical region that results from the complex interactions, through time, of the natural ecology (climate, mineral and other deposits, soil, vegetation, water and wildlife, etc.) and culture (distinctive customs, expressions of values, economic activities, forms of association, ideas for education, traditions, religion etc.) (Cole, 2010; Dias, 2015). The issues of scale concern space, time, system levels and the importance of designing solutions and management for sanitation services (Hoxie *et al.*, 2012). PaS is the starting point for articulating and designing sanitation solutions because it is linked to households, apartment buildings, neighbourhoods, political wards, schools, functional buildings (e.g. power plants) and local and urban communities (Tainter, 2012), and is more encompassing, identifying 'regions', urban centres, cities and very large-scale places like sani-sheds (Du Plessis, 2012; Svec *et al.*, 2012).

Instructively, sanitation problems can rarely be solved at the household level alone, from where materials such as wastewater, faecal sludge, etc. are exported to the neighbourhood, town or downstream population. In such cases, it is crucial that sanitation boundaries be extended to include larger spatial sections that take into account technology, management and governance components for storage, collection, transportation, treatment, reuse and minimal discharge (Zurbrügg & Tilley, 2009). Sani-sheds should be planned along hydrogeological/geographical boundaries or socio-cultural-political boundaries. This is logical because it constitutes homogeneity of cultures and customs, demographical information, livelihood activities, identification of felt needs and the removal of rural and urban dichotomies.

The main benefit of sani-sheds is the economics of scale for resource mobilization, as well as net-benefit-sharing: participation and partnership for efficient and effective management of sanitation systems. It has the added advantage of maximization of resources, enhanced resource recycling and safe reuse, respect for diversity, energy recovery, conservation and infrastructural rehabilitation and maintenance (McDonough & Braungart, 2002; Thayer, 1994).

Sanitation, from a sustainability approach, should involve environmental, cultural, social and local perspectives. Studies have shown that technological sanitation solutions that improve environmental and human health are not sufficient and adequate for concrete application. This is because solutions that emerge from local contexts, using appropriate technology that mimics the ecological systems of the 'place', will guarantee the widest local ownership, especially when they are supported by local authorities (Andersson & Minoia, 2017). The 'place and scale' principle ensures that sanitation systems function in close concert with natural and societal processes and within a range of local environmental and governance conditions to provide solutions at scale in sani-sheds and takes into account the psycho-social, cultural, economic, geographical and ecological systems to support resource recovery at source and safe reuse (Spaargaren *et al.*, 2007). A soil-based absorption sanitation system that works well on one soil type or geological structure may not work at all in another. Thus within a regional system there might be several different technologies in operation, with each responding to local conditions, but playing its part in the larger whole (Lyle, 1994). On the other hand, contexts relate to place in that common peculiarities of different specific locations are considered. Tilley *et al.* (2014a, b) noted that acceptance of any technology or practice is inherently contextual and not necessarily guaranteed. They noted that waterless technologies based on separation and dehydration have ecological benefits, but may not meet cultural expectations in communities that use water for anal cleansing.

For illustrative purposes only, the siting of the natural sewage treatment at the Lyle Center for Regenerative Studies of the California State Polytechnic University, Pomona, USA could be considered as related to PaS. This siting was due to unusual topographic conditions whereby the valley slopes and so water naturally settles there. In essence, wastewater can easily be channelled down the valley (rather than up) for treatment and the treated resource can be used for aquaculture, irrigation for rice paddies and other crops within the valley (Figure 2.7, Lyle, 1994).

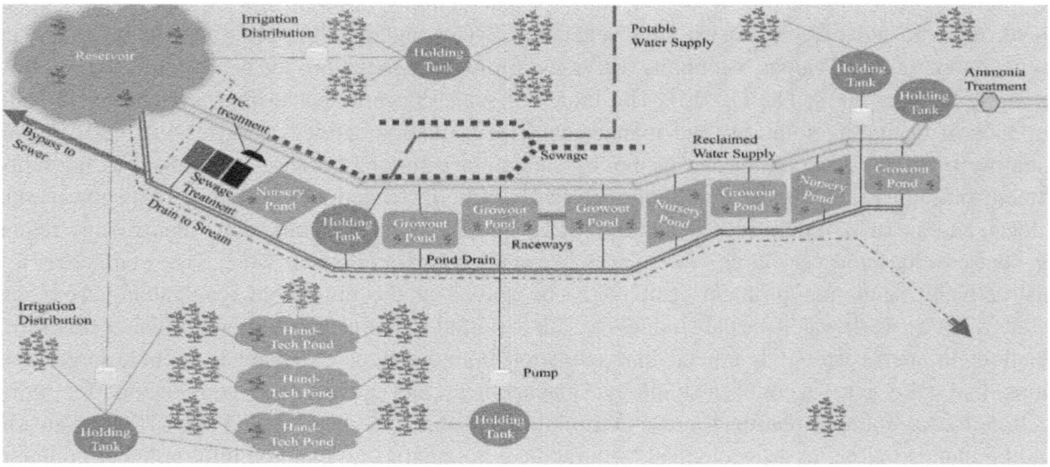

Figure 2.7 Siting of the natural sewage treatment at the Lyle Center for Regenerative Studies, California State Polytechnic University, Pomona, USA (*Source*: Lyle, 1994).

2.4.8 Participation and partnership (PaP)

The PaP principle is essentially a recipe for successful implementation of sanitation programmes because it advocates for needed solutions through the instruments of involvement, collaboration and engagement of all stakeholders from all spheres of the society (Aggestam, 2007). In relation to sanitation services, households and communities are increasingly considered as active participants in the development process and should be engaged in making decisions concerning type of infrastructural programmes, operations and maintenance, as well as ownership arrangement and cost recovery, they would support (Christmas & de Rooy, 1991; Kvarnström *et al.*, 2004; Najlis & Edwards, 1991; Rosenqvist *et al.*, 2016; White, 1981). Public participation of different stakeholders in the decision-making process introduces a range of ideas, experiences and expertise that motivates the development of alternative sanitation solutions and partnerships that provide a collaborative environment for actors from all spheres of the society such as the state, markets, businesses, civil societies, community-based organizations, etc. (Cookey *et al.*, 2016c; ILECF, 2005; van Huijstee *et al.*, 2007).

Participatory processes are fundamental, starting from respect for local cultures, traditions and norms and the recognition of users as agents of change rather than simply customers of solutions identified elsewhere (Andersson & Minoia, 2017). PaP is supported by Principle 10 in the Rio Declaration on Environment and Development (UNCED, 1992), which states that environmental issues are best handled with the participation of all concerned citizens, at the relevant level. This is further enhanced by the UN Convention on Access to Information, Public Participation in Decision Making and Access to Justice in Environmental Matters (UN Convention, 1998). Partnerships were also declared an important tool for implementing sustainable development at the 2002 World Summit on Sustainable Development (WSSD) in Johannesburg (Eweje, 2007; Hens & Nath, 2003; Norris, 2005); and the SDG no. 6 targets support strengthening stakeholders' participation and partnership as well as expanding international cooperation and capacity for improving technological and management solutions, especially in developing countries (United Nations, 2015).

The PaP principle aims to facilitate better implementation of sanitation solutions and ensures institutional effectiveness through support and cooperation of involved parties and the resulting input of knowledge (Cookey *et al.*, 2016c; Luyet *et al.*, 2012; Rowe & Frewer, 2000). It is proactive in creating an enhanced broad understanding of objectives, problems and solutions among various actors and stakeholders. PaP intends to create more informed operative decisions, provide a more solid base into policy outputs, enable interested parties to share knowledge and encourage government agencies to develop mechanisms that will blend the psycho-social, cultural, ecological, geographical, scientific and technological elements towards the aspirations of ReGenSan (Appelstrand, 2002; Nelson, 1995). PaP policies, practices and techniques should ensure that citizens and communities, individuals, groups and organizations have the opportunity to engage meaningfully in the process of decision-making as well as the design and management of their sanitation infrastructure and services.

The PaP principles could be illustrated in some cases in Sweden, where farmers were involved as stakeholders and partners in a sanitation recycling system that used excreta nutrients as fertilizer. The farmers participated in the initial planning and then transitioned to being business partners with the technology-provider and government agency to distribute and spread the recycled product on farmland. This was key to the success of seven Swedish urine-recycling sanitation systems. Two of these systems recycled blackwater (i.e. toilet wastewater), one system recycled a mixture of source-separated urine and blackwater and one system recycled sewage sludge (Box 2.12, Tidåker, 2007).

BOX 2.12 SWEDISH FARMERS' PARTICIPATION IN URINE-RECYCLING SANITATION SYSTEM

In 1999, 22 households using urine-separating toilets in the Haga-Ecovillage, Uppsala, initiated early contact with cereal production farmers within the neighborhood. The farmers accepted the urine only when paid the cost for the spreading (Jönsson *et al.*, 2010). In 2006, about 130 households had urine separation systems in Norrkoping, and in the same manner early contact was made with farmers, who stored the urine and reused it in the spring, but without any payment either way (Jönsson *et al.*, 2010; van Vliet *et al.*, 2011). Also, in 1998 Tanum municipality, Sweden, started to require urine separation for all new houses with on-site sanitation systems, and in 2006 there were more than 400 permits issued for urine separation systems. The municipality established contact with farmers at an early stage too, and seven farmers were engaged as entrepreneurs for collection, storage and spreading. The farmers charged the households for emptying the urine, and if households contacted the municipality a higher rate was charged, steering the households back to the farmers. The farmers were free to set their fee and thus the fees varied. Some of the farmers also collected and used blackwater from different households as fertilizer (Tidåker, 2007).

The farmers were required annually to submit documentation on which urine installation they had emptied (Jönsson *et al.*, 2010; Tidåker, 2007; van Vliet *et al.*, 2011). This illustration clearly shows how the PaP principle can synergize ReGenSan with regenerative agriculture and aquaculture. In addition, the farmers acquire a new line of business, which opens up job opportunities and other lines of ancillary businesses – even as they also get to access organic fertilizer for their farmland.

2.4.9 Recycling and safe reuse (RaSR)

The RaSR principle advocates for recovery and reuse of valuable resources from materials or constituents of human disposal such as urine, faeces, brownwater, beigewater (anal cleansing water) and greywater. The sanitation systems are expected to contribute to the provision of resources for livelihood support in communities, minimize natural resource depletion and environmental degradation and mainstream benefit-sharing into sanitation technological solutions and management. The principle is designed to achieve a vision for resource efficiency and cradle-to-cradle management as in the circular economy, and change how human waste is managed across the spectrum of the sanitation cycle, thus moving away from the 'cradle-to-grave' syndrome (Brands, 2014; Esrey, 2001, 2002; Lens *et al.*, 2001; McDonough & Braungart, 2002; SuSanA, 2008; Larsen & Gujer, 2013). The regenerative science cycle of 'cradle-to-cradle' is driven by the underlying desire to grow and develop, to generate and regenerate highly efficient systems of metabolism and flows of nutrients for which the concept of waste does not exist (de Macedo Guimarães, 2012; McDonough & Braungart, 2002). RaSR ensures that valuable products present in excreta and wastewater such as nutrients – N (nitrogen), P (phosphorus), K (potassium), S (sulphur) – water and soil amendment materials contribute to resource efficiency and food security as well as energy recovering and capturing from organic matter (Anderson *et al.*, 2016a, b; Polprasert & Koottatep, 2017; Rittmann, 2013; Tilley, 2013). It supports a transition from waste that needs disposal toward creating a valuable resource that can benefit farmers, create jobs and generate energy and funds to improve the sanitation service chain (Murray & Buckley, 2009; Rao *et al.*, 2016, 2017). To further advance and support resource recovery in sanitation spheres, the World Health

Organization (WHO) developed guidelines and standards for the use of sanitation resources (excreta, wastewater and greywater) in agriculture and aquaculture (USEPA, 2012; WHO, 2006a, b, c, d). In addition, this principle is in consonance with the SDG no. 6 target 3, which advocates for pollution reduction, eliminating dumping and halving the proportion of untreated wastewater as well as substantially increasing recycling and safe reuse technologies globally (United Nations, 2015).

RaSR ensures that resource flows provide multiple human and ecological benefits by maintaining and enhancing the quality of the input products (constituents of human waste) and output products (nutrients, energy, water resource and soil organics) to make sure that they do not degrade natural systems during generation, while they replenish and assimilate to build natural capital (Cole, 2012). RaSR is also a basic concept of ecological sanitation (EcoSan), resource-oriented sanitation (ROS), source separation and decentralization (SSD) and decentralization sanitation and reuse (DESAR), all of which contribute effectively and efficiently to pollution control by recycling pollutants into valuable products instead of discharging them into the environment (Brands, 2014; Esrey, 2002; Larsen *et al.*, 2009; Larsen & Gujer, 2013; Larsen & Maurer, 2011; Lens *et al.*, 2001; Tobias *et al.*, 2017). Thus, the next generation of sanitation infrastructure that provides improved sanitation facilities and expands access should strive toward greater efforts and sociotechnical innovations for resource recovery and safe reuse opportunities to users (Simha *et al.*, 2017).

The RaSR principles could be illustrated by the estimated economic benefits from resource recovery, generated in an exercise in Vientiane, Lao PDR (one of the CityBlues++ project, Figure 2.8). The city population was estimated to be 760,000 and the water-saving potential with low-flush/waterless urinals was about 13,700 m^3/day; the agricultural potential using biogas digestate and urine as fertilizers was 40,000 ha. of rice cultivation, reduced CO^2 emissions due to substitution of mineral fertilizer and diesel was 44,000 tons CO^2/year and energy potential for transport sector for organic waste was 10,000 km of bus travel per day (adjusted for energy consumption due to increased transport in waste collection) (Andersson *et al.*, 2016b).

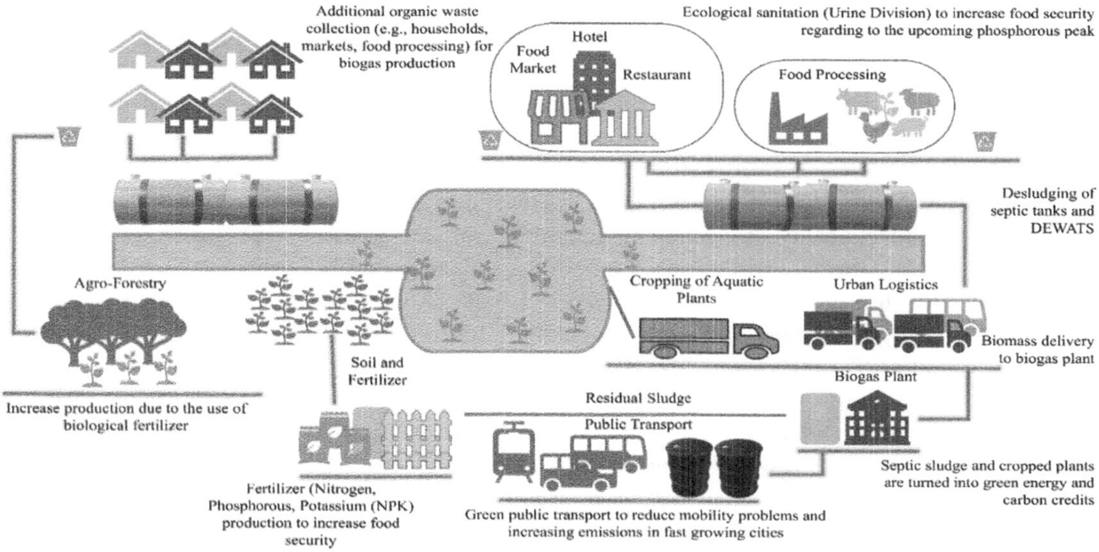

Figure 2.8 Potential added value of resource recovery in the city of Vientiane (*Source*: Andersson *et al.*, 2016b).

2.4.10 Rehabilitation of dysfunctional facilities (RoDF)

RoDF refers to the application of repair, renewal, retrofitting and replacement technologies in an effort to return functionality to sanitation infrastructures (USEPA, 2007, 2009). This principle aims at addressing upgrading and retrofitting of dysfunctional, disused and unimproved sanitation facilities and basic sanitation systems to safely manage sanitation. It addresses the challenges of faecal sludge management from on-site sanitation systems and retrofit resource recovery and energy-saving into conventional sanitation systems. It also includes the mitigation and remediation of sites that were polluted, degraded and damaged by existing sanitation practices (Mang & Reed, 2012a, b; Jenkin & Pedersen, 2009). RoDF became necessary in view of the large population (2.4 billion) without adequate and functional sanitation infrastructures, especially in developing countries of Latin America, Caribbean, Africa and Asia Pacific impacted by poor sanitation (JMP, 2015; WHO & UNICEF, 2017). Structural inadequacies, material degradation, poor construction and workmanship and ageing centralized sewage treatment facilities in developed countries, as well as lack of operations and maintenance cultures, are of great concern (Brikke, 2000; Brepolsa *et al.*, 2008; Cookey *et al.*, 2016d).

Clearly, rehabilitation of dysfunctional and unimproved sanitation infrastructure is crucial to meeting SDG no.6 and its targets, especially in addressing the backlog of people with inadequate sanitation systems in developing countries. The principle is capable of moving these huge numbers of people with inadequate sanitation facilities up the rung of the sanitation ladder to functional, safely managed systems (Kvarnström *et al.*, 2012; Exley *et al.*, 2014). A malfunctioning system is a risk to ecological systems, public health and environment and reoccurring problems create further complications if they are not quickly attended to. Effective and efficient rehabilitation systems require the recognition of symptoms of malfunctioning at an early stage, identification of the cause of the problems and repairs with appropriate measures (Ulrich *et al.*, 2009). The pressing goal is to understand better why these system failures occur and how they could be avoided in the future (Starkl *et al.*, 2013). Rehabilitation of sanitation infrastructures has been well defined and standardized with centralized systems, but poorly applied in sewer systems in developing countries. It is even less so for decentralized systems, which are poorly defined, non-standardized and in most cases totally absent, resulting in a large quantity of sanitation infrastructure that does not function adequately (Baum *et al.*, 2013; Starkl *et al.*, 2013; Cookey *et al.*, 2016d). The aim of this principle is to use the instrument of RoDF to expand access and improve sanitation service delivery and reduce environmental and public health hazards, as well as improve energy and water conservation (Dixon *et al.*, 2014; Eames, 2011).

Several studies highlight the necessity of the RoDF principles as they note high levels of sanitation infrastructures dysfunctionality in developing countries, with sub-Saharan Africa recording as high as 68% (Boulenouar & Gibson, 2015). Also, the results of an independent evaluation of case studies of sanitation infrastructure projects in India, Mexico and South Africa (most of them implemented since 2000) demonstrated an ongoing problem of failing infrastructure even in economically advanced developing countries (Starkl *et al.*, 2013). Perhaps this principle might prove more cogent for developed countries with a large quantity of ageing, inadequate and breaking sanitation infrastructure. These evident failures prompted the South African government to refocus not just on building infrastructure, but also on maintenance and rehabilitation of dysfunctional facilities. The result was the introduction of asset management policies and guidelines as a recognized approach to maintaining infrastructures and using sanitation facilities wisely in South Africa. These guidelines have been implemented by local governments with varying levels of precision and enforcement (Boulenouar & Gibson, 2015).

2.4.11 Systemic functioning of all parts

In concluding this section, it is necessary to restate that ReGenSan is more about a systemic and integrated whole with less emphasis on unit technology and system. Even in the process of innovating unit systems, it should still address the issue of how they fit into the bigger sani-shed system, while innovation or rehabilitation of sanitation systems must also fit into the systemic whole. In short, the principles of ReGenSan must operate in synergy with one another within the system of the 'place' and 'scale' in the sani-sheds. Taking the principles in isolation is not ReGenSan because the process requires a representation of each principle in symbiotic relationship with the others to fit the requirements of a 'place' and 'scale'.

In the sani-sheds, for instance, there can be several appropriate technologies, but they must all work systematically, integratively and synergistically to deliver smart services. Sani-sheds emphasize the whole system and not any individual technology or system. In fact, the illustrations and examples provided in the principles section do not exactly depict ReGenSan, but show the practicalities, possibilities, opportunities and usability of these principles for sanitation technologies and management solutions improvement. Consequently and reiteratively, these 10 principles are not designed to be used in isolation, but symbiotically and integratively, for sanitation systems and solutions to be termed ReGenSan. In this regard, they provide fodder for research into scenario development on the possibilities and opportunities of ReGenSan and, in addition, they can be used as indicators for measuring ReGenSan systems. However, the major challenge for ReGenSan will be the fact that sanitation practitioners and professionals are more comfortable with reductionist-mechanistic thinking than systemic, holistic and integrative thinking.

2.5 REGENERATIVE SANITATION PRACTICES

In the course of case explorations, we noticed that there are really no practices that could be said to be wholly ReGenSan, although some have some regenerative elements. As investigations continue into ReGenSan, core and particular practices shall be identified. Research and development will be needed to determine what will give significant effect in a ReGenSan system; for instance, retrofitting practices for on-site treatment, rehabilitation for dysfunctional infrastructure, upgrading of existing-improved facilities among others. Some practices with ReGenSan elements are presented in Box 2.13.

BOX 2.13 SOME PRACTICES WITH ReGenSan ELEMENTS

Some practices with ReGenSan elements:

(I) Resource recovery and source separation,
(II) Energy and water conservation,
(III) Recycling and reuse of by-products,
(IV) Hybridized sanitation solutions and management,
(V) Integrating sanitation with agriculture and aquaculture,
(VI) Systemic, holistic and integrated management,
(VII) Evaluating functionality and not just availability,
(VIII) Respect for nature and geography,
(IX) Fit-for-purpose governance mechanisms,

(X) Design and management based on peculiarities and conceptualities of specific locations (place) and sites (scale),
(XI) Revitalizing and renewing the ecosystem,
(XII) Public and environmental health protection (hygiene, water quality, solid waste, etc),
(XIII) On-site and at-source treatment – with minimal transportation, and
(XIV) Recognition of psycho-social, economic, cultural, religious and geographical considerations.

This is by no means a complete list, but these items (to a great extent) cover the range of regenerative practices based on the principles highlighted. There are many sanitation management practices being implemented; however, merging their rationale into a whole sanitation system can make a world of difference.

2.6 REGENERATIVE SANITATION POTENTIAL BENEFITS

ReGenSan could yield several benefits such as:

(I) Reduce water use and energy conservation,
(II) Revitalize and reintegrate ecosystems and social systems,
(III) Strengthen and accelerate the path towards 2030 SDG no. 6 targets on sanitation,
(IV) Innovate and reinvent sanitation technology, service and management solutions to be apt and fit for purpose,
(V) Improve national and local uptakes and selection of sani-solutions,
(VI) Support food sufficiency by contributing to regenerative agriculture,
(VII) Support the development of regenerative cities and even strengthen the cycle of circular economy through a sanitation economy,
(VIII) Ensure resource recycling and safe reuse at all levels,
(IX) Support local livelihoods and enhance the health, environment, comfort and well-being of served communities through net-benefit-sharing,
(X) Facilitate active stakeholders' engagement, psycho-social and cultural management, wealth creation and positive change in the communities (Cole *et al.*, 2012).

ReGenSan is expected to be a catalytic driving force towards progressive change that could positively impact the SDG no. 6 sanitation targets by 2030.

2.7 RETHINKING SANITATION

Approaches to providing sanitation solutions have been characterized with a high degree of fragmentation and institutional gaps, which neglect systemic, holistic and integrated approaches that seek to provide much-needed alternatives from solving sanitation problems at individual and/or household levels (which results in more problems at other scales). For example, ending 'individual open defecation' at household levels by building more toilets created more acute problems of 'public open defecation', that is, indiscriminate disposal of untreated faecal sludge into the environment (water and land). This makes it pertinent to approach sani-solutions from an integrated-systemic-holistic-hybridized perspective at sani-sheds that could yield inter-synergistic benefits beyond the current fragmented and haphazard service delivery norms. This implies that sanitation should no longer be reduced to just toilets and toilet systems, no longer seen as isolated structures for defecation where products and by-products are considered as waste and a source of nuisance that contaminate, pollute and degrade the ecological systems; but rather as a catalyst for the transformation of a 'place'.

Addressing sanitation regeneratively is viewing sanitation as a global system of human-societal-related components including physical, social and ecological ones as well as technological components and their interactions. The functional systemic perspectives of ReGenSan state that every aspect and level of sanitation service delivery has to work and households, neighbourhoods, rural communities, urban centres, cities, states, provinces and countries shall not be judged by the amount of available infrastructure, but by the functionality of such infrastructure (that is, a rejuvenation and revitalization of the entire spectrum of sanitation systems). Thus, ReGenSan involves integrating comprehensively political, scientific, social, economic, cultural and technological strategies in a holistic and systemic manner with the aim of providing solutions that are fit for particular locations (place) and sites (scale). The subsequent chapters present the ReGenSan framework and practical descriptions of its dimensions and implementation.

2.8 EXERCISES

(I) Create a scenario where culture, religion, status and income can impact sanitation technology and service delivery.
(II) What do you understand by 'place' and 'scale' in the contexts of sani-sheds in sanitation management?.
(III) In a maximum of 1,000 words, critically show the connections between ReGenSan theory, principles, practices and benefits and their level of practicality.
(IV) Critically assess in 500 words how the assumptions and theoretical framework guided the proposed theory and principles of ReGenSan. State your own theory of ReGenSan.
(V) Critically assess the ReGenSan paradigm and concept definitions in the light of the theory and principles and show how this could affect sanitation management perspectives in the future.
(VI) Critically examine five possible technology and management innovative solutions that could be derived from the principles and practices highlighted in this chapter.
(VII) Briefly expatiate on five of the ReGenSan principles and show what practices and innovations are core to each principle as well as the interconnections between them.
(VIII) Explain the symbiotic and integrative pattern of the 10 principles and how they can innovatively make up ReGenSan management.

2.9 RECOMMENDATIONS FOR FUTURE RESEARCH

The ReGenSan theory and principles are new and a lot of investigation and further research is still needed for better appreciation and understanding of the concept:

(I) Explore the role of culture and religion in sanitation technology design and service provision.
(II) Explore the combination and balance between centralized and decentralized sanitation system for ReGenSan hybridization programme.
(III) Explore the principles of ReGenSan and the kind of innovative solutions that could come out of it.
(IV) A critical analysis of ReGenSan paradigm, theory and concepts.
(V) Explore the ReGenSan principle of NToB, ApT, SHI and PaS and its possible impact on sanitation technological and management solutions.

(VI) Investigate possibilities of appropriate technology for place and scale levels in sanitation management.
(VII) Explore the fundamental connection of functionalism and system theory with complex-dynamic systems like sanitation management and different levels of technology, governance, pyscho-socio-cultural, economic, ecological systems and also holistically.
(VIII) Develop scenario management tool for ReGenSan systems for different types of sani-sheds.
(IX) Explore scenario development for ReGenSan at different levels, sani-sheds and scale.
(X) Explore ways that PaP and SHI in conjunction with FPG principles could impact community livelihood and economic status.

REFERENCES

Aggestam F. (2007). Stakeholder Analysis and Engagement: Reviewing projects of the United Nations Development Programme (UNDP) and International Institute for Applied Systems Analysis (IIASA). SERI Working paper. March 2007.

Ahluwalia P. (2012). Centralized and decentralized wastewater systems. Journal of Indian Water Works Association, January–March 2012, 53–58.

Akturk A. (2016). Regenerative Design and Development for a Sustainable Future: Definitions and Tool Evaluation. M. S. thesis, University of Minnesota, USA.

Anarow B., Greener C., Gupta V., Kinsley M., Henderson J., Page C. and Parrot K. (2003). Whole-Systems Framework for Sustainable Consumption and Production. Danish Ministry of the Environment, Denmark, vol. **807**, pp. 1–51.

Andersson K., Dickin S. and Rosemarin A. (2016a). Towards 'sustainable' sanitation: challenges and opportunities in urban areas. *Sustainability*, **8**, 1289.

Andersson K., Rosemarin A., Lamizana B., Kvarnström E., McConville J., Seidu R., Dickin S. and Trimmer C. (2016b). Sanitation, Wastewater Management and Sustainability: From Waste Disposal to Resource Recovery. United Nations Environment Programme and Stockholm Environment Institute, Nairobi and Stockholm. https://www.sei-international.org/mediamanager/documents/Publications/SEI-UNEP-2016-SanWWM&Sustainability.pdf (accessed 9 November 2017).

Andersson M. and Minoia P. (2017). Ecological sanitation: a sustainable goal with local choices. A case study from Taita Hills, Kenya. *African Geographical Review*, **36**(2), 183–199, doi: 10.1080/19376812.2015.1134336

Appelstrand M. (2002). Participation and societal values: the challenge for lawmakers and policy practitioners. *Forest Policy and Economics*, **4**(2002), 281–290.

Badarnah L. and Kadri U. (2015). A methodology for the generation of biomimetic design concepts. *Architectural Science Review*, **58**(2), 120–133, doi: 10.1080/00038628.2014.922458

Bakir H. A. (2001). Sustainable wastewater management for small communities in the Middle East and North Africa. *Journal of Environmental Management*, **61**(4), 319–328.

Balkema A. J., Preisig H. A., Otterpohl R. and Lambert F. J. D. (2002). Indicators for the sustainability assessment of wastewater treatment systems. *Urban Water*, **4**, 153–161.

Baum R., Luh J. and Bartram J. (2013). Sanitation: a global estimate of sewerage connections without treatment and the resulting impact on MDG progress. *Advances in Environmental Science and Technology*, **47**(4), 1994–2000, doi: 10.1021/es304284f

Baumeister D. (2012). Biomimicry Resource Handbook: A Seed Bank of Knowledge and Best Practices. Published by Biomimicry 3.8, Missoula, USA.

Benne B. and Mang P. (2015). Working regeneratively across scales-insights from nature applied to the built environment. *Journal of Cleaner Production*, **109**, 42–52. http://dx.doi.org/10.1016/j.jclepro.2015.02.037

Benyus J. M. (2002). Biomimicry: Innovation Inspired by Nature. rev. edn. Perennial, New York, USA.

Bergen S. D., Bolton S. M. and Fridley J. L. (2001). Design principles for ecological engineering. *Ecological Engineering*, **18**, 201–210.

Bieker S., Cornel P. and Wagner M. (2010). Semicentralised supply and treatment systems: integrated infrastructure solutions for fast growing urban areas. *Water Science and Technology*, **61**(11), 2905–2913.

Borsuk M. E., Maurer M., Lienert J. and Larsen T. A. (2008). Charting a path for innovative toilet technology using multicriteria decision analysis. *Advances in Environmental Science and Technology*, **42**, 1855e1862.

Boulenouar J. and Gibson J. (2015). Asset Management in Rural South Africa. A Case Study of Amatole and Chris Hani District Municipalities. IRC, Aguaconsult, The Hague, Netherlands. https://www.ircwash.org/resources/asset-management-rural-south-africa (accessed 13 November 2017).

Bracken P., Wachtier A., Panesar A. R. and Lange J. (2006). The road not taken: how traditional excreta and greywater management may point the way to a sustainable future. *Water Science & Technology: Water Supply*, **7**(1), 219–227, doi: 10.2166/ws.2007.025

Brands E. (2014). Prospects and challenges for sustainable sanitation in developed nations: a critical review. *Environmental Reviews*, **22**, 1–18, doi: 10.1139/er-2013-0082

Brepolsa C. H., Dorgeloh E., Frechen F. B., Fuchs W., Haider S., Joss A., de Korte K., Ruiken C. H., Schier W., van der Roest H., Wett M. and Wozniak T. H. (2008). Upgrading and retrofitting of municipal wastewater treatment plants by means of membrane bioreactor (MBR) technology. *Desalination*, **231**, 20–26.

Brikke F. (2000). Operation and Maintenance of Rural Water Supply and Sanitation Systems: A Training Package for Managers and Planners. World Health Organization Geneva, Switzerland. http://apps.who.int/iris/bitstream/10665/66716/1/WHO_SDE_WSH_00.2_p1-124.pdf (accessed 30 April 2017).

Brown & Root (2001). On-site Sewage Risk Assessment System. Prepared for New Department Local Government, Queensland, Australia. Brown & Root Services Asia Pacific Pty Ltd., Brisbane, Australia. http://www.dlg.nsw.gov.au/dlg/dlghome/dlg_osras.asp

Brown R., Farrelly M. and Keath N. (2009). Practitioner perceptions of social and institutional barriers to advancing a diverse water source approach in Australia. *International Journal of Water Resources Development*, **25**(1), 15–28.

Brown R., Ashley R. and Farrelly M. (2011). Political and professional agency entrapment: an agenda for urban water research. *Water Resources Management*, **25**, 4037–4050.

Burkhard R., Deletic A. and Craig A. (2000). Techniques for water and wastewater management: a review of techniques and their integration in planning. *Urban Water*, **2**, 197–221.

Butler D. and Parkinson J. (1997). Towards sustainable urban drainage. *Water Science and Technology*, **35**(9), 53–63, doi: 10.1016/S0273-1223(97)00184-4

Capra F. (1997). The Web of Life. Doubleday-Anchor Books, New York, USA.

Carabias W. J. V., Winistoerfer H. and Stuecheli A. (1999). Social aspects of public waste management in Switzerland. *Waste Management*, **19**(6), 417–425.

Carragher B. J., Stewart R. A. and Beal C. D. (2012). Quantifying the influence of residential water appliance efficiency on average day diurnal demand patterns at an end use level: A precursor to optimised water service infrastructure planning. *Resources Conservation and Recycling*, **62**, 81–90.

Charnley F. and Lemon M. (2011). Exploring the process of whole system design. *Design Studies*, **32**(2), 156–179. doi: 10.1016/j.destud.2010.08.002

Checkland P. (1997). Systems Thinking, Systems Practice. John Wiley & Sons Ltd, Chichester, UK.

Checkland P. B. (1981). Systems Thinking, Systems Practice. John Wiley Inc., Chichester, New York, USA.

Christian-Smith J. and Gleick P. H. (2012). Introduction: the soft path for water. In: A Twenty-First Century US Water Policy, J. Christian-Smith, P. H. Gleick, H. Cooley, L. Allen, A. Vanderwarker and K. A. Berry (eds), Oxford University Press, New York, USA.

Christmas J. and de Rooy C. (1991). The Decade and beyond: at a glance. *Water International*, **16**(3), 127–134.

Clark R. M. (1978). Analysis of Urban Solid Waste Services: A Systems Approach. Ann Arbor Science, Ann Arbor, USA.

Clarvis M. H., Allan A. and Hannah D. M. (2014). Water, resilience and the law: from general concepts and governance design principles to actionable mechanisms. *Environmental Science & Policy*, **43**, 98–110.

Cole R. J. (2010). New Context, New Responsibilities: Building Capability. Unpublished paper (Cited In: Mang, P. and Reed, B. (2012) Regenerative Development and Design. Regenesis Group and Story of Place, Encyclo. Sustainability Science & Technology). University of British Columbia, British Columbia, Canada..

Cole R. J. (2012). Regenerative design and development: current theory and practice. *Building Research & Information*, **40**(1), 1–6, doi: 10.1080/09613218.2012.617516

Cole R. J., Busby P., Guenther R., Briney L., Blaviesciunaite A. and Alencar T. (2012). A regenerative design framework: setting new aspirations and initiating new discussions. *Building Research & Information*, **40**(1), 95–111. doi: 10.1080/09613218.2011.616098.

Coley F. and Lemon M. (2008). Supporting multiple stakeholders in the design of more sustainable and systemic solutions: the case of the sustainable sports car. *Design Principles and Practices: An International Journal*, **2**, 27–33.

Cookey P. E. (2016a). Water Governance Performance in Songkhla Lake Basin, Thailand. PhD thesis, Prince of Songkla University, Thailand. http://kb.psu.ac.th/psukb/handle/2016/10677

Cookey P. E. (2016b). Adaptive Integrated Lake Basin Management: Measuring Lake Basin Governance Performance. LAP Lambert Academic Publishing, Saarbrucken, Germany, pp. 336, ISBN: 978-3-659-94249-5.

Cookey P. E., Darnswadi R. and Ratanachai C. (2015). Using Text Mining to Evaluate the Integrative and Adaptive Elements of Water Resource Institutions for Songkhla Lake Basin, Thailand. *Journal of Water Resource and Hydraulic Engineering*, **4**(4), 339–357, doi: 10.5963/JWRHE0404005

Cookey P. E., Koottatep T., van der Steen P. and Lens P. N. L. (2016a). Public health risk assessment tool: strategy to improve public policy framework for onsite wastewater treatment systems (OWTS). *Journal of Water, Sanitation and Hygiene for Development*, **06.1**, 74–88, doi: 10.2166/washdev.2016.081

Cookey P. E., Darnsawasdi R. and Ratanachai C. (2016b). A conceptual framework for assessment of governance performance of lake basins: towards transformation to adaptive and integrative governance. *Hydrology*, **3**(1), 12, doi: 10.3390/hydrology3010012

Cookey P. E., Darnswadi R. and Ratanachai C. (2016c). Local people's perceptions of Lake Basin water governance performance in Thailand. *Ocean & Coastal Management*, **120**, 11–28. http://dx.doi.org/10.1016/j.ocecoaman.2015.11.015

Cookey P. E., Darnswadi R. and Ratanachai C. (2016d). Performance evaluation of lake basin water governance using composite index. *Ecological Indicators*, **61**(2), 466–482. https://doi.org/10.1016/j.ecolind.2015.09.048.

Cookey P. E., Darnsawasdi R. and Ratanachai C. (2017). Text mining analysis of institutional fit of Lake Basin water governance. *Ecological Indicators*, **72**, 640–658. https://doi.org/10.1016/j.ecolind.2016.08.057

Cooper P. F. (2001). Chapter 2: Historical aspects of wastewater treatment. In: Decentralised Sanitation and Reuse – Concepts, Systems and Implementation, P. Lens, G. Zeeman and G. Lettinga (eds), IWA Publishing, the Hague, Netherlands.

Daigger G. T. (2009). Evolving urban water and residuals management paradigms: water reclamation and reuse, decentralization, and resource recovery. *Water Environment Research*, **81**, 809–823.

Daigger G. T. and Crawford G. V. (2007). Enhancing water system security and sustainability by incorporating centralized and decentralized water reclamation and reuse into urban water management systems. *Environmental Engineering and Management*, **17**, 1–10.

de Macedo Guimarães L. B. D. (2012). Sociotechnical design for a sustainable world. *Theoretical Issues in Ergonomics Science*, **13**(2), 240–269, doi: 10.1080/1463922X.2011.641230

Delleur J. W. (2003). The evolution of urban hydrology: past, present, and future. *Journal of Hydraulic Engineering*, **129**(8), 563–573, doi: 10.1061/(ASCE)0733-9429(2003)129:8(563)

DeWitt R. (2010). Worldviews: An Introduction to the History and Philosophy of Science. Wiley-Blackwell, Oxford, UK.

Dias B. D. (2015). Beyond sustainability – biophilic and regenerative design in architecture. *European Scientific Journal*, **1**, 1857–7881. http://dx.doi.org/10.19044/esj.2015.v11n9p%25p (accessed 12 October 2017).

Dijkema G. P. J., Reuter M. A. and Verhoef E. V. (2000). A new paradigm for waste management. *Waste Management*, **20**(8), 633–638.

Dixon T., Eames M., Britnell J., Watson G. B. and Hunt M. (2014). Urban retrofitting: Identifying disruptive and sustaining technologies using performative and foresight techniques. *Technological Forecasting & Social Change*, **89**, 131–144.

Dubrovsky V. (2004). Toward system principles: general system theory and the alternantive approach. *Systems Research and Behavioral Science*, **21**(2), 109–122. https://doi.org/10.1002/sres.572.

Du Plessis C. (2009). An Approach to Studying Urban Sustainability from within an Ecological Worldview. Unpublished doctoral thesis, University of Salford, UK.

Du Plessis C. (2012). Towards a regenerative paradigm for the built environment. *Building Research & Information*, **40**(1), 7–22, doi: 10.1080/09613218.2012.628548

Eames M. (2011). Developing Urban Retrofit Scenarios: An Outline Framework for Scenario Foresight and Appraisal, Retrofit 2050 Working Paper WP 2011/4. www.retrofit2050.org.uk (accessed 1 May 2017).

Esrey S. A. (2001). Towards a recycling society ecological sanitation – closing the loop to food security. *Water Science & Technology*, **43**(4), 177–187.

Esrey S. A. (2002). Philosophical, ecological and technical challenges for expanding ecological sanitation into urban areas. *Water Science and Technology*, **45**(8), 225–228, PMID:12019825

Eweje G. (2007). Strategic partnerships between MNEs and civil society: the post-WSSD perspectives. *Sustainable Development*, **15**(1), 15–27.

Exley J. L. R., Liseka B., Cumming O. and Ensink J. H. J. (2014). The sanitation ladder, what constitutes an improved form of sanitation? *Advances in Environmental Science and Technology*, **49**, 1086–1094, doi: 10.1021/es503945x

Fewtrell L., Kaufmann R. B., Kay D., Enanoria W., Haller L. and Colford J. M. (2005). Water, sanitation, and hygiene interventions to reduce diarrhoea in less developed countries: a systematic review and meta-analysis. *The Lancet Infectious Diseases*, **5**(1), 42–52, doi: 10.1016/S1473-3099(04)01253-8

Fewtrell L., Prüss-Üstün A., Bos R., Gore F. and Bartram J. (2007). Water, Sanitation and Hygiene: Quantifying the Health Impact at National and Local Levels in Countries with Incomplete Water Supply and Sanitation Coverage. World Health Organization, Geneva (WHO Environmental Burden of Disease Series No. 15).

Folke C., Pritchard Jr L., Berkes F., Colding J. and Svedin U. (2007). The problem of fit between ecosystems and institutions: ten years later. *Ecology and Society*, **12**(1), 30. http://www.eologyandscoiety.org/vol12/iss20/ (accessed 14 October 2017).

Galaz V., Olsson P., Hahn T. and Svedin U. (2008). The problem of fit among biophysical systems, environmental and resource regimes, and broader governance systems: insights and emerging challenges. In: Institutions and Environmental Change: Principal Findings, Applications and Research Frontiers, O. R. Young, L. A. King and H. Schroeder (eds), MIT Press, Cambridge, USA, pp. 147–186.

Gamage A. and Hyde R. (2012). A model based on Biomimicry to enhance ecologically sustainable design. *Architectural Science Review*, **55**(3), 224–235, doi: 10.1080/00038628.2012.709406

Gebeshuber I. C. and Drack M. (2008). An attempt to reveal synergies between biology and mechanical engineering. In Proceedings of the IMECHE, Part C. *Journal of Mechanical Engineering Science*, **222**(7), 1281–1287.

Geels F. (2005). Co-evolution of technology and society: the transition in water supply and personal hygiene in the Netherlands (1850–1930) – a case study in multi-level perspective. *Technology in Society*, **27**, 363–397.

Geisinger D. and Chartier G. (2005). Managed onsite/decentralised wastewater systems as long-term solutions. *Clearwaters*, **35**, 6–11.

Gikas P. and Tchobanoglous G. (2009). The role of satellite and decentralized strategies in water resources management. *Journal of Environmental Management*, **90**, 144–152, doi: 10.1016/j.jenvman.2007.08.016

Giné-Garriga R., Flores-Baquero O., de Palencia A. J. and Pérez-Foguet A. (2017). Monitoring sanitation and hygiene in the 2030 agenda for sustainable development: a review through the lens of human rights. *Science of the Total Environment*, **580**, 1108–1119.

Giri R., Takeuchi J. and Ozaki H. (2006). Biodegradation of domestic wastewater under the stimulated conditions of Thailand. *Water and Environment Journal*, **20**(3), 109–202.

Go E. and Demir I. (2006). Cost analysis of alternative methods for wastewater handling in small communities. *Journal of Environmental Management*, **79**(4), 357–363.

Griffiths H., Keirns N., Strayer E., Cody-Rydzewski S., Scaramuzzo G., Sadler T., Vyain S., Bry J. and Jones F. (2015). Introduction to Sociology, 12th Media Services. Suwanee, USA. https://openstax.org/details/introduction-sociology-2e (accessed 20 October 2017).

Hegger D. L. (2007). Greening Sanitary Systems: An End-user Perspective. PhD thesis, Wageningen University, Netherlands.

Helms M., Vattam S. S. and Goel A. K. (2009). Biologically inspired design: process and products. *Design Studies*, **30**(5), 606–622, doi: 10.1016/j.destud.2009.04.003

Henry R. K., Yongsheng Z. and Jun D. (2006). Municipal solid waste management challenges in developing countries – Kenyan case study. *Waste Management*, **26**(1), 92–100.

Hens L. and Nath B. (2003). The Johannesburg conference. *Environment, Development and Sustainability*, **5**(1–2), 7–39, Kluwer Academic. https://doi.org/10.1023/A:1025303511864

Hes D. and du Plessis C. (2014). Designing for Hope: Pathways to Regenerative Sustainability. Routledge, Abingdon, UK.

Heylighen F. (2006). The Newtonian World View. Principia Cybernetica Web. http://pespmc1.vub.ac.be/NEWTONWV.html (accessed 15 May 2014).

Hollick M. (1982). The appropriate technology movement and its literature: a retrospective. *Technology in Society*, **4**(3), 213–229. https://doi.org/10.1016/0160-791X(82)90019-7.

Hoxie C., Berkebile R. and Todd J. A. (2012). Stimulating regenerative development through community dialogue, Building. *Research & Information*, **40**(1), 65–80, doi: 10.1080/09613218.2011.628546

Hutton G., Haller L. and Bartram J. (2007). Economic and Health Effects of Increasing Coverage of Low Cost Household Drinking-Water Supply and Sanitation Interventions to Countries Off-Track to Meet MDG Target 10. World Health Organization, Geneva, Switzerland. http://www.irc.nl/page/38443 (accessed 15 March 2018).

IDRC, International Development Research Centre (2010). Cofounding Variables. http://www.idrc.ca/en/ev-1-201-1-DO_TOPIC.html (accessed 30 June 2010).

ILECF (International Lake Environmental Committee Foundation) (2005). Managing Lakes and their Basins for Sustainable Use: A Report for Lake Basin Managers and Stakeholders. Kusatsu, Japan. http://wldb.ilec.or.jp/Details/Download/10216 (accessed 15 October 2017).

Illich I. (1973). Tools for Conviviality. Harper and Row, New York, USA.

IWA (24:2016). Non-sewered sanitation systems – General safety and performance requirements for design and testing. https://www.iso.org/standard/70604.html (accessed 9 May 2017).

Jackson M. (2003). Systems Thinking: Creative Holism for Managers. John Wiley & Sons, Ltd, Chichester, UK.

Jenkin S. and Pedersen Z. M. (2009). Rethinking Our Built Environments: Towards a Sustainable Future. A Research Document. Ministry for the Environment, Manatu Mo Te Taiao, Wellington, New Zealand. https://www.researchgate.net/publication/261476724_Rethinking_our_built_environments_Towards_a_sustainable_future (accessed 10 February 2018).

Joint Monitoring Programme (JMP) (2015). Progress on Sanitation and Drinking Water: 2015 Update and MDG Assessment Geneva/New York.

Jones D. S., Efroymson R. A., Armstrong A. Q., Muhlheim M. D. and Carnes S. A. (2004). Integrated Risk Assessment for Individual Onsite Wastewater Systems. National Decentralized Water Resources Capacity Development Project. Oak Ridge National Laboratory, Oak Ridge, TN, USA, NDWRCDP Project Number: WU-HT-01-18. Washington University.

Jönsson H., Tidåker P. and Stintzing A. R. (2010). Role of farmers in improving the sustainability of sanitation systems. In: Social Perspectives on the Sanitation Challenge, B. van Vliet, G. Spaargaren and P. Oosterveer (eds), Springer, Dordrecht, Netherlands, pp. 31–47.

Kalbermatten J. M., Julius D. S., Mara D. D. and Gunnerson C. G. (1980). Appropriate Technology for Water Supply and Sanitation A Planner's Guide. The International Bank for Reconstruction and Development/the World Bank. https://www.ircwash.org/sites/default/files/201-80AP-19010.pdf (accessed 27 May 2017).

Kalbermatten J. M., Julius D. S. and Gunnerson C. G. (1982). Appropriate sanitation alternatives: a technical and economic appraisal. World Bank studies in Water Supply and Sanitation; no. 1. World Bank, Washington D.C., USA. http://documents.worldbank.org/curated/en/637021468740368977/Appropriate-sanitation-alternatives-a-technical-and-economic-appraisal (accessed 27 May 2017).

Katukiza A. Y., Ronteltap M., Oleja A., Niwagaba C. B., Kansiime F. and Lens P. N. L. (2010). Selection of sustainable sanitation technologies for urban slums – a case of Bwaise III in Kampala, Uganda. *Science of the Total Environment*, **409**, 52–62.

Katukiza A. Y., Ronteltap M., Niwagaba C. B., Foppen J. W. A., Kansiime F. and Lens P. N. L. (2012). Sustainable sanitation technology options for urban slums. *Biotechnology Advances*, **30**, 964–978.

Kellert S. (2012). Birthright: People and Nature in the Modern World. Yale University Press, New Haven, USA.

Kellert S. (2016). Biophilia and biomimicry: evolutionary adaptation of human versus nonhuman nature. *Intelligent Buildings International*, **8**(2), 51–56, doi: 10.1080/17508975.2014.902802

Kellert S. and Wilson E. O. (1993). The Biophilia Hypothesis. Island Press, Washington D.C., USA.

Kennedy E. B. and Marting T. A. (2016). Biomimicry: streamlining the front end of innovation for environmentally sustainable products. *Research-Technology Management*, **59**(4), 40–48. http://dx.doi.org/10.1080/08956308.2016.1185342

Kluge T. and Moser-Nørgaard P. M. (2008). Innovative water supply and disposal technologies as integral part of integrated water resources management: an example. *International Journal of Water*, **4**, 41–54.

Kollikkathara N., Feng H. and Stern E. (2009). A purview of waste management evolution: special emphasis on USA. *Waste Management*, **29**(2), 974–985.

Koltko-Rivera M. E. (2004). The psychology of worldviews. *Review of General Psychology*, **8**, 3–58.

Koottatep T., Polprasert C. and Laugesen C. (2007). Integrated eco-engineering design for sustainable management of fecal sludge and domestic wastewater. *Journal of Korean Wetland Society*, **9**, 69–78.

Kosek M., Bern C. and Guerrant R. L. (2003). The global burden of diarrhoeal disease, as estimated from studies published between 1992 and 2000. *Bull World Health Organ*, **81**, 197–204.

Kvarnström E., Bracken P., Karrman E., Finnson A. and Saywell D. (2004). People-centred approaches to water and environmental sanitation. Proceedings of the 30th WEDC international conference, Vientiane, Lao, PDR: WEDC.

Kvarnström E., McConville J., Bracken P., Johansson M. and Fogde M. (2011). The sanitation ladder – a need for a revamp? *Journal of Water Sanitation and Hygiene for Development*, **1**(1), 3–12.

Kvarnström E., Joep V., Mats N., Vishwanath S., Shubha R. and Karan S. (2012). The business of the honey-suckers in Bengaluru (India): the potentials and limitations of commercial faecal sludge recycling – an explorative study. Occasional paper 48 (online). The Hague: IRC (International Water and Sanitation Centre). http://www.hydratelife.org/wp-content/uploads/2012/12/The-Business-of-the-Honey-Suckers-inBengaluru-India-The-Potentials-and-Limitations-of-Commercial-FaecalSludge-Recycling.pdf (accessed 10 February 2018).

Larsen T. A. and Gujer W. (2013). Implementation of source separation and decentralization in cities. In: Source Separation and Decentralization for Wastewater Management, T.A. Larsen, K. Udert and J. Lienert (eds), IWA Publishing, London, UK, pp. 135–150.

Larsen T. A. and Maurer M. (2011). Source separation and decentralization. In: Treatise on Water Science, P. Wilderer (ed.), Elsevier, Oxford, UK, pp. 203–229.

Larsen T. A., Alder A. C., Eggen R. I., Maurer A. and Lienert J. (2009). Source separation: will we see a paradigm shift in wastewater handling. *Environmental Science & Technology*, **43**, 6121–6125.

Larsen T. A., Maurer M., Eggen R. I. L., Pronk W. and Lienert J. (2010). Decision support in urban water management based on generic scenarios: the example of NoMix technology. *Journal of Environmental Management*, **91**, 2676–2687.

Larsen T. A., Udert K. and Lienert J. (2013). Source Separation and Decentralization for Wastewater Management. IWA Publishing, London, UK.

Lebel L., Nikitina E., Pahl-Wostl C. and Knieper C. (2013). Institutional fit and river basin governance: a new approach using multiple composite measures. *Ecology and Society*, **18**(1), 1. http://dx.doi.org/10.5751/ES-05097-180101

Lens P., Zeeman G. and Lettinga G. (2001). Decentralized Sanitation and Reuse: Concepts, Systems and Implementation. IWA Publishing, London, UK, 650 p.

Leverenz H. and Tchobanoglous G. (2009). Satellite Systems for Enhanced Wastewater Management in Urban Areas. Proceedings, WEFTEC 2009, Orlando, FL, USA.

Libralato G., Ghirardini A. V. and Avezzu F. (2012). To centralize or to decentralize: an overview of the most recent trends in wastewater treatment management. *Journal of Environmental Management*, **94**(1), 61–68.

LIXIL (Link to Good Living), Oxford Economics and Water Aid (2016). The true cost of poor sanitation. Salt Communications. London, United Kingdom. See. https://www.ircwash.org/resources/true-cost-poor-sanitation (accessed 19 October 2017).

Lopes A. M., Fam D. and Williams J. (2012). Designing sustainable sanitation: involving design in innovative, transdisciplinary research. *Design Studies*, **33**(3), 298–317. https://doi.org/10.1016/j.destud.2011.08.005

Luyet V., Schlaepfer R., Parlange M. B. and Buttler A. (2012). A framework to implement Stakeholder participation in environmental projects. *Journal of Environmental Management*, **111**, 213–219.

Lyle J. T. (1994). Regenerative Design for Sustainable Development. John Wiley & Sons, Hoboken, USA.

Mang P. and Reed B. (2012). Chapter 3. Regenerative development and design: Regenesis Group and Story of Place Institute. In: *Encyclopedia Sustainability Science and Technology*, R. A. Meyers (ed.), Springer, New York, NY, USA, pp. 8855–8879. https://doi.org/10.1007/978-1-4419-0851-3.

Mang P. and Reed B. (2012a). Designing from place: a regenerative framework and methodology. *Building Research & Information*, **40**(1), 23e38.

Mang P. and Reed B. (2012b). Regenerative development and design. Regenesis Group and story of place. *Encyclopedia of Sustainability Science and Technology*.

Mang P., Haggard B. and Regenesis (2016). Regenerative Development and Design: A Framework for Evolving Sustainability. John Wiley & Sons, Inc, Hoboken, USA.

Mara D. (2003). Water, sanitation and hygiene for the health of developing nations. *American Journal of Public Health*, **117**, 452–456.

Mara D., Drangert J., Anh N. and Tonderski A. (2007). Selection of sustainable sanitation arrangements. *Water Policy*, **9**, 305–318.

Mara D., Lane J., Scott B. and Trouba D. (2010). Sanitation and Health. *PLoS Medicine*, **7**(11), e1000363, doi: 10.1371/journal.pmed.1000363

Marlow D. R., Moglia M., Cook S. and Beale D. J. (2013). Towards sustainable urban water management: a critical reassessment. *Water Research*, **47**, 7150–7161.

Marshall R. E. and Farahbakhsh K. (2013). Systems approaches to integrated solid waste management in developing countries. *Waste Management*, **33**, 988–1003. http://dx.doi.org/10.1016/j.wasman.2012.12.023

Massoud M. A., Tarhini A. and Nasr J. A. (2009). Decentralized approaches to wastewater treatment and management: Applicability in developing countries. *Journal of Environmental Management*, **90**, 652–659, doi: 10.1016/j.jenvman.2008.07.001

Mathers C. D., Lopez A. D. and Murray C. J. L. (2006). The burden of disease and mortality by condition: data, methods, and results for 2001. In: Global Burden of Disease and Risk Factors, A. D. Lopez, C. D. Mathers, M. Ezzati, D. T. Jamison and C. J. L. Murray (eds), Oxford University Press, New York, USA, pp. 45–240.

McDonough W. and Braungart B. (2001). The Next Industrial Revolution, William McDonough, Michael Braungart and the Birth of the Sustainable Economy. Earthome Productions, Stevenson, USA.

McDonough W. and Braungart M. (2002). Cradle to Cradle: Remaking the Way We Make Things. North Point Press, New York, USA.

McDougall F., White P. R., Franke M. and Hindle P. (2001). Integrated Solid Waste Management: A Lifecycle Inventory, 2nd edn. Blackwell Science, Oxford, UK.

Meadows D. (1999). Leverage Points: Places to Intervene in a System. The Academy for Systems Change. The Sustainability Institute. http://donellameadows.org/archives/leverage-points-places-to-intervene-in-a-system/ (accessed 20 September 2017).

Mele C., Pels J. and Polese F. (2010). A brief review of systems theories and their managerial applications. *Service Science*, **2**(1–2), 126–135. http://dx.doi.org/10.1287/serv.2.1_2.126

Melo J. C. (2005). The Experience of Condominial Water and Sewerage Systems in Brazil: Case Studies from Brasilia, Salvador and Parauebas. Water and Sanitation Program. World Bank, Washington D.C., USA. http://documents.worldbank.org/curated/en/505601468226737476/The-experience-of-condominial-water-and-sewerage-systems-in-Brazil-case-studies-from-Brasilia-Salvador-and-Parauebas (accessed 13 November 2017).

Miller M. E. and West A. N. (1993). Influences of world view on personality, epistemology, and choice of profession. In: Development in the Work-Place, J. Demick and P.M. Miller (eds), Erlbaum, Hillsdale, USA, pp. 3–19.

Mitsch W. J. (1993). Ecological engineering a cooperative role with the planetary life-support system. *Advances in Environmental Science and Technology*, **27**(3), 438–445.

Mitsch W. J. and Jørgensen S. E. (1989). Ecological Engineering: An Introduction to Ecotechnology. Wiley, New York, USA.

Mitsch W. J. and Jørgensen S. E. (2003). Ecological engineering: a field whose time has come. *Ecological Engineering*, **20**, 363–377.

Mohammed Shahalam A. B. (1985). An optimal approach for the selection of appropriate sanitation technologies for developing countries. In: Appropriate Waste Management for Developing Countries, K. Curi (ed.), Springer, Boston, MA, USA. https://doi.org/10.1007.978-1-4613-2457-7-3.

Mooney L. A., Knox D. and Schacht C. (2013). Understanding Social Problems, 8th edn. Wadsworth, California, USA.

Morrissey A. J. and Browne J. (2004). Waste management models and their application to sustainable waste management. *Waste Management*, **24**(3), 297–308.

Moss T. (2007). Solving problems of 'fit' at the expense of problems of 'interplay'? The spatial reorganisation of water management following the EU water framework directive. In: Integrated Water Resources Management: Global Theory, Emerging Practice and Local Needs, P. P. Mollinga, A. Dixit and K. Athukorala (eds), Sage Publications, New Delhi/Thousand Oaks/London.

Moss T. (2012). Spatial fit, from panacea to practice: implementing the EU Water Framework Directive. *Ecology and Society*, **17**(3), 2. http://dx.doi.org/10.5751/ES-04821-170302

Murphy H. M., McBean E. A. and Farahbakhsh K. (2009). Appropriate technology – a comprehensive approach for water and sanitation in the developing world. *Technology in Society*, **31**, 158–167, doi: 10.1016/j.techsoc.2009.03.010

Murray A. and Buckley C. (2009). Designing reuse-oriented sanitation infrastructure: the design for service planning approach. In: Wastewater Irrigation and Health Assessing and Mitigating Risk in Low-income Countries, P. Drechsel, C. A. Scott, L. Raschid-Sally, M. Redwood and A. Bahri (eds), Earthscan, London, UK, pp. 303–318.

Najlis P. and Edwards A. (1991). The international drinking water supply and sanitation decade in retrospect and implications for the future. *Natural Resources Forum*, **15**(2), 110–117.

Nansubuga I., Banadda N., Verstraete W. and Rabaey K. (2016). A review of sustainable sanitation systems in Africa. *Reviews in Environmental Science and Bio/Technology*, **15**, 465, doi: 10.1007/s11157-016-9400-3

Nelson R. H. (1995). The federal land management agencies. In: A New Century for Natural Resources Management, R. L. Knight and S. F. Bates (eds), Island Press, Covel, USA.

Nhapi I., Siebel M. A. and Gijzen H. J. (2006). A proposal for managing wastewater in Harare, Zimbabwe. *Water and Environment Journal*, **20**, 101–108.

Norris C. (2005). Partnerships for sustainable development. The role of type II agreements. In: Global Challenges: Furthering the Multilateral Process for Sustainable Development, A. C. Kallhauge, G. Sjostedt and E. Corell (eds), Greenleaf Publishing, Sheffield, UK, pp. 210–230.

North D. C. (1990). Institutions, Institutional Change and Economic Performance. Cambridge University Press, Cambridge, UK.

O'Reilly K., Dhanju R. and Goel A. (2017). Exploring 'The Remote' and 'The Rural': open defecation and latrine use in Uttarakhand, India. *World Development*, **93**, 193–205, 2017 0305-750X. https://doi.org/10.1016/j.worlddev.2016.12.022

Orth H. (2007). Centralised versus decentralised wastewater systems? *Water Science and Technology*, **56**(5), 259–266, doi: 10.2166/wst.2007.579

Ostrom E. (1990). Governing the Commons. Cambridge University Press, Cambridge, UK. http://dx.doi.org/10.1017/CBO9780511807763

Ostrom E. (2005). Understanding Institutional Diversity. Princeton University Press, Princeton, USA.

Ostrom E. (2010). Polycentric systems for coping with collective action and global environmental change. *Global Environmental Change*, **20**, 550–557. http://dx.doi.org/10.1016/j.gloenvcha.2010.07.004

Patton M. Q. (2002). Qualitative Research and Evaluation Methods (3rd edn). Sage Publications, Thousand Oaks, California, USA.

Pahl-Wostl C. (2009). A conceptual framework for analysing adaptive capacity and multi-level learning processes in resource governance regimes. *Global Environmental Change*, **19**(3), 354–365, doi: 10.1016/j.gloenvcha.2009.06.001

Pahl-Wostl C., Kabat P. and Möltgen J. (2008). Adaptive and Integrated Water Management. Coping with Complexity and Uncertainty. Springer Verlag, Heidelberg, Germany, 440p.

Pattnaik B. K. and Dhal D. (2015). Mobilizing from appropriate technologies to sustainable technologies based on grassroots innovations. *Technology in Society*, **40**, 93–110. http://dx.doi.org/10.1016/j.techsoc.2014.09.002

Persson C. P. (2017). 1,000-year-old Viking toilet uncovered in Denmark. http://sciencenordic.com/1000-year-old-viking-toilet-uncovered-denmark (accessed 3 July 2017).

Petts J. (2000). Municipal waste management: inequities and the role of deliberation. *Risk Analysis*, **20**(6), 821–832.

Polprasert C. and Koottatep T. (2017). Organic Waste Recycling – Technology, Management and Sustainability, 4th edn. IWA Publishing, the Hague, Netherlands.

Radcliffe-Brown A. R. (1952). Structure and Function in Primitive Society: Essays and Addresses. Cohen and West, London.

Rao K. C., Kvarnström E., Di Mario L. and Drechsel P. (2016). Business Models for Fecal Sludge Management. International Water Management Institute (IWMI), Colombo, Sri Lanka, CGIAR Research Program on Water, Land and Ecosystems (WLE). 80p. (Resource Recovery and Reuse Series 6), doi: 10.5337/2016.21

Rao K. C., Otoo M., Drechsel P. and Hanjra M. A. (2017). Resource recovery and reuse as an incentive for a more viable sanitation service chain. *Water Alternatives*, **10**(2), 493–512. http://www.water-alternatives.org/index.php/alldoc/articles/vol10/v10issue2/367-a10-2-17/file (accessed 18 October 2017).

Reed B. (2007). Shifting from 'sustainability' to regeneration. *Building Research & Information*, **35**(6), 674–680, doi: 10.1080/09613210701475753

Rijke J., Brown R., Zevenbergen C., Ashley R., Farrelly M., Morison P. and Van Herk S. (2012). Fit-for-purpose governance: a framework to make adaptive governance operational. *Environmental Science & Policy*, **22**, 73–84. https://doi.org/10.1016/j.envsci.2012.06.010 (accessed 10 February 2018).

Rittmann B. E. (2013). The energy issue in urban water management. In: Source Separation and Decentralisation for Wastewater Management, T. A. Larsen, K. M. Udert and J. Lienert (eds), IWA Publishing, London, UK, pp. 13–27.

Rocky Mountain Institute (2006). Hypercar design and technology. https://old.rmi.org/sitepages/pid390.php (Accessed on 7 February 2008).

Rosenqvist T., Mitchell C. and Willetts J. (2016). A short history of how we think and talk about sanitation services and why it matters. *Journal of Water, Sanitation and Hygiene for Development*, **06.2**, doi: 10.2166/washdev.2016.118

Ross K., Abeysuriya K., Mikhailovich N. and Mitchell C. (2014). Governance for decentralised sanitation: Global Practice Scan. A working document. Prepared by the Institute for Sustainable Futures, UTS as part of the Australian Development Research Award Scheme (ADRAS). http://communitysanitationgovernance.info/wp-content/uploads/2016/06/20140125-ADRAS-GDS-Global-Practice-Scan-a.pdf (accessed 14 October 2017).

Rowe G. and Frewer L. J. (2000). Public participation methods: a framework for evaluation. *Science Technology & Human Values*, **25**(1), 3–29.

Sankaran S., Abeysuriya K., Gray J. and Kachenko A. (2010). Closing the loop: a systems thinking led sustainable sanitation project in Australia. Proceedings of 54th Annual Conference of the International Society for the Systems Sciences 2010: Governance for a Resilient Planet, pp. 346–358. Waterloo, Ontario, Canada.

Sapkota M., Arora M., Malano H., Moglia M., Sharma A., George B. and Pamminger F. (2015). An overview of hybrid water supply systems in the context of urban water management: challenges and opportunities. *Water*, **7**, 153–174.

Sato T. (2011). Functionalism: Its axiomatics'. *Sociopedia.isa.*, doi: 10.1177/205684601332. http://www.sagepub.net/isa/resources/pdf/FunctionalismNew.pdf (accessed 31 October 2017).

Saubolle B. R. (1980). A women's toilet, Nepal. In: Experiences in appropriate technology, R. J. Mitchell (ed.), The Canadian Hunger Association, Ottawa, Canada, pp. 9–10.

Schumacher E. F. (1973). Small Is Beautiful: A Study of Economics as if People Mattered. Blond and Briggs, London, pp. 153–156.

Scott W. R. (2001). Institutions and Organizations: Ideas, Interests and Identities (2nd edn). Sage Publications, Thousand Oaks, California, USA.

Seadon J. K. (2006). Integrated waste management – looking beyond the solid waste horizon. *Waste Management*, **26**(12), 1327–1336.

Seadon J. K. (2010). Sustainable waste management systems. *Journal of Cleaner Production*, **18**(16–17), 1639–1651.

Seiffert M. E. B. and Loch C. (2005). Systemic thinking in environmental management: support for sustainable development. *Journal of Cleaner Production*, **13**, 1197–1202, doi: 10.1016/j.jclepro.2004.07.004

Semiyaga S., Okure M. A. E., Niwagaba C. B., Katukiza A. Y. and Kansiime F. (2015). Decentralized options for faecal sludge management in urban slum areas of sub-Saharan Africa: a review of technologies, practices and end-uses. *Resources, Conservation and Recycling*, **104A**, 109–119.

Sharma A., Burn S., Gardner T. and Gregory A. (2010). Role of decentralised systems in the transition of urban water systems. *Water Science and Technology*, **10**, 577–583.

Sharma A. K., Tjandraatmadja G., Cook S. and Gardner T. (2013). Decentralised systems – definition and drivers in the current context. *Water Science and Technology*, **67.9**, doi: 10.2166/wst.2013.093

Shuval H. I., Gunnerson C. G. and Julius D. S. (1981). Appropriate Technology for Water Supply and Sanitation: Night Soil Composting. International Bank for Reconstruction and Development/World Bank, Washington DC, USA.

Siegrist R. L., Tyler E. J. and Jenssen P. D. (2000). Design and Performance of Onsite Wastewater Treatment Soil Adsorption Systems. National Research Needs Conference Proceedings: Risk-Based Decision Making for Onsite Wastewater Treatment, EPRI, Palo Alto, CA, U.S. Environmental Protection Agency, and National Decentralized Water Resources Capacity Development Project: 1001446.

Simha P., Lalander C., Vinnerås B. and Ganesapillai M. (2017). Farmer attitudes and perceptions to the re-use of fertiliser products from resource–oriented sanitation systems – the case of Vellore, South India. *Science of the Total Environment*, **581–582**, 885–896, doi: 10.1016/j.scitotenv.2017.01.044

Simiyu S. (2017). Preference for and characteristics of an appropriate sanitation technology for the slums of Kisumu, Kenya. *International Journal of Urban Sustainable Development*, **9**(3), 300–312, doi: 10.1080/19463138.2017.1325366

Spaargaren G., Mommaas H., Van Den Burg S., Maas L., Drissen E., Dagevos H., et al. (2007). More sustainable lifestyles and consumption patterns: a theoretical perspective for the analysis of transition processes within consumption domains. Contrast Research Report, TMP project. Wageningen: Environmental Policy Group, Wageningen University; Telos: Tilburg University; RIVM; LEI, Wageningen, Netherlands.

Speck T., Harder D. L., Milwich M., Speck O. and Stegmaier T. (2006). Bionik: Die Natur als Innovationsquelle für Technische Textilien (Bionics: nature as a source of innovation for technical textiles). In: Technische Textilien, P. Knecht (ed.), Deutscher Fachverlag, Frankfurt, pp. 83–101.

Spencer H. (1998). The Principles of Biology. D. Appleton and Company, New York, USA.

Starkl M., Brunner N. and Stenstrom T. (2013). Why do water and sanitation systems for the poor still fail? Policy analysis in economically advanced developing countries. *Advances in Environmental Science and Technology*, **47**(12), 6102–6110, doi: 10.1021/es3048416

Starkl M., Brunner N., Feil M. and Hauser A. (2015). Ensuring sustainability of non-networked sanitation technologies: an approach to standardization. *Advances in Environmental Science and Technology*, **49**, 6411–6418, doi: 10.1021/acs.est.5b00887

Stasinopoulos P., Smith M., Hargroves K. and Desha C. (2009). Whole System Design: An Integrated Approach to Sustainable Engineering. EarthScan, London, UK.

Suriyachan C., Nitivattananon V. and Nurul Amin A. T. M. (2012). Potential of decentralized wastewater management for urban development: case of Bangkok. *Habitat International*, **36**, 85–92, doi: 10.1016/j.habitatint.2011.06.001

SuSan A (ed.) (2008). Towards More Sustainable Sanitation Solutions (Version 1). Sustainable Sanitation Alliance (SuSanA), Eschborn. https://www.susana.org/_resources/documents/default/3-267-7-1452594644.pdf (accessed 21 June 2017).

Svec P., Berkebile R. and Todd J. A. (2012). REGEN: toward a tool for regenerative thinking. *Building Research & Information*, **40**(1), 81–94, doi: 10.1080/09613218.2012.629112

Tainter J. A. (2012). Regenerative design in science and society. *Building Research & Information*, **40**(3), 369–372, doi: 10.1080/09613218.2012.671998

Tchobanoglous G. and Leverenz H. (2008). The Role of Onsite and Decentralized Wastewater Management in the Twenty-First Century. Proceedings of the Onsite and Decentralized Sewerage and Recycling Conference, Benalla, Victoria, Australia.

Tchobanoglous G. and Leverenz H. (2013). The rationale for decentralization of wastewater infrastructure. In: Source Separation and Decentralization for Wastewater Management, T. A. Larsen, K. M. Udert and J. Lienert (eds), IWA Publishing, London, UK, pp. 101–115.

Tchobanoglous G., Leverenz H. and Gikas P. (2009). Impacts of New Concepts and Technology on the Energy Sustainability of Wastewater Management. Proceedings of the Conference on Climate Change, Sustainable Development, and Renewable Energy Sources, Aristotle University of Thessaloniki, Thessaloniki, Greece.

Thayer R. L. G. (1994). World, Green Heart: Technology and the Sustainable Landscape. Wiley, New York, USA.

Thomas B. and McDougall F. (2005). International expert group on life cycle assessment for integrated waste management. *Journal of Cleaner Production*, **13**(3), 321–326.

Tidåker P. (2007). Integrating farming and wastewater management. Doctoral dissertation, Swedish University of Agricultural Sciences, Uppsala, Sweden.

Tilley E. (2013). Conceptualizing sanitation systems to accounts for new complexities in processing and management. In: Source Separation and Decentralization for Wastewater Management, T. A. Larsen, K. M. Udert and J. Lienert (eds), IWA Publishing, London, UK, pp. 227–239.

Tilley E., Strande L., Luthi C., Mosler H., Udert K. M., Gebauer H. and Hering J. G. (2014a). Looking beyond technology: An integrated approach to water, sanitation and hygiene in low income countries. *Environmental Science & Technology*, **48**(17), 9965–9970, doi: 10.1021/es501645d

Tilley E., Ulrich L., Luthi C., Reymond P. H. and Zurbrugg C. (2014b). Compendium of Sanitation Systems and Technologies. 2nd rev. edn. Swiss Federal Institute of Aquatic Science and Technology (EAWAG), Dübendorf, Switzerland.

Tobias R., O'Keefe M., Künzle R., Gebauer H., Gründl H., Morgenroth E., Pronk W. and Larsen T. A. (2017). Early testing of new sanitation technology for urban slums: the case of the Blue Diversion Toilet. *Science of the Total Environment*, **576**, 264–272.

Todd J. and Josephson B. (1996). The design of living technologies for waste treatment. *Ecological Engineering*, **6**, 109–136. https://doi.org/10.1016/0925-8574(95)00054-2 (accessed 10 February 2018).

Todd J. and Todd N. J. (1980). Tomorrow is Our Permanent Address: The Search for an Ecological Science of Design. Harper and Row, New York, USA.

Todd N. J. and Todd J. (1984). Bioshelters, Ocean Arks, City Farming: Ecology as the Basis of Design. Sierra Club Books, San Francisco, USA.

Todd N. J. and Todd J. (1994). From Eco-Cities to Living Machines: Principles of Ecological Design. North Atlantic Books, Berkeley, USA.

Ulrich A., Reuter S., Gutterer B., Sasse L., Panzerbieter T. and Reckerzügel T. (2009). Decentralised Wastewater Treatment Systems (DEWATS) and Sanitation in Developing Countries. WEDC, Loughborough, UK. https://www.wbdg.org/FFC/ARMYCOE/PWTB/pwtb_420_46_9.pdf (accessed 4 April 2018).

Unsal F. (2016). Sustainability, systems thinking and ecological worldview. Proceedings of the International Conference on Power, Energy Engineering and Management (PEEM 2016). https://doi.org/10.12783/peem2016/5050. ISBN 978-1-60595-324-3.

UN Millennium Project (2005). Health, Dignity, and Development: What Will it Take? Task Force on Water and Sanitation. http://www.unoosa.org/documents/pdf/psa/activities/2005/graz/watercomplete-lowres.pdf (accessed 10 November 2017).

UNCED (United Nations Conference on Environment and Development) (1992). Agenda 21, Rio Declaration. United Nations, New York, USA.

United Nations (1998). Convention on Access to Information, Public Participation in Decision-Making and Access to Justice in Environmental Matters. New York and Geneva.

United Nations (2015). Sustainable development goals. https://sustainabledevelopment.un.org/?menu=1300 (accessed 28 October 2015).

USEPA (U.S. Environmental Protection Agency) (2002). On-site wastewater treatment systems manual, EPA/625/R-00/008. Washington D.C., USA.

USEPA (U.S. Environmental Protection Agency) (2004). Primer for Municipal Wastewater Treatment Systems. Office of Wastewater Management and Office of Water, Washington D.C., USA.

USEPA (2007). Innovation and Research for Water Infrastructure for the 21st Century – Research Plan, U.S. Environmental Protection Agency, Office of Research and Development – National Risk Management Research Laboratory, Cincinnati, USA.

USEPA (2009). Rehabilitation of Wastewater Collection and Water Distribution Systems – White Paper. U.S. Environmental Protection Agency, Office of Research and Development – National Risk Management Research Laboratory, Cincinnati, USA. EPA/600/R-09/048.

USEPA (2012). Guidelines for Water Reuse. EPA/600/R-12/618. U.S. Environmental Protection Agency Office of Wastewater Management, Washington D.C, USA.

van de Klundert A. and Anschutz J. (2001). Integrated sustainable waste management – the concept: tools for decision-makers. In: Experiences from the Urban Waste Expertise Programme (1995–2001), A. Scheinberg (ed.), Urban Waste Expertise Programme, Gouda, Netherlands, 46pp. http://www.waste.nl/sites/waste.nl/files/product/files/tools_iswm_concept_eng1.pdf (accessed 10 February 2018).

Van Huijstee M. M., Francken M. and Leroy P. (2007). Partnerships for sustainable development: a review of current literature. *Biomedical and Environmental Sciences*, **4**(2), 75–89, doi: 10.1080/15693430701526336

van Vliet B., Spaargaren G. and Osterveer P. (eds) (2010). Social Perspectives on the Sanitation Challenge. Springer, Dordrecht, Netherlands. https://doi.org/10.1007/978-90-481-3721-3

van Vliet B. J. M., Spaargaren G. and Oosterveer P. (2011). Sanitation under challenge: contributions from the social sciences. *Water Policy*, **13**, 797–809.

Vattam S., Helms M. and Goel A. K. (2007). Biologically Inspired Innovation in Engineering Design: A Cognitive Study. Technical Report, Graphics, Visualization and Usability Center, Georgia Institute of Technology, USA, GIT-GVU-07-07.

Vester F. (2004). A Report to the Club of Rome. Deutscher Tashenbuch Verlag, Munich, Germany.

Von Bertalanffy L. (1956). General system theory. In: General System, Yearbook of the Society for the Advancement of General System Theory, F.E. Emery (ed.), 279pp. https://monoskop.org/images/7/77/Von_Bertalanffy_Ludwig_General_System_Theory_1968.pdf (accessed 10 February 2018).

Von Bertalanffy L. (1968). General System Theory: Foundations, Development, Applications. George Braziller, New York, USA.

Weber B., Cornel P. and Wagner M. (2007). Semi-centralised supply and treatment systems for (fast growing) urban areas. *Water Science and Technology*, **55**, 349–356.

WEDC (Water, Engineering and Development Centre) and DFID (Department for International Development) (1998). DFID Guidance Manual on Water Supply and Sanitation Programmes. Published by WEDC for DFID, London, UK.

Weinberg G. M. (2001). An Introduction to General Systems Thinking, 25th anniversary edn. Dorset House Publishing Company, Incorporated.

West S. (2001). The Key to Successful On-Site Sewerage Service. Sydney Water Corporation On-site '01 Conference Armidale, September 2001.

White A. (1981). Community Participation in Water and Sanitation: Concepts, Strategies and Methods (No. 17). IRC Technical Paper Series, Rijswijk, Netherlands.

WHO and UNICEF (2017). Progress on Drinking Water, Sanitation and Hygiene: Update and SDG Baselines. World Health Organization (WHO) and the United Nations Children's Fund (UNICEF), Geneva. http://www.who.int/mediacentre/news/releases/2017/launch-version-report-jmp-water-sanitation-hygiene.pdf?ua=1 (accessed 5 October 2017).

WHO (2006a). WHO Guidelines for the Safe Use of Wastewater, Excreta and Greywater/World Health Organization. Volume I Policy and Regulatory Aspects. Publications of the World Health Organization can be obtained from WHO Press, World Health Organization, 20 Avenue Appia, 1211 Geneva 27, Switzerland.

WHO (2006b). WHO Guidelines for the Safe Use of Wastewater, Excreta and Greywater. Volume II, Wastewater Use in Agriculture/World Health Organization. Publications of the World Health Organization can be obtained from WHO Press, World Health Organization, 20 Avenue Appia, 1211 Geneva 27, Switzerland.

WHO (2006c). WHO Guidelines for the Safe Use of Wastewater, Excreta and Greywater. Volume IV, Excreta and Greywater Use in Agriculture/World Health Organization. Publications of the World Health Organization can be obtained from WHO Press, World Health Organization, 20 Avenue Appia, 1211 Geneva 27, Switzerland.

WHO (2006d). WHO Guidelines for the Safe Use of Wastewater, Excreta and Greywater. Volume III, Wastewater and Excreta Use in Aquaculture/World Health Organization. Publications of the World Health Organization can be obtained from WHO Press, World Health Organization, 20 Avenue Appia, 1211 Geneva 27, Switzerland.

WHO (2007). Preventing Disease through Healthy Environments. The Contribution of Water, Sanitation and Hygiene. World Health Organization, Geneva.

Wilderer P. A. and Schreff D. (2000). Decentralized and centralized wastewater management: a challenge for technology developers. *Water Science and Technology*, **41**(1), 1–8.

Willoughby K. W. (1990). Technology Choice – A Critique of the Appropriate Technology Movement. Westview Press, London, UK.

Wilson E. O. (1984). Biophilia. Harvard University Press, Cambridge, USA.

Wolf M. and Störmer E. (2010). Decentralisation of wastewater infrastructure in eastern Germany. *Network India Quarterly*, **12**, 7–10.

World Bank (2003). Appraisal Report: Tanzania-Dar-es Salaam Water Supply and Sanitation Project. World Bank, Washington D.C., USA. http://documents.worldbank.org/curated/en/2003/04/2301336/tanzania-dar-es-salaam-water-supplysanitation-project (accessed 10 February 2018).

Young O. R. (2002a). The Institutional Dimensions of Environmental Change: Fit, Interplay and Scale. MIT Press, Cambridge, USA.

Young O. R. (2002b). Matching institutions and ecosystems: the problem of fit. Conference donnée dans le cadre du séminaire Économie de l'environnement et du développement durable, coorganisé par l'Iddri et le Medd. Institut du développement durable et des relations internationals (Conference given on the framework of the seminar economy of the environment and sustainable development, co-organized by Iddri and Medd. Institute of Sustainable Development and International Relations), Paris, France.

Young O. R., Agrawal A., King L. A., Sand P. H., Underdal A. and Wasson M. (1999). Institutional dimensions of global environmental change (IDGEC). Science Plan. International Human Dimensions Programme Report No. 9. IHDP, Bonn, Germany.

Zari M. P. and Jenkins S. (2010). Rethinking cutting edge sustainable design: from eco-efficiency to regenerative development. Paper presented at the Sustainable Building Conference 2010 (SB10, Wellington, New Zealand). https://www.branz.co.nz/cms_show_download.php?id=2ab15295f01d3078189c96855a6fa055226d58b4 (accessed 5 April 2017).

Zeldovich L. (2017). Reinventing the toilet. Mosaic Science Online. https://mosaicscience.com/story/poo-toilet-waste-energy-madagascar-loowatt-future (accessed 3 July 2017).

Zurbrügg C. and Tilley E. (2009). A system perspective in sanitation – Human waste from cradle to grave and reincarnation. *Desalination*, **248**, 410–417.

Chapter 3
Regenerative sanitation framework

'contrary to popular opinion, innovation without some standardized conceptual framework is tantamount to chaos'

Stephen Bush

3.1 INTRODUCTION

Improving services and expanding access to safely managed sanitation facilities requires a radical shift from technocentric and toilet-focused conceptual frameworks to a more comprehensive, inclusive, holistic, integrated and standardized framework with elements of technology, institutions, cultural values, resource recovery and reuse, users' expectations and rules and regulations; all embedded in a systemic pattern. The restrictive nature of existing frameworks does not depict the complex-dynamic interactions that are inherent in sanitation systems because they primarily focus on particular problem areas of individual aspects of sanitation. For example, the *service ladder framework* focuses on provision and access to toilets without regard to facility functionality (Kvarnström *et al.*, 2011; Potter *et al.*, 2011), the *service chain framework* focuses on analysis of physical flow of faecal sludge and its management (Trémolet & Rama, 2012; USAID, 2016) and the *decision support framework* focuses on technology selection, etc. (Kalbermatten *et al.*, 1980, 1982; Zurbrügg & Tilley, 2007) (Section 1.5); all focusing on technology.

These frameworks leave a lot of lacuna, which create challenges for sanitation management because by providing solutions to a few aspects, other aspects (such as acceptability) are left unattended, so whatever achievement could have been recorded is nullified when placed within the whole sanitation system. A comprehensive and systemic sanitation framework, however, that accommodates synergistic operations of these other aspects and allows for interdisciplinary and transdisciplinary research and practice in policy, technology, sociocultural, economic and ecological considerations, will produce more effective and inclusive results. The regenerative sanitation (ReGenSan) conceptual framework aims to fill this gap

© IWA Publishing 2019. Regenerative Sanitation: A New Paradigm for Sanitation 4.0
Editor(s): Thammarat Koottatep, Peter Emmanuel Cookey and Chongrak Polprasert
doi: 10.2166/9781780409689_0079

and seeks to connect the different sanitation subsystems, dimensions and components and synergize their interactions toward a common goal (Cumming, 2011; Pickett *et al.*, 2007). It would serve as a foundation for other elaborate models as it would guide a systemic understanding of the sanitation system towards solution provision (Pollack, 2005). The overall aim of ReGenSan is to refocus the perspectives of researchers, professionals, practitioners, advocacy-agents, policymakers and solution-providers from attacking a problem to providing services and solutions; and then ultimately expand access and improve service by 2030.

3.2 DEVELOPING AN INTEGRATED HOLISTIC FRAMEWORK

An iterative mixed methods approach was adopted in developing this framework by combining technical, experimental and experiential practical knowledge as well as a literature review that included previous related theories, research and concepts that were used to represent similar problems (Brewerton & Millward, 2001; Bocken *et al.*, 2014; Campbell *et al.*, 2015; Martínez-Jurado & Moyano-Fuentes, 2014; Novak & Canas, 2008; Seuring & Muller, 2008). A four-step methodological 'input-processing-output-outcome' detailed approach with rigorous sequential steps of activities was used (Levy & Ellis, 2006; Mayring, 2003; Sethi & King, 1998). The holistic integrated system research approach from Pahl-Wostl *et al.* (2011) was also adopted, together with evidence from scientific literature to assess the ongoing discourse for paradigm shift from technocentric sanitation systems to integrated holistic sanitation systems. To assess the current state of development in this direction, a bibliometric analysis of publications was conducted using the SCOPUS database, and different combination of keywords were applied to define search terms. The resulting dataset, shown in Table 3.1, records the number of publications in peer-reviewed journals in the search space (Pahl-Wostl *et al.*, 2010).

Table 3.1 highlights the fact that research on sanitation focuses primarily on technology and the impact of sanitation on health, hygiene and water. Publications on sanitation-related science show serious emphasis on technology over integrated systems and governance, even given the increase in publications on integrated systems and governance in recent years. In the case of integrated sanitation systems, the publications grew from 14 in 2002 to over 44 in 2015; while sanitation management and governance grew from two publications in 2002 to 15 in 2016. There was also an increase in research that focused on resource recovery and reuse.

The data showed that not much research focused on retrofitting, rehabilitation and repairs of sanitation infrastructures, suggesting the need for a holistic, integrated and systemic sanitation conceptual framework that is not technocentric, but embraces dynamic complexities and contextual issues while delivering context-dependent integrated solutions. This framework seeks to capture the essential characteristics of sanitation systems, elements and structures enunciated by several authors in literature, which constitute the ReGenSan paradigm.

Despite the recognition of the importance of psycho-socio-cultural-economic-ecological issues (including governance, resource recovery and reuse) to providing sanitation services, there has really been no inclusive and comprehensive conceptual framework that captures these dynamic complexities and answers the following questions:

(I) What are the essential elements needed to transit from rigid locked-in linear conventional (sewer and non-sewer) sanitation to more integrated, holistic and systemic approaches?

(II) What are the ingredients necessary to understand the complex dynamics of sanitation infrastructure and management solutions?

Table 3.1 Development of Holistic Integrated Sanitation System Research (Cell Value Indicates Number of Papers Matching Each Search Term in Each Year).

	Year																
Search Term	**2002**	**2003**	**2004**	**2005**	**2006**	**2007**	**2008**	**2009**	**2010**	**2011**	**2012**	**2013**	**2014**	**2015**	**2016**	**2017**	**Total**
Governance AND sanitation	6	9	9	12	13	18	24	39	21	27	36	31	44	30	40	15	374
Sanitation AND social AND science	11	14	11	14	9	9	14	15	22	15	17	30	26	24	14	9	254
Sanitation AND Stakeholders	8	13	17	22	15	27	32	37	29	37	27	59	48	54	46	22	493
Sanitation AND behaviour AND change	7	16	19	13	13	25	24	28	27	30	39	47	44	46	41	11	430
Sanitation AND management AND governance	2	7	6	5	6	10	11	19	10	11	18	14	22	15	15	0	171
Sanitation AND integration	9	5	15	14	15	17	15	19	23	24	26	34	33	24	29	6	308
Sanitation AND integrated AND system	14	19	19	11	23	37	25	38	27	33	29	39	30	44	32	12	432
Sanitation AND management AND integration	3	4	6	9	11	9	9	12	12	11	12	13	16	11	14	2	154
Sanitation AND Integrated AND management AND governance	0	1	1	0	0	3	4	6	0	3	8	3	2	6	4	0	41
Sanitation AND resource AND recovery	3	8	11	7	7	12	9	15	8	14	9	21	12	17	20	14	187
Sanitation AND Reuse	8	8	30	25	17	26	22	45	39	28	17	25	30	39	44	14	417
Sanitation AND retrofitting	0	0	1	1	1	0	3	1	1	4	0	1	0	1	1	0	15
Sanitation AND rehabilitation	8	6	11	10	11	16	11	12	9	9	11	18	15	18	10	2	177
Sanitation AND repairs	2	6	6	8	11	8	8	4	7	4	11	7	8	16	7	7	120
Sanitation AND maintenance	24	24	38	36	32	47	40	70	37	46	56	49	53	59	34	18	663
Sanitation AND water	368	391	554	561	457	618	586	714	659	778	742	882	811	868	766	283	10,038
Sanitation AND health	426	453	520	572	517	601	547	622	680	739	706	824	823	824	605	201	9660
Sanitation AND hygiene	123	167	186	205	180	208	211	251	249	266	253	367	313	380	274	107	3740
Sanitation AND toilet	45	62	50	55	74	93	67	95	104	108	100	124	124	200	97	40	1536
Sanitation AND technology	54	69	99	100	75	106	96	117	115	147	109	158	137	161	135	57	1438
Sanitation AND technology AND management	21	34	37	42	33	49	43	41	49	53	47	53	61	62	45	23	693

Source: Scopus database.

(III) Which kinds of guidance are required to restore unimproved and dysfunctional sanitation systems and incorporate resource recovery and reuse as well as ensure continuous upgrading of existing facilities and providing opportunities for nouveau sanitation systems, especially for populations without access to any sanitation facilities?

(IV) What are the appropriate methodologies to incorporate existing proven approaches such as governance, sociocultural, technologies, social marketing and mobilization and financing mechanisms, at scale, to rethink sanitation practices?

(V) What kind of guidance and tools are required for policymakers and practitioners to implement integrated, holistic and systemic approaches in sanitation spheres?

The ReGenSan framework is primarily based on the potential of holistic and systemic integration as highlighted above. This framework is an interdisciplinary and transdisciplinary conceptual framework that supports an understanding of the complex-dynamic, systemic, integrated and holistic characteristics of sanitation. It is based on the ecological worldview and the 'place-context' and 'scale' focus where comprehensive sanitation solutions are provided within sani-sheds (see Chapter 4). ReGenSan defines 'place-based' as activities, programmes, projects, plans, designs, solutions and such like that are focused on particular singular or groups of localities geographically connected within specific regions and/or areas with similar ethnicities and/or cultures, while context-specific refers to regions, areas and other places that have similar features without necessarily being geographically linked (e.g. coastal areas, arid/water-scarce areas, densely populated, shared religions, high-end, industrial, rural/urban, etc). In addition, scale could be based on population of users/inhabitants, number of required units and area of specific development (e.g. factory/industry, large office area, estates, apartment buildings, airports, schools, healthcare facilities, military bases) as well as levels of pilot study.

This framework aims to induce improvement and mind-change in sanitation planning, technology innovations and implementation. It provides a functional and progressive approach to improving and expanding access to sanitation services at a 'scale' by achieving net benefits for the society (Kvarnström *et al.*, 2011; Potter *et al.*, 2011). The framework combines and integrates sanitation processes and technologies into a systemic whole that addresses the interconnectedness of socio-cultural-economic, environmental, technical and institutional aspects of sanitation management (Marshall & Farahbakhsh, 2013). The aim is to ensure that sanitation systems do not transfer ecological and public health burdens, but provide net benefits to the people and environment. The framework embraces diagnostic (Ostrom, 2005), prescriptive (Cookey *et al.*, 2016a) and normative (Pahl-Wostl, 2009) analytical approaches designed to deliver comprehensive integrated and adaptive sanitation solutions (sani-solutions). This framework is essentially built on the ReGenSan principles (Section 2.4, Chapter 2) and theory (Section 2.3, Chapter 2).

3.3 SYSTEM ELEMENTS OF REGENSAN FRAMEWORK

The ReGenSan system elements include the subsystems, dimensions, components, their relationships, attributes and interactions. Seiffert and Loch (2005) suggested that the most important property of systems is that they are made up of several parts that are not isolated but closely interlinked, forming a complex structure (Chang *et al.*, 2011; Charnley & Lemon, 2011). Therefore, subsystems are part of the whole system, but can still be regarded as individual systems; indicative of the 'system of systems' concept, which depicts the collection of a few dedicated systems or subsystems that pool their resources and capabilities together to connect a more complex metasystem offering more functionality and performance than simply the sum of the constituent systems (Maier, 1998). ReGenSan fits the complex dynamic 'system of the systems' where subsystems (socio-ecological, technology and resource systems) are linked with each other (Chang *et al.*, 2011). The social-ecological subsystem (SES) is made up of two dimensions, governance and psycho-socio-ecophilia (PSEP), with six components; the technology subsystem (TeS) comprises three dimensions: existing improved design technology (EIDT), restorative design technology (ReDT) and nouveau design technology (NoDT) with fourteen components; and the resource subsystem (ReS) is made up of three dimensions: design for recovery and reuse (DeRaR), sanitation service chain raw materials (SSC-RaMs) and sanitation-derived products (SDPs) with 16 components (Figure 3.1 and Table 3.2).

The ReGenSan framework is unique in that it is applicable to both developed and developing countries, and to all stages, processes and cycles of delivering sanitation solutions. The framework has the potential to critically evaluate, analyze and provide credible, adequate and appropriate sanitation solutions by adopting diagnostic, prescriptive and normative analytical approaches (Cookey *et al.*, 2016a; Ostrom, 2005; Pahl-Wostl, 2009). The 'diagnostic' provides detailed existing status reports for interventions that are geared towards either restorative, existing or nouveau solutions. The 'prescriptive' is built upon the diagnostic and strategically determines the appropriateness and adequacy of the applied interventions and at the same time promotes continuous improvement of the sanitation system. The 'normative' is associated with the governance and PSEP considerations that provide the optimum enabling and operating environment for the uptake of scalable solutions. Together, they provide better understanding of the interconnections of knowledge and interactions among the sanitation subsystems, dimensions and components. The framework's overarching objectives are presented in Box 3.1.

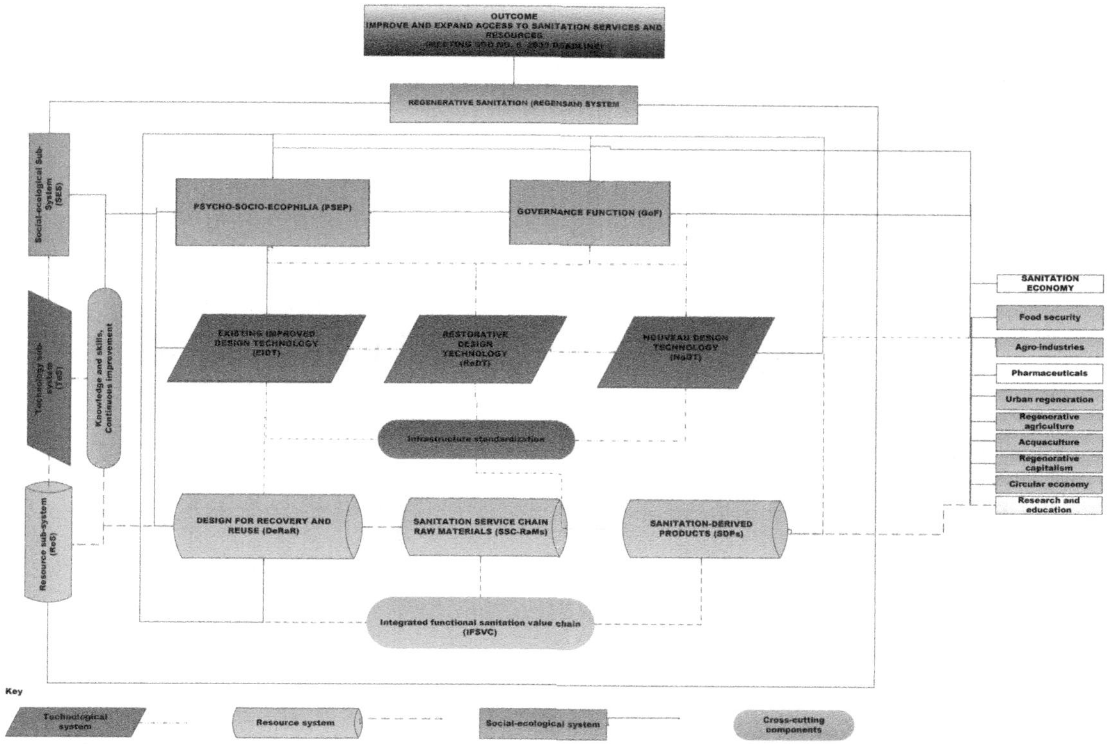

Figure 3.1 Conceptual framework of ReGenSan designed for the acceleration of sanitation coverage and service improvement.

Table 3.2 Summary of Regenerative Sanitation Subsystems, Dimensions and Components

Subsystems	Dimensions	Cross-Cutting Components	Components
Social-ecological system (SES)	**Psycho-socio-ecophilia (PSEP):** provides contextual and sociotechnical contributions for decision-making in sanitation infrastructure and management solutions as well as providing the needed socio-ecological-technology fit.	*(i) Continuous improvement*[†] *(CiM), (ii) knowledge and skills*[†] *(KaS), (iii) sanitation infrastructure standardization (SIS), (iv) integrated functional sanitation service chain (IFSVC)*	(*i*) *Psycho-social-cultural*: religion, norms, customs, availability, accessibility, acceptability, willingness to use, gender equity, inclusiveness, perceptions and attitude, health, hygiene and handwashing, preference, status, (*ii*) *Socio-economic:* educational level (formal and informal), income level (affordability, willingness to charge and willingness-to-pay), livelihood support (economic activities and occupational engagement), access to basic services (water supply, electricity, and healthcare), access to sanitation services and types (sewer, non-sewer, *shared and non-shared facilities*), geographical location (urban and rural), housing/dwelling types, household sizes and head of household, age, gender and race *(iii) Bio-geo-chemical:* nutrients and water (C (carbon), N, P, H_2O) cycling, ecosystem services and functions, fauna and flora
	Governance: provides an enabling environment, standards and regulations, funding mechanism and asset management procedures with the socio-ecological support for the uptake of functional and adaptable technologies and resource recovery fit for the 'place' and 'scale'.		*(iv) Institutions:* policies, legislation and regulations, informal institutions, international treaties and protocols *(v) Management:* sanitation management entities (SMEs), planning and implementation, asset management, clearly defined roles of actors, stakeholders' information and communication, regulatory compliance and enforcement, monitoring and evaluation *(vi) Sustainable financing:* Blending public and private funding sources, resource leveraging, cost recovery and lifecycle costing, supplementary income and social marketing and utilization of local/domestic private sector

Technological system (TeS)	**Existing improved design technology (EIDT):** focuses on continuous service improvement, expansion of access and coverage of the currently adjudged improved and safely managed sewer and non-sewer sanitation facilities estimated to be used by 5 billion people.	*(i) service chain functionality and serviceability, (ii) integrity, reliability, safety and quality, (iii) preventive and proactive maintenance, (iv) inspection and monitoring and (v) infrastructure standardization**
	Restorative design technology (ReDT): addresses the issues of upgrading and retrofitting of dysfunctional, aged, disused and unimproved, basic and limited sanitation facilities for both sewer and non-sewer to safely manage sanitation estimated to be used by 2.4 billion people.	*(vi) Rehabilitation, (vii) retrofitting and (viii) remediation and mitigation*
	Nouveau design technology (NoDT): focuses on the issues of emerging novel sanitation facilities solutions because of failure of current technologies to address the underlying challenges behind the lack of sanitation infrastructure and resources recovery as well as provide better alternative for new development area and the population exposed to open defecation estimated to be nearly 1 billion people.	*(ix) Design for service and products (DeSaP), (x) design for nature (DeN) and (xi) design for the base-of-the-pyramid (DeBoP).*

(*Continued*)

Table 3.2 Summary of Regenerative Sanitation Subsystems, Dimensions and Components (*Continued*).

Subsystems	Dimensions	Cross-Cutting Components	Components
Resource system (ReS)	**Design for recovery and reuse (DeRaR):** provides technological and infrastructural support for operationalization of ReS.		*(i) Recovery and reuse from existing facilities, (ii) alternative design for recovery and reuse, (iii) demand based-design and (iv) health and safety, (v) integrated functional sanitation value chain (IFSVC)*.*
	Sanitation service chain raw materials (SSC-RaMs): provides the input materials for ReS.		*(vi) faeces, (vii) urine, (vii) flushwater, (ix) greywater, (x) anal cleansing water and other materials and (xi) sludge (faecal and sewage)*
	Sanitation-derived products (SDPs): provide recoverable and reusable products and sanitation systems.		*(xii) used water, (xiii) sanitation-derived fertilizers (SDFs) (xiv) bioenergy, (xv) soil conditioners and (xvi) other output materials, e.g. protein feed, building materials, trace elements*

*Components applicable to all dimensions of technology subsystem.

†Components applicable to all subsystems.

BOX 3.1 OBJECTIVE OF ReGenSan FRAMEWORK

(I) Generate transdisciplinary system understanding for service improvement and access expansion;
(II) Develop and implement sani-shed specific methods, models, technological and resource recovery and reuse solutions as well as evolve contextual-dependent governance structures and strategies for 'place'- and 'scale'-based sanitation service delivery;
(III) Integrate science-policy-technology-practice research into sanitation space through a cyclical learning process; and
(IV) Indicate the potential for a sanitation economy that interacts with the circular economy.

ReGenSan identifies and integrates three subsystems and cross-cutting subsystems' components, seven dimensions and 26 components (Table 3.2). These subsystems and dimensions are described below (for more details see Chapters 4 to 6).

3.3.1 Social-ecological subsystem (SES)

SES seeks to provide an understanding of the multifaceted/multilevel interactions and intersections between ecology and society in the management of sanitation in sani-sheds. It acknowledges that the divide between ecosystem, human system and technological system is largely arbitrary; and the success of any sanitation intervention depends on the symbiotic-synergistic understanding of the SES and their linkages to other subsystems (Repella, 2014). The subsystem seeks to provide better understanding of the interactive effects of psycho-social, cultural, economic, geographical and ecological issues that influence the whole spectrum of sanitation science, governance, practice and technology guided by knowledge and skills (Hinkel *et al.*, 2014; Ostrom, 2009; Ostrom & Cox, 2010; Partelow, 2016; van Vliet *et al.*, 2011; Whittington *et al.*, 1993). SES consists of two dimensions: (psycho-socio-ecophilia and governance) and seven components (Figure 3.1 and Table 3.2).

3.3.1.1 Psycho-socio-ecophilia (PSEP)

PSEP provides contextual and behavioural perspectives to introduce psycho-socio-ecological-technical variables into assessment and decision-making for sanitation solutions. It emphasizes the importance of 'place' and 'scale' and focuses on providing sanitation solutions within 'sani-sheds', which ensures that sanitation systems connect with the peculiar characteristics of the locality (Kellert *et al.*, 2008; Mang & Reed, 2012a, b; Whittington *et al.*, 1993) to sustain the essential functions and processes of the ecosystem (Lyle, 1984, 1994; Van der Ryn & Cowan, 2007). Ambitious sustainability goals can only be fulfilled by co-design of sanitation technology with potential users in the 'system of the place' (Tobias *et al.*, 2017). ReGenSan argues that sanitation systems should be designed to suit 'place' because the diversity of social and ethnic groups existing in any locality greatly influences appropriate designs, accessibility, acceptability, affordability, construction and use of such systems (Domènech & Saurí, 2010; Hegger *et al.*, 2007; Larsen *et al.*, 2010; McConville *et al.*, 2014; Raven, 2007; Whittington *et al.*, 1993). PSEP could potentially become a strong driving force for the needed transition (Spaargaren *et al.*, 2007) in planning, operations, maintenance and cost recovery, where stakeholders are empowered to make choices about what kind of sanitation systems they wish to use and how it will be managed (See Box 3.2, Ferguson *et al.*, 2003).

3.3.1.2 Governance

Governance enhances the effectiveness and efficiency of sanitation operational responsibilities (i.e. the day-to-day functionality of the services delivered) and institutional arrangements (i.e. the formal and informal institutional contexts that help or hinder the successful delivery of day-to-day activities) (Joshi *et al.*, 2015; Kooiman, 2003; Maurer *et al.*, 2012; Ross *et al.*, 2014; Schertenleib, 2005; Starkl *et al.*, 2013). It encompasses arrangements for ownership, management and operations, institutional and regulatory frameworks (including setting and meeting requirements of standards, resource recovery, emission standards, pricing guidelines, funding mechanisms, cost recovery etc. (Galli *et al.*, 2014; Kooiman, 2003; Lüthi *et al.*, 2011a, b; Maurer *et al.*, 2012; Rodic, 2015; Ross *et al.*, 2014; Schertenleib, 2005; Starkl *et al.*, 2013; Tilley *et al.*, 2014a, b). It could also support and indicate technology selection and uptake as well as financial mechanisms for sanitation solutions (Figure 3.1 and Table 3.2).

3.3.2 Technological subsystem (TeS)

TeS is the configurations of infrastructures and services combined to treat by-products of human digestion from the point of generation to the final point of reuse or disposal. It includes an extensive range of possible combinations of technical and sociotechnical components driven by sanitation infrastructural standardization to deliver the most appropriate and best applicable solutions (Maurer *et al.*, 2012; Ramoa *et al.*, 2015; Tilley *et al.*, 2014a, b; Zurbrügg & Tilley, 2009). The ReGenSan TeS encompasses the entire sanitation service of *'users' interface-drop/or flush-store-treat-recover-utilize'*. The sociotechnical components are the economic, social, cultural and institutional contexts made up of multidimensional technological characteristics (place, scale and process), management and governance, financial arrangements and organizational set-up as well as active involvement of the end users (Altaf, 2011; Grübler, 1998; Oosterveer & Spaargaren, 2010; Starkl *et al.*, 2015). There are three dimensions to TeS: Existing Improved Design Technology (EIDT), Restorative Design Technology (ReDT) and Nouveau Design Technology (NoDT), and 12 components (Figure 3.1 and Table 3.2).

3.3.2.1 Existing Improved Design Technology (EIDT)

EIDT focuses on continuous improvement of the currently adjudged improved sanitation (safely managed, basic and limited services) facilities estimated to be used by 5 billion people (JMP, 2015; WHO/UNICEF, 2017) as well as ensuring an effective and efficient service chain management for the estimated 2.7 billion people using on-site sanitation worldwide (Nakagiri *et al.*, 2016; Strande *et al.*, 2014). The cardinal rule of EIDT is to ensure long-term sustainability of sanitation services and avoid the risk of 'slippage' – a term coined in India to refer to the slipping back of a community from 'fully covered with improved services' to 'partially or non-covered' (Lockwood & Smits, 2011; Rosenqvist *et al.*, 2016, UN-Water & WHO, 2012; Verhagen & Carrasco, 2013). This is achieved through standard operations and maintenance and enhanced advancement on the sanitation service ladder (Exley *et al.*, 2014; Ghosh & Elvira, 2011; Nakagiri *et al.*, 2016; Strande *et al.*, 2014). It aims to address the ageing problems of sewer infrastructures that do not deliver standard and quality effluent and sludge treatment (Baum *et al.*, 2013; Hutton *et al.*, 2007; Nansubuga *et al.*, 2016; Nhapi *et al.*, 2006; Starkl *et al.*, 2013, World Bank, 2003) as well as issues of actual use, cleanliness, quality of latrines/toilets, handwashing, hygiene promotion, toilet waste management (e.g. used tissue, menstrual pads, baby diapers, etc) behavioural change, facilities' integrity, quality control and assurance (Exley *et al.*, 2014; Gine *et al.*, 2011) (Figure 3.1 and Table 3.2).

3.3.2.2 Restorative design technology (ReDT)

ReDT seeks to retrofit unimproved and dysfunctional sanitation systems as well as support mitigation and remediation of polluted/degraded sites damaged by existing sanitation practices (Jenkin & Pedersen, 2009; Mang & Reed, 2012a, b). This strategy is necessary in view of the large population (2.4 billion) with inadequate and dysfunctional sanitation infrastructures (JMP, 2015). The pressing goal is to better understand the reasons for continued system failure and how they could be avoided in the future (Starkl *et al.*, 2013). Restorative designs arise due to structural inadequacies, material degradation and poor construction and workmanship, as well as ageing facilities and lack of operation and maintenance (Brepolsa *et al.*, 2008; Brikke, 2000; Cookey *et al.*, 2016a), which increase exposure to pathogens present in unconfined or poorly disposed human excreta (Boschi-Pinto *et al.*, 2006; Cairncross & Feachem, 1993; Ekane *et al.*, 2014). ReDT will ensure sustainability of sanitation systems to deliver services at an appropriate level of hygiene, quality, quantity, convenience, comfort, continuity, efficiency and reliability; and the institutionalization of complete coverage of operations, maintenance, administration and replacement strategies (Brikke, 2000). ReDT advocates for retrofitting and upgrading of existing wastewater treatment systems to meet stringent treatment requirements, increasing hydraulic and/or organic loading capacity, improving poor performance due to improper plant designs and/or operations, elongating expiry lifespan of ageing treatment plants and changing effluent limitations and wastewater treatment trends (Bozkurt *et al.*, 2015; Brepolsa *et al.*, 2008; El-Sheikh, 2011; USEPA, 1974). ReDT also aims to address the challenges of on-site systems, especially the unimproved systems that are poorly designed, constructed, operated and maintained (Giné-Garriga *et al.*, 2017; JMP, 2008, 2012, 2015; Mjoli, 2012; Pham & Kuyama, 2013) as well as improved systems that do not deliver, especially where local contexts were ignored (Starkl *et al.*, 2013) (Figure 3.1 and Table 3.2).

3.3.2.3 Nouveau design technology (NoDT)

NoDT addresses a spectrum of emerging technological and novel solutions based on the core principles of regenerative sanitation. It seeks to develop sanitation technologies that integrate eco-efficiency with ecologically based approaches that mimic nature to deliver improved functional systems peculiar to local contexts. The strategic role of NoDT is to contribute to the development of innovative solutions for the 700 million people who missed the MDG new development areas, as well as the 892 million people who have no facilities at all and are exposed to open defecation (JMP, 2015; WHO/UNICEF, 2017). Depending on the possibilities of the opportunity in the 'place', NoDT is to provide solutions at a scale in sani-sheds, taking into account the PSEP to deliver optimum services (Spaargaren *et al.*, 2006). Thus, within an SES, there might be different technologies in operation with each responding to local conditions, but playing its part in the larger whole (Lyle, 1994) (Figure 3.1 and Table 3.2). NoDT systems set an extensive range of possible combinations of SES and ReS to deliver sanitation technological infrastructural solutions for the provision of adequate products and services.

3.3.3 Resource subsystem (ReS)

ReS is concerned with 'closing the loop' by turning the linear resource management schemes of sanitation systems into a cyclical one that feeds the sanitation economy to eventually support the circular economy. Thus, ReS focuses on how to recover and reuse valuable resources of by-products of human digestion in order to ensure that sanitation systems minimize depletion of natural resources and environmental degradation. Recovering and reusing the valuable resources present in excreta, urine and wastewater contributes to resource efficiency and improved food security (Andersson *et al.*, 2016a, b). Closing these loops require synergies between technological, social, environmental and institutional systems and sound financing mechanisms, thus, making resource management systems major components of the overall sanitation system. There are three dimensions: design for recovery and reuse (DeRaR), sanitation service chain raw materials (SSC-RaMs) and sanitation-derived products (SDPs), with 16 components (Figure 3.1 and Table 3.2).

3.3.3.1 Design for recovery and reuse (DeRaR)

DeRaR facilitates a shift from 'design-for-disposal' to 'design-for-recovery' to ensure that SSC-RaMs are provided in the right quality and quantity for the needed transformational processes to SDPs. This approach effectively shifts the goal of sanitation from solely 'safe disposal' and/or 'safely managed' facilities to maximizing the extent to which embodied resources are safely captured for recovery and reuse as well as to deliver public and environmental health benefits. It has the potential to contribute productively to local economies and livelihoods (Murray & Buckley, 2009) and is characterized by the process of 'sanitise and recycle', based on preventing pollution through nutrient recovery for food production in the local communities (Haq & Cambridge, 2012) and producing water, nutrients, organic matter, energy and other valuable minerals from human waste and wastewater (Bahri, 1999; Jenkins & Sugden, 2006; Kujawa-Roeleveld & Zeeman, 2006; Lazarova *et al.*, 2001; Lens *et al.*, 2001; Maurer *et al.*, 2012; Simha *et al.*, 2017; Wielemaker *et al.*, 2016; Zeeman, 2012). DeRaR approaches are not technocentric or toilet-focused, but centre on resources and their management (Andersson *et al.*, 2016a, b) and have the capacity to act across the whole spectrum of the sanitation system (Brands, 2014; Esrey, 2002; Larsen *et al.*, 2013; Lens *et al.*, 2001; SuSanA, 2008) (Figure 3.1 and Table 3.2).

3.3.3.2 Sanitation service chain raw materials (SSC-RaMs)

SSC-RaMs provide the raw materials for the biophysical-chemical transformation to SDPs through DeRaR. These input materials such as faeces, urine, flushwater, greywater, cleansing water, sludge (faecal and sewage) and other materials. These materials require integrated and holistic management approaches to ensure that adequate resources are recovered and to maintain public health and environmental protection.

3.3.3.3 Sanitation-derived products (SDPs)

SDPs ensure that the transformed materials are delivered to target potential direct and indirect users and customers for utilization. For example, SDPs such as used water and struvite can contribute to improving agricultural productivity and soil quality (Andersson *et al.*, 2016a, b; Brands, 2014; Esrey, 2002; Larsen *et al.*, 2013; Lens *et al.*, 2001; SuSanA, 2008). Thus, a viable business model could emerge from used materials recovered from sanitation systems, which in turn helps ensure sustainable provision of safely managed sanitation facilities (Murray & Ray, 2010) (Figure 3.1 and Table 3.2).

3.3.4 Cross-cutting components

The ReGenSan framework has four major cross-cutting components, namely: continuous improvement (CiM), knowledge and skills (KaS), sanitation infrastructure standardization (SIS) and integrated functional sanitation value chain (IFSVC). These transversing features cut across the whole system and are vital and applicable to all the subsystems, dimensions and other components. They are essential for a successful implementation of ReGenSan solutions, practices, innovations and interventions. They serve as decussating (cross-cutting) parts that keep the wheels of the system well-oiled to guarantee unhindered progress and sustained maintenance. First, they interact and blend together in appropriate portions to form a brew that provides subsystems' dimensions and components with essential ingredients to operate and function as they should; and then they interact with actors in the subsystem along the

IFSVC to create a sanitation informion system that supports the entire system process as well as the robustness of the sanitation economy within and across sani-sheds. Thus, they enable a ReGenSan system to contribute to food security, agro-industries, agriculture, construction etc. (Figure 3.2).

In broad terms, knowledge of the processes involved in core sanitation is key to ensuring sustainability and effective comprehension of the intricate and complex dynamics that often play out in practice. One major challenge in the field of sanitation is that decisions are generally made on available information (e.g. number of toilets), but information is not knowledge. Knowledge is the logical assessment of information that produces concepts and direction for design, development, fabrication, installation, treatment, recovery, distribution, monitoring and evaluation etc. Subsequently, products and services are created for sanitation and support. When acquired and accumulated knowledge is put to practice, appropriate skills are developed to provide products and render services. Without adequate and required skills, ReGenSan will be stopped in its tracks because (contrary to past opinion) sanitation is a skilled occupational field. Most sanitation workers (particularly in the informal sector of developing countries) are unskilled and untrained and so are exposed to occupational hazards as well as putting public health at risk (Košturiak & Frolík, 2006). Therefore, KaS is essential in all subsystems, dimensions and components.

In addition, standards are key to ensure replicability, reliability and functionality across the subsystems. They also support continuous improvement (CiM), where KaS is required to perform and maintain acceptable quality output. These cross-cutting components form the backdrop for the structure of the ReGenSan framework and guarantee the sustainability of improved, enhanced and upgraded sanitation provisions in the system. Consequently, standardization interacts with all dimensions and components to ensure that sanitation facilities, technology and practices are regularized to create required uniformity, scalability and replicability that eventually leads to affordability, which encourages return on investment (see Section 5.3). In addition, sanitation infrastructural standardization (SIS) is an essential foundation for CiM because without standards there can be no strategic improvement and progressive processes for effective and efficient implementation of EIDT, ReDT, NoDT, DeRaR, SDPs, IFSVC and Governance (Mĺkva *et al.*, 2016). Integrating SIS and CiM ensures that sanitation systems operate at a reduced failure, abandonment and operational dysfunctional rate while also maintaining and rehabilitating aged infrastructure/facilities to reduce negative effects on public health and environment (Košturiak & Frolík, 2006). The IFSVC of ReS (see Section 6.3) is designed to drive the economic viability, livelihood support and supply chain within and across sani-sheds to boost the sanitation economy; and also interacts with other dimensions and components to ensure holistic and systemic integration between sanitation enterprises' operational functions such as primary productions, transformation, marketing and final sale with key actors/operators such as designers, producers, integrators, processors, traders, suppliers and distributors to deliver value-added SDPs to users and/or customers (Chapter 6). For details of CiM and KaS see the next two sections.

3.3.4.1 Continuous improvement (CiM)

ReGenSan proposes a CiM that will ensure sustained progressive strategies, practices and processes (Kiran, 2016; Yang *et al.*, 2016). This defines key performance indicators (KPIs) that would be used to measure the performance of sanitation solutions (Kiran, 2016) and provides opportunistic learning from past performance that could lead to future improvements. CiM operates with standards and certifications to guarantee performance quality, replicability, reliability, functionality, serviceability, acceptability, accessibility and contextuality. It will provide innovative measures for reducing abandonment, dysfunctionality and failures of existing sewered and non-sewered sanitation by increasing the information flow and decreasing the costs of poor-quality infrastructure (Robescu *et al.*, 2016; Silivestru *et al.*, 2014). KaS will deliver innovative management using best applicable and most appropriate techniques, tools and practice to ensure that the quality of products and services meet users/customers' requirements and long-term objectives. CiM will depend on effective record-keeping that builds up a sanitation information system database that is used to pursue environmental quality, technological specification and resource recovery among other things on a persistent loop in order to correct errors, problems and gaps as they occur or are exposed. This cross-cutting component is based on the ReGenSan principles of ApT (2.4.1), NToB (2.4.5), FPG (2.4.3) and RoDT (2.4.10) and is designed to incrementally improve sanitation products/services and access/delivery so as to meet set goals and targets (Anand *et al.*, 2009; Peng *et al.*, 2008; Yang *et al.*, 2016). ReGenSan proposes that CiM could support effective and efficient products and service and delivery (Robescu *et al.*, 2016) if:

(I) CiM is seen as a primary tool for positive change that will improve sanitation infrastructure technology;
(II) Sanitation services and products are designed and provided from users/customers' perspectives;
(III) System and subsystem processes are assessed externally and collaboratively to identify the issues for improvement;
(IV) Processes in the sanitation system are continuously improved (this will require constantly providing KaS to empower people to tackle problems as soon as they arise to prevent escalation; as early improvements will lead to proactive activities that produce positive effects later);
(V) Provide KaS training (for CiM) to equip operators, service providers, users/customers, regulators, auditors etc. with requisite understanding and capacity; and
(VI) Standards and key performance indicators (KPIs) are developed to direct CiM.

CiM strengthens SIS by KaS to support sanitation facility integrity, reliability and improved performance. This is dependent on knowledge, skills, competencies and significant information, which make it possible to visualize total quality sanitation management striving towards expanding access and improving service of safely managed and resource recovery facilities in the sani-sheds (Deighton, 2016).

3.3.4.2 Knowledge and skills (KaS)

Knowledge is acquired through the comprehension of information gained through learning that creates and enhances the skills, and the proven ability to apply acquired knowledge and skills (KaS) to technological management of solutions in sanitation is indispensable to successful and effective ReGenSan implementation (Grün *et al.*, 2009; Quendler *et al.*, 2013). Over the years, KaS for regenerative transformations in society has been in short supply, which has led to serious shortfall in the human resource requirement for disciplines related to regenerative development (Bagwell, 2007). In addition, there already exist skills gaps in traditional/conventional sanitation practice, which impacts the ability to deliver set goals and targets for the sector (Bos, 2006), especially in the services sectors that lack holistic perspective on skill development (IWA, 2014a). Countries such as Indonesia and the Netherlands among others are faced with systemic issues such as staff attrition, erosion of experience, the reluctance of new graduates to join the sanitation sector (WWAP, 2016) and (particularly in OECD) an ageing workforce (Wehn & Alaerts, 2013). The gruelling tasks of sanitation workers in the informal sectors of most developing countries are often carried out without requisite KaS to ensure occupational health and safety (OHS), public health and environmental protection. This negates the progress of the sanitation movement and it is imperative that ReGenSan target KaS upgrading and dissemination to guarantee effective programmes.

This challenge requires the design of adequate training tools and innovative learning approaches to enhance the competencies of staff as well as strengthen institutional capacity to deliver effective and efficient services (WWAP, 2016). Lower- and middle-income countries failed to ensure that the corresponding human resource base needed to design, construct, operate and maintain services was available and adequate for longer-term efforts in the pursuit of agreed goals and targets (IWA, 2014b). This could be related to:

(I) Lack of financial resources for hiring and retaining staff (salaries and benefits), especially in the public sector;
(II) Difficulty of attracting skilled workers to live and work in rural areas;
(III) Mismatch between courses offered and job requirements;
(IV) Inadequate funding for educational institutions;
(V) High cost of tuition;
(VI) Absence of continuous training systems in many countries;
(VII) Lack of government policies that create enabling environments;
(VIII) Image problems and stigma associated with the sanitation sector.

Optimum services satisfactory to users/customers will require new ways of working and new sets of skills for sector practitioners (Bagwell, 2007). Sanitation workers should have the requisite understanding of and ability in KaS in order to comprehend, operate and implement ReGenSan requirements, as it will take creative and systemic planning and designs to weave diverse factors (human and natural) together into one coherent whole. And then, within the whole, creative, functional, operational and logical skills are crucial for effective service delivery (Lyle, 1994). Subsequently, this could be achieved by:

(I) Creating enabling policy environments for collaborative frameworks between education, employers (public, private, NGOs), trade unions and employees in the sector;
(II) Developing incentives to attract and retain staff;
(III) Strengthening technical and vocational training; and
(IV) Giving attention to human resources capacity development in rural areas

A transition to sanitation 4.0, as proposed by the ReGenSan framework (see Sections 1.4.4 and 1.4.5) will entail critical support for KaS that provides the ability to expand access to safely managed facilities, devices and services as well as enhance resource capturing with environmental/natural resource protection. Accordingly, ReGenSan advocates that sanitation practitioners' generic and technical KaS (at high and low levels) be sufficient for effective and satisfactory ReGenSan practice that could deliver the required changes the sector craves (McGrath & Powell, 2016). Adequate and appropriate education and training are necessary so as to equip sanitation practitioners, workforce and the society with the core values, competences and KaS for the future of sanitation. ReGenSan draws from several disciplines such as environment, medicine, nutrition, agronomics, geography, engineering, architecture, citizenship, sociology, psychology, fishery/aquaculture, urban and rural development, political science, history, law, anthropology, economics and business; and so cannot be pursued in isolated disciplines (Annan-Diab & Molinari, 2017; Defries *et al.*, 2012). It therefore implies a revision of sanitation teaching and research contents to respond to global and local challenges. The idea is to promote teaching methods that generate KaS for interdisciplinary and systemic thinking, integrated planning, cooperative/collaborative decision-making and understanding the inherent complexities in the sanitation system, within sani-sheds. In other words, targeted and practical capacity-building for adequate and appropriate KaS across IFSVC (Quendler *et al.*, 2013). KaS that simultaneously cuts across through the subsystems (SES, TeS and ReS) and relevant external forces (sanitation economy, agriculture etc) will facilitate real-time problem-solving as it engenders better comprehension of each part to create solutions that tap in to different disciplines thereby blurring boundaries in the delivery of sustainable sanitation solutions. In essence, scientists (physical, medical, biological and social); engineers (sanitary, environmental, chemical, biological, mechanical, electrical/electronics etc), humanities professionals (philosophers, anthropologists, historians etc), builders (architecture, urban regeneration, rural development, planners, landscapers etc), finance and economics professionals and others should work together from local to global scales to combine knowledge, tools and approaches to develop ReGenSan infrastructural technological and management sanitation solutions (Annan-Diab & Molinari, 2017; Defries *et al.*, 2012).

3.4 SUBSYSTEMS' SYNERGISTIC INTERACTIONS

Figure 3.1 highlights the interactions, interrelationships and interdependencies of the three subsystems, seven dimensions and 27 components of ReGenSan for research, policy-making, planning, design, development and management of sanitation systems that are in harmony with the physical environment, human systems and the surrounding natural systems. The framework depicts a holistic and system thinking design for the integration of technological, ecological, economic, psycho-social and political systems for sanitation management.

The SES is the foundation of the ReGenSan system, in which the governance dimension provides the rules, guidelines and enabling environment for practice and service delivery based on the system of the 'place' and 'scale' (PSEP dimension). Governance and PSEP work in tandem to prescribe design and develop technologies for the 'place' and 'scale' and also influence resource recovery and reuse by determining appropriate approaches for adoption. The objective is to improve service delivery and expand access by meeting the goals and targets of SDG no. 6 by the year 2030. The overall outcome could support a sanitation economy where resources from sanitation support the circular economy through a symbiotic synergistic relationship between the subsystems, dimensions and other external actors via continuous feedback mechanisms. The governance and PSEP dimensions provide needed support to each other with governance directing behaviours and incorporateing peculiar features, norms, customs and practices, while PSEP assists governance processes and mechanisms with information, knowledge, skill, support and acceptability among other things. There is a positive and negative feedback loop between the two dimensions of SES. Previously, the inadequate considerations for PSEP and lack of synergy between all dimensions negatively impacted the successful implementation of sanitation intervention programmes, as illustrated in the National Sanitation Bucket Replacement programme in South Africa presented in Box 3.2.

BOX 3.2 SOUTH AFRICA'S NATIONAL SANITATION BUCKET REPLACEMENT PROGRAMME

The South African National Sanitation Bucket Replacement was a near classic case of RoDT aimed at replacing and upgrading an estimated 252–254 bucket toilets located across seven provinces. The bucket sanitation systems were unhygienic and expensive to maintain and violated the dignity of users and those responsible for collection and disposal (Mjoli, 2012). The national programme started in 2005 and by the end of 2008 more than 91% of the target number of bucket toilets had been replaced by waterborne sanitation systems and upgraded Ventilated Improved Pit (VIP) toilets (DWAF, 2008a). The programme, however, faced challenges that could be attributed to a disregard of the PSEP dimension because some municipalities' bulk sewers and wastewater treatment capacities were inadequate and in some areas the available water supply could not support the new waterborne sanitation systems (Mjoli, 2012). Also, in areas where upgraded VIP toilets were implemented, emptying trucks could hardly access the more than 60,000 VIPs due to the nature of the houses, especially in the informal settlements; and many resource recovery and reuse strategies also impacted negatively (Murray & Buckley, 2009 – Figure 3.2).

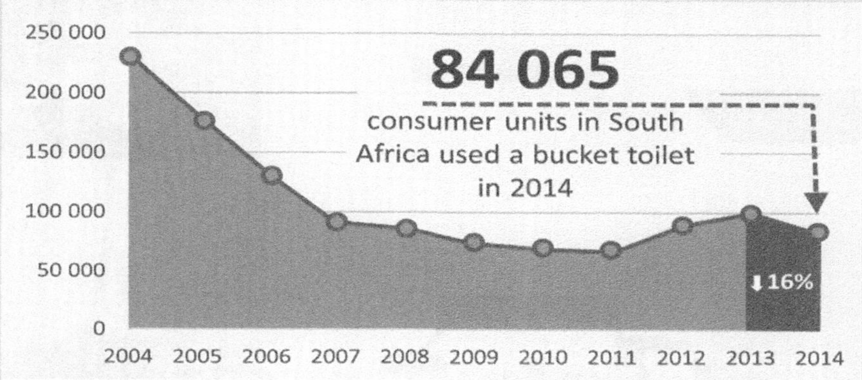

Figure 3.2 Use of bucket toilets in South Africa. (*Source*: Stats SA, 2015).

In addition, the governance dimension provides adequate guidance, standards, design criteria and asset management strategies for the technology dimensions of EIDT, RoDT and NoDT and resource system dimensions: design for recovery and reuse (DeRaR), sanitation service chain raw materials (SSC-RaMs) and sanitation-derived products (SDPs). Subsequently, the technology dimensions are adequately connected through negative feedback mechanisms, especially in situations where the three technology dimensions are required in a 'place' and 'scale' for appropriate sanitation solutions/interventions. The understanding of these complex dynamic interactions and their maximization could result in innovative sanitation solutions with output minimization and resource utilization, as in the case of the Asian I-nstitute of Technology's (AIT) solar septic tank, which increased recovery of biogas and inactivation of pathogens in the effluents (Koottatep *et al.*, 2014, 2015, 2016; Pussayanavin *et al.*, 2015) as presented in Box 3.3.

BOX 3.3 ASIAN INSTITUTE OF TECHNOLOGY (AIT) SOLAR SEPTIC TANK

The unique feature of this on-site sewage treatment system is that it draws energy directly from the sun, which makes it suitable for places with abundant sunlight and inadequate electricity supply. In addition, the increased temperature of the influent inside the septic tank rapidly and effectively destroys pathogens and reduces sludge formation while increasing biogas production. This biogas can then be used for domestic purposes, while the effluent, further purified by constructed wetland, can be used for irrigation, aquaculture and horticulture as well as redirected to the sanitation system as flush water (Koottatep *et al.*, 2015, 2016, Pussayanavin *et al.*, 2015 – Figure 3.3, Table 3.3).

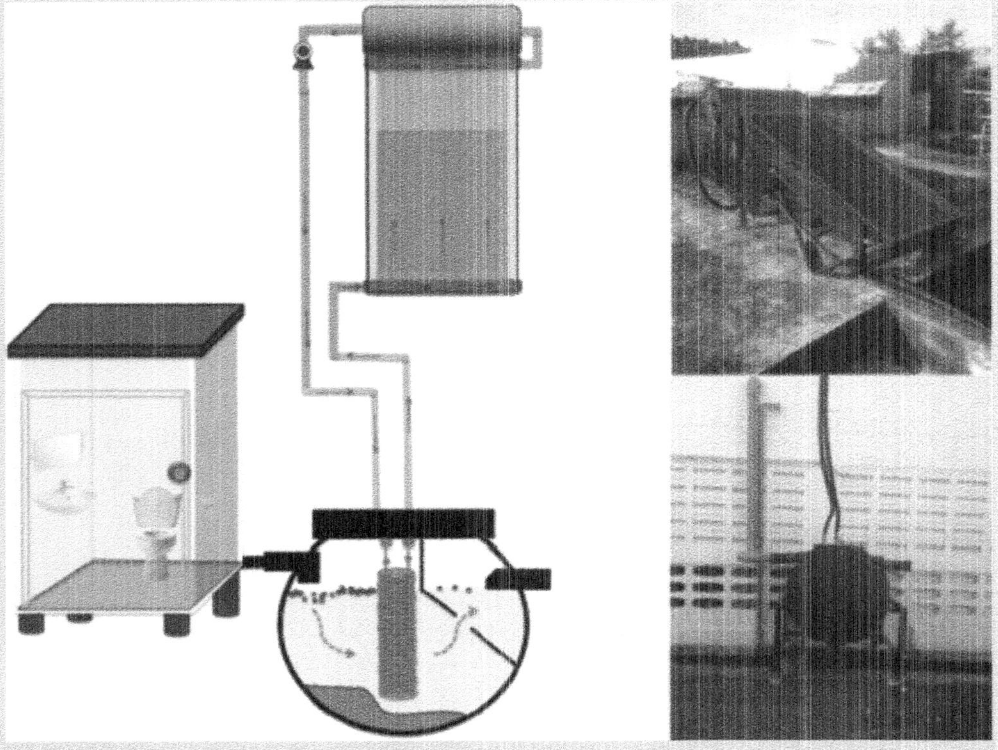

Figure 3.3 Field testing of the solar septic tank at the Asian Institute of Technology, Thailand (*Source*: AIT 2015).

Table 3.3 Comparative Performance of Laboratory-Scale Solar Septic Tanks and Conventional Septic Tanks (Koottatep *et al.*, 2015, 2016; Pussayanavin *et al.*, 2015).

Parameters	**Solar Septic Tank (40–50°C)**	**Conventional Septic Tank (30°C)**
Sludge accumulation (m^3/year)	0.5	1
TVS accumulation in sludge (g/g TVS input)	0.26	0.50
CH_4 production (L/d)	1.5	0.9
CH_4 production (L/g TVS input)	0.027	0.01
E.coli in effluents (MPN/mL)	1.6×10^3 12.3 (50°C)	4.6×10^3

The TeS dimensions can also independently provide the needed inputs for each dimension in the ReS and the ReS in turn could provide necessary feedbacks into the technology design system. For example, EIDT delivers SSC-RaMs (e.g. urine) required by the fit-for-purpose infrastructure of DeRaR, then DeRaR in turn delivers the SSC-RaMs in the right quality and quantity to transform them to SDPs and deliver products and services to the customers in the IFSVC (e.g. struvite to farmers), which could then be used in food production and subsequently returned by users (through toilets – EIDT) as urine. South Africa showcases an example where well-intentioned sanitation resource-recovery and reuse-intervention projects produce low return on investment due to weak synergy and poor understanding of the SES, TeS and ReS (Box 3.4).

BOX 3.4 POOR SES AND TECHNOLOGY SUBSYSTEMS SYNERGY EXPERIENCE, SOUTH AFRICA

A recovery and reuse strategy in the Mnini district of the eThekwini Municipality, South Africa, which involved treating wastewater via a pond system that was then used for irrigation of a 2-ha. banana and mango farm, was stalled at 75% completion in spite of its technical viability. The system was never commissioned because it did not comply with ReGenSan's PSEP and governance even though it fitted with the ReDT component of replacement. Particularly, lack of prior permissions from local traditional leaders for land use, no prior permission from local households to install irrigation systems and utilize land for agriculture, as well as no approval from relevant government agencies, shot down the scheme before it could begin (Murray & Buckley, 2009).

Consequently, the entire process could have a cyclical feedback that creates beneficial outcomes between ReGenSan interactions and other aspects of the society. For example, recovered products and materials for reuse will serve agriculture, aquaculture, horticulture and other circular economy industries within and outside the sanitation economy, and in the same way such agro-industries will also feed ReGenSan through value chain services and products.

3.5 DELIVERING INTEGRATED SANITATION SOLUTIONS

ReGenSan does not seek to replace existing frameworks, but integrates all aspects of sanitation holistically and systematically to interact with related issues (e.g. water, health, hygiene, gender etc.) and creates a comprehensive scheme for the best approach and solution in particular contexts. The ReGenSan framework shows for the first time how socio-ecological, resource recovery and technology considerations can be inter-synergistically linked in a systemic whole to deliver comprehensive integrated sani-solutions. ReGenSan brings four innovative perspectives into the sanitation sphere (Box 3.5).

BOX 3.5 ReGenSan FRAMEWORK INNOVATION

(I) Offers a comprehensive, holistic, integrated and systemic outlook for sanitation technological and management solutions rather than a singular focus,
(II) Embraces the dynamic complexities and contextual peculiarities of 'place and scale' in delivering sanitation solutions,
(III) Introduces the concept of sani-sheds in sanitation management, and
(IV) Applies the regenerative concept to the field of sanitation for the first time

The intention is to improve management, planning and technological innovations to help guide the activities of practitioners, policymakers and academics working in the sanitation space. ReGenSan framework advocates for comprehensive integration of all the components of the sanitation systems to ensure the capacity of sustained service delivery as well as improve participatory processes that enhance users' ownership. The diversity that often exists in social and ethnic groups of any locality greatly influences the design, access, acceptability, affordability, construction and use of sanitation systems. Therefore, this framework will strengthen understanding of the unique dynamics and potential of a location to influence types and processes of technologies and determine systems' capacity to improve their own performance.

This ReGenSan framework has the potential to induce improvement and mind-change in sanitation system-thinking, management, planning and technology innovations without compromising ecosystem sustainability. It is also broad enough to accommodate other pre-existing frameworks and perspectives, as it provides functional and progressive approaches towards improving and expanding access to sanitation services at a 'scale'. The framework also advocates comprehensive, explorative, experimental and descriptive assessments using qualitative and quantitative datasets for better understanding of sanitation subsystems, dimensions and components. ReGenSan shall serve as a guide for funding/donor and intervention agencies by helping them to identify where and when to fund and support for more effective and practical outcomes. It could also help to strengthen global planning and implementation towards expanding access and improving service delivery of sanitation solutions. At the same time, it could really serve as a pathway or portal for innovative designs and development of sanitation technologies and social solutions.

The major constraint on the application of the ReGenSan framework is the disregard for the use of conceptual frameworks in most sanitation research. This trend is changing as framework perspectives and paradigm research in the sanitation space increase. The perception of sanitation only through the lens of toilets as isolated structures for defecation with products and by-products only considered as waste and nuisances that contaminate, pollute and degrade the ecological system must change. *Sanitation should rather be seen as a catalyst for systemic community regeneration and transformation.*

3.6 EXERCISES

(I) Give examples of NoDT, RoDT and EIDT and show possible complex-dynamic interactions between the three technology dimensions using a system diagram applicable to your home town.

(II) Describe the following ReGenSan subsystems: (a) SES, (b) TeS and (c) ReS and their application to help your country meet the SDG no. 6 by the year 2030.

(III) Explore the relationship between the ReGenSan principles and the framework.

(IV) Briefly describe the interactions between the cross-cutting components and the subsystems shown in Figure 3.1 and Table 3.2.

(V) Using scenario development, create sani-solutions for the following options based on the ReGenSan framework: (a) upgrading of urban slum areas' bucket sanitation systems, (b) community transformation from open defecation and (c) provision of sanitation to new development areas in coastal, valley and arid regions.

(VI) Using the ReGenSan framework, propose sanitation solutions at a scale of 5,000 population in high-, middle- and low-income urban areas.

3.7 RECOMMENDATIONS FOR FUTURE RESEARCH

Future research could look at the following:

(I) Explore PSEP pre-considerations that could lead to practical, effective and acceptable technological solutions.

(II) Using the ReGenSan framework, design sani-solution to match a 'place' and 'scale'.
(III) Using the ReGenSan framework, design sani-solution for retrofitting dysfunctional and failing infrastructure and facilities to fit a particular place and scale.
(IV) Using the ReGenSan framework, design sani-solutions to replace dysfunctional and failing infrastructure and facilities with NDT that fit with a chosen location and particular scale type.
(V) Using the ReGenSan framework, explore options for a governance and institutional framework that fits a particular chosen place towards improving sanitation solutions and service delivery.
(VI) Using the ReGenSan framework, explore options to incorporate ReGenSan into the design and plans for new cities, areas, estates and apartments, which include reducing transfer of burden at final disposal (i.e. resource recovery/reuse and treatment within point of generation) as much as possible.
(VII) Explore the need and potential for capacity development and training for relevant government officials and regulators in sanitation management.

REFERENCES

AIT (Asian Institute of Technology) (2015). Innovative DEWAT technologies. Reinventing the Toilet – Progress Report 3, submitted to Bill & Melinda Gates Foundation.

Altaf S. W. (2011). So Much Aid, So Little Development: Stories from Pakistan. Johns Hopkins University Press, Baltimore, USA.

Anand G., Ward P., Tatikonda M. and Schilling D. (2009). Dynamic capabilities through continuous improvement infrastructure. *Journal of Operations Management*, **27**(6), 444–461.

Andersson K., Dickin S. and Rosemarin A. (2016a). Towards 'sustainable' sanitation: challenges and opportunities in urban areas. *Sustainability*, **8**, 1289.

Andersson K., Rosemarin A., Lamizana B., Kvarnström E., McConville J., Seidu R., Dickin S. and Trimmer C. (2016b). Sanitation, Wastewater Management and Sustainability: From Waste Disposal to Resource Recovery. UNEP/GPA/SEI, Stockholm, Sweden, Sanitation,_wastewater_management_and_sustainability_From_waste_disposal_to_resource_recovery-2016sanitation_wastewater_management_and_sustainabilit.pdf?sequence=2&isAllowed=y (accessed 15 March 2018).

Annan-Diab F. and Molinari C. (2017). Inter-disciplinarity: practical approach to advancing education for sustainability and for the Sustainable Development Goals. *The International Journal of Management Education*, **15**, 73–83, http://dx.doi.org/10.1016/j.ijme.2017.03.006

Bagwell S. (2007). Skills needs for regeneration professionals. *Journal of Urban Regeneration and Renewal*, **1–2**, 191–204.

Bahri A. (1999). Agricultural reuse of wastewater and global water management. *Water Science and Technology*, **40**(4), 339–346.

Baum R., Luh J. and Bartram J. (2013). Sanitation: a global estimate of sewerage connections without treatment and the resulting impact on MDG progress. *Advances in Environmental Science and Technology*, **47**(4), 1994–2000, doi: 10.1021/es304284f

Bocken N. M. P., Short S. W., Rana P. and Evans S. (2014). A literature and practice review to develop sustainable business model archetypes. *Journal of Cleaner Production*, **65**, 42–56.

Bos A. (2006). Capacity Building in the Water and Sanitation Sector at times of the MDGs. Discussion paper prepared for the Round Table Meeting organised by WaterLinks and PSO, January 2006. UNESCO-IHE, December 2005–January 2006. http://www.bibalex.org/Search4Dev/files/314800/145385.pdf (accessed 15 July 2018).

Boschi-Pinto C., Lanata C. F., Mendoza W. and Habte D. (2006). Diarrheal diseases. Disease and mortality in Sub-Saharan Africa. In: Disease and Mortality in Sub-Saharan Africa, D. T. Jamison, R. G. Feachem, M. W. Makgoba, E. R. Bos, F. K. Baingana, K. J. Hofman and K. O. Rogo (eds), 2nd edn, World Bank, Washington D.C., USA, http://www.ncbi.nlm.nih.gov/books/NBK2302/

Bozkurt H., Gernaey K. and Sin G. (2015). Computer-Aided Framework for Synthesis, Design and Retrofit of Wastewater Treatment Plants. Danmarks Tekniske Universitet (DTU), Kgs. Lyngby. http://orbit.dtu.dk/files/110810194/Hande_Bozkurt_CAP15_08_01_1_.pdf (accessed 30 April 2017).

Brands E. (2014). Prospects and challenges for sustainable sanitation in developed nations: a critical review. *Environmental Reviews*, **22**, 1–18, doi: 10.1139/er-2013-0082

Brepolsa C. H., Dorgeloh E., Frechen F. B., Fuchs W., Haider S., Joss A., de Korte K., Ruiken C. H., Schier W., van der Roest H., Wett M. and Wozniak T. H. (2008). Upgrading and retrofitting of municipal wastewater treatment plants by means of membrane bioreactor (MBR) technology. *Desalination*, **231**, 20–26.

Brewerton P. and Millward L. (2001). Organisational Research Methods. Sage, London, UK.

Brikke F. (2000). Operation and Maintenance of Rural Water Supply and Sanitation Systems: A Training Package for Managers And planners. World Health Organization Geneva, Switzerland, http://apps.who.int/iris/bitstream/10665/66716/1/WHO_SDE_WSH_00.2_p1-124.pdf (accessed 30 April 2017).

Cairncross S. and Feachem R. G. (1993). Environmental Health Engineering in the Tropics: An Introductory Text. 2nd edn. Wiley, Chichester, UK, http://eu.wiley.com/WileyCDA/WileyTitle/productCd-0471938858.html

Campbell O. M. R., Benova L., Gon G., Afsana K. and Cumming O. (2015). Getting the basic rights – the role of water, sanitation and hygiene in maternal and reproductive health: a conceptual framework. *Tropical Medicine and International Health*, **20**(3), 252–267, doi: 10.1111/tmi.12439

Chang N.-B., Pires A. and Martinho G. (2011). Empowering systems analysis for solid waste management: challenges, trends, and perspectives. *Critical Reviews in Environmental Science and Technology*, **41**(16), 1449–1530, https://doi.org/10.1080/10643381003608326

Charnley F. and Lemon M. (2011). Exploring the process of whole system design. *Design Studies*, **32**, 156–179. doi: 10.1016/j.destud.2010.08.002

Cookey P. E., Darnsawasdi R. and Ratanachai C. (2016a). A conceptual framework for assessment of governance performance of Lake Basins: towards transformation to adaptive and integrative governance. *Hydrology*, **3**(1), 12, doi: 10.3390/hydrology3010012

Cumming G. S. (2011). Spatial Resilience in Social-Ecological Systems. Springer, Dordrecht, Heidelberg, London, New York, doi:10.1007/978-94-007-0307-0

DeFries R. S., Ellis E. C., Chapin F. S., III, Matson P. A., Turner B. L., II, Agrawal A. et al. (2012). Planetary opportunities: a social contract for global change science to contribute to a sustainable future. *BioScience*, **62** (6), 603e606, http://dx.doi.org/10.1525/bio.2012.62.6.11

Deighton M. G. (2016). Facility Integrity Management: Effective Principles and Practices for the Oil, Gas and Petrochemical Industries. Gulf Professional Publishing, Elsevier, Amsterdam, the Netherlands, eBook ISBN: 9780128018323.

Domènech L. and Saurí D. (2010). Socio-technical transitions in water scarcity contexts: public acceptance of greywater reuse technologies in the Metropolitan Area of Barcelona. *Resources Conservation and Recycling*, **55**, 53–62.

DWAF (Department of Water Affairs and Forestry) (2008a). the National Sanitation Bucket Replacement Programme – Lessons Learnt. http://www.ewisa.co.za/misc/WIN_SA_Facts/WIN%20SA/TheNSRPLessonLearnt.pdf (accessed 12 July 2017).

Ekane N., Nykvist B., Kjellén M., Noel S. and Weitz N. (2014). Multi-level sanitation governance: understanding and overcoming the challenges in the sanitation sector in Sub-Saharan Africa. Working paper no. 2014–4. Stockholm Environment Institute. https://www.sei-international.org/mediamanager/documents/Publications/Water-sanitation/sei-workingpaper-ekane-multi-level-sanitation.pdf (accessed 8 May 2017).

El-Sheikh M. A. (2011). Optimization and upgrading wastewater treatment plants. *Journal of Engineering Sciences, Assiut University*, **39**(4), 697–713. http://www.aun.edu.eg/journal_files/82_J_8366.pdf (accessed 30 April 2017).

Esrey S. A. (2002). Philosophical, ecological and technical challenges for expanding ecological sanitation into urban areas. *Water Science and Technology*, **45**(8), 225–228, PMID:12019825.

Exley J. L. R., Liseka B., Cumming O. and Ensink J. H. J. (2014). The sanitation ladder, what constitutes an improved form of sanitation? *Environmental Science & Technology*, **49**, 1086–1094, doi: 10.1021/es503945x

Ferguson G., Dakers A. and Gunn I. (2003). Sustainable Wastewater Management A Handbook for Smaller Communities. Ministry for the Environment, Wellington, New Zealand, http://www.mfe.govt.nz/sites/default/files/summary_0.pdf (accessed 11 May 2017).

Galli G., Nothomb C. and Baetings E. (2014). Towards systemic change in urban sanitation. (IRC Working Paper) The Hague: IRC.

Ghosh B. S. and Elvira M. (2011). Africa's Water and Sanitation Infrastructure: Access, Affordability, and Alternatives. Directions in Development; Infrastructure. World Bank, Washington D.C., USA, http://documents.worldbank.org/curated/en/712211468202191672/Africas-water-and-sanitation-infrastructure-access-affordability-and-alternatives (accessed 24 June 2017).

Gine R., Jimenez A. and Perez-Foguet A. (2011). The Future of Water, Sanitation and Hygiene: Innovation, Adaptation and Engagement in a Changing World – A Closer Look at the Sanitation Ladder: Issues of Monitoring the Sector. Water, Engineering and Development Centre International Conference. 35th WEDC International Conference: The Future of Water, Sanitation and Hygiene in Low-Income Countries: Innovation, Adaptation and Engagement in a Changing World, Loughborough, UK. http://hdl.handle.net/2117/14199 (accessed 15 March 2018).

Giné-Garriga R., Flores-Baquero O., de Palencia A. J. and Pérez-Foguet A. (2017). Monitoring sanitation and hygiene in the 2030 Agenda for Sustainable Development: a review through the lens of human rights. *Science of the Total Environment*, **580**, 1108–1119.

Grübler A. (1998). Technology and Global Change. Cambridge University Press, Cambridge, UK.

Grün G., Tritscher-Archan S. and Weiß S. (2009). Guidelines for the Description of Learning Outcomes. http://ibw4.mservices.at/zoom/pdf/wp2/Guidelines_EN_final_2.pdf (accessed 3 June 2013).

Haq G. and Cambridge H. (2012). Exploiting the co-benefits of ecological sanitation. *Current Opinion in Environmental Sustainability*, **4**(4), 431–435.

Hegger D. L. T., van Vliet J. and van Vliet B. J. M. (2007). Niche management and its contribution to regime change: the case of innovation in sanitation. *Technology Analysis and Strategic Management*, **19**, 729–746.

Hinkel J., Bots P. W. G. and Schluter M. (2014). Enhancing the Ostrom social–ecological system framework through formalization. *Ecology and Society*, **19**(3), 51, 1–19, doi: 10.5751/ES-06475-190351

Hutton G., Haller L. and Bartram J. (2007). Economic and Health Effects of Increasing Coverage of Low Cost Household Drinking-Water Supply and Sanitation Interventions to Countries Off-Track to Meet MDG Target 10. World Health Organization, Geneva, Switzerland, http://www.irc.nl/ page/38443 (accessed 15 March 2018).

IWA (2014a). An Avoidable Crisis: WASH Human Resource Capacity Gaps in 15 Developing Countries. IWA Publishing, London, http://www.iwa-network.org/downloads/1422745887-an-avoidable-crisis-wash-gaps.pdf (accessed 15 March 2018).

IWA (2014b). YWP Conference: Key Message from IWA International Young Water Professional Conference hosted in Taipei December 2014.

Jenkin S. and Pedersen Z. M. (2009). Rethinking Our Built Environments: Towards a Sustainable Future. Ministry for the Environment, Manatu Mo Te Taiao, Wellington, New Zealand, https://www.researchgate.net/publication/261476724_Rethinking_our_built_environments_Towards_a_sustainable_future (accessed 10 February 2018).

Jenkins M. and Sugden S. (2006). 'Rethinking sanitation: lessons and innovation for sustainability and success in the new millennium'. in Human Development Report 2006, United Nations Development Programme, New York, p. 37.

Joint Monitoring Programme (JMP) (2015). Progress on Sanitation and Drinking Water: 2015 Update and MDG Assessment Geneva/New York.

Joshi D. K., Hughes B. B. and Sisk T. D. (2015). Improving governance for the post-2015 sustainable development goals: scenario forecasting the next 50 years. *World Development*, **70**, 286–302, 2015 0305-750X, http://dx.doi.org/10.1016/j.worlddev.2015.01.013 (accessed 15 March 2018).

Kalbermatten J. M., Julius D. S., Mara D. D. and Gunnerson C. G. (1980). Appropriate Technology for Water Supply and Sanitation A Planner's Guide. The International Bank for Reconstruction and Development/the World Bank. https://www.ircwash.org/sites/default/files/201-80AP-19010.pdf (accessed 27 May 2017).

Kalbermatten J. M., Julius D. S. and Gunnerson C. G. (1982). Appropriate sanitation alternatives: a technical and economic appraisal. World Bank studies in Water Supply and Sanitation; no. 1. World Bank, Washington D.C.,

USA. http://documents.worldbank.org/curated/en/637021468740368977/Appropriate-sanitation-alternatives-a-technical-and-economic-appraisal (accessed 27 May 2017).

Kellert S. R., Heerwagen J. and Mador M. (2008). Biophilic Design: The Theory, Science and Practice of Bringing Buildings to Life. John Wiley & Son Inc, New York, USA.

Kiran D. R. (2016). Total Quality Management. Butterworth-Heinemann-Elsevier, Oxford, UK.

Kooiman J. (2003). Governing as Governance. Sage Publications, London, UK.

Koottatep T., Phuphisith S., Pussayanavin T., Panuvatvanich A. and Polprasert C. (2014). Modeling of pathogen inactivation in thermal septic tanks. *Journal of Water Sanitation and Hygiene for Development*, **4**(1), 81–88.

Koottatep T., Chapagain S. K., Watanatanachart J. and Polprasert C. (2015). Application of nano-silver coated granular activated carbon for inactivation of septic tank effluent. *Journal of Water Sanitation and Hygiene for Development*, **5**(4), 632–637.

Koottatep T., Chapagain S. K., Polprasert C., Panuvatvanich A. and Ahn K.-H. (2016). Sanitation situations in selected Southeast Asian countries and application of innovative technologies. *Environ Dev Sustain*, doi: 10.1007/s10668-016-9892-6

Košturiak J. and Frolík Z. (2006). Lean and Innovative Company. Alfa Publishing, s. r. o., Prague, Czech Republic.

Kujawa-Roeleveld K. and Zeeman G. (2006). Anaerobic treatment in decentralized and source-separation-based sanitation concepts. *Reviews in Environmental Science and Bio/Technology*, **5**, 115–139.

Kvarnström E., McConville J., Bracken P., Johansson M. and Fogde M. (2011). The Sanitation Ladder – a Need for a Revamp? *Journal of Water Sanitation and Hygiene for Development*, **1**(1), 3–12.

Larsen T. A. and Gujer W. (2013). Implementation of source separation and decentralization in cities. In: Source Separation and Decentralization for Wastewater Management, T. A. Larsen, K. Udert and J. Lienert (eds), IWA Publishing, London, UK, pp. 135–150.

Larsen T. A., Maurer M., Eggen R. I. L., Pronk W. and Lienert J. (2010). Decision support in urban water management based on generic scenarios: the example of NoMix technology. *Journal of Environmental Management*, **91**, 2676–2687.

Larsen T. A., Udert K. and Lienert J. (2013). Source Separation and Decentralization for Wastewater Management. IWA Publishing, London, UK.

Lazarova V., Levine B., Sack J., Cirelli G., Jeffrey P., Muntau H., Salgot M. and Brissaud F. (2001). Role of water reuse for enhancing integrated water management in Europe and Mediterranean countries. *Water Science and Technology*, **43**(10), 25–33.

Lens P., Zeeman G. and Lettinga G. (2001). Decentralized Sanitation and Reuse: Concepts, Systems and Implementation. IWA Publishing, London, UK.

Levy Y. and Ellis T. J. (2006). A systems approach to conduct an effective literature review in support of information systems research. *Informing Science: The International Journal of an Emerging Transdiscipline*, **9**, 181–212.

Lockwood H. and Smits S. (2011). Supporting Rural Water Supply: Moving Towards a Service Delivery Approach. Practical Action Publishing Ltd, Warwickshire, UK.

Lüthi C., Morel A., Tilley E. and Ulrich L. (2011a). Community-Led Urban Environmental Sanitation Planning (CLUES); Swiss Federal Institute of Aquatic Science and Technology (EAWAG) Dübendorf, Switzerland. http://www.susana.org/_resources/documents/default/2-1300-cluesguid.pdf (accessed 14 May 2017).

Lüthi C., Panesar A., Schütze T., Norström A., McConville J., Parkinson J., Saywell D. and Ingle R. (2011b). Sustainable Sanitation in Cities: A Framework for Action. Papiroz Publishing House, Netherlands.

Lyle J. T. (1984). Designing Human Ecosystems. John Wiley & Sons, Hoboken, USA.

Lyle J. T. (1994). Regenerative Design for Sustainable Development. John Wiley & Sons, Hoboken, USA.

Maier M. W. (1998). Architecting principles for systems-of-systems. *Bioprocess and Biosystems Engineering*, **1**, 267–284.

Mang P. and Reed B. (2012a). Designing from place: a regenerative framework and methodology. *Building Research & Information*, **40**(1), 23–38.

Mang P. and Reed B. (2012b). Regenerative Development and Design. Regenesis Group and Story of Place, Encyclo. Sustainability Science & Technology.

Marshall R. E. and Farahbakhsh K. (2013). Systems approaches to integrated solid waste management in developing countries. *Waste Management*, **33**, 988–1003, http://dx.doi.org/10.1016/j.wasman.2012.12.023 (accessed 15 March 2018).

Martínez-Jurado P. J. and Moyano-Fuentes J. (2014). Lean management, supply chain management and sustainability: a literature review. *Journal of Cleaner Production*, **85**, 134–150.

Maurer M., Bufardi A., Tilley E., Zurbrügg C. and Truffer B. (2012). A compatibility-based procedure designed to generate potential sanitation system alternatives. *Journal of Environmental Management*, **104**, 51–61.

Mayring P. (2003). Qualitative Inhaltanalyse – Grundlagen und Techniken. (Qualitative Content Analysis), 8th edn. Beltz Verlag, Weinheim, Germany.

McConville J., Kunzle R., Messmer U., Udert K. M. and Larsen T. A. (2014). Decision support for redesigning wastewater treatment technologies. *Environmental Science & Technology*, **48**(20), 12238–12246.

McGrath S. and Powell L. (2016). Skills for sustainable development: transforming vocational education and training beyond 2015. *International Journal of Educational Development*, **50**, 12–19, http://dx.doi.org/10.1016/j.ijedudev.2016.05.006

Mjoli N. P. (2012). Evaluation of Sanitation Upgrading Programmes – The Case of the Bucket Eradication Programme. Report to the Water Research Commission. WRC Report no. 2016/1/12.

Mĺkva M., Prajová V., Yakimovich B., Korshunov A. and Tyurin I. (2016). Standardization – One of the Tools of Continuous Improvement. International Conference on Manufacturing Engineering and Materials, ICMEM 2016, Nový Smokovec, Slovakia, 6–10 June 2016, Procedia Engineering 149, pp. 329–332. https://ac.els-cdn.com/S1877705816311845/1-s2.0-S1877705816311845-main.pdf?_tid=aa028662-9f87-4594-9932-77b079fe27db&acdnat=1531597772_5e793fff7675c6f2103c88d09db8b0bc (accessed 15 July 2018).

Murray A. and Buckley C. (2009). Designing reuse-oriented sanitation infrastructure: the design for service planning approach. In: Wastewater Irrigation and Health Assessing and Mitigating Risk in Low-income Countries, P. Drechsel, C. A. Scott, R.-S. Liqa, M. Redwood and A. Bahri (eds), Earthscan, London, UK, pp. 303–318.

Murray A. and Ray I. (2010). Commentary: back-end users: the unrecognized stakeholders in demand-driven sanitation. *Journal of Planning Education and Research*, **30**(1), 94–102.

Nakagiri A., Niwagaba C. B., Nyenje P. M., Kulabako R. N., Tumuhairwe J. B. and Kansiime F. (2016). Are pit latrines in urban areas of SubSaharan Africa performing? A review of usage, filling, insects and odour nuisances. *BMC Public Health*, **16**, 120, doi: 10.1186/s12889-016-2772-z

Nansubuga I., Banadda N., Verstraete W. and Rabaey K. (2016). A review of sustainable sanitation systems in Africa. *Reviews in Environmental Science and Bio/Technology*, **15**, 465, doi: 10.1007/s11157-016-9400-3

Nhapi I., Siebel M. A. and Gijzen H. J. (2006). A proposal for managing wastewater in Harare, Zimbabwe. *Water and Environment Journal*, **20**, 101–108.

Novak J. and Canas A. (2008). The Theory Underlying Concept Maps and How to Construct and Use Them. Florida Institute for Human and Machine Cognition. http://cmap.ihmc.us/Publications/ResearchPapers/TheoryUnderlyingConceptMaps.pdf

Oosterveer P. and Spaargaren G. (2010). Meeting social challenges in developing sustainable environmental infrastructures in East African Cities. In: Social Perspectives on the Sanitation Challenge, B. van Vliet (ed.), Springer Science+Business Media B.V., Dordrecht, Netherlands. doi: 10.1007/978-90-481-3721-3_2

Ostrom E. (2005). Understanding Institutional Diversity. Princeton University Press, Princeton, USA.

Ostrom E. (2009). A general framework for analyzing sustainability of social-ecological systems. *Science 24*, **325** (5939), 419–422. https://doi.org/10.1126/science.1172133

Ostrom E. and Cox M. (2010). Moving beyond panaceas: a multi-tiered diagnostic approach for social-ecological analysis. *Environmental Conservation*, **37**, 451–463, doi: 10.1017/S0376892910000834

Pahl-Wostl C. (2009). A conceptual framework for analysing adaptive capacity and multi-level learning processes in resource governance regimes. *Global Environmental Change*, **19**(3), 354–365, doi: 10.1016/j.gloenvcha.2009.06.001

Pahl-Wostl C., Holtz G., Kastens B. and Knieper C. (2010). Analyzing complex water governance regimes: the management and transition framework. *Environmental Science & Policy*, **13**, 571–581.

Pahl-Wostl C., Jeffrey P., Isendahl N. and Brugnach M. (2011). Maturing the new water management paradigm: progressing from aspiration to practice. *Water Resources Management*, **25**, 837–856, doi: 10.1007/s11269-010-9729-2

Partelow S. (2016). Coevolving Ostrom's social–ecological systems (SES) framework and sustainability science: four key co-benefits. *Sustainability Science*, **11**, 399–410, doi: 10.1007/s11625-015-0351-3

Peng D. X., Schroeder R. G. and Shah R. (2008). Linking routines to operations capabilities: a new perspective. *Journal of Operations Management*, **26**(6), 730–748.

Pham N. B. and Kuyama T. (2013). Decentralized domestic wastewater management in Asia – challenges and opportunities. Policy Brief – Series 1, Water Environment Partnership in Asia (WEPA), Institute for Global Environmental Strategies (IGES). http://www.wepa-db.net/pdf/1403policy_brief/WEPA_PB1_2013.pdf (accessed 30 April 2017).

Pickett S. T. A., Jones C. and Kolasa J. (2007). Ecological Understanding: The Nature of Theory and the Theory of Nature. Academic Press, New York, USA.

Pollack H. N. (2005). Uncertain Science … Uncertain World, 1st edn. Cambridge University Press, Cambridge, UK.

Potter A., Klutse A., Snehalatha M., Batchelor C., Uandela A., Naafs A., Fonseca C. and Moriarty P. (2011). Assessing sanitation service levels. IRC International Water and Sanitation Centre, Working Paper 3, 2nd edn. https://www.ircwash.org/sites/default/files/Potter-2011-Assessing.pdf (accessed 18 May 2017).

Pussayanavin T., Koottatep T., Eamrat R. and Polprasert C. (2015). Enhanced sludge reduction in septic tanks by increasing temperature. *Journal of Environment Science and Health, Part A*, **50**(1), 81–89.

Quendler E., Van der Luit J., Monteteone M., Aguado P., Pfeiffenscheider M., Wagner K., Valente F. and Cunha-Queda C. (2013). Employers' needs on competences, knowledge and skills for sustainable development as a reference framework for higher education in life sciences. *Procedia – Social and Behavioral Sciences*, **106**, 1063–1085, https://doi.org/10.1016/j.sbspro.2013.12.121 (last accessed 05/04/2018).

Ramoa A., Matos J. and Luthi C. (2015). System-Based Decision Trees for the Selection of Sanitation Technologies. 38th WEDC International Conference, Loughborough University, UK. http://wedc.lboro.ac.uk/resources/conference/38/Ramoa-2250.pdf (accessed 26 June 2017).

Raven R. (2007). Co-evolution of waste and electricity regimes: multi-regime dynamics in the Netherlands (1969–2003). *Energy Policy*, **35**, 2197–2208.

Repella A. (2014). Framework Perspectives for Water: a small sample of the range of frameworks for addressing questions about water. Water Diplomacy Network Blog. http://blog.waterdiplomacy.org/2014/01/framework-perspectives-for-water/ (accessed 6 June 2017).

Robescu L. D., Silivestru C., Presura A., Pana A. and Mihai R. (2016). Application of Continuous Improvement Strategy for reducing environmental impact of a wastewater treatment plant. *Water Utility Journal*, **14**, 29–40, https://www.ewra.net/wuj/issue_14.htm (accessed 14 July 2018).

Rodic L. (2015). Waste governance. In: Global Waste Outlook, D. C. Wilson (ed.), ISWA/UNEP, Nairobi, Kenya, http://www.unep.org/ietc/Portals/136/Publications/Waste%20Management/GWMO%20report/GWMO%20full%20report.pdf (accessed 12 September 2016).

Rosenqvist T., Mitchell C. and Willetts J. (2016). A short history of how we think and talk about sanitation services and why it matters. *Journal of Water, Sanitation and Hygiene for Development*, **06.2**, doi: 10.2166/washdev.2016.118

Ross K., Abeysuriya K., Mikhailovich N. and Mitchell C. (2014). Governance for decentralised sanitation: Global Practice Scan. A working document. Prepared by the Institute for Sustainable Futures, UTS as part of the Australian Development Research Award Scheme (ADRAS). http://communitysanitationgovernance.info/wp-content/uploads/2016/06/20140125-ADRAS-GDS-Global-Practice-Scan-a.pdf (accessed 14 October 2017).

Schertenleib R. (2005). From conventional to advanced environmental sanitation. *Water Science and Technology*, **51**(10), 7–14.

Seiffert M. E. B. and Loch C. (2005). Systemic thinking in environmental management: support for sustainable development. *Journal of Cleaner Production*, **13**, 1197–1202. doi: 10.1016/j.jclepro.2004.07.004

Sethi V. and King W. R. (1998). Organizational Transformation thought Business Process Reengineering. Prentice Hall, Upper Saddle River, USA.

Seuring S. and Muller M. (2008). From a literature review to a conceptual framework for sustainable supply chain management. *Journal of Cleaner Production*, **16**, 1699–1710.

Silivestru C., Ifrim D., Gugu A. M. and Mihalache R. (2014). Define Manual, MIS ETC 2177.

Simha P., Lalander C., Vinnerås B. and Ganesapillai M. (2017). Farmer attitudes and perceptions to the reuse of fertiliser products from resource-oriented sanitation systems – The case of Vellore, South India. *Science of the Total Environment*, **581–582**, 885–896, https://doi.org/10.1016/j.scitotenv.2017.01.044

Spaargaren G., Mommaas H., Van Den Burg S., Maas L., Drissen E. and Dagevos H. (2007). More sustainable lifestyles and consumption patterns: a theoretical perspective for the analysis of transition processes within consumption domains. Contrast Research Report, TMP project. Wageningen: Environmental Policy Group, Wageningen University; Telos: Tilburg University; RIVM; LEI, Wageningen, Netherlands.

Spaargaren G., Oosterveer P., Van Buuren J. and Mol A. P. J. (2006). Mixed Modernities: Towards Viable Urban Environmental Infrastructure Development in East Africa. Wageningen University, Wageningen, Netherlands.

Starkl M., Brunner N. and Stenstrom T. (2013). Why do water and sanitation systems for the poor still fail? Policy analysis in economically advanced developing countries. *Advances in Environmental Science and Technology*, **47**(12), 6102–6110, doi: 10.1021/es3048416

Starkl M., Brunner N., Feil M. and Hauser A. (2015). Ensuring sustainability of non-networked sanitation technologies: an approach to standardization. *Advances in Environmental Science and Technology*, **49**, 6411–6418, doi: 10.1021/acs.est.5b00887

Stats SA (Statistic South Africa) (2015). Bucket toilets: an update. http://www.statssa.gov.za/wp-content/uploads/2015/09/BucketToilet_Slider.jpg (accessed 22 June 2018).

L. Strande, M. Ronteltap and D. Brdjanovic (2014). Faecal Sludge Management: Systems Approach for Implementation and Operation. IWA Publishing, London, UK.

SuSanA (ed.) (2008). Towards More Sustainable Sanitation Solutions. (Version 1). Sustainable Sanitation Alliance (SuSanA), Eschborn. https://www.susana.org/_resources/documents/default/3-267-7-1452594644.pdf (accessed 21 June 2017).

Tilley E., Strande L., Luthi C., Mosler H., Udert K. M., Gebauer H. and Hering J. G. (2014a). Looking beyond technology: an integrated approach to water, sanitation and hygiene in low income countries. *Advances in Environmental Science and Technology*, **48**(17), 9965–9970, doi: 10.1021/es501645d

Tilley E., Ulrich L., Luthi C., Reymond P. H. and Zurbrugg C. (2014b). Compendium of Sanitation Systems and Technologies, 2nd rev. edn. Swiss Federal Institute of Aquatic Science and Technology (EAWAG), Dübendorf, Switzerland.

Tobias R., O'Keefe M., Künzle R., Gebauer H., Gründl H., Morgenroth E., Pronk W. and Larsen T. A. (2017). Early testing of new sanitation technology for urban slums: the case of the Blue Diversion Toilet. *Science of the Total Environment*, **576**, 264–272.

Trémolet S. and Rama M. (2012). Tracking National Financial Flows into Sanitation, Hygiene and Drinking Water. World Health Organization.

UN-Water and WHO (2012). GLAAS 2012 Report. UN-Water Global Analysis and Assessment of Sanitation and Drinking-Water. The Challenge of Extending and Sustaining Services. WHO, Geneva, Switzerland.

USAID (2016). Global Sanitation Service Chain Technology Manual: An analysis of sanitation technology innovations. http://www.path.org/publications/files/PATH_dt_ssd_tech_manual.pdf (accessed 20 June 2017).

USEPA (1974). Process design manual for upgrading existing wastewater treatment plants. U.S. Environmental Protection Agency, Technology Transfer. http://www.un.org/waterforlifedecade/pdf/glaas_report_2012_eng.pdf (accessed 28 June 2017).

Van der Ryn S. and Cowan S. (2007). Ecological Design, 10th anniversary edn. Island Press, Washington D.C., USA.

van Vliet B. J. M., Spaargaren G. and Oosterveer P. (2011). Sanitation under challenge: contributions from the social sciences. *Water Policy*, **13**, 797–809.

Verhagen J. and Carrasco M. (2013). Full-Chain Sanitation Services that Last: Non-Sewered Sanitation Services. IRC International Water and Sanitation Centre, Netherlands, https://www.ircwash.org/sites/default/files/201306_sanitation_framework_final.pdf (accessed 17 June 2017).

Wehn U. and Alaerts G. (2013). Leadership in knowledge and capacity development in the water sector: a status review. *Water Policy*, **15**, 1–14.

Whittington D., Lauria D. and Choe K. (1993). Household sanitation in Kumasi, Ghana: a description of current practices, attitudes, and perceptions. *World Development*, **21**(5), 733–748. https://doi.org/10.1016/0305-750X(93)90030-D (accessed 15 March 2018).

WHO/UNICEF (2017). Progress on Drinking Water, Sanitation and Hygiene: 2017 Update and SDG Baselines. World Health Organization (WHO) and the United Nations Children's Fund (UNICEF), Geneva, Licence: CC BY-NC-SA 3.0 IGO.

Wielemaker R. C., Weijma J. and Zeeman G. (2016). Harvest to harvest: recovering nutrients with New Sanitation systems for reuse in Urban Agriculture. *Resources, Conservation and Recycling*, **128**, 426–437, http://dx.doi.org/10.1016/j.resconrec.2016.09.015 (accessed 15 March 2018).

World Bank (2003). Appraisal report: Tanzania-Dar-es Salaam water supply and sanitation project. World Bank, Washington D.C., USA, http://documents.worldbank.org/curated/en/2003/04/2301336/tanzania-dar-es-salaam-water-supplysanitation-project (accessed 15 March 2018).

WWAP (United Nations World Water Assessment Programme) (2016). The United Nations World Water Development Report 2016: Water and Jobs. UNESCO, Paris, http://unesdoc.unesco.org/images/0024/002439/243938e.pdf (accessed 16 July 2018).

Yang Y., Lee P. K. C. and Cheng T. C. E. (2016). Continuous improvement competence, employee creativity, and new service development performance: a frontline employee perspective. *International Journal of Production Economics*, **171**, 275–288, http://dx.doi.org/10.1016/j.ijpe.2015.08.006 (accessed 15 March 2018).

Zeeman G. (2012). New Sanitation: Bridging Cities and Agriculture. Wageningen University, Wageningen, Netherlands.

Zurbrügg C. and Tilley E. (2007). Evaluation of Existing Low Cost Conventional as Well as Innovative Sanitation System and Technologies – NETSSAF Deliverable D22&23. Swiss Federal Institute of Aquatic Science and Technology (EAWAG), Dübendorf, Switzerland.

Zurbrügg C. and Tilley E. (2009). A system perspective in sanitation – Human waste from cradle to grave and reincarnation. *Desalination*, **248**, 410–417.

Chapter 4

Social-ecological system

'everything affects everything else in one way or another, whether you are aware of that or not does not change the fact that this is what is happening'

Murray Gell-Mann

4.1 INTRODUCTION

ReGenSan nests sanitation services and products within the social ecological system (SES) of the 'place' to enhance an understanding of the systemic, holistic and integrated (SHI) interactions in the sanitation subsystems (Andries *et al.*, 2004; Binder *et al.*, 2013; Ostrom & Cox, 2010). This reinforces the established fact that to achieve adequate and sustainable sanitation services and proper institutional frameworks as well as adoption of appropriate technology requires sanitation solutions to be embedded in local physical and socio-economic, cultural, spatial and decision-making structures (Elledge *et al.*, 2002; Letema, 2012; Seghezzo, 2004; WECD, 1987). These complex-dynamic interactions have the capacity to facilitate sanitation infrastructural innovation by reinforcing technological, institutional and sociocultural transformations (Geels, 2005). The SES indicates a commitment to adopt SHI perspectives towards human and non-human elements of sanitation systems (Halliday & Glaser, 2011) intricately linked with and affected by one or more social systems and mediated through interactions with biophysical systems (Andries *et al.*, 2004). This SES dynamic shows the interactions and connections between technology and resource systems' dimensions and components of the whole sanitation system as well as the sanitation economy. Any change within the SES will affect all other aspects of the whole system. For example, technology cannot be separated from the economic and social contexts from which it evolves because issues of affordability, acceptability and willingness-to-pay should determine technology selection within an SES. A successful implementation of ReGenSan requires better understanding of the SES (Altaf, 2011; Grübler, 1998; Nilsson & Olsson, 2014) to remove the largely unknown 'black box' of interactions between the sanitation system components such as users and service

© IWA Publishing 2019. Regenerative Sanitation: A New Paradigm for Sanitation 4.0
Editor(s): Thammarat Koottatep, Peter Emmanuel Cookey and Chongrak Polprasert
doi: 10.2166/9781780409689_0107

providers. Worldwide, toilets and sanitation practices are sealed-off from the society and culturally loaded with notions of dirt, shame and waste (Curtis & Biran, 2001) and this negatively impacts improving services and expanding access to sanitation (Hegger *et al.*, 2007; van Vliet *et al.*, 2011). Therefore, a proper understanding of the interactions between sanitation entities within the SES could transform the sanitation burden into something more positive and sustainable.

Evidence has shown the inadequacy of looking at sanitation only as a matter of toilets (infrastructure) rather than the whole integrated cycle of management (including other psycho-socio-cultural factors) and this could be responsible for the slow progress experienced in sanitation service delivery. Working from within the SES ensures that infrastructure is embedded as an interactive part of the system that connects users and other SES entities (van Vliet *et al.*, 2011). Generally, even where users are knowledgeable about sanitation technology and governance, they are hardly ever considered in design and planning through the sanitation value chain; technology has been prioritised over users and the ecosystem. Some have even argued '*Let us sort out technology first; if we have proven that it works, we should implement it in society*' (Hegger *et al.*, 2007). But the implementation of sustainable decentralized sanitation technology is hampered by many non-technical barriers and it is increasingly becoming apparent that users are key to the success of technology selection, especially, when certain new well-designed sanitation technologies fail at implementation phase because they did not fit the standards of comfort that users uphold or were incompatible with cultural beliefs or religious codes. Domestic end users might be vital co-producers of change in sanitation if they are taken seriously as system users and invited to rethink sanitation practices (Hegger *et al.*, 2007; Spaargaren *et al.*, 2007; van Vliet *et al.*, 2011). This disconnection and the inadequate consideration for local contexts wherein technology is nested are largely responsible for the slow progress towards sanitation goals and targets. A classic example is presented in Box 4.1 as reported in the case of Gafumba, Karangara and Cyahi cells, Burera District, Rwanda, where traditional pit toilets upgraded to urine-diversion dry toilets (UDDT) are still used as a traditional pit toilets (Ekane *et al.*, 2012);

BOX 4.1 UDDT USED AS TRADITIONAL PIT TOILETS, GAFUMBA, KARANGARA AND CYAHI CELLS, BURERA DISTRICT, RWANDA

In Gafumba, Karangara and Cyahi cells, Burera District, Rwanda, traditional pit toilets were upgraded to urine-diversion dry toilets (UDDT), but the people still used them as pit toilets and none of these UDDTs collected urine separately. Also, in a bid to assist the UDDTs, some water was poured into both the urine and faeces compartments and sometimes urine even went into the faeces compartment in households where men are in the habit of drinking excessive amounts of alcohol (Ekane *et al.*, 2012).

This is caused by lack of understanding by the users because promoters of improved sanitation, in most cases, have completely different perspectives from the would-be users, which often results in incompatibility between technologies and the values, beliefs and experiences of the users. Also, lack of proper understanding of these dynamic interactions between the entities of the SES could lead to poor institutional and policy frameworks in the sector, which impacts negatively on issues of affordability, acceptability, willingness to pay for services and active participation in infrastructural operations and maintenance (Banerjee & Duflo, 2011; Rogers & Hall, 2003).

Furthermore, sanitation systems have been known to exert some pressures on the SES that could lead to negative impacts, especially when they are not adequately understood and incorporated early into sanitation technological design and management solutions (Golden & Earp, 2012; Stokols, 1992, 1996). For example, failed and disused infrastructures could pollute ecosystems, degrade environmental aesthetics and affect public health. Sociotechno-ecological (STE) sanitation problems facing the world today, particularly developing countries, are both systemic and managerial in character. They are systemic because they arise from deep-rooted, complex, interrelated, multi/transdisciplinary processes that cannot be resolved by technocentric approaches alone. They are management and governance problems because their solutions require sustained, integrated, coordinated and goal-driven responses across disciplines and professions and there are no quick fixes (Halliday & Glaser, 2011). Therefore, understanding specific SES (of a 'place') ensures that complementary issues such as resource recovery, psycho-socio-cultural and institutional frameworks among others are incorporated and addressed early in the design of sanitation systems to ensure fit and improve feedback mechanisms between all the subsystems, dimensions and components of the ReGenSan system. Figure 4.1 models the SES dynamics of ReGenSan and is adopted from Andries *et al.* (2004), Halliday and Glaser (2011) and Liehr *et al.* (2017). Table 4.1 shows descriptions of entities, exogenous and endogenous drivers, and Table 4.2 shows the interactions involved in the SES.

Figure 4.1 clearly depicts the fact that better understanding of other entities of SES beyond technology is a requirement to fix the infrastructural challenges of sanitation as against previous assumptions of 'fix

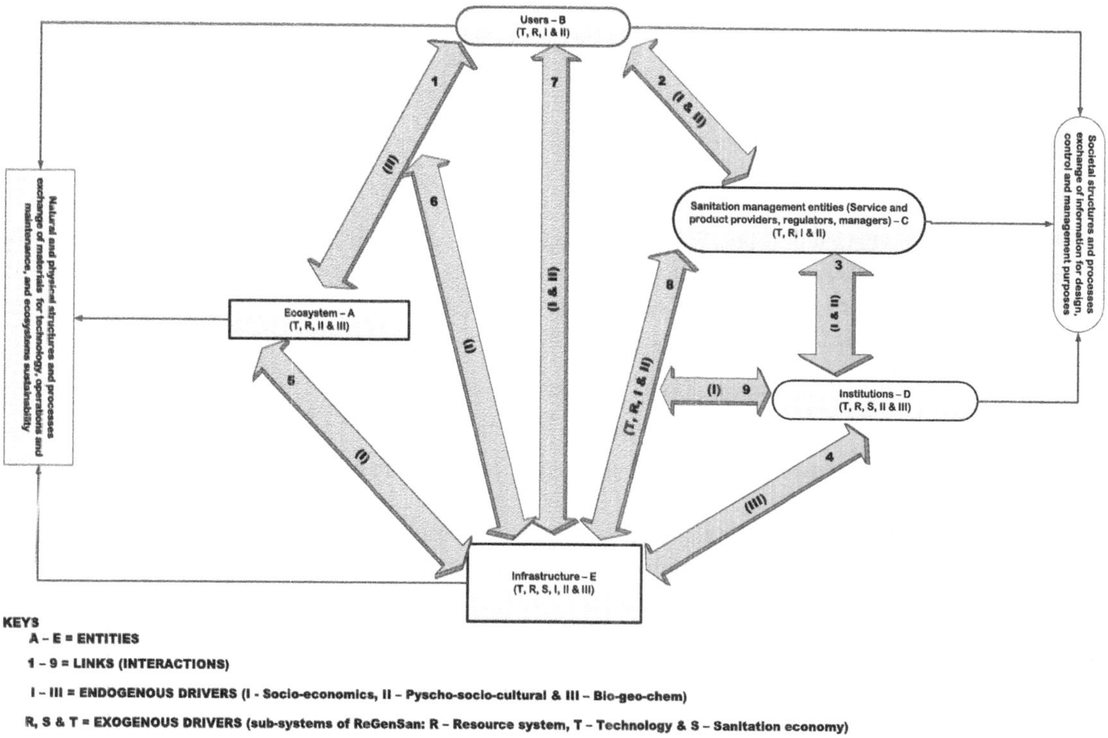

Figure 4.1 Model of social-ecological subsystem dynamics interactions of ReGenSan (Author figure inspired by Andries *et al.*, 2004.)

Table 4.1 Entities descriptions, exogenous and endogenous drivers.

Entities		Descriptions	Exogenous Drivers	Endogenous Drivers
A	Ecosystem	Provides services and functions	Technology and resource systems	Socio-economic, psycho-socio-cultural and bio-geo-chemical and hydrological
B	Users	Makes use of infrastructures and services	Technology and resource systems	Socio-economic, psycho-socio-cultural
C	Service and product providers, regulators, managers etc.	Provides infrastructures, services and products	Technology and resource systems as well as sanitation economy	Socio-economic, psycho-socio-cultural
D	Institutions	Provides the rules and regulations	Technology and resource systems as well as sanitation economy	Socio-economic, psycho-socio-cultural and bio-geo-chemical and hydrological
E	Infrastructure	Sanitation facilities/infrastructure	Technology and resource systems as well as sanitation economy	Socio-economic, psycho-socio-cultural and bio-geo-chemical and hydrological

technology first and others will follow'. Previous literature on sanitation had focused on technology, with limited interactions with users of the technology and, where users and technology were considered, the focus was on how to make users accept provided technology and not necessarily on exploring the kinds and types of technology that users will accept (see equation 4.1–4.3).

$$E \rightarrow A, B, C, D \quad (E \text{ determines all other entities without feedback}) \tag{4.1}$$

$$B, C, D, E \rightarrow A \quad (\text{All other entities impact } A \text{ without feedback}) \tag{4.2}$$

where: $A =$ ecosystem, $B =$ users, $C =$ service and product providers, regulators, managers, $D =$ institutions and $E =$ infrastructure.

ReGenSan, therefore, proposes that real solutions to various sanitation challenges lie in better understanding and utilization of the complex dynamic interactions and linkages between the entities of the SES.

$$A \leftrightarrow B, C, D, E \quad (\text{Systemic holistic integration of entities with feedback}) \tag{4.3}$$

where: $A =$ ecosystem, $B =$ users, $C =$ service and product providers, regulators, managers, $D =$ institutions and $E =$ infrastructure

It is indicative that one model cannot capture all potential interactions in the SES, but an attempt is made to present a broad structure with aligned processes for overall improvement of sanitation service delivery,

Table 4.2 Interactions involved in the SES.

Links		Exogenous Drivers	Endogenous Drivers
1	Ecosystem and users	Technology and resource systems	Socio-economic, psycho-socio-cultural and bio-geo-chemical and hydrological
2	Users and service and product providers	Technology and resource systems as well as sanitation economy	Socio-economic and psycho-socio-cultural
3	Service and product providers, regulators, managers and institutions	Technology and resource systems as well as sanitation economy	Socio-economic, psycho-socio-cultural and bio-geo-chemical and hydrological
4	Institutions and infrastructures	Technology and resource systems as well as sanitation economy	Socio-economic, psycho-socio-cultural and bio-geo-chemical and hydrological
5	Infrastructures and ecosystem	Technology and resource systems as well as sanitation economy	Bio-geo-chemical and hydrological
6	Users and infrastructures	Technology and resource systems	Socio-economic and psycho-socio-cultural
7	Ecosystems, users and infrastructures	Technology and resource systems	Socio-economic, psycho-socio-cultural and bio-geo-chemical and hydrological
8	Service and product providers, regulators, managers and infrastructures	Technology and resource systems as well as sanitation economy	Socio-economic, psycho-socio-cultural and bio-geo-chemical and hydrological
9	Service and product providers, regulators, managers, infrastructures and institutions	Technology and resource systems as well as sanitation economy	Socio-economic, psycho-socio-cultural and bio-geo-chemical and hydrological

product design and access expansion. The ReGenSan SES model of interactions presented in this chapter indicates the dissonance between users, service and product providers, institutions, ecosystems and infrastructure and the way it could hinder progress in global and local sanitation pursuits. In essence, all sanitation entities should work together within the given SES to determine types/kinds of sanitation technology and style/range of services suitable for the local context. This SES model will, therefore, contribute to policy, practice and research by providing information and input data for planning and design. This will also generate in-depth insight into these dynamic interactions towards evolving sanitation policies and regulations (Hinkel *et al.*, 2014; Ostrom, 2009; Ostrom & Cox, 2010; Partelow,

2016; van Vliet *et al.*, 2011; Whittington *et al.*, 1993) as well as appropriate/innovative technology provision. The five entities of this ReGenSan SES model are grouped into two dimensions: psycho-socio-ecophilia (PSEP) and governance (Figure 3.1 and Table 3.2).

4.2 DIMENSIONS OF REGENSAN SES

4.2.1 Psycho-Socio-Ecophilia (PSEP)

PSEP is a core dimension of ReGenSan whichever specific SES is being focused on. It stipulates that psychological (individual sensibilities, concerns, health, safety, privacy and preferences), social (community, status, acceptability, perception, participation/involvement, regulations, norms, behaviours, hygiene etc), cultural (traditions, customs, belief systems, religion, family/ancestry), economic (affordability, willingness-to-pay, livelihood/entrepreneurial opportunity, financing etc), physiological (urination and defecation) and geographical/geological/ecological (place, scale and ecosystem, etc) conditions should be incorporated into the design, planning and modelling of sanitation technologies and management, which could result in a most appropriate fit (van Vliet *et al.*, 2011). It recognizes that ReGenSan system is embedded within a SES (Dias, 2015; Mang & Reed, 2012b). This dimension operates from the place and scale (PaS) principle of ReGenSan, which focuses on 'sani-sheds' for sanitation management and planning that connects with the socio-ecological characteristics of local contexts (Kellert *et al.*, 2008; Mang & Reed, 2012a). It is designed to explore interlinkages between psychology, sociology, ecology, physics, political science, public administration, economics, biology, chemistry, anthropology, management, geography, geology, medical and health sciences, food production/security and engineering disciplines (among others) for multi/transdisciplinary solutions to sanitation challenges (Mang & Reed, 2012b).

PSEP is key to ReGenSan because a diversity of social and ethnic groups often exists in a locality and individual preferences cannot just be overlooked within any group (Lyle, 1984, 1994; Van der Ryn & Cowan, 2007) and all of these greatly influence acceptability, willingness to use and pay, affordability, accessibility, design, construction and usage of sanitation systems. Also, the normative contents of the human rights to sanitation (i.e. the human rights obligations of countries related to sanitation) are encapsulated in the psycho-socio-cultural and socio-economic components of PSEP (Giné-Garriga *et al.*, 2017; UNGA, 2009; UNHRC, 2011). For the economic, social and cultural rights (e.g. the right to sanitation), the content of these obligations is clarified and operationalized under the criteria of availability, accessibility, quality, affordability and acceptability (de Albuquerque & Roaf, 2014; UNESCO, 2002).

PSEP will characteristically determine system design, installation and operations in ReGenSan. In other words, user interface, collection, treatment, recovery and disposal processes should be determined by the variety of these within local situations (Chirisa *et al.*, 2017) in order to understand how the unique dynamic-potentials of a location could cardinally influence types and processes of technology (USLG, 1996). The primary intent of PSEP is to systematically integrate psychological, social, economic, cultural, health, hygiene, behaviour, environment, land use, ecosystem and natural resources concerns as well as gender equity, dignity and stakeholder management into design, technology selection and management of sanitation systems (Marshall & Farahbakhsh, 2013; Schübber, 1996).

PSEP becomes crucial as failure of technology/infrastructure due to sociocultural and socio-economic factors like lack of demand and acceptance or unwillingness/inability to pay for services is increasingly evident. This could then lead to environmental and public health risks (Starkl *et al.*, 2013) as well as negatively impacting resource recovery due to differences in social groups' values and attitudes. In this case, it is essential to drive interactions within the SES with users as the focal point rather than technology/infrastructure as was the case in the unsuccessful past (Spaaragaren *et al.*, 2007; van Vliet *et al.*, 2011).The key components of PSEP are presented in Box 4.2.

BOX 4.2 KEY COMPONENTS OF PSEP

(1) Psycho-socio-cultural: religion, norms, customs, availability, accessibility, acceptability, willingness to use, gender equity, inclusiveness, perceptions and attitude, health, hygiene and handwashing, preference, status
(2) Socio-economic: educational level (formal and informal), income level (affordability, willingness-to-get and willingness-to-pay), livelihood support (economic activities and occupational engagement), access to basic services (water supply, electricity and healthcare), access to sanitation services and types (sewer, non-sewer, shared and non-shared facilities), geographical location (urban and rural), housing/dwelling types, household sizes and head of household, age, gender and race
(3) Bio-geo-chemical (BGC): nutrients and water (C, N, P, H_2O) cycling, ecosystem services and functions, fauna and flora

4.2.1.1 Psycho-Socio-Cultural (PSC)

The psycho-socio-cultural component is the meso, macro and other contextual factors that influence sanitation design, technology, practices and management solutions. ReGenSan is framed to ensure that PSC is incorporated into sanitation infrastructural designs and management solutions increase access expansion and service improvement and help users meet their sanitation needs without any psycho-social stress (Siegrist & Marmot, 2004). PSC seeks to explore the various interactions of users with their sanitation systems to determine the willingness to accept, pay for and use the facilities provided as well as determine their level of involvement and participation (Egan *et al.*, 2008). This stems from the fact that sanitation choices are based on psycho-socio-cultural and political factors, and most service and product providers need to recognize this and include dialogue with all parties in their strategies, especially vulnerable groups (Andersson & Minoia, 2017; Jewitt, 2011) because several sanitation failures arise from poor attention to sociocultural conditions and lack of local ownership. In essence, sanitation solutions cannot be provided in uniform fashion, but should be addressed at the local level and in differentiated ways, aligned with the cultural norms and socio-economic status (Andersson & Minoia, 2017; O'Reilly & Louis, 2014).

Cultural and social sustainability is not simply an acceptance of technical propositions or even awareness-raising and capacity-building. It is important to acknowledge certain general attitudes towards faeces that are part of specific cultures and influenced by religions. Cultural acceptability in sanitation refers to pre-adaptation to behaviours, beliefs and organization of daily lives and it contributes to transformative development that impacts well-being, resource use and sustainable production (Andersson & Minoia, 2017). Culturally, humans have been known to interact with and/or relate to issues of excrement in two ways: faecophobic and faecophilic. Faecophobic cultures are common among Hindus, some South-East Asian and African sub-Saharan countries (Winblad & Simpson-Hebert, 2004). For example, Pinsem and Vinnerås (2003) observed that getting people in Thailand to accept the application of human urine as fertilizer was not easy due to sociological difficulties where human excreta is commonly viewed as being dirty and a pathway for disease transmission (Duncker *et al.*, 2007). Also, in some Nigerian ethnicities tradition prohibits collection of urine by strangers for fear that it may be used against the people through 'black magic' or 'evil spirits' (Duncker *et al.*, 2007; Sridhar *et al.*, 2005). In Rwanda, faeces are called *'amazirantoki'*, meaning 'do not touch' or 'untouchable' and defecation inside or very close to living space is not allowed. Therefore, toilets constructed in most rural communities in Rwanda are located away from the houses and outside the household fences, usually at the exit of the compound. In addition, toilets should be constructed away from the kitchen (Ekane *et al.*, 2012; Jain, 2011).

On the other hand, faecophilic cultures are common among Chinese societies, where human excreta is seen as a valuable product and has for several millennia been used as fertilizer (Winblad & Kilama, 1985). Most world cultures occupy a position somewhere in between these two extremes (Rosenqvist, 2005) and will determine the response to sanitation choices. This fear (though somewhat rational) often hinges on the fact that human excreta is malodorous and potentially dangerous and this has proven to be one key obstacle for resource recovery and reuse of sanitation materials (Rosenqvist, 2005; Winblad & Simpson-Hebert, 2004). Of course, it is often a case of denial because most people would rather not talk about sanitation and cultures are mostly indifferent to sanitation issues – ironically seeing as every human engages in faecal and urine excretion at least four times a day (Loudon, 1977; Rosenqvist, 2005). It is vital to understand PSC issues in order to manage perception of by-products of human digestion, as is evidenced in Sweden where it became popular practice to mix cow dung with human faeces to make them less offensive and repulsive (Duncker *et al.*, 2007).

To this effect, ReGenSan shifts the focus from just building toilets to engaging the PSC component to determine the right kind of design (Devine, 2009; O'Reilly & Louis, 2014), community involvement (Kar & Chambers, 2008), involvement of the state (Black & Fawcett, 2008) and understanding people's perceptions, ideas and values around sanitation (Drangert & Bahadar, 2011; O'Reilly & Louis, 2014; Rheinländer *et al.*, 2010; Simha *et al.*, 2017b). Therefore, the concept of PSC advocates for adequate information and communication strategies that encourage people to show interest in, talk about and act on, as well as get involved in, the management of their sanitation facilities. The trend is to ensure that ReGenSan is responsive to sanitation approaches that emphasize the physical, social and environmental needs of users. Recognizing the intimate connection between the sanitation environment and the psychological, social and mental well-being of users further underscores the need for a contextualized understanding of sanitation-related psycho-socio-cultural stress (Sahoo *et al.*, 2015). This can be clearly illustrated by the TepozEco ecological toilet project in San Juan, Mexico, where previous attempts to provide ecological toilets proved to be unsuccessful. The criticisms of this earlier project were lack of consultation with recipients in its design, scanty operation and maintenance education and no follow-up technical support; all of these resulted in a quick slide into disrepair that compromised the use of the toilets. Lessons from this earlier project were applied in the new project to increase its chances of success (Davies-Colley & Smith, 2012). Box 4.3 presents this t project, with adequate considerations for the psycho-socio-cultural, economic and governance issues.

BOX 4.3 TepozEco ECOLOGICAL TOILET PROJECT IN SAN JUAN, MEXICO

The socio-economic profile of San Juan showed a population of 2,000 predominantly rural residents, mostly earning an average income of USD 90/month and heavily involved in agriculture, particularly maize and nopal cactus (a staple vegetable crop). The mean rainfall is 1,200 mm and virtually all rain falls between May and August, with little in the remaining eight months. There is no reticulated water supply, so residents capture rainwater during the rainy season, but for the rest of the year must carry water from springs or purchase it from water-tanker trucks for around USD 60 per m^3.

This is very expensive and makes water a precious commodity and because of the lack of piped water supply, the price of water, and the absence of any form of reticulated sewerage, the community relies on waterless toilets, which have been poorly constructed and maintained (Davies-Colley & Smith, 2012). The first 12 months of the project were dedicated to planning, feasibility studies and raising the community's awareness of local environmental issues and the potential of ecological toilets. In 2002, a community meeting was held, and although many who attended initially wanted flush toilets, they conceded that they were impractical due to the lack of reticulated water and sewerage. The community accepted ecological toilets as the best potential solution. Being dissatisfied with their existing toilets and the possibility of getting toilets that did not require water (and was not a pit latrine) were important initial motives. For most recipients, knowing that their toilets would not contaminate the environment and would allow the reuse of urine and faeces on their home gardens were major additional benefits (Davies-Colley & Smith, 2012).

The governance aspect of the TepozEco ecological toilet project revolved around the formation of the community group (informal institution) and a formal letter to Morelos State Water and Environmental Commission (CEAMA) (formal institution) requesting assistance. It should be noted that state governments in Mexico are responsible for the provision of sanitation to communities: CEAMA agreed to provide a subsidy sufficient for the construction of the 30 toilets. The community itself then decided which of more than 100 interested families would get the toilets.

The education programme endeavoured to address the participants' fears and doubts and ensure they understood all aspects of the toilets. The programme included a workshop to promote proper hygiene, to provide information on the transmission of faecal pathogens, how to disinfect dehydration toilets, operation and maintenance and the value of the urine-fertilizer and faeces-compost. During this phase, project staff encouraged participants to talk about defecation, faeces and urine, which helped break their psycho-socio-cultural inhibitions. By the completion of the toilets' construction and the educational programme, the participants had become experts in the operations and maintenance of the toilets. All the toilets used urine-diverting dehydration systems, the same basic plans but specifically designed architecturally for the needs and wants of individual families; the final cost to the recipient was necessarily heavily subsidized. TepozEco trained five local youths in operations and maintenance to provide follow-up support and assist toilet recipients (Davies-Colley & Smith, 2012).

Between 2004 and 2006, 30 dehydrating-style ecological toilets were constructed in the rural community of San Juan Tlacotenco (San Juan), near the town of Tepoztlan, approximately 80 km south of Mexico City. The objective was to establish a fully functioning example of urban ecological sanitation, with a long-term goal of encouraging the adoption of ecological sanitation systems in Latin America.

4.2.1.2 Socio-Economic (SE)

The socio-economic (SE) component of ReGenSan (PSEP) is a complex phenomenon predicted by a broad spectrum of variables conceptualized as a combination of financial, occupational and educational influences that impact on sanitation access expansion and service improvement (Duncan, 1961; Green, 1970; Hollingshead, 1971; Mueller & Parcel, 1981; Winkleby *et al.*, 1992). Although these variables are interrelated, they reflect different individual and societal forces associated with sanitation. For example, income level reflects affordability and willingness to pay for sanitation services as well as access to other basic services; and educational level indicates willingness to use and the possibility of improving income that will enhance the ability to access other essential services due to increased opportunities to acquire positive social, psychological and economic resources (Antonovsky, 1967; Susser, 1985; Winkleby *et al.*, 1992). If users understood the economic and income-generating opportunities embedded in sanitation, especially through resource recovery and reuse, they would most likely be more interested in sanitation management even at household levels. This relates to the fact that humans are generally more actively interested in things relating to livelihood support. Studies have shown that using SE factors as incentives can be a good strategy to influence users to adopt dry toilets, especially in most water-scarce regions. This is due to cost-saving from the volume of water that it is no longer necessary to supply, treat, distribute, collect and treat again, as well as lower capital investment costs that this can imply (Anorve, 1994; Cordova & Knuth, 2005; Costner *et al.*, 1990; Del Porto & Steinfeld, 1999; Esrey *et al.*, 1998; Otterpohl *et al.*, 1997; Pollard *et al.*, 1997; Van der Ryn, 1995).

Disadvantages in SE have negative sanitation-related consequences for the society because poor sanitation has been known to cause economic losses associated with the direct costs of treating sanitation-related illnesses and lost income through reduced or lost productivity (Minh & Hung, 2011). Also, better SE conditions imply improved living situations, thus resulting in better access and improved sanitation services (Malik *et al.*, 2012; Sedhain, 2014). This could also explain the worsening sanitation situation in rural areas compared with urban areas due to variations in socioeconomic status and geographical locations (WHO/UNICEF, 2017). Therefore, ReGenSan advocates that SE factors be key drivers in sanitation policy design, development, planning and management as well as infrastructure/technology/service design, development and provision as this could affect allocation of sanitation-based resources, infrastructure provisions and sanitation pricing (Rhodes & McKenzie, 2015).

4.2.1.3 Bio-geo-Chemical (BGC)

The PSEP component of ReGenSan is designed to ensure that the elemental and molecular nutrients cycle inherent in the by-products of human digestion during degradation-stabilization and management processes is not broken. It is the main sanitation treatment process that can be enhanced to mine the elements (nutrients) converted during these cyclical processes for effective and efficient resource recovery as well as support the ecosystem services and functions of the 'place', that is, the SES. BGC draws upon the ReGenSan principles of ApT, NToB, RaSR, SHI, HS and RoDF (see Chapter 2, Section 2.4) to guarantee a system of symbiotic integration of the bio-geo-chemical cycle process within the contextual sanitation system and the ecosystem. BGC will enable appropriate technology that reduces burden transfer, encourages and supports resource recovery/reuse and operates systematically and holistically in any type of combination as well as guarantees performance quality and functionality. When sanitation systems/technology/infrastructure fail, it causes a disruption in the BGC cycling of urine and faeces whereby nutrients rich in nitrogen and phosphorous could travel on to local SES, leading to pollution and degradation of natural resources and aesthetics. To this effect, policymakers and regulators, service providers, technology designers/developers and users alike should have an understanding of how the BGC cycle concerns sanitation, its impact on technology, recovery/reuse, ecosystem protection (public health and water quality goals) and benefits as well as operations and maintenance.

Consequently, the primary concern of this component is the circulation, removal and transformation of organic and inorganic materials derived from by-products of human digestion and related activities (faeces, urine, flush/anal cleansing water and dry cleaning materials) processed in sanitation facilities. It also focuses on the operating system that regulates the physical, chemical and biological processes of the elemental (carbon, nitrogen, phosphorus and water) cycle, which supports ecosystem services and processes as well as fauna and flora diversity in the sani-shed (Reddy & Delaune, 2008). One major aim of this component is to enhance cyclical flows and continuous resource recovery/reuse of energy while ensuring that sanitation operational processes mutually integrate with natural processes to guarantee that waste volume is within the carrying capacity of the SES (Lyle, 1994). BGC cycling processes in sanitation are the result of a few major constituents of by-products of human digestion (faeces/urine) in treatment systems with controlled complex interactions of the oxidation (aerobic) and reduction (anaerobic) reactions of these organic and inorganic constituents of human digestion (Wilhelm *et al.*, 1994) leading to inactivation of pathogens and detoxification of toxic compounds (heavy metals and micropollutants) actively contributing to the global BGC elemental (C, N, P, H_2O) cycling in nature (Reddy & Delaune, 2008). The septic system is used here to illustrate the bio-geo-chemical reaction of domestic waste according to Wilhelm *et al.* (1994).

Wolgast (1993) reported that an average individual excretes around 500 kg of urine and 50 kg of faeces (dry matter content of 20%) each year, with the total nutrient composition of excreta (faeces + urine) as follows: 5.7 kg N, 0.6 kg P and 12 kg K. However, 90% of the tot-N, 60–65% of tot P and 50–80% of K is partitioned by the human body and excreted in the urine (Simha & Ganesapillai, 2017a). In relation to the ReGenSan principle of PaS, Feachem *et al.* (1983) distinguished the faecal production in developing countries (130–520 g person^{-1} d^{-1}) and North America and Europe (100–200 g person^{-1} d^{-1}) and Polprasert *et al.* (1981) estimated urine production from an average Vietnamese person at 0.82–12 kg person^{-1} d^{-1} and faeces as 130–140 g person^{-1} d^{-1}), respectively (Guyton; 1992; Jönsson *et al.*, 2005; Lentner *et al.*, 1981; Schouw *et al.*, 2002; Simha and Ganesapillai, 2017a; Strande *et al.*, 2014; Vinnerås *et al.*, 2006). This indicates that understanding the BGC cycle for sanitation management purposes should be contextual and technology/service delivery must follow suit. Ironically, several studies have explored the subsurface fate of wastewater constituent by constituent, but only a few have considered sanitation treatment infrastructure as BGC reservoirs. It could make more sense to begin to look beyond targeting specific constituents and recognize that sanitation systems (e.g. septic systems) are true BGC systems whereby wastewater constituents react with each other and with subsurface gases and porous media (Reneau *et al.*, 1989; Wilhelm *et al.*, 1994). Box 4.4 presents two major redox environments in which wastewater is transformed by the bio-geo-chemical reactions to their individual nutrient and elemental cycle.

BOX 4.4 IMPORTANT BIO-GEO-CHEMICAL REACTIONS IN SEPTIC SYSTEMS (WILHELM *ET AL.*, 1994)

Anaerobic zone (septic tank and biological mat):

Organic molecule hydrolysis:

$$\begin{aligned}&\text{Proteins} + H_2O \rightarrow \text{Aminoacids}\\&\text{Carbohydrates} + H_2O \rightarrow \text{Simplesugars}\\&\text{Fats} + H_2O \rightarrow \text{Fatty acids and glycerol}\end{aligned} \tag{4.4}$$

Ammonium release:

$$\text{Urea}\left[CO(NH_3^+)_2\right] + H_2O \rightarrow 2NH_4^+ + CO_2 \tag{4.5a}$$

$$\text{Amino acids} + H_2O \rightarrow NH_4^+ + \text{Organic compounds} \tag{4.5b}$$

Fermentation:

$$\text{Amino acids, simple sugars} \rightarrow H_2,\ \text{acetate}\,(CH_3OO^-),\ \text{other organic acids} \tag{4.6}$$

Anaerobic oxidation:

$$\text{Fatty acids} + H_2O \rightarrow H_2, CH_3OO^- \quad (4.7)$$

Sulphate reduction:

$$SO_4^{2-} + 2CH_3O^a + 2H^+ \rightarrow H_2S + CO_2 + 2H_2O \quad (4.8)$$

Methanogenesis:

$$CH_2OO^- (\text{acetate}) + H^+ \rightarrow CH_4 + CO_2 \quad (4.9a)$$

$$CO_2 + 4H_2 \rightarrow CH_4 + 2H_2O \quad (4.9b)$$

Aerobic zone (unsaturated zone, and saturated zone to lesser extent):

Organic matter oxidation:

$$CH_2O + O_2 \rightarrow CO_2 + H_2O \quad (4.10)$$

Nitrification:

$$NH_{4+} + 2O_2 \rightarrow NO_3^- + 2H^+ + H_2O \quad (4.11)$$

Sulphide oxidation:

$$H_2S(\text{or organic sulphide}) + 2O_2 \rightarrow SO_4^{2-} + 2H^+ \quad (4.12)$$

Carbonate buffering:

$$H^+ + HCO_3^- \rightarrow H_2CO_3 \quad (4.13a)$$

$$CaCO_3 + H_+ \rightarrow Ca^{2+} + HCO_3^- \quad (4.13b)$$

$$CaCO_3 + CO_2 + H_2O \rightarrow Ca^{2+} + 2HCO_3^- \quad (4.13c)$$

Second anaerobic zone (saturated or near-saturated conditions)[b]:

Denitrification:

$$4NO_3^- + CH_2O + 4H^+ \rightarrow 2N_2 + 5CO_2 + 7H_2O \quad (4.14)$$

[a]organic matter is simplified as CH_2O throughout. Actual organic matter contains C of various oxidation state and other elements such as N, P, and S, and therefore actual reaction products vary.
[b]Nitrate reduction can also be accomplished via oxidation of reduced sulphur compounds (Howarth & Stewart, 1992).

BGC interactions could improve the quality and performance of the next generation of resource recovery/reuse facilities and technology to boost the sanitation economy. A whole world of research is urgently needed in this area. In essence, BGC processes in urine separation technologies could learn from human physiology (which excretes faeces and urine separately) to design an engineered system that captures nutrients for return to the soil, thereby improving sanitation (Elser & Bennett, 2011; Simha & Ganesapillai, 2017a). In another light, pit latrines (used by an estimated 1.77 billion people concentrated primarily in rural areas of developing- and middle-income countries) pose huge environmental challenges through global methane (CH_4) emissions and groundwater contamination (Graham & Polizzotto, 2013; Reid *et al.*, 2014). Meanwhile, a better understanding of the BGC reactions within pit latrine systems could provide much-needed solutions: for example, pit latrines are estimated to emit ~14 Tg $CH_4\ y^{-16}$, or >4% of global anthropogenic emissions, but mitigations for wastewater treatment technologies do not apply to them (Reid *et al.*, 2014). The limited studies on pit latrines and groundwater only focus on a few indicator contaminants; much is not known about the relevant elemental nutrients and molecular BGC cyclical interactions and their behaviours. Furthermore, although groundwater is frequently observed downstream of latrines, contaminant transport distances, empirically based recommendations and guidelines are varied and not well aligned (Graham & Polizzotto, 2013). The ability to effectively predict the behaviours of BGC constituents' cycles and the changes that take place in the systems can help to boost resource recovery and reuse in ReGenSan.

ReGenSan further advocates that the BGC cycle could play a critical role in buffering the impacts of nutrients and other contaminants through the process of biodegradation, transformation and assimilation and ultimately contribute to ecosystem services and functions as well as fauna and flora diversity. Further studies in this regard are crucial because a well-functioning BGC process can enhance exchange of materials between living and non-living components of the sanitation system, while also supporting systems biodegradation processes, preventing system failures and improving resource recovery/reuse (Wilhelm *et al.*, 1994). Consequently, overlooking the specific BGC operations and mechanisms in different sanitation facilities and contexts could cause 'broken bio-geo-chemical cycles' and could then lead to large release of nutrients from failed sanitation systems to air, soil and water, causing enormous devastation to ecosystem services and functions (Elser & Bennett, 2011).

These major BGC interactions show the behaviour of specific contaminants and the impact of these changes in septic systems (sanitation systems) (Wilhelm *et al.*, 1994).

4.2.2 Governance Function (GoF)

The ReGenSan dimension of governance function (GoF) is predicated on the fact that adequate intervention and service/solution provision in any local context is predominantly dependent on prevailing policy, legislations, regulations and acceptable standards. As acknowledged in the 2006 United Nations World Water Report, the 'global water crisis' and by extension the 'sanitation crisis' is primarily a 'crisis of governance' (UNESCO–WWAP, 2006), a fundamentally ethical challenge that the sector faces in the twenty-first century (Castro & Heller, 2009; Davis & McGinn, 2001; Delli Priscoli *et al.*, 2004; EUWATER Network, 2005). According to the World Bank, governance includes public sector management, accountability, legal framework, transparency and information availability as well as equitability and responsiveness to people's needs (Larmour, 1998). Governance should also be seen as the enforcement of law, accountability, transparency and implementation of government policies (Mwanza, 2005). Without adequate governance backing, sanitation management will be weak and the SDG targets a pipe dream. Governance is the structure and process that society uses to make decisions and share power. It is also the interaction of laws, norms, institutions and processes through which society exercises powers and responsibilities to make and implement decisions and ensure accountability, as well as set the rules under which management operates (Cookey *et al.*, 2016; Folke *et al.*, 2005; Moench *et al.*, 2003; Moore & Unsworth, 2006; Pahl-Wostl, 2009; Pahl-Wostl *et al.*, 2012; Rogers & Hall, 2003; Tropp, 2007). Balancing competing interests about who is entitled to what services, how services are provided, who pays and how competing interests are balanced, as well as decisions about how to recover resources and reuse from sanitation, are governance responsibilities (Moriarty *et al.*, 2007).

The ideal governance takes account of the different actors and networks that help formulate and implement sanitation policies. Therefore, sanitation governance entails the ongoing dialogue between public and private sanitation stakeholders to discern expectations for what results to achieve; translating these expectations including other information as well as perspectives and values of stakeholders into written criteria that is institutions and policies; and checking to see that criteria are met, that is through monitoring (Ekane *et al.*, 2013). GoF of sanitation is multi-level in nature because decision-making and service delivery involve different actors via multi-level processes within which different roles and powers are dispersed among different actors (Ekane *et al.*, 2014; Pahl-Wostl, 1995; Rotmans *et al.*, 2001). In between national policies (at the macro level) and individual households (the micro level) is the meso level, which includes the web of actors ranging from government employees (e.g environment and health inspectors) to private sector formal and informal service providers, civil society organizations and homeowners. The basic functions of ReGenSan governance are presented in Box 4.5.

BOX 4.5 BASIC FUNCTIONS OF ReGenSan GOVERNANCE

(1) Provision of enabling environment, political will and effective management, essential for long-term provision of sanitation services as well as ensuring good administration, planning and decision-making systems (ISF & Stone Environmental, 2009; Yeager *et al.*, 2006),
(2) Provision of institutional and regulatory framework for adequate and appropriate guidelines and standards for sanitation facilities operations and performance, with strict enforcement to ensure compliance (IWA 24:2016; Rodic, 2015),
(3) Institutional arrangements that support technological innovative interventions, operational and maintenance responsibilities of physical systems, make provisions for restoration of dysfunctional and unimproved sanitation technologies (Crous *et al.*, 2012; Eales *et al.*, 2012/2013; Roma & Jeffrey, 2010; Washcost, 2012) as well as technological institutional arrangement that clearly defines roles and responsibilities for service delivery (Montgomery *et al.*, 2009; Oosterveer & Spaargaren, 2010; Parkinson & Tayler, 2003; Roma & Jeffrey, 2010; Sansom, 2011),
(4) Provision of financial arrangements for infrastructural components and cost recovery from service delivery as well as financing operational and maintenance responsibilities (ORs) and institutional arrangements (IAs) (Fonseca *et al.*, 2011; Fonseca & Verhoeven, 2013; IRC & WSUP, 2012; Trémolet & Rama, 2012), and
(5) Psycho-socio-cultural acceptance such as ensuring that prevailing attitudes are compatible and match users' perceptions, preferences and level of commitment with the technologies or practices to be implemented (Lüthi *et al.*, 2011a, b; Tilley *et al.*, 2014).

These multi-level actors, their roles and interactions constitute sanitation governance. The meso-level actors operate in relation to the (macro-level) policies, plan and strategies of the national government and donor agencies. The micro-level actors de facto have much responsibility in realizing sanitation (Ekane *et al.*, 2014). GoF provides the effective and efficient sanitation operational responsibilities (ORs) (i.e. the day-to-day functionality of the service delivery) and fit-for-purpose institutional arrangements (IAs) (i.e. the formal and informal institutional contexts that help or hinder the successful delivery of the day-to-day activities) (Kooiman, 2003; Ross *et al.*, 2014; Starkl *et al.*, 2013).

GoF is a foundational component of ReGenSan in that it provides the enabling environment for effective and efficient service delivery, technology selection, financing and sustainability of sanitation systems and technologies (Maurer *et al.*, 2012; Schertenleib, 2005). This PSEP component draws extensively on the PaS principle as well as the PaP, SHI and NToB principles. It is required to define boundaries of sani-sheds and scale levels as well as set targets for source separation and resource recovery. Generally, sanitation governance encompasses arrangements for ownership, management and operations, institutional contexts and regulatory frameworks (i.e. requirements on service standards, resource recovery, emission standards, pricing guidelines and cost recovery) within 'place' and 'scale' (sani-sheds) of SESs (Ross *et al.*, 2014). It also provides strategic goals, guiding principles, policy instruments, institutional arrangements and capacities, stakeholders' interactions, financial mechanisms and technology selections towards sanitation management (Rodic, 2015). The key components of GoF are presented in Box 4.6.

BOX 4.6 KEY COMPONENTS OF GoF

(1) *Institutions: policies, legislation and regulations, informal institutions, international treaties and protocols*
(2) *Management: sanitation management entities (SMEs), planning and implementation, asset management, clearly defined roles of actors, stakeholders' information and communication, regulatory compliance and enforcement, monitoring and evaluation*
(3) *Sustainable financing: blending public and private funding sources, resource leveraging, cost recovery and lifecycle costing, supplementary income and social marketing, and utilization of local/domestic private sector.*

But, sanitation is really more about service delivery than technology provisions and deals with complex systems of interactions between social (intra and inter), institutions, ecosystems and physical infrastructure – all operating at various governance levels (Galli *et al.*, 2014), therefore, GoF of ReGenSan embodies particularly contextual institutional/policy frameworks (direct regulation, standards and guidelines, supportive economic and social instruments and guidelines for technology selection/sani-shed description and delineation), stakeholders' engagement and involvement, planning and management, operations and maintenance, financing and full cost recovery and compliance and enforcement (Kooiman, 2003; Lüthi *et al.*, 2011a, b; Ross *et al.*, 2014; Starkl *et al.*, 2013; Tilley *et al.*, 2014). For instance, Japan's achievement of universal access and coverage in sanitation thrived on an enabling environment characterized by the enactment of legislative instruments prioritizing sanitation; a robust regulatory regime; establishment of institutional arrangements and coordination mechanisms among stakeholders, where functions and responsibilities are clear-cut and delineated; and initial financing from governments at the forefront of allocating funds for sanitation interventions of varying complexities and scale (ADB, 2016).

4.2.2.1 Institutions

Institutions are rules, constraints and norms of behaviour, conventions and self-imposed codes of conduct and their enforcement characteristics (North, 1990; Obeng & Agyenim, 2011) as well as regularized behaviours that have turned into standard routines (Cookey *et al.*, 2017; Ekane *et al.*, 2012; North, 1990, 1998; Ostrom, 2005). Sanitation institutions are the configurations of various legal, policy and organization elements involved in sanitation technological design and development, ownership, resource allocation, recovery and use as well as management (Saleth, 2004, 2010; Saleth & Dinar, 2004). There are sets of rights, rules and decision-making procedures (Young *et al.*, 1999), which are often referred to as the rules of the game in a society, and the humanly devised constraints that shape human actions (North, 1990). These rules provide information about how agents (actors) are expected to act in certain situations and around rules to which others conform (Amable, 2003; Ekane *et al.*, 2012). Agents/actors (i.e. individuals and organizations) use institutions as instruments for coordination, reducing uncertainty and implementing sanitation programme strategies as well as responses to various challenging situations. Cases abound, where large public and private investments are made in off-site and on-site sanitation solutions, that are seldom backed by adequate institutional and regulatory frameworks at the critical stages of operations, maintenance and monitoring, thereby allowing service deficiencies to persist (Srinivasan, 2015). Thus, sanitation institutions must be relevant to all actors, providing them with an enabling environment and a common understanding of how the tasks are performed (Ekane *et al.*, 2012).

North (1990) separates institutions into two sets of rules or norms: formal or informal. Formal institutions are linked to the official channels of governmental bureaucracies. They are codified in regulatory frameworks or any kind of legally binding documents and they can be enforced by legal procedures; informal institutions refer to socially shared rules such as social or cultural norms. In most cases, they are not codified or written down. They are enforced outside of legally sanctioned channels (Pahl-Wostl, 2009). Government have the legal powers to use formal institutions to adopt, tolerate or outlaw informal laws and/or constraints based on their cultural acceptability, technical feasibility, suitability and practicability. For example, many developing countries have enacted laws to ban the use of the pan or bucket latrine, which has been practised by some communities for several years (See Box 3.2 – Obeng and Agyenim, 2011). In the sanitation sector, informal institutions and constraints are major determinants of the commitments of various stakeholders to the enforcement of and compliance with formal institutions (Vogler, 2003). They influence such critical factors as attitudes, willingness to get and pay for services, compliance with sanitation laws and participation in resource recovery/reuse.

The impact of informal institutions and constraints on the sanitation sector are generally more pronounced in developing countries (Obeng & Agyenim, 2011). For example, in the Burera District, Rwanda, a mismatch was found between the prevailing sanitation behaviour practices and the provisions of the national guidelines and standards for hygiene and sanitation (Ekane *et al.*, 2012; Ekane, 2013 – see Box 4.8). Therefore, it is worth noting that the functionality of institutions depends on the degree of match/fit between the formal and informal institutions (Khan, 1995; Kjellén, 2006) because in many developing societies informal rules have a tendency to override formal rules, making the enforcement of formal rules very difficult and thereby affecting performance (Bandaragoda & Firdousi, 1992). Sanitation institutions have their own institutional structure and environment (Saleth & Dinar, 1999, 2000) and are characterized not only by factors determining the overall institutional environment, but also by those related to the conditions of infrastructural and management solutions, as well as to other sanitation-related sectors such as water, health, agriculture, environment, forest and urban development. The sanitation institutional structure is defined interactively by three institutional components: sanitation law, policy and administration (or sanitation-related organizations) (Ekane *et al.*, 2012, 2013; Saleth & Dinar, 1999). The important role of institutions could best be illustrated in the case of Thailand presented in Box 4.7; in 1897 the country promulgated its first specific sanitation law, which was responsible for the remarkable success of Thailand attaining 91.94% sanitation access coverage as long ago as 1999 (Luong *et al.*, 2000).

The biggest governance challenge of most countries that retard sanitation coverage and service improvement is absence of policy. Elledge *et al.* (2002) described the absence of supportive policies for planning and implementing sanitation projects and programmes as a missing link to improving sanitation coverage. ReGenSan advocates that specific and contextual fit-for-purpose policies, statements, guidelines and standards should make provisions and accommodate all the subsystems of sanitation

BOX 4.7 THAILAND'S FIRST SANITATION LAW

The Bangkok Sanitation Law of 1897 was aimed at curbing the spread of communicable diseases due to poor environmental sanitation. The campaign then targeted garbage collection and building public latrines. A hookworm eradication project initiated in 1918 emphasized using pit latrines. In 1926, the Ministry of the Interior issued regulations banning defecation in rivers and canals. All the ramshackle latrines along the banks of rivers/canals were destroyed. Establishment of the Ministry of Public Health in 1942 led to the integration of Environmental Sanitation into overall health development in the country. The Village Health and Sanitation Project initiated in 1960 to combat the prevalent water/faecal borne diseases gave impetus to expand the sanitation programme nationwide. Rural Environmental Sanitation programmes of the National Health Development Plan were incorporated into successive governments' five-year National Economic and Social Development Plans starting from the first five-year plan in 1961 (Luong *et al.*, 2000).

(SES, TeS and ReS). This could correct the current technocentric institutions responsible for poor performance of sanitation goals and targets. Moving forward will also require modifications to the current institutions of sanitation, which are more focused on micro-level (site and household) and counting of toilets (technology) without adequate attention paid to functionality (Kvarnström *et al.*, 2011).

There will be a need to extend the scope of institutions to cover the meso and macro-levels (sani-sheds) since it has already been established that sanitation solutions cannot be comprehensively provided at household level (Zurbrügg & Tilley, 2009). Thus, extending the scope will cover the neglected aspect of faecal sludge management (Graham & Polizzotto, 2013; Reid *et al.*, 2014) and adequate provisions for resource recovery and reuse for all categories of sanitation infrastructure.

4.2.2.2 Management

The management component refers to the actual operational and institutional responsibilities of organizations to provide sanitation services based on adopted business, legal and sociocultural orientations. The commonly applied approaches are public-sector management with emphasis on the role of local authorities/municipalities, private-sector management and cooperative and community-driven approaches for implementation of sanitation programmes at the local, regional or national level (Castro & Heller, 2009). Currently, sanitation management in most countries is unclear and lacks an organizational home and a champion for investment and service delivery, particularly regarding institutional mandates to decentralize activities that support promoting consumer demands. This lack of a responsible home-base can add to the complexity of the sector-management and resource-allocation processes while limiting organizational accountability for progress (Mehta & Knapp, 2004; WSP, 2011). Presently, sanitation service delivery is overseen by different government ministries, departments and agencies based on their priorities (e.g. health, environmental protection) with administrative decentralization for project implementation at regional and local government levels (Rouse, 2007). This often results in conflicts of the roles and responsibilities of different actors and weak coordination of projects and programmes among different levels of government (Ekane *et al.*, 2014). Different management structures and many layers of policy interpretation usually make it difficult to ensure that clear messages are conveyed, in particular to the local or household levels where the implementation of sanitation mainly occurs (Ekane *et al.*, 2016; Morella *et al.*, 2008). In most parts of the world, responsibilities for sanitation management generally fall on local governments/municipalities, communities and households (Gutterer *et al.*, 2009; Ross *et al.*, 2014), who more often than not have little or no capacity for effective implementation or monitoring of sanitation access expansion and service improvement. Box 4.8 presents sanitation organizational management in Rwanda (Ekane *et al.*, 2016).

BOX 4.8 SANITATION ORGANIZATIONAL MANAGEMENT IN RWANDA

Sanitation main management responsibilities in Rwanda are shared between five ministries, central government service management and regulatory agencies as well as other private-sector and local-level actors. The Ministry of Infrastructure (MININFRA) is responsible for national policies, guidelines and strategies, enhancing human resource capacity at the district level and the monitoring and implementation of government policies. The Ministry of Health (MINISANTE) leads primarily on the promotion of sanitation at the community level and provides preventive, curative and rehabilitative services as well as promoting behaviour change. The Ministry of Education (MINEDUC) is involved in the implementation of hygiene programmes at the local level. The Ministry of Finance and Economic Planning (MINECOFIN) is responsible for budgeting and financing as well as coordination of sanitation external funds and aid and the Ministry of Local Government (MINALOC) is responsible for the coordination of local actors for effective service delivery as well as funding small-scale sanitation projects. The Water and Sanitation Corporation (WASAC), a profit-making company owned by the government, is responsible for managing and regulating water and sanitation services. The districts (local government) and community-based organizations (CBOs) are responsible for providing access to basic sanitation services at the local level. Private-sector actors operate under contracts with the district government and play a major role at the local level. Thus, coordination of actors and other key stakeholders at different levels of the society becomes a major challenge in transforming the sanitation sector and translating policies into practice (Ekane *et al.*, 2016).

Apart from the problem of service coverage, especially in developing countries with low access and poor service delivery, the sanitation sector has long suffered from inadequate management and in most cases total absence of management (Mwanza, 2005; Schwartz, 2008; WHO/UNICEF, 2006). Previous attempts designed to address poor sector performance occasioned by inadequate management was focused on providing 'hardware' because it was thought that if service providers were technically equipped it would improve service delivery.

Unfortunately, the results of the last two decades of sanitation development (Drinking Water Decade, 1981–1990; MDG, 1990–2015) proved disappointing, with almost as many people still without access or improved service delivery (Schwartz, 2008). It has become obvious that providing improved and sustained services is not limited to technical capabilities, but requires more managerial and institutional strengthening because sustainability of sanitation services cannot happen in the absence of credible sanitation management systems (Spiller & Savedoff, 1999; Pahl-Wostl, 2007). This necessitates a shift from project-based approaches to planning, construction and management as well as changes in the role of the government from service provider to policymaker and decentralized service deliverer (Ekane *et al.*, 2016). A lack of a supportive institutional framework that establishes effective and efficient management systems has often resulted in fragmented organizational and conflicting mandates, poor coordination of actors and activities and much emphasis on technological solutions (hardware).

Sanitation management is essential to enable long-term service improvement and access expansion through effective and efficient administration, planning, organizing, directing, controlling and decision-making processes (Grigg, 1986; ISF and Stone Environmental, 2009; Ross *et al.*, 2014; Yeager *et al.*, 2006). Effective management should address all aspects of sanitation subsystems (SES, TeS and ReS) and not focus only on infrastructure, to ensure comprehensive system standardization for cost-effective solutions. It could also be a key to achieving best performance of sanitation facilities, from the simplest 'traditional pit toilets and septic systems' to the most complex treatment and recovery systems (WERF, 2009). This is best illustrated in Box 4.9, on sanitation management in Japan (ADB, 2016).

BOX 4.9 SANITATION MANAGEMENT IN JAPAN

Japan's Ministry of Land, Infrastructure, Transport, and Tourism (MLIT) is responsible for the management of off-site sanitation (sewer systems), while the Ministry of the Environment (MOE) handles on-site sanitation (i.e. Johkasou, septage management with scheduled desludging and sludge treatment facilities) and both ministries are expected to work together. To properly promote sewerage works subsidized by the national government, the Sewerage Law of Japan stipulates structural criteria and standards for effluent quality; guidelines for planning, construction and installation of pre-treatment facilities; household connections; user fees; national government financial support; and the respective roles of national and local governments. In areas where construction of a sewerage system is difficult, the households or small communities, located mostly in rural or peri-urban areas, use an on-site treatment system called Johkasou (a packaged aerated wastewater treatment plant installed in individual houses, buildings or small communities for collection and treatment of night soil and domestic greywater). The Johkasou Law mandates the owner to engage a desludging contractor for the facility at least once a year, with the owner paying the associated charge. The effluent water quality of the Johkasou is monitored by the authorized inspection agency once a year in accordance with the law. The Waste Management and Public Cleaning Law mandates local governments to create household wastewater treatment plans for their municipalities, including sludge disposal plans. Sludge treatment facilities are constructed, operated and maintained by municipalities and cities (ADB, 2016).

Sanitation management in the context of ReGenSan is sets of actions taken to guide the overall sanitation systems towards achieving desired goals and targets for the day-to-day operational and institutional responsibilities of the SMEs; within the context of the strategies, policies, processes and procedures that have been established by governance instruments (Lertzman, 2009; Pahl-Wostl, 2007). Its deals with the implementation of a comprehensive lifecycle series of elements and activities that address the structure of SMEs, public education and participation, planning and financing, facility standardization, construction, operations and maintenance, compliance/enforcement, asset management, monitoring and evaluation as well as establishing distinct roles and responsibilities of actors involved in ensuring proper sanitation system management (USEPA, 2003).

ReGenSan advocates for an integrated hybrid management system that provides opportunities for coordination of multi-technological systems, scales, institutional arrangements and multiple providers of sanitation services in sani-sheds (Letema, 2012). It aligns with the principle of decentralized flexible SMEs' organizational/operational structures with active engagement of stakeholders to decide on the types of business and management models to be adopted in sani-sheds. This approach will ensure that SMEs are structured to conform to the sociocultural norms within sani-sheds. The purpose is to adequately and appropriately integrate existing technological scales and types as well as using different management and institutional arrangements to ensure a synergy for effective and efficient access expansion and service improvement.

4.2.2.3 Sustainable Financing

The sustainable financing component of the GoF dimension provides innovative funding strategies that support sanitation institutional arrangements, technology and resource recovery as well as creating a sanitation economy (Badu *et al.*, 2015; Lerner & Tufano, 2011), with the ultimate goals being to improve services and expand access. The investment needed to generate these benefits is enormous because of sector under-funding that results in huge financing gaps, particularly in developing countries where the challenge of increasing access is substantial (OECD, 2011). Studies have identified major financial challenges impacting the sanitation sector as being inadequate resources for sanitation service improvement and expansion, low or non-existent cost recovery from service delivery and lack of financial sustainability of existing solutions (ADB, 2007; van Dijk *et al.*, 2014). These challenges arise from inappropriate institutional frameworks, ineffective and inefficient financing policies and public resources inadequate to meet the costs of sustained, enhanced coverage and improved service delivery (Badu *et al.*, 2015; World Bank, 2003). Previous experiences have shown that investments tended to concentrate on the construction of infrastructure (capital expenditure) with insufficient attention paid to the systems needed to make sanitation function properly, such as regulations, policies, monitoring, institutions and the human resources that provide and deliver services. For instance, having clear institutional roles and legal frameworks that define assets ownership is a major component of attracting private and public finance. If assets ownership is not legally defined, no comprehensive maintenance and rehabilitation can take place and no loans will be provided due to a lack of guarantee (IRC & Water.org, 2017).

Financial flows in the sanitation sector are derived from a large range of sources including overseas development assistance, loans, grants, international private-sector investments, domestic small-scale private service providers' investments, public-sector expenditures, individual households and community investments. In practice, there are only a few options that are actually available and used to finance sanitation investments. Much funding is required to finance the operational responsibilities and institutional arrangement activities critical for upgrading sanitation services and expanding access (Ross *et al.*, 2014). According to Hutton and Varughese (2016) an estimated figure of USD 68.5 billion per year, exclusive of costs to reach the unserved and sustain the gains, is required to meet sanitation SDG no. 6. Sustainable financing also requires proper understanding of lifecycle perspectives on sanitation costing, a vital key to determining recurrent costs and necessary planning for adequate resources to fund long-term operations (Fonseca *et al.*, 2011; IRC & WSUP, 2012; Trémolet & Rama, 2012).

Lifecycle costs approach (LCCA) is particularly important in clarifying the cost of service delivery to be able to arrive at a unit cost for each socio-economic status of the society (IRC & WSUP, 2012). Adequate financing requires adequate costing because when reoccurring and capital costs including maintenance, capacity-building and policy development are not accounted for in budgets, then systems cannot be properly maintained, replaced or expanded (Badu *et al.*, 2015; Dijk, 2003, 2008). This is fundamental, since existing data on costs often refer to the service as designed, with no exploration of the real costs that people actually pay for real services received (IRC & WSUP, 2012). Determining the real funding gaps and service quality that can be delivered at particular funding levels will help to facilitate the supply of adequate funding to service providers and households for optimum services. Recent World Bank studies to track expenditure in seven infrastructure sectors found that households were actually the largest source of finance in the sanitation sector, ahead of governments and international donors, in sub-Saharan Africa (Banerjee *et al.*, 2011). Sustainable financing is designed to provide a proper mix of funding sources that will deliver the required level of services as well as considering sanitation activities as a value chain where private actors play an important role (WSP, 2004; van Dijk, 2012). These strategies are as follows:

(I) Blending public and private funding sources to finance capital expenditure, repairs, renewal and expansion of infrastructure and service improvement. This mechanism addresses affordability constraints at household levels and expansion constraints for service providers (IRC & Water.org, 2017; OECD, 2011);
(II) Resource leveraging from additional sources such as private/domestic and public-sector participation, local investment through local credit markets (microfinance) and resource enhancement from households and community sources;
(III) Cost recovery through service delivered to finance operational and institutional responsibilities as well as life cycle costing approaches to ensure comprehensive costing of sanitation service delivery;
(IV) Supplementary income and social marketing, especially from the sales of materials recovered from sanitation systems and participation in carbon trading; and
(V) Utilization of local/domestic private businesses to provide a range of services such as latrine construction, emptying and management (Badu *et al.*, 2015; IRC & Water.org, 2017; IRC & WSUP, 2012; OECD, 2011; Ross *et al.*, 2014).

The main objective is to attract funds that would otherwise not be attracted by ensuring that basic public policy goals, such as increasing access and service improvement, are met (OECD, 2011). ReGenSan advocates for the use of mixed financing strategies to stimulate the growth of the sanitation sector and encourage local sanitation private-sector service providers to adopt policies and business models that would enhance affordable service delivery. Public finance and development aid alone will not be sufficient to achieve the ambitious targets of SDG no. 6. The sector requires significantly more funding to ensure that sanitation infrastructure investments are properly maintained to deliver quality services (IRC & Water.org, 2017). ReGenSan supports financing solutions that take a blended approach – strategically and intentionally combining aid, public and private finance – because universal access to sanitation cannot be achieved without increased resources to the sector. Classical examples are presented in Box 4.10.

BOX 4.10 BLENDED FINANCIAL APPROACHES

Bangladesh market-based instrument: National social mobilization campaigns combined with market creation approaches supported the emergence of more than 4,000 private workshops producing over one million latrines per year in response to demand promotion and gradual government withdrawal from direct subsidy and production. This led to an annual household investment of USD 10 million in sanitation improvement (Froehlich, 1999; Mehta & Knapp, 2004; SDC, 2003; Urs & Polak, 2000; WSP-SA, 2000).

Lesotho experiences in scaling up sanitation promotion: The National Sanitation Programme in Lesotho dates back 20 years and shows that determined government leadership, limited subsidies and involvement of the local private sector can lead to large increases in national latrine coverage, from approximately 20% to 53% of the population. It focuses on demand and supply and is basically not subsidized. Outstanding features of the programme are that it is an established item on the government budget independent of external support and has simple and clear financing rules including zero direct subsidies for building individual household latrines. Households directly employ private-sector latrine builders while the government concentrates on promotion and training (Pearson, 2002; Mehta & Knapp, 2004).

Ghana's revolving funds for sanitation: Significant challenge for increased coverage of sanitation in Ghana's rural communities is insufficient up-front cash to pay for latrine construction materials. Previous credit funding schemes for latrines offered by local banks or microfinance banks proved unsuccessful. Simavi and its local partner INTAGRAD developed an innovative loan structure where one-third was meant for the construction of a latrine and the remaining two-thirds for income-generating activities at an affordable interest rate. To ensure that funding was actually spent on latrine construction, the credit was disbursed in three tranches; the second tranche was released only after the latrine was built. To improve the financial viability of the loans, they were provided to community savings groups. The profit from the income generation activities allowed latrine loan repayment. The payback rates were as high as 98% over a period of one and a half years. The credit scheme offered adequate sanitation to 1,400 people with a net investment of USD 1,200 (https://rsr.akvo.org/en/project/4279/#report in IRC and Water.org, 2017).

4.3 SES APPROACH TO REGENSAN SOLUTIONS

Solving global sanitation challenges requires system thinking, partnership and the use of transdisciplinary skills. Transdisciplinary approaches can improve sanitation infrastructural and management solutions by reducing discrepancies and mismatch between the SES entities of ReGenSan. Nevertheless, transdisciplinary approaches are often difficult and challenging to perform in the real world because integration efforts are complex to facilitate due to human resource limitation, institutional biases and lack of adequate networking and communication among SES entities (Virapongse *et al.*, 2016). However, it could be critical to fostering better relationships between social, natural sciences, engineering, technology and governance issues as well as providing complex feed-back systems and dependencies.

SES enables proper understanding of the complex interactions between the users, service providers, institutions, infrastructure and the ecosystem (Figure 4.1). The ReGenSan SES subsystem will help address the 'sanitation system blindness' – which describes the focus on tangible infrastructure without attention to the support systems needed to ensure functional adequacy (IRC & Water.org, 2017). The SES consideration has broadened previous definitions of sanitation problems; this implies that they can always be fixed by technology, which has been shown, in case studies both presented in this book and found everywhere around the globe, to be inadequate.

The interactive interconnectedness of the SES entities is a fundamental trait of ReGenSan; no system can be managed effectively and efficiently by focusing on only a few entities, as no problem exists in isolation. All entities are part of the larger system's interacting networks of social, biogeophysical-chemical, political and economy considerations. This perspective provides better insights into infrastructure as an entity that links users, service providers, institutions and the ecosystem, thus providing an avenue to think about sanitation as a whole system. It proves that a sanitation system is more than a chain of actions that starts with faecal matter and urine with its organic and mineral compounds, flush water, wastewater, washing machines and showers. A sanitation system includes psycho-social-cultural, economic and ecological systems as well as the biogeophysical-chemical processes of converting faecal matter and urine into valuable resources to support livelihood and it also needs supportive institutional arrangements that provide the enabling environment (van Vliet *et al.*, 2010).

Users' engagement is well emphasized in the SES of ReGenSan because the isolation of users from different sanitation initiatives is often the underlying cause of many current challenges in achieving successful, sustained results in service improvement and access expansion. Providing a sanitation framework that integrates both social and biophysical components into one management approach ensures that SES of ReGenSan offers great potential for sustainable sanitation interventions and solutions.

4.4 EXERCISES

(I) How could understanding the SES in ReGenSan improve service and expand access?
(II) How do the SES entities and their interactions affect technology selection?
(III) In what ways can understanding the ReGenSan SES affect the success of sanitation infrastructure provisions and wastewater/faecal sludge management?
(IV) Why is governance key to ReGenSan?
(V) Design a sani-shed with a hybrid sanitation system and proffer adequate and appropriate financial and management strategies.
(VI) Describe the institutional arrangement and organizational system of sanitation in your country.
(VII) What are the stages of BGC reactions in septic systems and their corresponding chemical reactions and equations?

4.5 RECOMMENDATIONS FOR FUTURE RESEARCH

(I) Explore three scenarios in sani-management whereby users are the central focus in a systemic and cyclical relationship.
(II) Explore three scenarios where users are not consulted and infrastructure is the focus.
(III) From the perspective of this SES, how is the ecological systemic worldview preferred over the mechanistic-reductionist worldview?
(IV) Explore the BGC recycling process of human digestion by-products in sanitation systems.
(V) Explore the governance challenges of coordinating different stakeholders and actors across societal levels.
(VI) Explore the challenges of integrated hybrid management system in the coordination of multi-technological systems, types, scales, institutional arrangements and multiple providers of sanitation services.
(VII) Explore and advance the concept of sani-sheds as the most appropriate scale of sanitation management.
(VIII) Explore the appropriate mix of blending various financial sources based on systems, types, scales, management and institutional arrangements.

REFERENCES

ADB (Asian Development Bank) (2007). Smarter Sanitation, How to Clean Up Your Sanitation and Waste Water Mess. Asian Development Bank, Manila.

ADB (Asian Development Bank). (2016). Sanitation and Sustainable Development in Japan. Publication Stock No. RPT168293-2. Asian Development Bank, Manila.

Altaf S. W. (2011). So Much Aid, so Little Development: Stories from Pakistan. Johns Hopkins University Press, Baltimore, USA.

Amable B. (2003). Diversity of Modern Capitalism. Oxford University Press, Oxford, UK.

Andersson M. and Minoia P. (2017). Ecological sanitation: a sustainable goal with local choices. *A case study from Taita Hills, Kenya. African Geographical Review*, **36**(2), 183–199, doi: 10.1080/19376812.2015.1134336

Andries J. M., Janssen M. A. and Ostrom E. (2004). A framework to analyze the robustness of social-ecological systems from an institutional perspective. *Ecology and Society* **9**(1), 18, http://www.ecologyandsociety.org/vol9/iss1/art18/ (accessed 15 March 2018)

Anorve C. (1994). Sociedad Civil y Tecnología Sanitaria Alternativa. El caso del excusado seco ecológico en Morelos, México (Civil Society and Alternative Health Technology. The Case of the Ecological Dry Toilet in Morelos, Mexico). Habitat International Coalition/UNDP, (in Spanish), México DF.

Antonovsky A. (1967). Social class, life expectancy and overall mortality. *The Milbank Memorial Fund Quarterly*, **45**, 31–73.

Badu E., Owusu-Manu D., Donkor-Hyiaman K. A., Conway C. and Edwards D. J. (2015). Innovative Financing Mechanisms for Delivering Sanitation Infrastructure. *Mindanao Journal of Science and Technology*, **13**, 238–258, https://www.researchgate.net/publication/292970897 (accessed 15 January 2018).

Bandaragoda D. J. and Firdousi G. R. (1992). Institutional Factors Affecting Irrigation Performance in Pakistan: Research and Policy Priorities. IMMI Country Paper – Pakistan No. 4. International Irrigation Management Institute, Colombo, Sri Lanka.

Banerjee S. G. and Morella E. (2011). Africa's Water and Sanitation Infrastructure: Access, Affordability, and Alternatives. Directions in Development; Infrastructure. World Bank, Washington D.C., USA. http://documents.worldbank.org/curated/en/712211468202191672/Africas-water-and-sanitation-infrastructure-access-affordability-and-alternatives (accessed 27 May 2017).

Banerjee S. G. and Morella E. (2011). Africa's Water and Sanitation Infrastructure: Access, Affordability, and Alternatives. Directions in Development; Infrastructure. World Bank, Washington DC, USA. http://documents.worldbank.org/curated/en/712211468202191672/Africas-water-and-sanitation-infrastructure-accessaffordability-and-alternatives (accessed 27 May 2017).

Binder C. R., Hinkel J., Bots P. W. G. and Pahl-Wostl C. (2013). Comparison of frameworks for analyzing social-ecological systems. *Ecology and Society*, **18**(4), 26.

Black M. and Fawcett B. (2008). The Last Taboo: Opening the Door on the Global Sanitation Crisis. Earthscan, London, UK; Sterling, USA.

Castro J. E. and Heller L. (2009). Water and Sanitation Services: Public Policy and Management. Earthscan, London, UK.

Chirisa I., Banduko E., Matamanda A. and Mandisvika G. (2017). Decentralised domestic wastewater systems in developing countries: the case study of Harare (Zimbabwe). *Applied Water Science*, **7**(3), 1069–1078. https://doi.org/10.1007/s13201-016-0377-4.

Cookey P. E., Koottatep T., van der Steen P. and Lens P. N. L. (2016). Public health risk assessment tool: strategy to improve public policy framework for onsite wastewater treatment systems (OWTS). *Journal of Water, Sanitation and Hygiene for Development*, **06**(1), 74–88, doi: 10.2166/washdev.2016.081

Cookey P. E., Darnsawasdi R. and Ratanachai C. (2017). Text mining analysis of institutional fit of Lake Basin water governance. *Ecological Indicators*, **72**, 640–658, https://doi.org/10.1016/j.ecolind.2016.08.057 (accessed 18 April 2018).

Cordova A. and Knuth B. A. (2005). User satisfaction in large-scale, urban dry sanitation programs in Mexico. *Urban Water Journal*, **2**(4), 227–243.

Costner P., Gettings H. and Booth G. (1990). We All Live Downstream. A Guide to Waste Treatment that Stops Water Pollution. Waterworks Publishing Company, Eureka Springs, USA.

Crous P., Haarhoff J. and Buckley C. (2012). Can Shared Facilities be Sustainable? Experience from Communal Ablution Blocks in eThekwini, South Africa. 36th WEDC International Conference, Delivering water, Sanitation and Hygiene in an Uncertain Environment, Nakuru, Kenya. https://wedc-knowledge.lboro.ac.uk/resources/conference/36/Crous-1809.pdf (accessed 18 April 2018).

Curtis V. and Biran A. (2001). Dirt, disgust, and disease. Is hygiene in our genes? *Perspect Biol Med.* 2001 Winter, **44** (1), 17–31.

Davies-Colley C. and Smith W. (2012). Implementing environmental technologies in development situations: the example of ecological toilets. *Technology in Society*, **34**(1), 1–8. https://doi.org/10.1016/j.techsoc.2011.002

Davis C. K. and McGinn R. E. (eds) (2001). International Perspectives on Ethical Dilemmas in the Water Industry. American Water Works Association, Denver, USA.

de Albuquerque C. and Roaf V. (2014). Realising the Human Rights to Water and Sanitation: A Handbook by the UN Special Rapporteur. UN Special Rapporteur on the Human Right to Safe Drinking Water and Sanitation, Lisbon, Portugal.

Delli Priscoli J., Dooge J. and Llamas R. (2004). Water and ethics: Overview, in Water and Ethics, United Nations Educational Scientific and Cultural Organization (UNESCO). International Hydrological Programme and World Commission on the Ethics of Scientific Knowledge and Technology, Paris.

Del Porto D. and Steinfeld C. (1999). The Composting Toilet System Book. A Practical Guide to Choosing, Planning and Maintaining Composting Toilet Systems, an Alternative to Sewer and Septic Systems. The Center for Ecological Pollution Prevention (CEPP), Concord, USA.

Devine J. (2009). Introducing SaniFOAM: A Framework to Analyze Sanitation Behaviors to Design Effective Sanitation Programs, Water and Sanitation Program Working Paper, Report 72205, vol. 1, World Bank, Washington D.C., USA http://documents.worldbank.org/curated/en/272351468334778050/Introducing-SaniFOAM-a-framework-to-analyze-sanitation-behaviors-to-design-effective-sanitation-programs (accessed 18 April 2018).

Dias B. D. (2015). Beyond sustainability – biophilic and regenerative design in architecture. *European Scientific Journal*, 1857–7881, http://eujournal.org/index.php/esj/article/view/5385 (accessed 12 October 2017).

Dijk M. P. V. (2003). Liberalization of Drinking Water in Europe and Developing Countries. UNESCO-IHE Institute for Water Education, Delft, Netherlands, ISBN 10: 9073445108/ISBN 13: 9789073445109.

Dijk M. P. V. (2008). Role of small-scale private operators in water and sanitation. *International Journal of Water*, **4** (3/4), 275–290.

Drangert J. and Bahadar N. (2011). A cultural-spatial analysis of excreting, recirculation of human excreta and health – the case of North West Frontier Province, Pakistan. *Health Place*, **17**(1), 57–66.

Duncan O. D. (1961). A socioeconomic index for all occupations. In: Occupation and Social Status, A.J. Reiss, Jr, O.D. Duncan, P.K. Hatt and C.C. North (eds) The Free Press of Glencoe, New York, USA, pp. 109–138.

Duncker L. C., Matsebe G. N. and Moilwa N. (2007). The Social/Cultural Acceptability of Using Human Excreta (Faeces and Urine) for Food Production in Rural Settlements in South Africa, Report No. TT 310/07. Water Research Commission, Pretoria, South Africa. ISBN 978-1-77005-592-6.

Eales K., Siregar R. and Febriani E. (2012). Review of Community Managed Decentralized Wastewater Treatment Systems in Indonesia, Report and Recommendations for the Government of Indonesia, World Bank Water and Sanitation Program https://openknowledge.worldbank.org/handle/10986/17751?locale-attribute=es (accessed 18 April 2018).

Eales K., Siregar R., Febriani E. and Blackett I. (2013). Review of Community Managed Decentralized Wastewater Treatment Systems in Indonesia, Final Report, World Bank Water and Sanitation Program http://documents.worldbank.org/curated/en/181941468260140100/pdf/851920WSP00Box0esia0Technical0Paper.pdf (accessed 18 April 2018).

Egan M., Tannahill C., Petticrew M. and Thomas S. (2008). Psycho-social risk factors in home and community settings and their associations with population health and health inequalities: a systematic meta-review. *BMC Public Health*, **88**:239, https://doi.org/10.1186/1471-2458-8-239 (systematic meta-review. *BMC Public Health* **88**:239, https://doi.org/10.1186/1471-2458-8-239 (accessed 18 April 2018).

Ekane N. (2013). Sanitation Policy and Practice in Rwanda: Tackling the Disconnect. SEI Policy Brief. Stockholm Environment Institute, Stockholm, Sweden. http://www.sei-international.org/publications?pid=2289 (accessed 18 April 2018).

Ekane N., Kjellén M., Noel S. and Fogde M. (2012). Sanitation and Hygiene: Policy, Stated Beliefs and Actual Practice A Case Study in the Burera District, Rwanda. Working Paper 2012-07. Stockholm Environment Institute, Stockholm, Sweden https://www.sei-international.org/mediamanager/documents/Publications/SEI-WorkingPaper-Ekane-SanitationHygieneRwanda.pdf (accessed 18 April 2018).

Ekane N., Kjellén M., Noel S. and Fogde M. (2013). Sanitation Governance viewed through different lenses. https://www.sei-international.org/publications?pid=2344 (accessed 30 December 2017).

Ekane N., Nykvist B., Kjellén M., Noel S. and Weitz N. (2014). Multi-level sanitation governance: understanding and overcoming the challenges in the sanitation sector in Sub-Saharan Africa. Working Paper no. 2014–4. Stockholm Environment Institute, https://www.sei-international.org/mediamanager/documents/Publications/Water-sanitation/sei-workingpaper-ekane-multi-level-sanitation.pdf (accessed 08 May 2017).

Ekane N., Weitz N., Nykvist B., Nordqvist P. and Noel S. (2016). Comparative Assessment of Sanitation and Hygiene Policies and Institutional Frameworks in Rwanda, Uganda and Tanzania. Working Paper 2016-05. Stockholm Environment Institute, Sweden.

Elledge M. F., Rosensweig F., Warner D. B., Austin J. H. and Perez E. A. (2002). Guidelines for the Assessment of National Sanitation Policies, Strategic Report 2, Prepared for the Office of Health, Infectious Diseases and Nutrition, Bureau for Global Health, U.S. Agency for International Development, under EHP Project 26568/CESH.SANITATION. POLTOOL (accessed 31 December 2017).

Elser J. and Bennett E. (2011). Phosphorus cycle: a broken biogeochemical cycle. *Nature*, **478**, 29–31, doi: 10.1038/478029a

Esrey S., Gough J., Rapaport D., Sawyer R., Simpson-Hebert M., Vargas J. and Winblad U. (1998). Ecological Sanitation. Swedish International Development Cooperation Agency (SIDA), Stockholm).

EUWATER Network (2005). European Declaration for a New Water Culture. New Water Culture Foundation, Saragossa.

Feachem R. G., Bradley D. J., Garelick H. and Mara D. D. (1983). Sanitation and Disease: Health Aspects of Excreta and Wastewater Management. John Wiley and Sons, Chichester, UK.

Folke C., Hahn T., Olsson P. and Norberg J. (2005). Adaptive governance of social-ecological systems. *Annual Review of Environment and Resources*, **30**, 441–473. https://doi.org/10.1146/annurev.energy.30.050504.144511.

Folke C., Pritchard Jr L., Berkes F., Colding J. and Svedin U. (2007). The problem of fit between ecosystems and institutions: ten years later. *Ecology and Society*, **12**(1), 30, https://www.ecologyandsociety.org/vol12/iss1/art30/ (accessed 14 October 2017).

Fonseca C., Franceys R., Batchelor C., Mcintyre P., Klutse A., Komives K. and Snehalatha M. (2011). Briefing Note 1a Life-Cycle Costs Approach Costing Sustainable Services. IRC International Water and Sanitation Centre, The Hague, Netherlands. http://www.washcost.info/page/1557 (accessed 18 April 2018).

Fonseca C. and Verhoeven J. (2013). Monitoring the Finance Needed for Sustainable Service Delivery. IRC Symposium 2013, Addis Ababa, Ethiopia. http://www.irc.nl/page/77982 (accessed 18 April 2018).

Froehlich U. (1999). Private Sector – Just a (New) Hope? Report on the 15th AGUASAN Workshop. SKAT. https://www.ircwash.org/sites/default/files/Frolich-1999-Private.pdf (accessed 16 January 2018).

Galli G., Nothomb C. and Baetings E. (2014). Towards Systemic Change in Urban Sanitation. IRC Working Paper, The Hague, Netherlands.

Geels F. (2005). Co-evolution of technology and society: the transition in water supply and personal hygiene in the Netherlands (1850–1930) – a case study in multi-level perspective. *Technology in Society*, **27**, 363–397.

Giné-Garriga R., Flores-Baquero O., de Palencia A. J. and Pérez-Foguet A. (2017). Monitoring sanitation and hygiene in the 2030 agenda for sustainable development: a review through the lens of human rights. *Science of the Total Environment*, **580**, 1108–1119.

Golden S. D. and Earp J. A. L. (2012). Social ecological approaches to individuals and their contexts: twenty years of health education & behavior health promotion interventions. *Health Education & Behavior*, **39**(3), 364–372, doi: 10.1177/1090198111418634

Graham J. P. and Polizzotto M. L. (2013). Pit Latrines and their impacts on groundwater quality: a systematic review. *Environ Health Perspect*, **121**(5), 521–530, doi: 10.1289/ehp.1206028, https://doi.org/10.1289/ehp.1206028 https://www.ncbi.nlm.nih.gov/pubmed/23518813 (accessed 18 January 2018).

Green L. (1970). Manual for scoring socioeconomic status for research on health behavior. *Public Health Report*, **85**, 815–827.

Grigg N. S. (1986). Urban Water Infrastructure: Planning, Management and Operations. John Wiley and Sons, New York, USA.

Grübler A. (1998). Technology and Global Change. Cambridge University Press, Cambridge, UK.

Gutterer B., Sasse L., Panzerbieter T. and Reckerzu gel T. (2009). In: Decentralised Wastewater Treatment Systems (DEWATS) and Sanitation in Developing Countries – A Practical Guide. A. Ulrich, S. Reuter and B. Gutterer (eds), BORDA and WEDC: Water, Engineering and Development Centre (WEDC), Loughborough, UK.

Guyton A. C. (1992). Human Physiology and Mechanisms of Disease. W.B. Saunders Co, Philadelphia, USA.

Halliday A. and Glaser M. (2011). A management perspective on social ecological systems: a generic system model and its application to a case study from Peru. *Human Ecology Review*, **18**(1), 1–18, http://www.humanecologyreview.org/pastissues/her181/halliday.pdf (accessed 12 December 2017).

Hegger D. L. T., van Vliet J. and van Vliet B. J. M. (2007). Niche management and its contribution to regime change: the case of innovation in sanitation. *Technology Analysis and Strategic Management*, **19**, 729–746.

Heierli U. and Polak P. (2000). The Market Creation Approach to Development. Swiss Agency for Development and Co-operation, Division Industry, Vocational Training and Urban Development, Zurich, Switzerland. http://www.poverty.ch/wp-content/uploads/2017/10/poverty.pdf (accessed 16 January 2018).

Hinkel J., Bots P. W. G. and Schluter M. (2014). Enhancing the Ostrom social-ecological system framework through formalization. *Ecology and Society*, doi: 10.5751/ES-06475-190351

Hollingshead A. B. (1971). The indiscriminate state of social class measurement. *Social Forces*, **49**, 563–567.

Howarth R. W., Stewart J.W.B. and Hunter M. V. (1992). The Interactions of Sulphur Cycling on the Continents: Wetlands, Terrestrial Ecosystems, and Associated Waterbodies. Marcel Dekker, Inc., New York, USA, pp. 67–84.

Hutton G. and Varughese M. (2016). The Costs of Meeting the 2030 Sustainable Development Goal Targets on Drinking Water, Sanitation, and Hygiene. Water and Sanitation Program (WSP): Technical paper – 103171. http://documents.worldbank.org/curated/en/415441467988938343/pdf/103171-PUB-Box394556B-PUBLIC-EPI-K8543-ADD-SERIES.pdf (accessed 5 October 2017).

IRC and Water.org (2017). Financing WASH: How to Increase Funds for the Sector While Reducing Inequities. Position paper for the Sanitation and Water for All Finance Ministers Meeting, pp. 23. https://www.ircwash.org/resources/financing-wash-how-increase-funds-sector-while-reducing-inequalities-position-paper (accessed 15 January 2018).

IRC and WSUP. (2012). Financing Water and Sanitation for the Poor: Six Key Solutions. Discussion paper DP#003, https://www.ircwash.org/sites/default/files/Norman-2012-financing.pdf (accessed 15 January 2018).

ISF and Stone Environmental. (2009). Establishing Successful Responsible Management Entities (RMEs) Guidance Fact Sheets. Prepared for Water Environment Research Foundation by the Institute for Sustainable Futures and Stone Environmental Inc., http://www.onsiteconsortium.org/Linked%20files/WERF%20Establishing%20Successful%20RMEs.pdf (accessed 18 April 2018).

IWA. (2016). Non-Sewered Sanitation Systems – General Safety and Performance Requirements for Design and Testing. https://www.iso.org/standard/70604.html (accessed 9 May 2017).

Jain N. (2011). Getting Africa to Meet the Sanitation MDG: Lessons From Rwanda. Water and Sanitation Program Case Study. World Bank, Washington DC. https://www.wsp.org/sites/wsp/files/publications/wsp-rwanda-sanitations-lessons.pdf (accessed 18 April 2018).

Jewitt S. (2011). Geographies of shit: Spatial and temporal variations in attitudes towards human waste. *Progress in Human Geography*, **35**, 608.

Jönsson H., Baky A., Jeppsson U., Hellström D. and Kärrman E. (2005). Composition of Urine, Faeces, Greywater and Biowaste for Utilization in the URWARE Model. Urban Water Report 2005: 6. Chalmers University of Technology, Gothenburg, Sweden. ISSN 1650-3791.

Kar K. and Chambers R. (2008). Handbook on Community-Led Total Sanitation. Plan International (UK), London.

Kellert S. R., Heerwagen J. and Mador M. (2008). Biophilic Design: The Theory, Science and Practice of Bringing Buildings to Life. John Wiley and Sons, New York, USA.

Khan M. (1995). State failure in weak states: A critique of new institutionalist explanations. In: The New Institutional Economics and Third World Development, J. Harriss, J. Hunter and C.M. Lewis (eds), Routledge, London, UK, pp. 71–86.

Kjellén M. (2006). From Public Pipes to Private Hands: Water Access and Distribution in Dar es Salaam, Tanzania. PhD thesis. Stockholm University, Stockholm, Sweden.

Kooiman J. (2003). Governing as Governance. Sage Publications, London, UK.

Kvarnström E., McConville J., Bracken P., Johansson M. and Fogde M. (2011). The sanitation ladder – a need for a Revamp? *Journal of Water Sanitation and Hygiene for Development*, **1**(1), 3–12.

Larmour P. (1998). State Society and Governance in Melanesia: Making Sense of GOOD GOVERNANCE. Australian National University, Research School of Pacific and Asian Studies, Canberra, Australia.

Lentner C. and Wink A. (1981). Units of Measurement, Body Fluids, Composition of the Body, Nutrition. Geigy Scientific Tables. Ciba-Geigy, Basel, Switzerland.

Lerner J. and Tufano P. (2011). The consequences of financial innovation: a counterfactual research agenda. *Annual Review of Financial Economics*, **3**(1). https://doi.org/10.1146/annurev.financial.050808.114326. https://www.researchgate.net/publication/228291432_The_Consequences_of_Financial_Innovation_A_Counterfactual_Research_Agenda. Accessed 08 January 2018.

Lertzman K. (2009). The paradigm of management, management systems, and resource stewardship. *Journal of Ethnobiology*, **29**(2), 339–358, http://rem-main.rem.sfu.ca/forestry/publications/PDFs.of.Papers/Lertzman09 ParadigmofManagement.pdf (accessed 5 January 2018).

Letema S.C. (2012). Assessing Sanitary Mixtures in East African Cities. PhD thesis, Wageningen University, Wageningen, Netherlands, https://doi.org/10.3920/978-90-8686-769-1 (accessed 08 May 2017).

Liehr S., Röhrig J., Mehring M. and Kluge T. (2017). How the social-ecological systems concept can guide transdisciplinary research and implementation: addressing water challenges in Central Northern Namibia. *Sustainability*, **9**, 1109.

Loudon J. B. (1977). On body products. In: The Anthropology of the Body, J. Blacking (ed.), A.S.A. Monographic Press, London, UK.

Luong T. V., Chanacharnmongkol O. and Thatsanatheb T. (2000). Universal Sanitation – Thailand Experiences. 26th WEDC Conference, Dhaka, Bangladesh. https://wedc-knowledge.lboro.ac.uk/resources/conference/26/Luong.pdf (accessed 31 December 2017).

Lüthi C., Morel A., Tilley E. and Ulrich L. (2011a). Community-Led Urban Environmental Sanitation Planning (CLUES). Swiss Federal Institute of Aquatic Science and Technology (EAWAG), Dübendorf, Switzerland, http://www.susana.org/_resources/documents/default/2-1300-cluesguid.pdf (accessed 14 May 2017).

Lüthi C., Panesar A., Schütze T., Norström A., McConville J., Parkinson J., Saywell D. and Ingle R. (2011b). Sustainable Sanitation in Cities: A Framework for Action. Papiroz Publishing House, Rijswijk, Netherlands.

Lyle J. T. (1984). Designing Human Ecosystems. John Wiley and Sons, Hoboken, USA.

Lyle J. T. (1994). Regenerative Design for Sustainable Development. John Wiley and Sons, Hoboken, USA.

Malik A., Yasar A., Tabindha A. B. and Abubakar M. (2014). Waterborne diseases, cost of illness and willingness to pay for diseases interventions in rural communities of developing countries. *Iranian J Publ Health*, **41**(6), 39–49.

Mang P. and Reed B. (2012a). Designing from place: a regenerative framework and methodology. *Build. Res. Inform,*, **40**(1), 23–38.

Mang P. and Reed B. (2012b). Chapter 3. Regenerative development and design: Regenesis Group and Story of Place Institute. In: Encyclopedia Sustainability Science and Technology, R. A. Meyers (ed.), Springer, New York, NY, USA, pp. 8855–8879. https://doi.org/10.1007/978-1-4419-0851-3.

Marshall R. E. and Farahbakhsh K. (2013). Systems approaches to integrated solid waste management in developing countries. *Waste Management*, **33**, 988–1003, http://dx.doi.org/10.1016/j.wasman.2012.12.023 (accessed 15 March 2018).

Maurer M., Bufardi A., Tilley E., Zurbrügg C. and Truffer B. (2012). A compatibility-based procedure designed to generate potential sanitation system alternatives. *Journal of Environmental Management*, **104**, 51–61.

Mehta M. and Knapp A. (2004). The Challenge of Financing Sanitation for Meeting The Millennium Development Goals. This Paper was Commissioned as a Think Piece by the Norwegian Ministry of the Environment, for the 12th Session of the United Nations Commission on Sustainable Development (CSD-12), April 14–30 2004, New York, 35pp, http://www.wsp.org/sites/wsp.org/files/publications/af_finsan_mdg.pdf (accessed 10 January 2018).

Minh H. V. and Hung N. V. (2011). Economic aspects of sanitation in developing countries. *Environmental Health Insights*, **5**, 63–70, doi: 10.4137/EHI.S8199, https://www.ncbi.nlm.nih.gov/pmc/articles/PMC3212862/pdf/ehi-5-2011-063.pdf (accessed 19 December 2017).

Moench M., Dixit A., Janakarajan M., Rathore S. and Mudrakartha S. (2003). The fluid mosaic, water governance in the context of variability, uncertainty and change: A Synthesis Paper. Nepal Water Conservation Foundation, Kathmandu, and the Institute for Social and Environmental Transition. Boulder, USA, 71pp, https://www.issuelab.org/resources/23415/23415.pdf (accessed 18 April 2018).

Montgomery M. A., Bartram J. and Elimelech M. (2009). Increasing functional sustainability of water and sanitation supplies in rural Sub-Saharan Africa. *Environmental Engineering Science*, **26**(5), 1017–1023, doi: 10.1089/ees.2008.0388

Moore M. and Unsworth S. (2006). Britain's new white paper: making governance work for the poor. *Development Policy Review*, **24**(6), 707–715.

Morella E., Foster V. and Ghosh Banerjee S. (2008). Climbing the Ladder: The State of Sanitation in Sub-Saharan Africa. Background paper 13 (Phase 1). Africa Infrastructure Country Diagnostic. Ministry of Health/WSP (2004). Strengthening budget mechanisms for sanitation in Uganda, http://www.wsp.org/sites/wsp.org/files/publications/af_strengtheningbudget_execsum.pdf (accessed 18/April 2018).

Moriarty P., Batchelor C., Laban P. and Fahmy H. (2007). The EMPOWERS Approach to Water Governance: Background and Key Concepts, 40p. INWRDAM on behalf of the EMPOWERS Partnership, Amman, Jordan. https://www.ircwash.org/sites/default/files/Moriarty-2007-EMPOWERS.pdf (accessed 18 April 2018).

Mueller C. W. and Parcel T. L. (1981). Measures of socioeconomic status: alternatives and recommendations. *Child Development*, **52**, 13–30.

Mwanza D. (2005). Promoting good governance through regulatory frameworks in African water utilities. *Water Science and Technology*, **51**(8), 71–79.

Nilsson M. and Olsson L. (2014). Sanitation in an informal settlement – A user-oriented study on sanitation solutions and factors influencing decisions for implementation in Cochabamba, Bolivia. MS thesis, Chalmers University of Technology, Gothenburg, Sweden, http://publications.lib.chalmers.se/records/fulltext/207030/207030.pdf

North D. C. (1990). Institutions, Institutional Change and Economic Performance. Cambridge University Press, Cambridge, UK.

North D.C. (1998). Five Propositions about institutional change. In: Explaining Social Institutions, J. Knight and I. Sened (eds), University of Michigan Press, Ann Arbor, USA, 12 pp.

Obeng P.A. and Agyenim J. B. (2011). Institutional matrix for sustainable waste management. In: Integrated Waste Management vol. I, S. Kumar (ed.) doi:10.5772/16423, https://www.intechopen.com/books/integrated-waste-management-volume-i/institutional-matrix-for-sustainable-waste-management (accessed 30 December 2017).

OECD (2011). Meeting the Challenge of Financing Water and Sanitation: Tools and Approaches. OECD Studies on Water, OECD Publishing. http://dx.doi.org/10.1787/9789264120525-en (accessed 15 January 2018.

Oosterveer P. and Spaargaren G. (2010). Meeting Social Challenges in Developing Sustainable Environmental Infrastructures in East African Cities. In: Social Perspectives on the Sanitation Challenge, B. van Vliet, G. Spaargaren and P. Oosterveer, (eds), Springer Science+Business Media B.V., Netherlands doi:10.1007/978-90-481-3721-3_2

O'Reilly K. and Louis E. (2014). The toilet tripod: understanding successful sanitation in rural India. *Health & Place*, **29**, 43–51.

Ostrom E. (2005). Understanding Institutional Diversity. Princeton University Press, Princeton, USA.

Ostrom E. (2009). A general framework for analyzing sustainability of social-ecological systems. *Science*, **325**(5939), 419–422. https://doi.org/10.1126/science.1172133.

Ostrom E. and Cox M. (2010). Moving beyond panaceas: a multi-tiered diagnostic approach for social-ecological analysis. *Environmental Conservation*, **37**, 451–463, doi: 10.1017/S0376892910000834

Otterpohl R., Grottker M. and Lange J. (1997). Sustainable water and waste management in urban areas. *Water Science Tech.*, **35**, 121–133.

Pahl-Wostl C. (1995). The Dynamic Nature of Ecosystems: Chaos and Order Entwined. Wiley, Chichester, p. 288.

Pahl-Wostl C. (2007). Transitions towards adaptive management of water facing climate and global change. *Water Resour Manage*, **21**, 49–62, doi: 10.1007/s11269-006-9040-4

Pahl-Wostl C. (2009). A conceptual framework for analysing adaptive capacity and multi-level learning processes in resource governance regimes. *Global Environmental Change*, **19**(3), 354–365, doi: 10.1016/j.gloenvcha.2009.06.001

Pahl-Wostl C., Lebel L., Knieper C. and Nikitina E. (2012). From applying panaceas to mastering complexity: toward adaptive water governance in river basins. *Environmental Science & Policy*, **23**, 24–34, http://dx.doi.org/10.1016/j.envsci.2012.07.014

Parkinson J. and Tayler K. (2003). Decentralized wastewater management in peri-urban areas in low-income countries. *Environment and Urbanization*, **15**(1), 75–90, doi: 10.1177/095624780301500119

Partelow S. (2016). Coevolving Ostrom's social–ecological systems (SES) framework and sustainability science: four key co-benefits. *Sustain Sci.*, **11**, 399–410, doi: 10.1007/s11625-015-0351-3

Pearson I. (2002). The National Sanitation Programme in Lesotho: How Political Leadership Achieved Long-Term Results. WSP-AF. Blue Gold series. 5pp, http://www.wsp.org/sites/wsp.org/files/publications/af_bg_lesotho.pdf (accessed 16 January 2018).

Pinsem W. and Vinnerås B. (2003). Composting with Human Urine: Plant Fertilizer Approach. Paper Presented at the 2nd International Symposium on Ecological Sanitation, Lübeck, Germany, April 2003.

Pollard R., Kohlenberg A. and Davison L. (1997). Effectiveness and User Acceptance of Composting Toilet Technology in Lismore, NSW. Proceedings of Environmental Technologies for Wastewater Management UNEP Conference, Murdoch University, Perth, WA, Murdoch University, Perth, Australia, 4–5 December.

Polprasert C., Lohani B. N. and Chan C. B. (1981). Human Faeces, Urine and Their Utilization. Asian Institute of Technology, Bangkok, Thailand (as cited in Schouw, 2002).

Reddy K. R. and DeLaune R. D. (2008). Biogeochemistry of Wetlands: Science and Applications. CRC Press, Taylor & Francis Group, New York, USA.

Reid M. C., Guan K., Wagner F. and Mauzerall D. L. (2014). Global methane emissions from pit latrines. *Environmental Science & Technology*, **48**, 8727–8734, doi: 10.1021/es501549h

Reneau S. L., Dietrich W. E., Rubin M., Donahue D. J. and Jull A. J. T. (1989). Analysis of Hillslope erosion rates using dated colluvial deposits. *Journal of Geology*, **97**(1), 45–63. doi:10.1086/629280.

Rheinländer T., Samuelsen H., Dalsgaard A. and Konradsen F. (2010). Hygiene and sanitation among ethnic minorities in Northern Vietnam: does government promotion match community priorities? *Social Science & Medicine* , **71**(5), 994–1001.

Rhodes B. and McKenzie T. (2015). To what extent does socio-economic status still affect household access to water and sanitation services in South Africa?. *Journal of Economic and Financial Sciences*, **11**(1), 1–9 doi:10.4102/jef.v11i1.173, https://hdl.handle.net/10520/EJC-10c2e15a74 (accessed 18 April 2018).

Rodic L. (2015). Waste Governance. In: Global Waste Outlook, David C. Wilson (ed.), ISWA/UNEP, UK, Blackwell Publishing Ltd., http://www.unep.org/ietc/Portals/136/Publications/Waste%20Management/GWMO%20report/GWMO%20full%20report.pdf (accessed 12 September 2016).

Rogers P. and Hall A.W. (2003). Effective Water Governance. TEC Background papers no. 7, Global Water Partnership, Technical Committee, Stockholm, Sweden.

Roma E. and Jeffrey P. (2010). Evaluation of community participation in the implementation of community-based sanitation systems: a case study from Indonesia. *Water Science and Technology: A Journal of the International Association on Water Pollution Research*, **62**(5), 1028–36, doi: 10.2166/wst.2010.344

Rosenqvist L. E. D. (2005). A psychosocial analysis of the human-sanitation nexus. *Journal of Environmental Psychology*, **25**, 335–346, doi: 10.1016/j.jenvp.2005.07.003

Ross K., Abeysuriya K., Mikhailovich N. and Mitchell C. (2014). Governance for decentralised sanitation: Global Practice Scan. A working document. Prepared by the Institute for Sustainable Futures, UTS as part of the Australian Development Research Award Scheme (ADRAS), http://communitysanitationgovernance.info/wp-content/uploads/2016/06/20140125-ADRAS-GDS-Global-Practice-Scan-a.pdf (accessed 14 October 2017).

Rotmans J., Kemp R. and van Asselt M. (2001). More evolution than revolution: transition management in public policy. *Foresight*, **3**(1), 15–31. https://doi.org/10.1108/14636680110803003.

Rouse M. J. (2007). Institutional Governance and Regulation of Water Services. IWA Publishing, London, UK.

Sahoo K. C., Hulland K.R.S., Caruso B. A., Swain R, Freeman M.C., Panigrahi P. and Dreibelbis R. (2015). Sanitation-related psychosocial stress: a grounded theory study of women across the life-course in Odisha, India. *Social Science & Medicine*, **139**, 80–89. https://doi.org/10.1016/j.socscimed.2015.06.031.

Saleth R. M. (2004). Strategic Analysis of Water Institutions: An Application of a NewParadigm for India, Research Report 97, International Water Management Institute, Colombo, Sri Lanka.

Saleth R. M. (2010). Institutional Response as an Adaptation to Water Scarcity. Madras Institute of Development Studies, 79-Second Main Road Gandhi Nagar Chennai-600020 India.

Saleth R. M. and Dinar A. (1999). Water Challenge and Institutional Response: A Cross-Country Perspective, World Bank Policy Research. Working paper no. 2045, World Bank, Washington D.C., USA.

Saleth R. M. and Dinar A. (2000). Institutional changes in global water sector: trends, patterns and implications. *Water Policy*, **2**(3), 175–199.

Saleth R. M. and Dinar A. (2004). The Institutional Economics of Water: A Cross-Country Analysis of Institutions and Performance. Edward Elgar, Cheltenham, UK.

Sansom K. (2011). Complementary roles? NGO-government relations for community-based sanitation in South Asia. *Public Administration and Development*, **31**, 282–293, doi: 10.1002/pad

Schertenleib R. (2005). From conventional to advanced environmental sanitation. *Water Science and Technology*, **51** (10), 7–14.

Schouw N. L., Danteravanich S., Mosbaek H. and Tjella J. C. (2002). Composition of human excreta - a case study from Southern Thailand. *The Science of the Total Environment*, **286**, 155–166.

Schübeler P. (1996). Conceptual framework for municipal solid waste management in low-income countries. In: UNDP/UNCHS/World Bank/SDC Collaborative Programme on Municipal Solid Waste Management in Low-Income Countries, K. Wehrle and J. Christen (eds), St. Gallen, Switzerland.

Schwartz K. (2008). The new public management: the future for reforms in the African water supply and sanitation sector? *Utilities Policy*, **16**, 49–58, doi: 10.1016/j.jup.2007.07.001

SDC (Swiss Agency for Development Co-operation). (2003). Water and Development: Experiences of the Swiss Agency for Development Cooperation (SDC) in the Water Sector. Bern, Switzerland.

Sedhain P. (2014). Water, Sanitation, Socioeconomic Status and Prevalence of Waterborne Diseases: A Cross Sectional Study at Makwanpur District, Nepal. Master's thesis, UiT The Arctic University of Norway, https://munin.uit.no/handle/10037/6503 (accessed 08 May 2017).

Seghezzo L. (2004). Anaerobic Treatment of Domestic Wastewater in Subtropical Regions. PhD thesis, University of Wageningen, Netherlands, https://www.researchgate.net/publication/34736397_Anaerobic_treatment_of_domestic_wastewater_in_subtropical_regions (accessed 08 May 2017).

Siegrist J. and Marmot M. (2004). Health inequalities and the psychosocial environment – two scientific challenges. *Social Science & Medicine*, **58**, 1463–1473, doi: 10.1016/S0277-9536(03)00349-6

Simha P. and Ganesapillai M. (2017a). Ecological Sanitation and nutrient recovery from human urine: how far have we come? A review. *Sustainable Environment Research*, **27**, 107–116, doi: 10.1016/j.serj.2016.12.001

Simha P., Lalander C., Vinnerås B. and Ganesapillai M. (2017b). Farmer attitudes and perceptions to the reuse of fertiliser products from resource-oriented sanitation systems – The case of Vellore, South India. *Science of the Total Environment*, **581–582**, 885–896.

Spaargaren G., Mommaas H., Van Den Burg S., Maas L., Drissen E. and Dagevos H. (2007). More Sustainable Lifestyles and Consumption Patterns: A Theoretical Perspective for the Analysis of Transition Processes within Consumption Domains, Contrast Research Report, TMP project. Wageningen: Environmental Policy Group, Wageningen University; Telos: Tilburg University; RIVM; LEI, Wageningen, Netherlands.

Spiller P. and Savedoff W. (1999). Chapter 1: Government opportunism and the provision of water. In: Spilled Water: Institutional Commitment in the Provision of Water Services, Spiller P. and Savedoff W. (eds), Inter-American Development Bank, Washington D.C., USA. pp 1–34. ISBN: 1886938563, see https://publications.iadb.org/handle/11319/331 (accessed 20 September 2017).

Sridhar M. K. C., Coker A. O., Adeoye G. O. and Akinjogbin I. O. (2005). Urine Harvesting and Utilization for Cultivation of Selected Crops: Trials from Ibadan, South West Nigeria. Paper presented at the 3rd International Ecological Sanitation Conference, Durban, South Africa, 23–26 May 2005.

Srinivasan S. (2015). Performance-based contracting in sanitation delivery. IFRM LEAD, http://www.susana.org/_resources/documents/default/3-2284-22-1437031496.pdf (accessed 4 January 2018).

Starkl M., Brunner N. and Stenstrom T. (2013). Why do water and sanitation systems for the poor still failPolicy analysis in economically advanced developing countries. *Environmental Science and Technology*, **47**(12), 6102–6110, doi: 10.1021/es3048416

Stokols D. (1992). Establishing and maintaining healthy environments: toward a social ecology of health promotion. *American Psychologist*, **47**(1), 6–22.

Stokols D. S. (1996). Translating social ecological theory into guidelines for community health promotion. *American Journal of Health Promotion*, **10**(4), 282–298. https://doi.org/10.4278/0890-1171-10.4.282.

Strande L., Ronteltap M. and Brdjanovic D. (eds) (2014). Faecal Sludge Management: Systems Approach for Implementation and Operation. IWA Publishing, London, UK.

Susser M.W. (1985). Watson W, Hopper K. Sociology in Medicine. Oxford University Press, New York, USA.

Tilley E., Strande L., Lüthi C., Mosler H., Udert K. M., Gebauer H. and Hering J. G. (2014). Looking beyond technology: an integrated apparoach to water, sanitation and hygiene in low income countries. *Environmental Science and Technology*, **48**(17), 9965–9970, doi: 10.1021/es501645d

Trémolet S. and Rama M. (2012). Tracking national financial flows into sanitation, hygiene and drinking water. World Health Organization. ISBN: 978-92-4-151356-2. Licence: CC BY-NC-SA 3.0 IGO.

Tropp H. (2007). Water governance: trends and needs for new capacity development. *Water Policy*, **9**(Supplement 2), 19–30, doi: 10.2166/wp.2007.137

UNESCO–WWAP (United Nations Educational Scientific and Cultural Organization–World Water Assessment Programme) (2006). Water, a Shared Responsibility: The United Nations World Water Report 2. UNESCO and Berghahn Books, Paris, New York.

United Nations Economic and Social Council. (2002). The Right to Water. General Comment No. 15. Resolution E/C.12/2002/11.

United Nations General Assembly. (2009). Report of the Independent Expert on the Issue of Human Rights Obligations Related to Access to Safe Drinking Water and Sanitation, Catarina de Albuquerque. Resolution A/HRC/12/24.

United Nations Human Rights Council. (2011). The Human Right to Safe Drinking Water and Sanitation. Resolution A/HRC/RES/16/2.

USEPA. (2003). Voluntary National Guidelines for Management of Onsite and Clustered (Decentralized) Wastewater Treatment Systems. EPA 832-B-03-001. Office of Water, Office of Research and Development U.S. Environmental Protection Agency, Cincinnati, USA, https://nepis.epa.gov/Exe/ZyPDF.cgi/20009NAM.PDF?Dockey=20009NAM.PDF (accessed 18 January 2018).

Van der Ryn S. (1995). The Toilet Papers. Recycling Waste and Conserving Water. Ecological Design Press, Sausalito, USA.

Van der Ryn S. and Cowan S. (2007). Ecological Design, 10th anv. ed. Island Press, Washington D.C., USA.

Van Dijk M. P. (2012). Sanitation in developing countries: innovative solutions in a value chain framework. In: H. Sun (ed.), Management of Technological Innovation in Developing and Developed Countries, Intech, Rijkeka, Netherlands, pp. 65–83.

van Dijk M. P., Etajak S., Mwalwega B. and Ssempebwa J. (2014). Financing sanitation and cost recovery in the slums of Dar es Salaam and Kampala. *Habitat International*, **43**, 206–213, http://dx.doi.org/10.1016/j.habitatint.2014.02.003

Van Vliet B., Spaargaren G. and Oosterveer P. (eds). (2010). Social Perspectives on the Sanitation Challenge. Springer, Dordrecht, Netherlands.

van Vliet B. J. M., Spaargaren G. and Oosterveer P. (2011). Sanitation under challenge: contributions from the social sciences. *Water Policy*, **13**, 797–809.

Vinnerås B., Palmquist H., Balmér P., Weglin J., Jensen A., Andersson Å and Jönsson H. (2006). The characteristics of household wastewater and biodegradable waste – a proposal for new Swedish norms. *Urban Water*, **3**, 3–11.

Virapongse A., Brooks S., Metcalf E. C., Zedalis M., Gosz J., Kliskey A. and Alessa L. (2016). A social-ecological systems approach for environmental management. *Journal of Environmental Management*, **178**, 83–91, http://dx.doi.org/10.1016/j.jenvman.2016.02.028

Vogler J. (2003). Taking institutions seriously: how regime analysis can be relevant to multilevel environmental governance. *Global Environmental Politics*, **3**(2), 25–39.

Washcost (2012). Providing a Basic Level of Water and Sanitation Services That Last: Cost Benchmarks. WashCost Infosheet 1. IRC, The Hague, Netherlands. https://www.ircwash.org/sites/default/files/IRC-2012-Providing.pdf (accessed 18 April 2018).

WECD (World Commission on Environment and Development) (1987). Our Common Future. Oxford University Press, Oxford, UK.

WERF (Water Environment Research Foundation) (2009). Establishing successful RMEs: Guide to the fact sheets. http://www.werf.org/i/c/KnowledgeAreas/DecentralizedSystems/RMEsite/RMEs_2.aspx (accessed 5 January 2018).

Whittington D., Lauria D. and Choe K. (1993). Household sanitation in Kumasi, Ghana: a description of current practices, attitudes, and perceptions. *World Development*, **21**(5), 733–748. https://doi.org/10.1016/0305-750X(93)90030-D.

WHO/UNICEF (2006). WHO/UNICEF Joint Monitoring Programme for Water Supply and Sanitation. Water for Life: Making it Happen. World Health Organization and United Nations Children's Fund, Geneva and New York.

WHO/UNICEF (2017). Progress on Drinking Water, Sanitation and Hygiene: 2017 Update and SDG Baselines. World Health Organization (WHO) and the United Nations Children's Fund (UNICEF), Geneva, Licence: CC BY-NC-SA 3.0 IGO.

Wilhelm S. R., Schiff S. L. and Cherry J. A. (1994). Biogeochemical evolution of domestic wastewater in septic systems: 1. *Conceptual model. Groundwater Groundwater*, **34**(6), 853–64, https://doi.org/10.1111/j.1745-6584.1996.tb02080.x

Winblad U. and Kilama E. rev.edn. (1985). Sanitation without Water, 159 pp Macmillan. Printed in Hong Kong.

Winblad U. and Simpson-Hebert M. (2004). Ecological sanitation – Revised and enlarged edition, 2nd edn. Stockholm Environmental Institute, Sweden. ISBN 91-88714-98-5.

Winkleby M. A., Jatulis D. E., Frank E. and Fortmann S. P. (1992). Socioeconomic status and health: how education, income, and occupation contribute to risk factors for cardiovascular disease. *American Journal of Public Health*, **82**(6), 816–820, https://www.ncbi.nlm.nih.gov/pmc/articles/PMC1694190/pdf/amjph00543-0042.pdf (accessed 19 December 2017).

Wolgast M. (1993). Rena vatten. Om tankar i kretslopp (Clean water. About thoughts in circuits), Crenom HB, Uppsala, Sweden, 186 pp.

World Bank. (2003). Meeting the Financing Challenge for Water Supply and Sanitation: Incentives to Promote Reforms, Leverage Resources, and Improve Targeting, Report for Water and Sanitation Program, http://www.wsp.org/sites/wsp.org/files/publications/46200735636_MeetingFinancingChallenge (accessed on 20 February 2015).

WSP. (2004). Sanitation is a Business, Approaches for Demand-Oriented Policies. Water and Sanitation Program, Nairobi, 38pp. Published by Swiss Agency for Development and Cooperation (SDC), Bern, Switzerland https://www.ircwash.org/sites/default/files/Heierli-2004-Sanitation.pdf (accessed 18 April 2018).

WSP (2011). The Political Economy of Sanitation: How can we increase investment and improve service for the poor? https://openknowledge.worldbank.org/bitstream/handle/10986/17276/599330V20WSP0R10Box358311B01PUBLIC1.pdf?sequence=1&isAllowed=y (accessed 8 January 2018).

WSP-SA (Water and Sanitation Program, South Asia). (2000). New Approaches to Promoting Sanitation in Rural Bangladesh. Field Note, https://www.wsp.org/sites/wsp.org/files/publications/sa_approach.pdf (accessed 16 January 2018).

Yeager T., Ehrhard R. and Murphy J. (2006). Business Attributes of Successful Responsible Management Entities. Water Environment Research Foundation (WERF), http://www.ndwrcdp.org/research_project_04-DEC-4SG.asp (accessed 18 April 2018).

Young O. R., Agrawal A., King L. A., Sand P. H., Underdal A. and Wasson M. (1999). Institutional Dimensions of Global Environmental Change (IDGEC), Science Plan. International Human Dimensions Programme Report no. 9. IHDP, Bonn, Germany.

Zurbrügg C. and Tilley E. (2009). A system perspective in sanitation – Human waste from cradle to grave and reincarnation. *Desalination*, **248**, 410–417.

Chapter 5

Technological system

'We are stuck with technology when what we really want is just stuff that works'

Douglas Adams

5.1 INTRODUCTION

ReGenSan technology system (TeS) refers to a functional and standardized infrastructure across the sanitation service chain (SSC) that is seamlessly networked (i.e. conventional/non-conventional sewerage) and loosely networked (i.e. non-sewered systems), designed to deliver reliable safely-managed sanitation infrastructure aimed at access expansion and service improvement (Kaminsky & Javernick-Will, 2015; Kvarnström *et al.*, 2011; Markard, 2011; Thomas *et al.*, 2018; WHO/UNICEF, 2017). These sanitation infrastructures are the configurations with which sanitation technologies are delivered to users/customers in a sani-shed; while sanitation technologies broadly cover the products, services and processes that operate within, around and on the network of systems that makes up the infrastructure. They should all be built on rigorous formulated standards that ensure uniformity, replicability, scalability, reliability, specificity and functionality as well as design, construction, installation, operation, maintenance, manufacturing and service delivery (Gasiorowski-Denis, 2018; Lazarte, 2016). This infrastructural standardization is in consonance with the principles and theory of ReGenSan (see Section 2.4, Box 2.4 and Figure 2.2), which aims to pursue total revitalization and rejuvenation of the complex streams of sanitation technological systems by ensuring continuous improvement of safely managed existing facilities and rehabilitation/retrofitting of dysfunctional and aged facilities as well as contextual, innovative, transparent and functional criteria for development of nouveau facilities. Studies have pointed to inadequate, 'unrecognized' and 'unacceptable' standards for design, production, installation, construction, operation, maintenance and testing as being responsible for the 'non-functionality' of most adjudged improved facilities (Starkl *et al.*, 2013), in spite of the deployment of strategic and efficient technology selection decision support tools.

© IWA Publishing 2019. Regenerative Sanitation: A New Paradigm for Sanitation 4.0
Editor(s): Thammarat Koottatep, Peter Emmanuel Cookey and Chongrak Polprasert
doi: 10.2166/9781780409689_0141

According to the International Social and Environmental Accreditation and Labelling Alliance (ISEAL), a standard is a 'document that provides, for common and repeated use, rules, guidelines or characteristics for products or related processes and production methods, with which compliance is not mandatory' (ISEAL, 2006); and refers to particular products, services and value chain (Springer-Heinzer, 2017). In recent global discourse, there is an increasing recognition of the need for standards in sanitation infrastructure, with some already existing (mainly for sewered systems) and others being proposed (for non-sewered systems) (Jean-Marie, 2013). However, most of these standards are built on broad and generic templates that may not be able to deliver the anticipated infrastructural revolution; particularly as most developing countries with predominantly latrine and septic tank sanitation infrastructure lack any kind of standards and have no particular responsible government agencies, which results in almost no attention ever being accorded to sanitation networks. It may then be difficult for these broad and generic standards to be adopted and implemented in most countries that need urgent improvement in sanitation access and services.

The non-conventional (condominial sewerage) sewer infrastructure, designed then as an alternative to sewer systems in response to challenges posed by expanding services into the peri-urban neighborhoods of Brazil during the 1980s (Melo, 2005), was substantially institutionalized and adopted in Brazil and Latin America under national condominial sewerage design standards and codes of practice (Jean-Marie, 2013; Satterthwaite *et al.*, 2006). However, due to inadequate operating standards the same non-conventional sewer infrastructure was not successfully implemented in sub-Saharan Africa (Pham & Kuyama, 2013). For instance, in Zambia, a 1991 study conducted by the Water and Engineering Development Centre (WEDC), Loughborough University, United Kingdom (UK), that assessed several condominial sewerage systems discovered that all infrastructure had serious maintenance issues traceable to inadequately adopting standards that were not tailored to suit the specific needs of the country (Jean-Marie, 2013; Reed & Vines, 1992). It is highly indicative then that innovative and contextual infrastructural standards are necessary for decentralized and centralized sanitation technologies that could resolve the current global challenges of failure, dysfunction, deterioration and abandonment, especially in countries with huge sanitation infrastructural burden (CSIR, 2007; Kaminsky & Javernick-Will, 2015; Rodgers *et al.*, 2007; WSP, 2007). Standards could remarkably reduce the difficulties associated with technology replicability, functionality and scalability, while also making it easier to determine causes of technologies' successes and failures (Gasiorowski-Denis, 2018; Lazarte, 2016; Starkl *et al.*, 2015). ReGenSan's TeS approaches the issue of sanitation technologies from the angle of design, development, use and service regimen and specifications that interpret national systems and indigenous knowledge of specific SES to determine required deliverables as well as the SSC and sanitation service ladder (SSL) levels. To this effect, product and service quality and provision will be guided by global, regional/state and national standards/regulations that are context-specific. Therefore, it is essential to first grasp an understanding of the considerations and framework of sanitation infrastructure and the decussating role of standardization (a cross-cutting component of the ReGenSan framework).

5.2 SANITATION INFRASTRUCTURE

Infrastructure could be described as sociotechnical system installations that act as the foundation for any operation or system as well as the stock of basic facilities, activities, instruments and interrelated systems required for the effective functioning of a nation, city or area to condition and support its citizens. Infrastructure can be hard assets which refers to physical networks and facilities required for the process to function, or soft assets, which may be services, interactions, institutions and standards required to maintain efficient operations and functioning (Lewis, 2008; Niskanen, 1991). They make up interdependent and self-contained public and privately funded facilities that provide essential services for citizens to enjoy good quality of life; and include, among others, sewers, sewage treatment plants, water supply systems, reservoirs, solid waste disposal and housing (AGCA, 1982; Finger *et al.*, 2005; Kaijser, 2004; Kwiakowski, 1986; Uddin *et al.*, 2013). They are more than physical artefacts (products of technology) and include people, services, organizations and knowledge as well as economic and legal conditions (Hughes, 1987; Jönsson, 2005; Kaijser, 2004). Infrastructure assets are those technology systems and people that provide vital services to communities and businesses such as energy, water supply, solid waste disposal, wastewater treatment, faecal sludge management (FSM) etc. (Uddin *et al.*, 2013). The implication is a network of hard and soft assets of entire interrelated systems of public/private facilities and activities defined by specific standards that continually replace and rehabilitate individual and connected asset components indefinitely to ensure continuous optimum performance (Fulmer, 2009).

Sanitation infrastructure, therefore, includes hard and soft assets of sociotechnical systems used to provide sanitation solutions at public and private levels such as sewers, septic tanks, latrines/toilets, treatment plants, disposal trucks etc. which must take into account all the technical aspects of the 'solution', and all economic, social, organizational, institutional and environmental aspects (Schertenleib, 2005; Maurer *et al.*, 2012; Chang *et al.*, 2011). Thus, ReGenSan infrastructure is the conglomeration of networked systems of sanitation technologies, institutions and standards designed to deliver effective and practical services and solutions that conform to the principles of ReGenSan (2.4). It is an aggregation of sanitation technologies as physical artefacts, which are arranged together in a service chain synergy to deliver effective and practical solutions within a standardized framework upon which technologies are placed to function in the sani-sheds, mimic natural systems and incorporate indigenous knowledge.

5.2.1 Sanitation service chain (SSC)

SSC focuses on the long-term provision of sanitation services for all and at scale as well as describing the ways in which sanitation services are provided. The application of the SSC provides many possible sanitation technology configurations, ranging from self-contained on-site systems to neighbourhood-scale treatment facilities to hybrid systems in which some functions (separation, collection, storage of urine) are performed on site and others (e.g. greywater treatment) are at centralized locations – see Figure 5.1 (Langergrabera & Muellegger, 2005; Larsen *et al.*, 2009; Magid *et al.*, 2006; Tchobanoglous & Leverenz, 2013; Vinnerås & Jönsson, 2013).

Figure 5.1 Sanitation infrastructure service chain (Reprinted from Zakaria F., Garcia H. A., Hooijmans C. M. and Brdjanovic D. (2015). Decision support system for the provision of emergency sanitation. *Science of the Total Environment*, 512–513, 645–658, with permission from Elsevier.)

Studies have shown that sanitation technologies exist at different spatial levels, each with specific technical, management, operations and maintenance conditions (Zurbrügg & Tilley, 2007). There are two broad categories of the wet and dry technologies. The terms 'wet' and 'dry' indicate the presence of flushing water for the transport of excreta and imply how wet or dry the collected waste materials will be. Although flushing water might not be used and it would not therefore qualify as a dry system, a system may nevertheless contain anal cleansing water, urine flushing water or even greywater. Also, wet systems are characterized by the production of a parallel product: faecal sludge. In wet systems then, the faecal sludge flow stream must be taken into account and treated accordingly, with its own set of processes and product-specific technologies until the point of ultimate disposal (Zurbrugg & Tilley, 2009).

Table 5.1 Overview of Existing Sanitation Technologies and Corresponding Input/Out Materials.

Existing Sanitation Technologies	Input/Out Materials
Wet mixed blackwater and greywater with off-site treatment technology	Blackwater mixed with greywater Faecal sludge
Wet mixed blackwater and greywater with on-site treatment technology	Blackwater mixed with greywater Faecal sludge
Wet blackwater technology (blackwater separated from greywater)	Blackwater Faecal sludge Greywater
Wet urine-diversion technology	Urine/yellow-water Brownwater mixed with greywater Faecal sludge
Dry greywater-separate technology	Excreta Greywater
Dry urine- and greywater-diversion technology	Urine Faeces Greywater
Dry all mixed technology	Excreta mixed with greywater

(Reprinted from Tilley E., Ulrich L., Lüthi, C., Reymond, P.H. and Zurbrugg C. (2014b). Compendium of Sanitation Systems and Technologies, 2nd rev. edn. p.108, with permission from Swiss Federal Institute of Aquatic Science and Technology (EAWAG).)

Sanitation technologies include five major components: collection and storage, transport, treatment on/off site, recovery/reuse and disposal; and they are often referred to as the sanitation service chain (SSC) (Wirmer, 2014; Zakaria *et al.*, 2015), or sanitation value chain (SVC) (which views human excreta as a potential resource for valuable products such as biogas or fertilizer (CSTEP, 2016; Van Dijk, 2012) or sanitation service delivery chain (SSDC) (which refers to access to toilets, emptying, transport and disposal or reuse) (Rao *et al.*, 2016) or functional groups (i.e. technologies that have similar functions, which, when selected together, can be used to build a sanitation infrastructure) (Tilley *et al.*, 2014a, b; Zurbrügg & Tilley, 2007, 2009).

The concept of SSC is actually embedded in every sanitation technology, whether sewered (centralized or off site) or non-sewered (decentralized or on site). The off-site infrastructure consists of a sequence of several individual sanitation processes where each individual process represents a step of the SSC that is carried out by a separate technology. However, the on-site infrastructure usually combines several sanitation processes/steps of the SSC in a single sanitation technology (Figure 5.2) (CAWST & EAWAG-Sandec, 2016; Wirmer, 2014; Zakaria *et al.*, 2015).

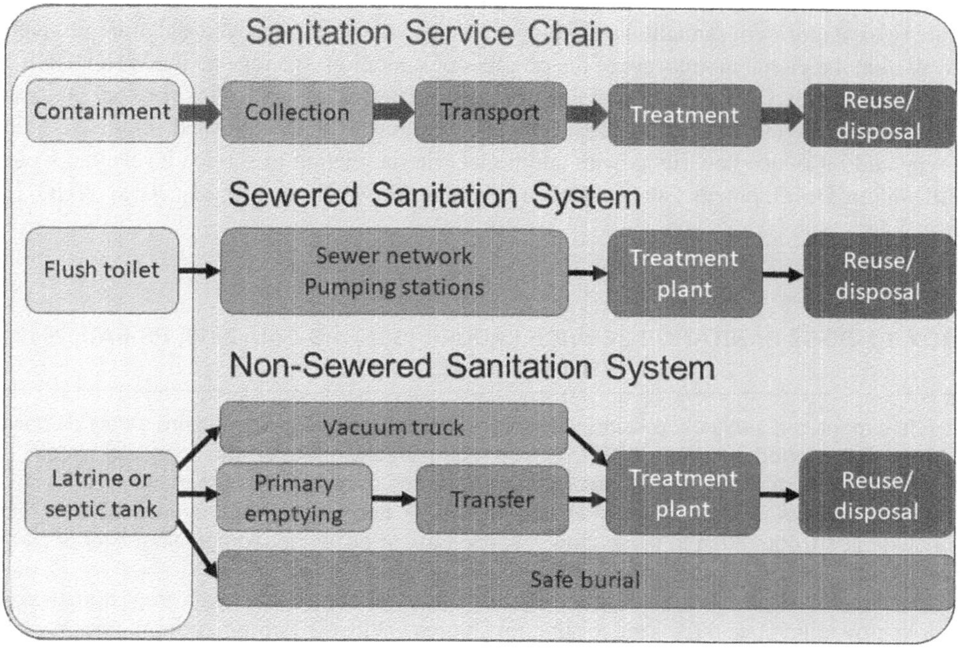

Figure 5.2 Sanitation service chain (*Source*: Blackett & Hawkins, 2015).

The SSC is the series of steps/processes involved in the provision of sanitation infrastructure (Sparkman, 2012; Verhagen & Ryan, 2008; Zakaria *et al.*, 2015), comprising:

(I) Processes for excreta disposal and collection from the user interface – production and capture,
(II) Processes for excreta conveyance – collection and transport, and
(III) Processes for treatment until final disposal or reuse – treatment or disposal and reuse (CAWST & EAWAG-Sandec, 2016; Jean-Marie, 2013; Wirmer, 2014; Zakaria *et al.*, 2015).

For instance, in Kenya, an organization called Sanergy is proving that sanitation is about more than just providing people with toilets. Sanergy has applied a service model that accounts for the entire sludge management train, from toilet construction, to emptying and sludge transport, to final waste treatment and conversion of the sludge into renewable energy or fertilizer (Templeton, 2015) (see Box 6.2). Hence, sanitation technology development and practices are strongly rooted in the SSC and sanitation service ladder (SSL).

5.2.2 Sanitation service ladder (SSL)

The SSL is based on an incremental toilet/latrine provision progression as it goes up the rungs of the ladder and corresponds to levels of service delivery (Gine *et al.*, 2011; Kvarnström *et al.*, 2011; Lenton *et al.*, 2005; Wood *et al.*, 1998), often used in varied situations as a tool to choose technology types in community-based sanitation projects. The lowest rung is improved traditional practice and hygiene (which is not usually considered sustainable over a long period) and the highest level is 'safely managed' toilets/latrines, also referred to as 'improved sanitation facility', which are not shared and where excreta is safely disposed of in-situ or treated off site (Exley *et al.*, 2014; Kvarnström *et al.*, 2011; Potter *et al.*, 2011; Thomas *et al.*, 2018; WHO/UNICEF, 2017).

There are two categories of sanitation technology related to the SSL: improved and unimproved facilities; terms now used in the global monitoring of the progress of sanitation coverage by the WHO/UNICEF Joint Monitoring Programme (JMP). The Sustainable Development Goals (SDGs) for the sanitation Joint Monitoring Programme (JMP) were built on the classifications of the established improved/unimproved facilities type and introduce new rungs with additional criteria relating to service levels that were lacking in the Millennium Development Goals (MDGs) (pS-Eau, 2018; Thomas *et al.*, 2018; WHO/UNICEF, 2017 – Box 5.1).

BOX 5.1 SDGS SANITATION SERVICE LADDER (SSL) (PS-EAU, 2018; PS-EAU, 2018; THOMAS *ET AL.*, 2018).

(I) Safely managed services: private improved facility where faecal wastes are safely disposed on site or transported and treated off site; plus a handwashing facility with soap and water;
(II) Basic service: private improved facility that separates excreta from human contact;
(III) Limited service: improved facility shared with other households. Improved sanitation facilities include flush/pour flush to piped sewer, septic tank or pit latrine; composting toilet or pit latrine with slab;
(IV) Unimproved service: unimproved facility that does not separate excreta from human contact. Unimproved sanitation facilities include pit latrines without a slab or platform, hanging latrines and bucket latrines.
(V) No service: open defecation

The SDGs address the hygiene targets that were not included in the MDGs, while the JMP also developed a specific ladder for handwashing with soap and water. Hygiene is here defined as 'the conditions and practices that help maintain health and prevent diseases including handwashing, menstrual hygiene management and food hygiene' (Thomas *et al.*, 2018; WHO/UNICEF, 2015a, b). The new SSL reveals the change in approach between the MDGs and SDGs, which focuses on the entire sanitation service chain (SSC) (Figure 5.3). Based on the improved and unimproved facilities categorization and national survey data, WHO/UNICEF JMP estimated that 5 billion people were using improved sanitation facilities in 2015, whereas 2.4 billion were using unimproved sanitation facilities (Baum *et al.*, 2013; WHO/UNICEF, 2017). Studies have shown that most of the improved technologies of countries of the southern hemisphere are not designed, constructed, installed, operated and maintained based on standards and there are no adequate codes of practice put in place to ensure regenerative infrastructure.

However, these classified technologies raised certain criticisms within the sector due to the exclusion of certain technologies like composting and urine-diverting toilets (UDT) from the list of approved technologies, although composting toilets were later added (Kvarnström *et al.*, 2011). In addition, these classifications do not indicate specific standard requirements for improved facilities such as design, construction, installation, etc. and do not easily allow for the inclusion of new, innovative and functional technology options (Kvarnström *et al.*, 2011; Potter *et al.*, 2011). For instance, globally, septic tanks account for 56% of on-site improved sanitation facilities in urban settings and 38% in rural areas. While septic tanks have certain defining design features (including watertight walls and floors, multiple chambers separated by baffles and an outlet pipe leading to a soakpit or leachfield), many on-site systems lack these features, and should actually be classified as simple vaults or cesspools (WHO/UNICEF, 2017). The SSL also specifies the level of services that are provided across the SSC from containment to end-disposal or processing and use.

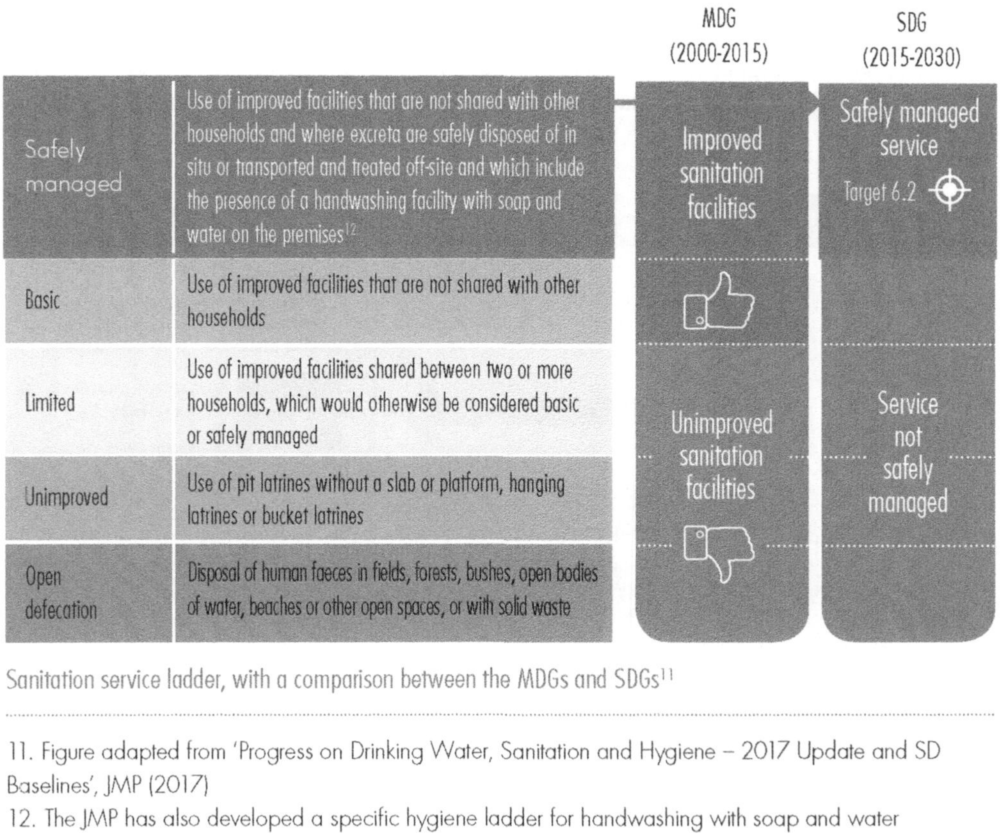

Figure 5.3 Sanitation service ladder, with a comparison between the MDGs and SDGs (Reprinted from The Sustainable Development Goals for Water and Sanitation Services. Interpreting the Targets and Indicators, 2018, with permission from Programme Solidarité Eau, Paris: pS-Eau. www.pseau.org/en/agenda-2030).

The combination of SSL and SSC produces or leads to a framework that shows compositions and functionalities of sanitation infrastructure in many towns and cities, especially in developing countries, as having a mixture of non-sewered and sewered facilities and services (Hawkins *et al.*, 2013; UN-Water, 2015). This is illustrated by using the 'faecal waste flow diagram' (FWFD) –often referred to as a shit flow diagram (SFD) – developed by Peal *et al.* (2014a, b), which shows the different pathways faecal waste takes along the SSC – a pictorial representation that summarizes city-level outcomes and highlights bottlenecks in faecal waste management (CAWST & EAWAG-Sandec, 2016; WSP, 2014). The application of SFD provides a robust overview of the challenges of sanitation infrastructural standardization in Dhaka, Bangladesh caused by inadequate and poorly managed sanitation facilities.

Figure 5.4 illustrates these problems where 20% of faecal waste is sewered and 79% goes to non-sewered containment (Peal *et al.*, 2014a, b; UN-Water, 2015). The width of the arrows and the percentages in the diagram represent the proportion of the population whose faecal waste takes each route. Although nearly all waste is effectively contained at the household level, poorly designed, constructed, operated and unsafely managed on-site facilities combined with highly inadequate sewerage and wastewater treatment mean that faecal waste is distributed throughout the urban environment of Dhaka. This is because the downstream end of the SSC is still deficient and only a tiny percentage (2%) of the waste generated is treated adequately prior to disposal (Peal *et al.*, 2014a, b; UN-Water, 2015; WSP, 2014). The Dhaka case strongly supports the argument for extending boundaries of sanitation management from households to sani-sheds as it could contribute greatly to service improvement and better sanitation infrastructural management.

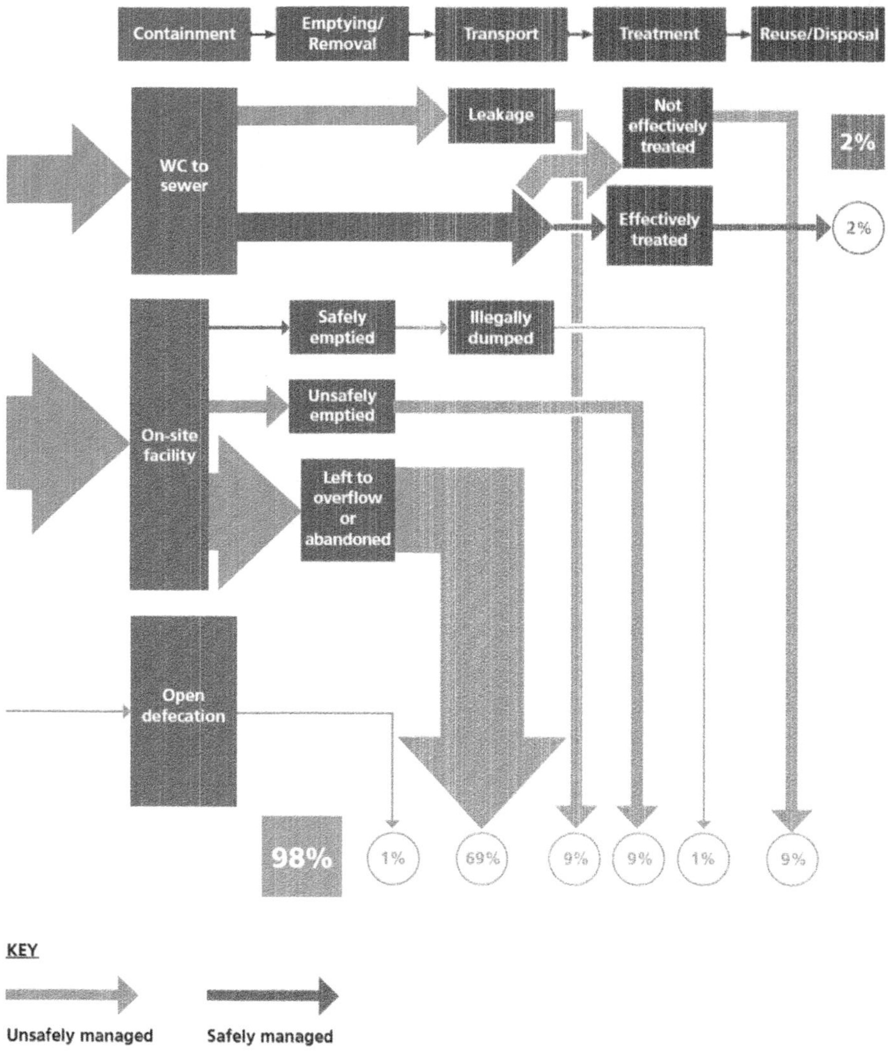

Figure 5.4 Faecal waste flows in Dhaka, Bangladesh (*Source*: Peal *et al.*, 2014a, b).

5.2.3 Sanitation technology selection support tools

Sanitation technology selection support tools are products that combine information on users' given situations with information on available technologies and options helping practitioners to select the best available technology or approach (Palaniappan *et al.*, 2008). However, it has been noted that decision support tools may not be able to enhance applicability and scalability of promising technologies and service models; certainly not without adequate and appropriate standards. The main challenge facing these policies, especially in sub-Saharan African countries, is a lack of specification for minimum technology options, designs and services standards (Nakagiri *et al.*, 2016; Potter *et al.*, 2011; WEDC, 2005). Just as has been noted in the previous section, the SSL has no service indicators, is restricted to only 'listed' technological options, is not open to innovations and, even though it could be useful for monitoring service improvement and coverage expansion, it is not backed up with technical and structural standard specifications. This will pose a great hindrance to technology selection as it is basically propped upon the SSL (Potter *et al.*, 2011). It is, therefore, not surprising that similar technological solutions implemented within the same geographical region may not produce similar results owing to a lack of infrastructure standards. A study that reviewed sanitation policies of nine countries noted that only South Africa, Mozambique and Ghana had adopted VIPs as their minimum sanitation technology, with no specific technical and industrial standards issued by these countries' national standards bodies to drive technological innovations (Nakagiri *et al.*, 2016; Potter *et al.*, 2011; WEDC, 2005). In essence, the use of selection support tools only assisted decision-makers in these countries to select facilities such as the VIP option, but obviously could not guarantee functionality, quality or performance or even inspire innovations.

For instance, in South Africa, several studies observed that the upscaling and delivery of the VIPs and their derivatives, as well as the urine-diversion technologies, were challenged by inadequate technical specification standards for design and construction (Bhagwan *et al.*, 2008; Still *et al.*, 2010; Still & Foxon, 2012a–c). In addition, conflicting advice on what should be put into the pits to keep them operating well, poor construction resulting in structural integrity and lack of the necessary anaerobic activity in the pit all indicated non-adherence to standards so the pits filled up much faster than their design life. The study called for an urgent and thorough understanding of the VIP technology so that it could be more effective and its potential optimized. Unfortunately, there is still little or no research done on VIPs' standardization, especially on the degradation mechanism or processes occurring in VIP latrines (Bhagwan *et al.*, 2008).

In Nigeria, with an estimated sanitation infrastructure of latrines (37.98%), septic tank (17.83%) and sewer (9.9%) (WHO/UNICEF, 2017), 30.1% of Nigerian households use improved sanitation facilities that are not shared with other households (25.1% in rural and 36.6% in urban areas), while 39.9% and 15.5% of rural and urban households, respectively, still lack access to sanitation facilities (Abubakar, 2017). Within the same period, the proportion of the population defecating in the open has increased from 24% to 25% (UNICEF/WHO, 2015) and, sadly, access to improved sanitation declined from 38% of the population in 1990 to 29% in 2015, while several studies still reveal an alarming rate of infrastructural depreciation due to non-existent nationally recognized and adopted standards. There are no acceptable technical specifications and industry standards for the design, construction and installation of sanitation infrastructures in Nigeria (Adelegan & Ojo, 1999; Cookey, 2013; Cookey *et al.*, 2016; NDHS, 2008; NISER, 2010; Sridhar, 1995). Particularly, a study in the oil and gas city of Port Harcourt (a metro-cosmopolitan city with a population of about 2 million people) reported that technical specifications for products, services and systems performance as well as regulations, guides and codes of practice for septic tanks and latrines were completely absent (Cookey, 2013; Cookey *et al.*, 2016).

Similar situations were reported in Indonesia and Vietnam, where 85% to 90% of urban citizens rely on septic tanks and latrines for human waste disposal, as well as in the Philippines where 85% of households have septic tanks, in Sri Lanka where it's 77% and Indonesia where it's 62%. These studies revealed that national standards for design and construction of septic tanks and other on-site sanitation systems were inadequate, and in most cases not known and never applied, which often resulted in sub-standard sanitation infrastructure (AECOM & SANDEC, 2010; WSP, 2015). Therefore, to ensure replicability and scalability ReGenSan TeS advocates for the development and adoption of internationally, regionally and nationally recognizable technical standard specifications for products, services and systems to drive sanitation infrastructural improvement and access expansion that go beyond the SSC and SSL.

5.2.4 Non-sewered sanitation infrastructure

Non-sewered sanitation is a technology in which excreta and used water are collected and stored in the location where they are produced. Usually, the faecal sludge generated from the facilities has to be transported off site for treatment, use or disposal and is often called on-plot, on-site or decentralized sanitation (CAWST & EAWAG-Sandec, 2016; Verhagen & Carrasco, 2013) (see Section 2.3.4). It is the predominant sanitation service model in developing countries. Globally, inadequate non-sewered infrastructure still provides services to most developing countries and evidence suggests that many constructed on-site sanitation systems (OSS) have extremely high abandonment rates of up to 42% due to failure that leads to the attendant public and environmental health impacts (CSIR, 2007; Fewtrell *et al.*, 2007; Kaminsky & Javernick-Will, 2015; Rodgers *et al.*, 2007; WSP, 2007). For instance, a study in Ghana shows that many of the existing sanitation facilities are in disrepair and only 10% of the wastewater and faecal sludge treatment plants function as planned (Murray & Drechsel, 2009; Murray & Buckley, 2009).

These technologies are obviously not yet considered to be fully fledged alternatives to centralized infrastructure because of the deficiencies of the currently available configurations, as well as uncertainty about functionality, reliability and cost-effectiveness in the near future (Truffer *et al.*, 2013). A households study on latrine coverage in some rural communities of Ethiopia showed that 56.9% of the latrines were dysfunctional and required maintenance, even though 52.6% of these latrines were constructed only two or more years prior to the time of the study (Awoke & Muche, 2013). Sanitation infrastructure in developing countries faces numerous challenges, with very few focus researches targeted towards resolving it.

For example, septic tanks, currently adjudged as one of the improved facility, have essentially retained their original design from the 1860s (Bradley *et al.*, 2002) and in the global South, the designs are still rudimentary and poorly constructed (Lüthi & Panesar, 2013). There are, however, some pockets of scaled-up technological innovations such as the Chinese three-chamber septic tanks (Cheng *et al.*, 2017; Fan *et al.*, 2017; Hu *et al.*, 2016); the anaerobic baffled reactor (ABR) that uses baffles to optimize treatment could be referred to as an 'improved' septic tank, but is not widely applied (Foxon *et al.*, 2004; Tilley *et al.*, 2014b; Ulrich *et al.*, 2009); the solar septic tank that utilizes solar energy to inactivate the pathogens and increase sludge biodegradability (Koottatep *et al.*, 2014, 2015, 2016; Pussayanavin *et al.*, 2015 – see section Box 3.3) and septic tank effluent drainage/pumped (STED/STEP) infrastructure that provides effluent sewer services treatment in a communal facility to millions of small communities in Australia and USA (Gardner & Sharma, 2013; Lyle, 1994; Orenco System, 2017).

The only upgrades to traditional pit latrines are the VIP latrines and biogas latrines such as the Sulabh biogas toilet system (India), Kibera biogas toilet (Kenya) (Katukiza *et al.*, 2012; Nakagiri *et al.*, 2016) (see Figure 7.2) and the Chinese biogas production technology for domestic sewage treatment in rural communities, which has been used to replace septic tanks in treating domestic sewage and faecal sludge (Cheng *et al.*, 2013, 2017; Jiang *et al.*, 2011) (see Table 5.2). Despite this progress, understanding is still deficient in issues of odour as well as insect, reptile and rodent nuisances, and rapid filling of facilities such as pit latrines, cesspits and septic tanks that defy their design lifespan (Bhagwan *et al.*, 2008; Medland *et al.*, 2016; Nakagiri *et al.*, 2016; Still & Foxon, 2012a–c). Design characteristics and other factors that cause some pits to fill more quickly than others are not well understood and very little is known of the physical characteristics of the faecal sludge that accumulates in pit latrines (Templeton, 2015 – see Section 4.2.3).

As a way of solving odour problems, studies have recommended ventilation criteria for odourless conditions in VIP latrines from two air changes per hour (2 ACH) in the superstructure (ASHRAE, 2005), six ACH (6 ACH) (Mara & Ryan, 1983) and 8-20 ACH (Viessman, 1964). For instance, the value of 6 ACH fell within the indoor air quality standards for a toilet room in the UK, at the time Mara and Ryan (1983) carried out their study. Therefore, considering the fact that the superstructures of the VIP are generally larger in air volume than those that were used to determine 6 ACH as a reasonable rate, how could such designs be achieved? What is the actual ACH value required for odourless conditions in VIP latrines? (Dumpert, 2008).

Management or treatment of faecal sludge from these 'improved' on-site latrines/toilets technologies is a major challenge even though it was not included in the definition of what constitutes improved sanitation facilities (Dodane *et al.*, 2012). Most developing countries do not have faecal sludge treatment standards and it is estimated that 1.2 billion urban residents are in need of FSM services and this number will increase (Templeton, 2015). Furthermore, it is unrealistic to adopt standards from industrialized countries that were established based on their peculiar conditions, technological availabilities and institutional climates, which are quite different from less developed countries' situations (Ingallinella *et al.*, 2002). Hence, comprehensive FSM from on-site sanitation (OSS) installations remains a major challenge for developing countries (Dodane *et al.*, 2012). This is best illustrated by Figure 5.5a, which shows a typical example of the FSM gap of incomplete pit-based sanitation infrastructure; missing conveyance, treatment and safe disposal, resulting in uncontrolled discharge of untreated faecal sludge. Figure 5.5b shows the infrastructure missing in Figure 5.5a based on the SSC and the roles of various entities (see Chapter 4, Figure 4.1, Tables 4.1 and 4.2), with specific reference to SMEs (see Section 4.3.2), which are currently lacking in most developing countries (Tilley, 2013).

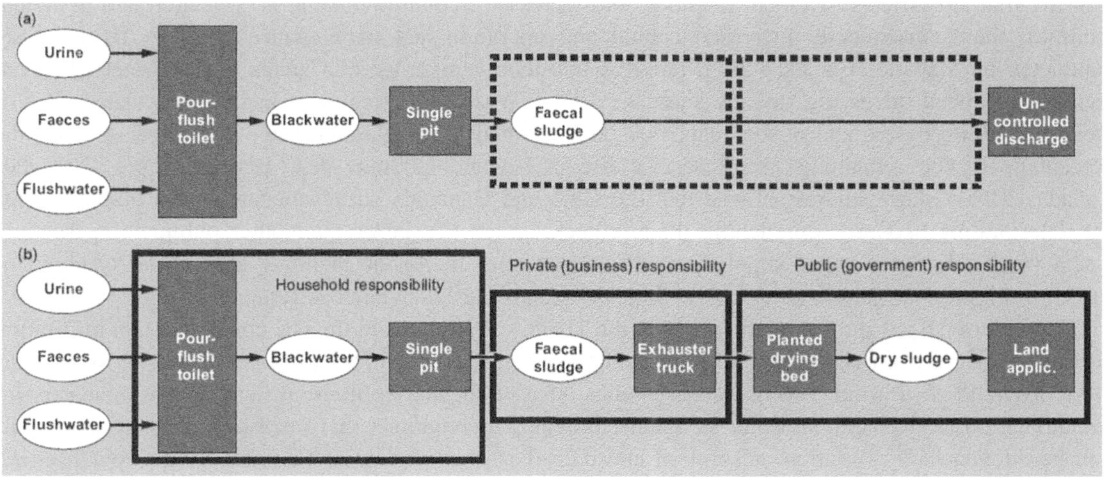

Figure 5.5 (a) A typical example of incomplete pit-based sanitation infrastructure, missing conveyance, treatment and safe disposal; (b) A complete pit-based sanitation infrastructure with appropriate technologies identified for each SSC (*Source*: Tilley, 2013).

However, Figure 5.1b is impeded at scale levels of households, which are limited by the boundary restrictions of individual facilities, which do not complete the service chain. To this effect, the concept of ReGenSan 'sani-sheds' provides opportunities to extend the boundaries of sanitation infrastructure to accommodate the processes of the pit-based SSC. Several studies have confirmed that BGC cycling processes in latrine infrastructures are not properly understood and this may be responsible for breakage of the complex processes of the SSC and the resultant inadequate service delivery (Bhagwan *et al.*, 2008; Lyle, 1994; Medland *et al.*, 2016; Nakagiri *et al.*, 2016; Still, 2010) (see Section 4.2.3).

5.2.5 Sewered sanitation infrastructure

Sanitation systems that transport wastewater through a pipe network (such as a simplified sewer, solid free sewer or conventional sewer) to another location for treatment, use or discharge are often called a sewer system, sewerage system, sewers, connected sanitation or networked sanitation (CAWST & EAWAG-Sandec, 2016) (see Section 2.3.3). They consist of three basic systemic components: (i) wastewater collection comprising large pipes, pumping systems and various access routes that cost more than 60% of the total budget of the whole wastewater management system budgets (Zaharia, 2017), (ii) wastewater treatment, a complex set of concurrent mechanical, physico-chemical and biological processes that can be broadly categorized into: activated sludge, rotating biological contactor (RBC) and trickling filter (Bayo *et al.*, 2006; Boujelben *et al.*, 2017; Cardot, 1999) and (iii) treated wastewater disposal or reuse, which has particularly attracted increasing attention because wastewater can be treated and reused for a variety of purposes (Boujelben *et al.*, 2017; Grant *et al.*, 2012; Wilderer & Schreff, 2000; Zaharia, 2017).

Studies have shown that sewered infrastructure faces challenges in both developed and developing countries. In developed countries, ageing sewered infrastructure is primarily responsible for deterioration and failure because most of the sewers date back over 100 years to the nineteenth century, when brick sewers were common (Ahmadi *et al.*, 2015; Ashley & Hopkinson, 2002; Brown & Caldwell, 1984; Corcoran *et al.*, 2010; Diagne, 2013; El-Sheikh, 2011; Fenner & Saward, 2004; Kemp, 2000; Miszta-Kruk, 2016; Qasem *et al.*, 2009; Reynolds & Barrett, 2003; USEPA, 1991a, b). When sewer systems deteriorate, water from excessive infiltration and inflow enters the system, resulting in decreased capacity that also affects treatment facilities, and increased hydraulic loading at collection and treatment facilities, then consequently increased capital and operation and maintenance costs. A recent study conducted by Statistics Canada (2008) reported that the average age of Canada's wastewater treatment facilities has been increasing steadily since the 1970s. Wastewater treatment facilities exceeded 63% of their service life by the end of the year 2007 on a national scale (Qasem *et al.*, 2009) and in 2015, the percentage of the remaining useful service life of sewage facilities decreased to 62.6% (Statistics Canada, 2015). Many studies showed that the USA and Canada's wastewater treatment infrastructures are deteriorating and are approaching their project service life faster than other public infrastructural assets (roads, bridges, water supply systems, wastewater treatment facilities and sewer systems) as shown in Figure 5.6 (Qasem *et al.*, 2009; Statistics Canada, 2008; Vanier & Rahman, 2004).

A survey of 27 sewage-treatment works in the Thames Water area in the UK concluded that infiltration normally accounts for over 50% of all flows and, in some cases, up to 75% of flows (Fiddes & Simmonds, 1981; Reynolds & Barrett, 2003). Recent studies have implicated exfiltration (leakage) from sewers for contamination of deep groundwater as against previous assumptions that microbiological contaminants will be attenuated by filtration, adsorption and die-off at shallow levels (Fairbain, 1996; Powell *et al.*, 2000, 2001). This has serious implications for the potential utilization of urban groundwater resources and for rural boreholes in the vicinity of sewage sources (Reynolds & Barrett, 2003). In the same way, sewered infrastructure in Japan faced high-level threats from ageing and recorded a high number of failed sewers, which were responsible for about 4000 cases of sinkholes annually (JSWA, 2016).

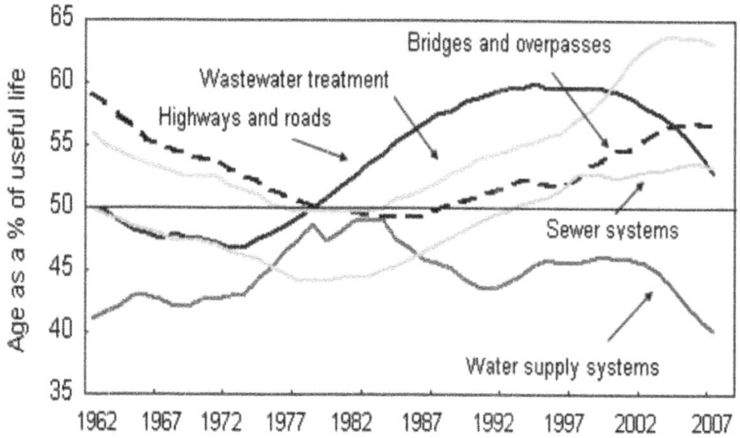

Figure 5.6 Canada's infrastructures remaining service life (*Source*: Qasem *et al.*, 2009).

Meanwhile, in Europe the proportion of population connected to wastewater treatment facilities ranges from 7% (Macedonia) to 99% (UK), with the majority of the EU nations being between 50% and 80% connected (Brands, 2014; Eurostat, 2011). Regulators in the UK reported that about 23% of the sewers in England and Wales are in Grades 3 to 5. Grade 3 sewers have deformations of up to 5%, i.e. cracking, fractures, joint defects or badly made connections, while Grade 5 sewers are in a state of collapse (Ashley & Hopkinson, 2002; OFWAT, 2000). In the United States of America (USA), there are about 16,000 wastewater treatment systems that serve approximately 70% of the population, with the remainder serviced by decentralized systems such as septic tanks (Tchobanoglous & Leverenz (2013); USEPA, 2010). There are over 459,815 km of sewer pipes and more than 12 million manhole fixtures associated with public wastewater systems, and another 310,686 km of lateral sewer pipes connecting private buildings and industries to public sewers in the USA (USEPA, 2010). A report from the USA wastewater infrastructure system grade had dropped from [C] in 1988 to [D-] in 2005 (Brands, 2014; Qasem *et al.*, 2009). The most recent grading from the American Society of Civil Engineers is of [D+]: that is, the infrastructure is in poor to fair condition and mostly below standard, which may compromise service reliability (ASCE, 2017).

In developing countries, sewered treatment infrastructure failure may not necessarily be attributed to age, but is most likely due to inadequate implementation strategies (Cuppens *et al.*, 2013; IWA, 2006). Implementation of sewered infrastructure in developing countries puts more emphasis on technology than on the process of solution identification, technology transfer and capacity-building mechanisms, which are crucial factors for the long-term success of sanitation infrastructure (Murphy *et al.*, 2009). Experience in Brazil has shown that when technologies with proven efficiency in developing countries are selected, there is still no guarantee of a satisfying performance of wastewater treatment plants in practice (Von Sperling & Oliveira, 2007). The observed poor performance and failure of treatment facilities cannot be attributed to the treatment processes per se but rather to design, operations and maintenance problems as well as poor standards and regulatory environment (Oliveira & Von Sperling, 2008). Even when new infrastructure are constructed, too little attention is given to why existing ones failed and/or are not performing well. The stories behind these poorly functioning facilities hardly reach decision-makers, researchers and consultants involved in sanitation (Cuppens *et al.*, 2013).

In a classic case, a limited sewer collection bottleneck was reported from the Kaliti wastewater treatment in Addis Ababa, the capital of Ethiopia. The Kaliti treatment plant was designed in 1982 to serve 50,000 people and, after 30 years of development, was only able to serve 13,000 people. One of the attributed reasons is insufficient connection from houses to municipal sewerage pipelines. The rate of connection is still very low with no improvement since 1993, and only 3% of wastewater of the city is treated in the facility (Abiye *et al.*, 2009; Wang *et al.*, 2014). This is also similar to China, which had more than 3437 wastewater treatment plants in 2015; however, quite a large number of them are not operating because of incomplete drainage networks (especially in rural areas) to connect households to the treatment infrastructure (Cheng *et al.*, 2017; Giovanni *et al.*, 2012). Another case is the Kisumu district in Kenya, where all three of the existing pump stations are broken down, resulting in overflow of sewage directly into Lake Victoria (Parkman *et al.*, 2008; Wang *et al.*, 2014).

The cost implications of sanitation infrastructure failures, dysfunctionalities and inadequacies are enormous for both developed and developing countries. For example, the cost of dysfunctional non-sewered infrastructure to 18 African countries is estimated to be USD 5.5 billion per year (Bartram & Cairncross, 2010; Bos & Gijzen, 2005; Cheng *et al.*, 2012; Ekane *et al.*, 2014; Fewtrell *et al.*, 2005; Pruss *et al.*, 2002; Prüss-Üstün *et al.*, 2008; UNDP, 2006; WSP, 2014; Yardley, 2010), while an estimated 1 billion euros are invested annually for sewer rehabilitation in the Netherlands (Korving *et al.*, 2003). In the USA, the capital investment needed for repairs of existing sewered systems is estimated to be nearly USD 300 billion for the next decades (Brands, 2014; USEPA, 2008b), with an annual expenditure of approximately USD 10 billion required for upgrading existing wastewater collection systems and USD 45 billion for controlling combined sewer overflows (Ahmadi *et al.*, 2015; ASCE/WEF, 1994). In France, about 3 billion euros is required annually for rehabilitation of ageing sewer infrastructure by 2020 (Ahmadi, 2015; Le Gauffre & Cherqui, 2009). Hence, to safeguard proper functioning of the sanitation infrastructure over time, appropriate strategies to address the challenges that hinder sustainability are crucial.

5.3 SANITATION INFRASTRUCTURE STANDARDIZATION (SIS)

Essentially, standards create uniformity, scalability and replicability that eventually lead to affordability, and encourage return on investment (NOWRA, 2007; Starkl *et al.*, 2015). If properly executed, standards will take sanitation infrastructural solutions and management way beyond 2030, particularly since non-standardized sanitation infrastructure will not be able to meet the required standard quality for effluents discharged into the environment. It is indicative, then, that adequate and appropriate standards for centralized and decentralized sanitation technology, design, development, delivery and use (which are broad and specific, global and localized) are crucial for the achievement of the SDG targets. Already, centralized and decentralized sanitation infrastructure and service delivery in most developed countries are based on agreed and acceptable standards, even though there are still issues for review in the face of current challenges (HES, 2014; USEPA, 2008a, b); however, developing countries continue to struggle with both centralized and decentralized options, mainly due to no standards and/or sub-standard designs, installations, construction, usage and policy/legislative instruments (CSIR, 2007; Kaminsky & Javernick-Will, 2015; Rodgers *et al.*, 2007; WSP, 2007).

Standards are key to institutional frameworks for sanitation management at all levels, specifically because they are required to support regulatory instruments and are necessary for policy/legislative implementation as well as compliance monitoring. For standards to be effective and practical, they need to reflect specific local, contextual, ecological and indigenous conditions; and so should

(I) Incorporate global standard frameworks,
(II) Understand the local SES and sani-sheds,

(III) Incorporate local natural systems and indigenous ecological knowledge, and
(IV) Conform to national, state and local policy/legislative framework on sanitation.

This implies that sanitation infrastructure standards should draw considerably from local adaptations in specific applicable contexts. For instance, in the case of greywater recycling, each installation has to be optimized to the specific building context and the targeted water reuse options (Truffer *et al.*, 2013). Consequently, the objectives for sanitation infrastructure standardization in ReGenSan TeS include, but are not limited to, the following:

(I) Translate the SDG 6 targets of 'safely-managed' facilities at local levels to affordable and quality technologies for sanitation infrastructure;
(II) Facilitate the use of a uniform classification process for components of sanitation infrastructure;
(III) Ensure applicability, scalability and replicability of technological solutions that maintain the same quality irrespective of locations;
(IV) Develop and promote industrial and technical standards for entrepreneurs and manufacturers that support innovative and regenerative solutions to enhance return on investments;
(V) Enhance recognizable products, services and certifications for the much-needed global improved sanitation (Gasiorowski-Denis, 2018; Lazarte, 2016; NOWRA, 2007; Starkl *et al.*, 2015).

Generally, standardization involves a process of collaborative production, dissemination and implementation of technology to create 'standards' (Feng, 2003; Russell, 2005). The information is then collated in a document that provides requirements, specifications, guidelines and characteristics, which could consistently be used to ensure that materials, products, processes and services are fit for purpose (Hatto, 2010; ISO, 2018; Starkl *et al.*, 2015). These standards can then be included in international, regional and local normative documents developed and established from a consensus agreement between stakeholders in the field of sanitation and related disciplines and occupations. Standards could include test methods, technical specifications for products, services, system performance, technical regulations, guides and codes of practice, terminology and organization management systems (Feng, 2003; ISO, 2018; Russell, 2005), according to the following descriptions:

(I) Terminology – a set of terms and definitions to help harmonise the use of language in the sanitation sector,
(II) Specification standards – define characteristics of a product (product standards), or service (service activities standards) and their performance thresholds such as fitness for use, interface and interoperability, health, safety and environmental protection;
(III) Test methods and analysis standards – are highly prescriptive, setting out an agreed way of measuring, testing or specifying what is reliably repeatable in different circumstances and places, wherever it needs to be applied;
(IV) Codes of practice – recommend sound good practice as currently undertaken by competent and conscientious practitioners in the field. They are drafted to incorporate a degree of flexibility in application, while offering reliable indicative benchmarks,
(V) Treatment standards – define the minimum acceptable parameters treatment systems are required to comply with by setting the point prior to release to the receiving environment; and,
(VI) Management systems standards – define the policies, processes and procedures that organizations use to manage the interrelated parts of its business in order to achieve its objectives. These objectives can relate to a number of different topics, including product or service quality, operational efficiency, environmental performance, occupational health and safety and many more (Hatto, 2010; ISO, 2018).

The most widely and commonly used standards in the sanitation sector are 'test methods' and 'treatment' standards. Others are seriously lacking, especially in developing countries, where standardized test methods are used to determine treatment regimens for non-standardized infrastructure. It should also be noted that undue haste in adopting standards that are too high or too low can lead to the use of inappropriate technology or unaffordable objectives. It is better to set appropriate and affordable sanitation-related standards and to have a phased approach for improving these standards that will allow for enough time to develop capacity and skills to deliver on these standards. In addition, it is important to allow for the maturation of these standards' requirements into suitable regulatory frameworks with matching institutional instruments and appropriate enforcement mechanisms (Gutterer & Reuter, 2016).

Standards and regulations should work together because standards are voluntary codes with which there are no legal obligations to comply, while regulations specify legally enforceable requirements and non-compliance leads to sanctions. In addition, standards can be used to demonstrate compliance with regulations (Hatto, 2010). Standards, policies and regulatory frameworks are needed for effective sanitation technologies and infrastructural advancement and practice, which seems to elude most countries with a huge sanitation burden, especially countries of central and southern Asia, Oceania and sub-Saharan Africa. Even though these countries have attained some levels of progress in the development of national sanitation norms and policies, they still have weak prescriptive and performance-based regulatory frameworks, including the absence of technical standards for material, product, process and service specifications. Thus, the anticipated infrastructural and service level improvement from the combined effect of policies and standards may not be effective since sanitation policies have no standard specifications to support their actual progress.

Previous efforts for SIS were targeted at sewered technologies (Jean-Marie, 2013), based on the assumption that non-sewered provisions were primarily temporary stop-gaps (USEPA, 1997). In 1947, 10 states in the Great Lakes-Upper Mississippi River region of the USA (Illinois, Indiana, Iowa, Michigan, Minnesota, Missouri, New York, Ohio, Pennsylvania and Wisconsin) came together to review and upgrade existing design standards for sewer works and produced a joint standards publication in 1951. These standards have since influenced the upscaling of sewer works and have been revised several times up to 2014 (EUWATER Network, 2005; HES, 2014; USEPA, 2008a; USEPA, 2009). Europe also has several standards for design, construction and operations of sewer systems, driven by the European Union and used by all the 18 countries of the European Committee for Standardization (CEN). These standards set out basic provisions, common to all user countries, and include annexes with additional information that references each country's national standards, codes, guides and other relevant technical information (CPDA, 2001). In spite of these advances in the standardization of sewered infrastructure, there are still increasing records of huge infrastructural operations and maintenance costs and high rates of ageing and deterioration, as well as engineering inefficiencies with regard to nutrient losses, energy demand, resource conservation and recovery and reuse (Brands, 2014). This indicates an urgent need for innovative and sustainable standard frameworks to retrofit existing infrastructure.

5.3.1 Sanitation infrastructure standards (SIs)

Standardizing sanitation infrastructure provides a framework for design, development and use of sanitation technologies as well as the systemic interactions/linkages that make up the whole infrastructure. The process of standardization enhances consensus among engineers, scientists, policymakers and users at all affected levels and encourages participation in setting and applying agreed rules of practice, where options for development and maintenance of 'safely managed' infrastructure and the costs benefits associated with such options are clearly understood (NOWRA, 2007). It demands manufacturing competence that goes beyond plumbing, piping and tank productions, essentially, for decentralized plant manufacturers (Olsson, 2013). The technical basis of a new standard is usually established through the preparation and submission of a proposal to relevant standard organizations, in this case the International Standards Organization (ISO). The proposal is then circulated to members for evaluation – three questions are normally asked:

(I) Is there a need?
(II) Are there existing standards that are relevant?, and
(III) Will you participate?

If approved, a technical committee is formed and drafts are produced over time, such as:

(I) Working draft,
(II) Committee draft,
(III) Draft standard,
(IV) Final draft standard, and
(V) Published standard.

The whole standard development process normally takes about three years of completion through all stages, with a mandatory review after five years (Ellison, 2017). There are some notable formal and informal standards in the sanitation sector, which could serve as foundations or guidelines towards developing specific and far-reaching standards that deal with current global and local sanitation challenges. The ***formal sanitation standards (FSS)*** are approved or adopted by national, regional or international standard bodies (Box 5.2). International standard bodies such as the ISO comprise members of approximately 160 countries' recognized National Standard Bodies (NSBs). Regional standard bodies are regional countries' organizational standard bodies, such as the European Union Committee for Standardization (Comité Européen de Normalisation – CEN), the Pacific Area Standards Congress (PASC), the Pan American Standards Commission (COPANT), the African Organization for Standardization (ARSO), the South Asian Regional Standards Organization (SARSO), etc.

BOX 5.2 SOME FORMAL INTERNATIONAL STANDARDS FOR SANITATION TECHNOLOGY SOLUTIONS

(I) ISO 24521: 2016 – Guidelines for the management of basic on-site domestic water services: guidelines for managing basic on-site domestic wastewater services and using technologies in their entirety at any level of development, as applicable (ISO, 2016);
(II) ISO 24516-3:2017 – Specific guidelines for technical aspects, tools and good practices for the managing of assets of wastewater networks to maintain value from existing assets. Applicable to all types and sizes of organizations and/or utilities operating wastewater networks, and all different roles/functions for the management of assets within a utility (e.g. asset owner/responsible body, asset manager/operator, service provider/operator) (ISO, 2018);
(III) ISO 24516–4:2017 – Specific guidelines for technical management of the following individual assets: wastewater pumping stations, retention and detention in the network, wastewater treatment plants and sludge treatment facilities (ISO, 2018);
(IV) ISO 22094:2017 – Non-Sewered sanitation systems – General safety and performance requirements for design and testing of existing on-site sanitation systems: This standard

specifies technical requirements, test methods and sustainability considerations for non-sewered sanitation systems. A non-sewered system for the purpose of this standard is a sanitation system that: (i) is not connected to a networked sewer system, and (ii) collects, conveys and fully treats the specific input to allow for safe reuse or disposal of the generated output (IWA, 24:2016);

(V) ISO 30500:2018 – Non-sewered sanitation systems – Prefabricated integrated treatment units – General safety and performance requirements for design and testing. This standard enhances the effectiveness and availability/appropriateness of the next generation of on-site plants by providing guidance on designing and testing (ISO/DIS, 2018)

(VI) **ISO/NP 23457: 2018** – Community-scale resource oriented sanitation treatment systems: the standard under-development on completion will define requirements and test methods to ensure safety, performance and sustainability of community-scale resource-oriented faecal sludge treatment units that serve approximately 1000 to 100,000 people. The standard will apply to treatment units that (a) primarily treat human excreta, (b) are able to operate in non-sewered and off-grid environments and (c) are prefabricated. The standard will not apply to sanitation treatment units requiring sewer infrastructure or electric grid access. Additionally, treatment units to which the standard will apply exhibit resource recovery capability (e.g. energy, drinking water, fertilizer) and are capable of being energy-independent or energy net positive. The standard is intended to ensure the general performance, safety and sustainability of such units (ISO/NP, 2018).

(VII) EN 12566 – Small wastewater treatment systems for up to 50 population total (PT) refers to a set of European standards that specifies the general requirements for packaged and/or site assembled wastewater treatment plants used for domestic wastewater treatment for up to 50 PT (BS EN, 2016).

The ***informal sanitation standards (ISS)*** are published by other standard development organizations (SDO) such as ASTM International (previously American Society for Testing Materials), NSF International (formerly National Sanitation Foundation), and International Social and Environmental Accreditation and Labeling (ISEAL) (Box 5.3). For instance, most states in the US have adapted the ANSI/NSF International standard for certifying sanitation technologies, which require that products are tested at a national testing laboratory. The certification ensures that only safe designs be permitted and some commercial composting toilets models are ANSI/NSF International certified (Anand & Apul, 2014). These standards are, however, not context-specific and may be unable to address and meet the design, construction, installation and user needs of most challenged countries, particularly in developing countries.

BOX 5.3 SOME INFORMAL INTERNATIONAL STANDARDS FOR ON-SITE SANITATION SOLUTIONS DEVELOPED BY NSF INTERNATIONAL AND AMERICAN NATIONAL STANDARDS INSTITUTE (ANSI)

(I) NSF/ANSI Standard 350 and 350-1: on-site reuse treatment systems – establishes material, design, construction and performance requirements for on-site residential and commercial water reuse treatment systems, including greywater treatment systems (NSF/ANSI, 2011);

(II) NSF/ANSI Standard 40: residential wastewater treatment systems – establishes standards for residential wastewater treatment systems with rated capacities between 400 and 1500 gallons (1514 and 5678 litres) per day (NSF/ANSI, 2010a);

(III) NSF/ANSI Standard 360: Field performance verification – establishes wastewater treatment systems' field performance verification processes such as: consistent site selection, sampling, laboratory analysis and data evaluation methods for obtaining field performance results for on-site wastewater treatment systems (NSF/ANSI, 2010b);
(IV) NSF/ANSI Standard 41: non-liquid saturated treatment systems: certifies composting toilets and similar treatment systems that do not use liquid saturated media as primary means of storing or treating wastes (NSF/ANSI, 2016);
(V) NSF/ANSI Standard 46: Water treatment system components: establishes minimum material, design, construction, product literature and performance requirements for components and devices used in handling, treating, recycling, reusing or disposing wastewater. Components and devices covered by this standard are intended for use with greywater, blackwater or both. It also evaluates the performance of wastewater treatment system components and devices such as grinder pumps, septic tank effluent filters, chlorination devices and UV disinfection devices (NSF/ANSI, 2017).
(VI) (Source: NSF International, 2012)

The national standard bodies (NSBs) of countries' recognized standard agencies are responsible for publishing ***formal national sanitation standards (FNSS)*** and are members of ISO. Some of them are, ANSI, Standards Australia, British Standards Institution (BSI), Thai Industrial Standards Institute (TISI), Standards Organisation of Nigeria (SON), South African Bureau of Standards (SABS) etc. (Box 5.4).

BOX 5.4 SOME FORMAL NATIONAL STANDARDS FOR ON-SITE SANITATION SOLUTIONS DEVELOPED BY AUSTRALIA/NEW ZEALAND STANDARDS BODIES

(I) AS/NZS 1546.1:2008: on-site domestic wastewater treatment units (septic tanks) – establishes performance requirements and performance criteria for septic tanks and enables manufacturers to comply with the performance requirements and criteria. These will enable certification bodies to check that a product conforms to the Standard (AS/NZS 1546.1:2008);
(II) AS/NZS 1546.2:2008: on-site domestic wastewater treatment units (waterless composting toilets) – establishes performance requirements for domestic waterless composting toilet (WCT), performance evaluation test against which any conventional or innovative domestic WCT may be assessed, and provides guidelines for the operation and maintenance of a domestic WCT so that public health and environmental requirements will be met (AS/NZS 1546.2:2008);
(III) AS/NZS 1546.3:2008: on-site domestic wastewater treatment units (aerated wastewater treatment systems) – sets out performance and design requirements, means of compliance and specifications for testing aerated wastewater treatment systems and associated fittings. It also provides guidance on installation, operation and maintenance requirements (AS/NZS 1546.3:2008).
(IV) AS/NZS 1547:2012: on-site domestic wastewater management – establishes performance objectives, requirements and criteria to achieve sustainable and effective on-site domestic wastewater management, and to protect public health and the environment (AS/NZS 1547:2012);
(V) AS/NZS 1546.4:2016: domestic greywater treatment systems – sets out requirements for the design, commissioning, performance, installation and conformity testing of domestic greywater treatment systems (DGTSs) and associated fittings for single domestic premises.

5.3.2 Global challenge of SIS

To address the global challenge of inadequate standardization of sanitation infrastructures and practices, especially for developing countries with huge deficits, international standard bodies such as ISO, ASTM international and NSF International etc. are developing new sanitation and wastewater technical standard specifications for products, services and resource recovery/reuse for sewer and non-sewer infrastructure. SIs will enhance and support innovation of commercially produced sewered and non-sewered technologies as well as help influence adoption of standards into alternative sanitation infrastructures produced locally or by users themselves (which constitutes the bulk of the sanitation infrastructure in developing countries). The role of international standards will be to provide the guide and overall global architecture for the development of more specific and minimum mandatory national and local technical standards for all categories of sanitation products, materials and services. Evidently, countries with appropriate national SIs that are supported with sound regulatory frameworks for sewered and non-sewered technologies, coupled with strong incentives for compliance and enforcement, performed better in the global JMP sanitation coverage monitoring reports (WHO/UNICEF, 2017).

A report on the performance of decentralized wastewater treatment plants (systems serving 1–100 persons) in EU countries showed that the quality performance of sanitation infrastructure was strongly influenced by the European Urban Wastewater Treatment Directives (UWWTD: 91/271/EEC), European Union Standards on Small Wastewater Treatment Systems and specific countries' sanitation infrastructure standards, and well supported with each country's regulatory framework (EU, 1991; BS EN, 2016). A particularly interesting case study is Austria, with 96.78% safely managed and 0.03% unimproved sanitation infrastructure. The country has developed adequate and appropriate standards for the design, installation, operations and maintenance of sewered and non-sewered infrastructure. Austria's decentralized wastewater treatment plants are guided by norms for application, size, building and usage of individual treatment plants with standards specified in ONORM B 2502-1:2001. The norm does not specify binding discharge limits, but states that wastewater will be suitable when the design, installation, operations and maintenance are carried out according to the rules, and it should be possible to obtain the required effluent characteristics (Bisschops *et al.*, 2006).

Ireland, with 87.73% safely managed, 8.36% limited, 2.93% basic and 0.98% unimproved sanitation infrastructure, illustrates the importance of SIs that are backed up with appropriate legislative and regulatory frameworks. The country's legislative and regulatory frameworks for on-site domestic wastewater treatment systems (DWWTS), used by one-third of the population, have undergone a major change with the introduction of registration and inspection regimes (Dubber & Gill, 2014). For instance, before 1991 the soil attenuation system, often referred to as soakaway or soakpit, was traditionally used. From 1991 the leach-field infiltration area became a common practice; here, effluent is split and distributed via parallel subsoil percolation trenches (Gill *et al.*, 2007). Sometimes, a secondary treatment system (packaged treatment plant) has been installed as an alternative to a septic tank or as an addition to provide aerobic treatment of septic tank effluent prior to discharge to subsoil (Dubber & Gill, 2014). The code of practice provides guidelines to ensure DWWTSs are situated and designed in a way that water bodies will be sufficiently protected (IrishEPA, 2009). The Water Services (Amendment) Act 2012 ensures that homeowners with a DWWTS register their system and make sure it does not constitute a risk to human health and the environment through compliance with standards for performance and operations (IrishEPA, 2012, 2013). Also, local authorities are required to establish and maintain a registration system and to undertake inspections to regulate discharges from these systems, while the Environmental Protection Agency of Ireland is the supervisory authority for these inspections (Dubber & Gill, 2014).

Infrastructural standards with strong regulatory enforcement contributed to Australia's progress to achieving 74.21% of safely managed and 25.79% basic sanitation infrastructure. The country has strong sewered and non-sewered SIs that encourage continuous improvements of the sanitation infrastructures. These standards and guidelines assist in design details, development and use for all types of on-site and off-site wastewater treatment and disposal systems available in Australia (see Box 5.4). A recent study reported the use of performance-based Joint Australia and New Zealand on-site wastewater management standards by the Department of Health Western Australia (DOHWA) to address poorly performing or failing on-site wastewater treatment systems (OWTS) (Beal *et al.*, 2005; Bremer & Harter, 2012; Gunady *et al.*, 2015; Ormiston & Floyd, 2004; USEPA, 1997).

Furthermore, Japan, with 99.80% safely managed and 0.20% basic sanitation infrastructure, showcases a successful system of hybridized sanitation infrastructure with standardizations in consonance with ReGenSan Principles (Section 2.4). The government of Japan sets the acceptable sanitation infrastructural technical, industrial, regulatory and compliance enforcement standards (ADB, 2016; Mizuochi, 2011) and the Japanese Industrial Standards agency determines determine capacity and structural standards for all sanitation infrastructure, and prescribes for future innovations. All sanitation technologies are produced, installed and managed under strict Japanese industrial and regulatory guidelines and standards (ADB, 2016), according to type, facility size and administrative support. Sanitation infrastructure can be classified into sewerage system, rural sewerage system and Johkasou system (the Japanese word for on-site sanitation treatment system, mainly used where there is no sewer and in high-population-density areas for on-site wastewater treatment including water reclamation) (Gaulke, 2006; Morita, 2015) (Figure 5.7).

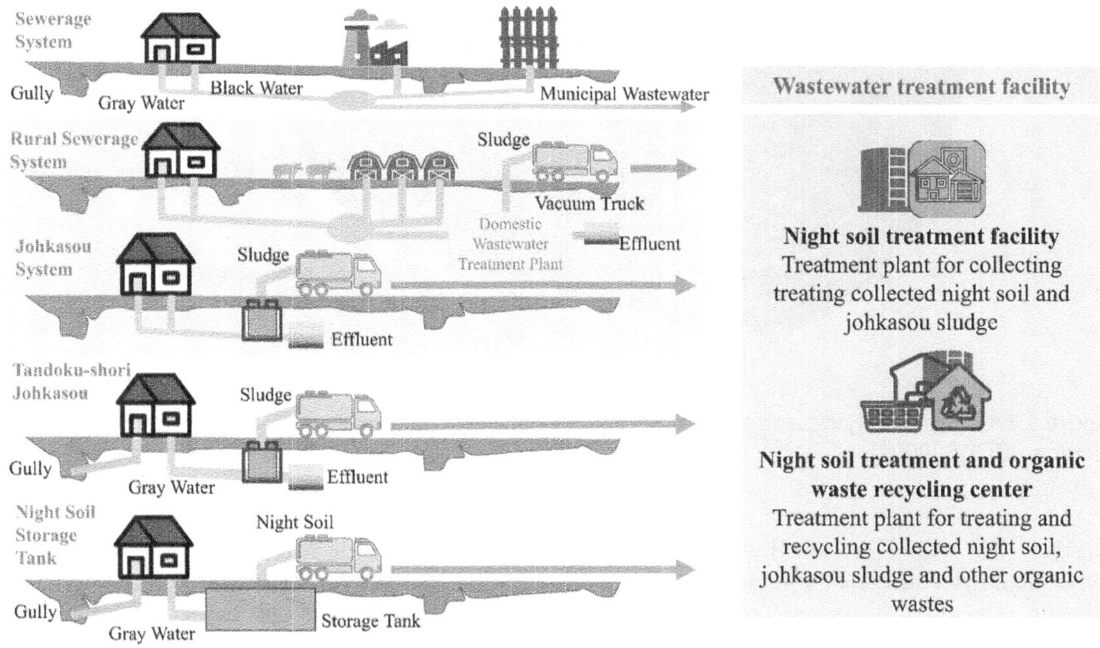

Figure 5.7 Sanitation technological infrastructure in Japan (adapted from JECES, 2009).

The Johkasou industrial structure and capacity standards were enacted in 1960 and 1969 respectively. In 1980, the Johkasou structural standards were revised and upgraded from 'Tandoku-shori Johkasou', which treated only blackwater, to 'Gappei-shori johkasou' (Figure 5.8), which treats all categories of wastewater (blackwater and greywater) from households (ADB, 2016; Mizuochi, 2011; Yang, 2012). Further to ensuring compliance, in 2000 the Johkasou Law was revised and Tandoku-shori johkasou deleted from the definition in the law; the revised law requires that newly installed johkasou be of the Gappei-shori johkasou variety (JECES, 2009; Mizuochi, 2011). The Tandoku-shori johkasou and night storage tanks, used for night soil treatment in the past, are no longer installed and are gradually being phased out. The johkasou system is divided into two major types: (i) in terms of sizes, small and middle (large) scales, and (ii) structural treatment standards, standard designed types based on specific standards and certified structure types that are designed and manufactured freely by other Johkasou manufacturers; but their treatment performance is still evaluated and certified.

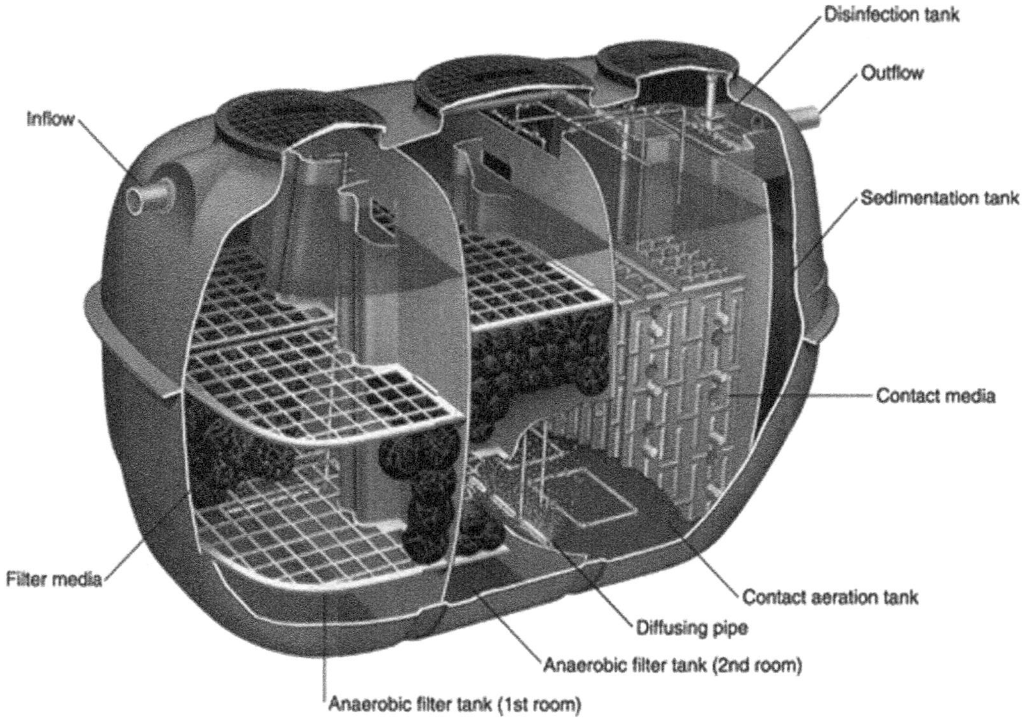

Figure 5.8 Gappei-shori johkasou system structure employing an anaerobic filter – contact aeration process (*Source*: JECES, 2009).

Figure 5.8 shows the small scale Gappei-shori johkasou made of two treatment processes: the conventional activated sludge process and the extended aeration process (JECES, 2009). The system is generally composed of a primary treatment unit (including a primary settling tank or anaerobic filter tank or both); an aerobic biological treatment unit (for example: an activated sludge process, aerobic submerged biofilter, fluidized bed reactor or biofilm filtration reactor) and a disinfection chamber

(JECES, 2009; Oladoja, 2017). Wastewater first flows into the anaerobic filter tank, where plastic filters are submerged and solid matter is separated from wastewater, which then flows into the contact aeration tank before discharging to the environment. The Johkasou treatment is able to reduce pathogens, but performance is dependent on BOD, temperature and recycle rates. It is considered sufficiently reliable and disinfection is typically accomplished with calcium hypochlorite tablets (Gaulke, 2006; JECES, 2009; Kaneko *et al.*, 2001; Magara, 2003; Mizuochi, 2011; Morita, 2015; Nakajima *et al.*, 1999; Yang *et al.*, 2001). Johkasou systems that include nutrient removal processes have also been developed (Nakajima *et al.*, 1999), and a membrane bioreactor for small-scale plants has been tested (Ratanatamskul *et al.*, 1995).

Also, China with a sanitation coverage of 73.43% safely managed, 12.80% basic, 7.99% and 7.17% unimproved sanitation facilities (WHO/UNICEF, 2017), through the 'Toilet Revolution' programme adopted six standardized sanitation infrastructure for rural areas designed to improve coverage of 'safely managed' facilities (Figure 5.9, Table 5.2).

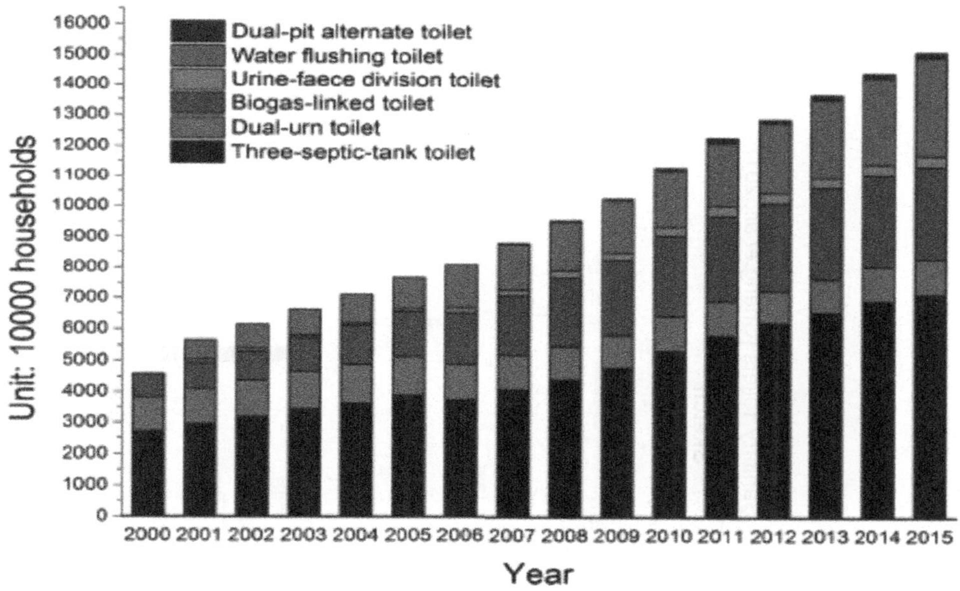

Figure 5.9 Number of six different harmless sanitary toilets installed annually in Chinese rural areas from 2000 to 2015 (*Source*: Cheng *et al.*, 2017).

The standardization of sanitation infrastructure in China has contributed to the provision of appropriate non-sewered sanitation technologies that are reliable, functional and resilient for rapid acceleration of progress in sanitation coverage, especially for rural households (Hu *et al.*, 2016; NHFPC, 2015, 2016 – Table 5.2). For instance, the development of standards for the biogas project for domestic sewage treatment (BPDST) by the Ministry of Agriculture (MOA) in 2014 was responsible for the advancement of biogas sanitation technologies. The specifications issued provided sets of standards for design drawings of biogas digesters for domestic sewage treatment, construction and regulations, and operations and maintenance specification for BPDST (Cheng *et al.*, 2017; Fan *et al.*, 2016; Hu *et al.*, 2016; MOA, 2009, 2014a, b, c, d; Yang, 2004).

Table 5.2 Scenarios of Six Different Harmless Sanitary Toilets (Cheng *et al.*, 2017).

Type of Toilet	Characteristic	Sketch	Suitable Area
Three-septic-tank toilet	It is composed of three septic tanks. The wastewater flows one by one. The first chamber is designed as settler, the second as post-settler and scum separator and the third as storage tank. The size of the last tank depends on seasonal reuse pattern or other available emptying services.	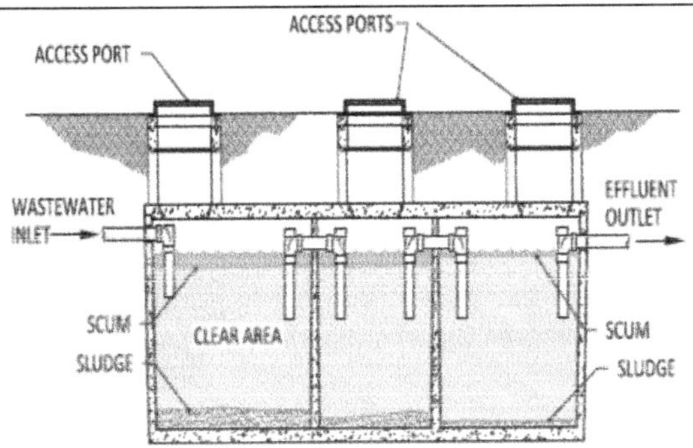	Mostly rural areas, mainly in the south of China. If in the north of China, freezing protection should be considered. In some areas, the tanks are emptied by vacuum trucks and the content is centrally treated.
Dual-urn toilet	The vault is prefabricated by ceramic, cements or composite. The installation is easy. Needs post-treatment of liquid before used as fertilizer or only used as soil improvement before planting seasons.	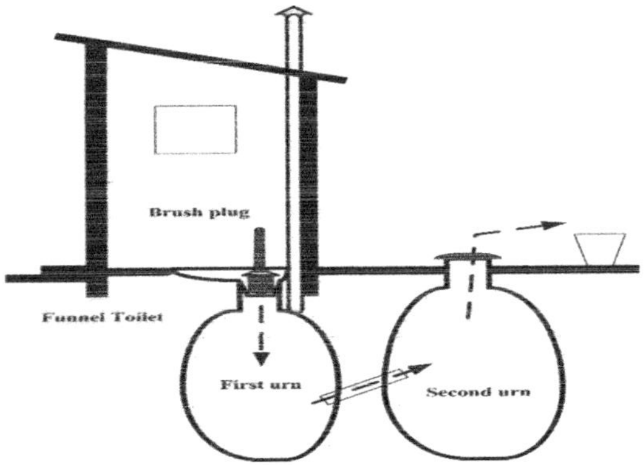	Huai River basin, middle and lower streams of Yangtze River, North China Plain, north-west and south-west arid areas short of rainfall.

Dual-pit alternate toilet	The two pits work alternately. When one is used, the other is stored for composting by correcting the C/N ratio.	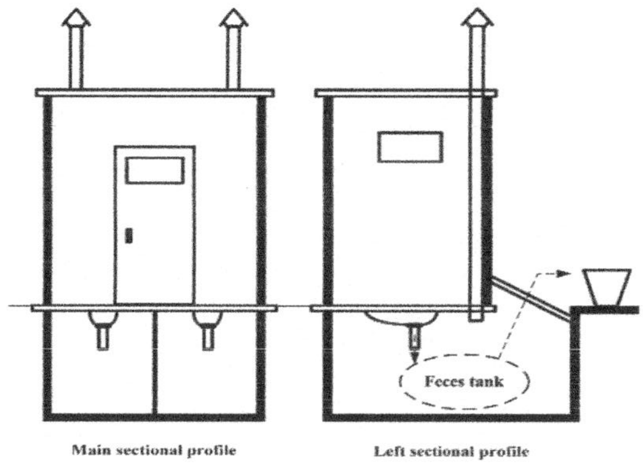	North-west and south-west arid areas short of rainfall.
Biogas-linked toilet	The toilet is built next to livestock sheds. Human waste and animal waste are both used for biogas production, which is used mainly for cooking and lighting.	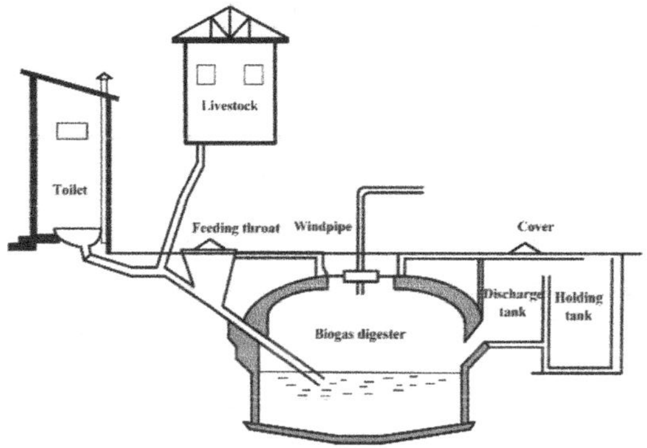	Nationwide, especially south and west of China where climate is warm.

(Continued)

Table 5.2 Scenarios of Six Different Harmless Sanitary Toilets (Cheng *et al.*, 2017) (*Continued*).

Type of Toilet	Characteristic	Sketch	Suitable Area
Urine-faeces division toilet	Urine and faeces are collected separately; urine can be diluted with water and used for fertilizer directly. Faeces are dehydrated for harmless treatment.	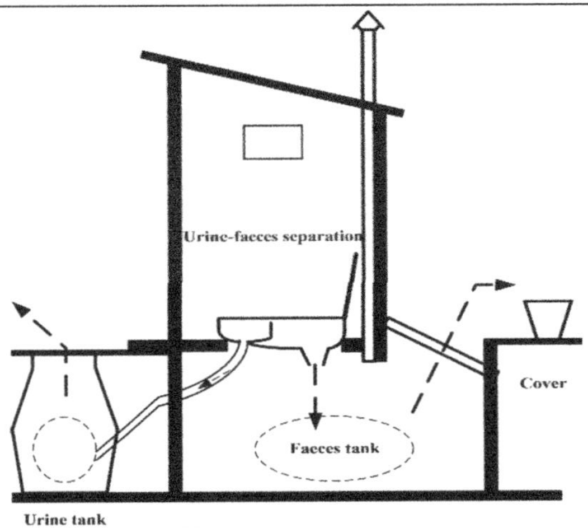	Nationwide, especially arid and water-deficient areas.
Water flushing toilets	The toilet is connected with fresh water supply pipe and complete drainage connection for sewage treatment. It requires building sewage pipeline and treatment facilities, so the overall cost is high.		Nationwide, rich rural areas or suburbs that are connected to sewer pipeline network combined with central or decentralized treatment station.

Table 5.2 illustrates support for national minimum mandatory standards for sanitation infrastructure with technical specifications for products, materials and services. To this effect, it is indicative that countries need to take responsibilities for their SIS needs because even though the global and regional standards may be helpful, they may not adequately fit into the peculiarities of the PSEP/SES of individual countries. Therefore, ReGenSan advocates that critical components of SIs should incorporate the following elements:

(I) Performance requirement and general safety for design, development and testing that covers the whole spectrum of the SSC;
(II) Performance requirements for NoTB that ensures human health and environment are protected;
(III) Compliance monitoring and enforcement to ensure system performance is achieved and maintained;
(IV) Technical specification guidelines for site evaluation, design, construction, operations and acceptable prescriptions for specific site conditions and use;
(V) Education/training for all practitioners, planners, owners and users;
(VI) Certification/licensing for all practitioners/professionals to maintain standards of competence and conduct;
(VII) Programme reviews to identify knowledge gaps, implementation shortcomings and necessary corrective actions;
(VIII) Certifications for products, services, operations and systems; and
(IX) Development of uniform terminologies for use in the sector (NOWRA, 2007).

ReGenSan goes further to prescribe integrative and comprehensive infrastructural standardization solutions for existing improved sanitation (especially for the long-neglected non-sewer infrastructure in developing countries'), retrofitting of unimproved-dysfunctional facilities, ageing and deteriorating sewer infrastructures and the development of nouveau technological solutions, as there is increasing failure of current technologies and underlying challenges of sanitation infrastructure and resource recovery (Gasiorowski-Denis, 2018; Lazarte, 2016).

5.4 REGENSAN TECHNOLOGICAL SYSTEM

Traditionally, technology was referred to as the discourse on or verbal expression of how things are gained or created (Vermaas *et al.*, 2011): that is, the study and development of products and services; but contemporary usage has narrowed it down to just tools and machines or products/services that are meant for human development or progress (Arthur, 2009; Mesthene, 1970). These broad descriptions could define technology as a combination of human knowledge, skills, abilities and processes to design and develop products and services using tools, procedures and raw materials to provide solutions, pleasures, needs and insights for people and communities. Sanitation technologies are, therefore, assets of sanitation infrastructure that work together to provide solutions, whether as individual components (e.g. toilets) or systems of components (e.g. connecting pipes). ReGenSan TeS is, then, the system of products and services (individually and synergistically) that work together to provide sani-solutions and needs (as well as additional solutions) within the sani-sheds for individuals and groups of individuals at place and scale. TeS presents a multi/transdisciplinary SSC process (Van Dijk, 2012; Seetharam, 2015; Tilley *et al.*, 2014b) that consists of three dimensions (Existing Improved Design Technology – EIDT, Restorative Design Technology – ReDT and Nouveau Design Technology – NoDT) and 11 components, all knitted together by the cross-cutting component of infrastructural standardization and supported by sociotechnical regimes (Box 5.5, Figure 5.10).

BOX 5.5 TES DIMENSIONS AND COMPONENTS

(I) *Existing Improved Design Technology (EIDT)* aims at continuous service improvement, expansion of access and coverage of currently adjudged improved and safely managed sewer and non-sewer sanitation facilities estimated to be used by 5 billion people; and consists of four components: (i) service chain functionality and serviceability, (ii) integrity, reliability, safety and quality, (iii) preventive and proactive maintenance, and (iv) inspection and monitoring.

(II) *Restorative Design Technology (ReDT)* contributes to upgrading and retrofitting of dysfunctional, disused, aged and unimproved, basic and limited sanitation facilities for both sewer and non-sewer to improved and safely managed facilities estimated to be used by 2.4 billion people. It consists of three components: (i) rehabilitation, (ii) retrofitting, and (iii) remediation and mitigation.

(III) *Nouveau Design Technology (NoDT)* focuses on issues of emerging contextual novel sanitation solutions due to failure of current technologies to address the underlying challenges behind the lack of sanitation infrastructure and resource recovery as well as provide better alternatives for new development areas and populations exposed to open defecation estimated to be nearly 1 billion; it comprises four components: (i) designs for service and products (DeSaP), (ii) design for nature (DeN), and (iii) design for the base-of-the-pyramid

(IV) (DeBoP).

(a) *Cross-cutting component*: (i) infrastructural standardization – international, regional and national standards, (ii) knowledge and skills, (iii) continuous monitoring.

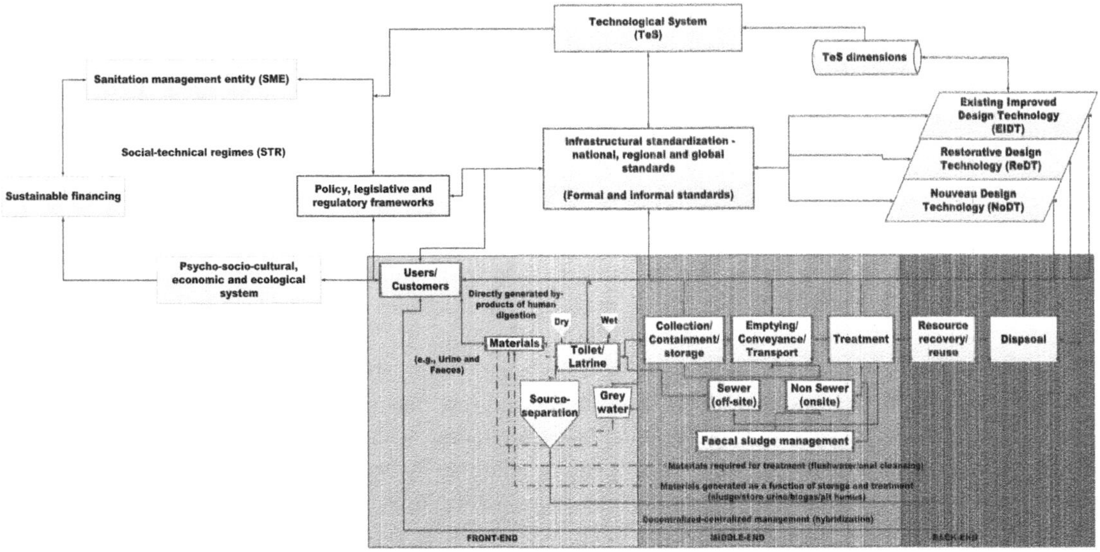

Figure 5.10 Linkages between the TeS dimensions service chain driven by infrastructural standardization.

ReGenSan TeS is driven by standards striving towards safely managed infrastructure that allow for treatment and minimal disposal of treated materials to occur on site, or for them to be collected and managed off site or transported off site by a sewer; and ensures resource recovery/reuse in any of the chosen options within sani-sheds (Orner & Mihelcic, 2018). ReGenSan technological and management solutions align with designs, structures and solutions that let nature do the work and respect the peculiarities of place by drawing on natural processes of decomposition to reintegrate materials into the ecosystem (Figure 2.2) as well as indigenous knowledge of local users and experts. TeS is an integrative-systemic-holistic-hybridized infrastructural (centralized and decentralized) network that works in harmonious relationship within the sani-sheds.

The social-technical regimes (STR) are the technological, organizational and institutional configurations that influence process variation and selection of sanitation infrastructure within the economic and social contexts from which TeS evolved (Altaf, 2011; Hegger *et al.*, 2007; Nilsson & Olsson, 2014; Markard, 2011; Sarkilahti *et al.*, 2017). STR is crucial to achieving effective and acceptable sanitation technological solutions in particular societies (Sarkilahti *et al.*, 2017; Smith *et al.*, 2010); and they consist of material and technical elements, networks of actors and the rules that guide activities (Sarkilahti *et al.*, 2017; Verbong & Geels, 2010). The users/customers are part of the STR and the SSC, and they are critical for improvement and implementation of sanitation infrastructure as they are co-producers of sanitation technologies (Hegger *et al.*, 2007; Spaargaren *et al.*, 2007; van Vliet *et al.*, 2011). Thus saying, '*Let us sort out technology first; if we have proven that it works, we should implement it in society*' (Hegger et al., 2007 – see Section 4.1) is no longer tenable. Users/customers and other stakeholders should be involved in the ideation/conceptualization stage in order to test the feasibility and acceptance of basic designs, technologies and project principles, test-run facilities in target areas and continuously interact with lead users at every project phase, even at the commission stage and follow-up (Lettl, 2007; Tobias *et al.*, 2017). TeS does not simply consider toilets and household level infrastructure/technologies, but also addresses sanitation issues across the spread of sani-sheds (catchment/community safety). The processes and structures are expected to consist of (Figures 5.1 and 5.10 and Table 5.3):

- (I) A safe, convenient, functional, replicable, reliable, scalable, quality and hygienic facility for urination and defecation such as: squatting platform or pedestal seat, optional urinal, provision for anal cleansing, superstructure (safe, well-lit, ventilation to prevent odours, aesthetic), and place for handwashing, bathing, disposal of urination and defecation waste material. This is often referred to as user interface, where users are allowed to access the sanitation infrastructure;
- (II) Collection, storage and treatment of waste materials, which may occur either on site or via transport off site and/or conveyance (sewers) to centralized facility and, sometimes, treating the products generated at the user interface (Figure 5.5);
- (III) Resource recovery/reuse and minimal disposal of treated materials; and
- (IV) Conveyance/emptying/transport, treatment and disposal or reuse of greywater (Nelson & Murray, 2008; Tilley, 2013).

The main goal of TeS is to provide appropriate and suitable treatment and management facilities for by-products of human digestion across the entire scope of the SSC to ensure that infrastructure does not transfer burden to public health, maintains environmental quality and recovers/reuses valuable resources with minimal final disposal of treated materials within sani-sheds in an ecologically acceptable manner (Nelson & Murray, 2008). Table 5.3 shows system boundaries, all inputs, immediate and output products and the specific technologies necessary for each functional group. Within one boundary, households, private sector (business) and the public sector (government) may each have different financial responsibilities (Tilley, 2013). ReGenSan demands that the SSC and its necessary functional groups should be conceived in consultation with all relevant responsible entities (Figure 4.1 and Table 4.1). In other words, system boundaries should be inclusive and drawn from all perspectives of all responsible entities. Therefore, sanitation infrastructure should consider all components required for adequate

Table 5.3 SSC Functional Groups for Describing Sanitation Technologies (Tilley *et al.*, 2014b).

Group	Stakeholders	Technologies	Key Functions
User interface	Homeowner	Toilets (urine separating), urinals, taps, appliances	Determine volume, concentration and composition of products requiring storage and treatment
Collection and storage	Homeowner	Pipes, storage tanks	Hold intermediate products, hygienization (of urine)
Conveyance	Public utility or private enterprise	Pipes, trucks	Move products between sanitation 'nodes'
Treatment	Homeowner, public utility or private enterprise	Filtration, sedimentation, precipitation, stabilization	Change characteristics of products (e.g. dehydration, disinfection)
Use and disposal	Public utility and (or) private enterprise	Solid spreading equipment, pelletizer, urine applicator, incinerator, biogas plant	Distribute products (e.g. urine as fertilizer)

(Reprinted from Tilley E., Ulrich L., Lüthi C., Reymond P. H. and Zurbrugg C. (2014). Compendium of Sanitation Systems and Technologies, 2nd rev. edn. p. 108, with permission from Swiss Federal Institute of Aquatic Science and Technology (EAWAG).)

management of human wastes (Zurbrugg & Tilley, 2009) because sanitation problems can rarely be solved at household levels alone. Households 'export' waste to the neighbourhood, town or downstream population, which implies that TeS should not be restricted to just the front end (user interface) and back end (collection and conveyance). This concept is supported by the newly adopted SDGs JMP of 'safely managed' sanitation facilities that go beyond 'household safety' to 'community safety', and indicates how well sanitation infrastructure separates excreta from human contact beyond the household (Thomas *et al.*, 2018). Starkl *et al.* (2015) also noted that toilet technology comprises of the front end, collection/transport, back end and other components. It becomes crucial, then, that sanitation infrastructural boundaries be extended to include processes and activities that take place in these large spatial sections (sani-sheds) (Cheng *et al.*, 2017; Mang *et al.*, 2016).

In summary, ReGenSan TeS applies engineering, scientific, social science and indigenous knowledge to practically and purposefully design and develop products and services that provide standard-driven sanitation infrastructural solutions to be safely managed for individuals and groups at scale and place. It includes an extensive range of possible combinations of technical components such as means of storage, conveyance and treatment in both sewer and non-sewer systems to deliver the most appropriate, best applicable and standardized technologies compatible with the entities and processes of specific SES (Maurer *et al.*, 2012; Ramoa *et al.*, 2015; Tilley *et al.*, 2014a, b; Zurbrugg & Tilley, 2009). ReGenSan postulates that TeS has 'front-end' (user interface), 'middle-end' (collection, conveyance, treatment) and 'back-end' (resource recovery/reuse and final minimal disposal of treated materials) processes that should operate in an ecologically sustainable manner within the sani-sheds.

5.4.1 ReGenSan technology system dimensions

ReGenSan TeS is made up of three dimensions, namely:

(I) Existing Improved Design Technology (EIDT) aimed at continuous improvement of the currently adjudged 'safely managed' sewered and non-sewered infrastructure;

(II) Restorative Design Technology (ReDT), which deals with upgrading and retrofitting of dysfunctional and unimproved sanitation infrastructure; and
(III) Nouveau Design Technology (NoDT), which supports further research and development of emerging novel sanitation technological solutions based on the core principles of ReGenSan.

NoDT intends to strategically promote and stimulate research and development for innovative solutions to meet the needs of the millions of people without sanitation provision at all as well as provide for new development areas like new estates, apartments, campuses, towns, cities etc. and critical areas such as coastal and arid regions. ReDT, on the other hand, focuses research and development to reduce cases of facility/infrastructure abandonment and investments loss, while EIDT targets the reduction of dysfunction and failure of available facilities/infrastructure. The expectation for these ReGenSan TeS dimensions is that empirical, contextual, analytical and real-life data and information shall be sought and made available for design, manufacture, service delivery, governance, standards formulation and compliance monitoring; all to ensure acceptability and sustainability as well as the targets of the SDG no. 6.

5.4.1.1 Existing improved design technology (EIDT)

EIDT presents comprehensive strategies that assure functionality, serviceability, resilience, reliability and continuous performance improvement of existing sewered and non-sewered infrastructure (Ghosh & Elvira, 2011). The overriding purpose is to support the implementation of practical, efficient and effective hygiene practices and faecal sludge management of on-site sanitation facilities that are used by an estimated 2.7 billion people worldwide (Exley *et al.*, 2014; Nakagiri *et al.*, 2016; Strande *et al.*, 2014). It seeks to reinforce concerns in actual use, cleanliness and overall quality of these facilities (Exley *et al.*, 2015). When non-sewered sanitation systems are improperly operated and inadequately maintained (as most often is the case), it does not matter how many facilities are counted, hygiene quality will still be compromised and disease transmission could ensure (Fewtrell *et al.*, 2007; Gine *et al.*, 2011; WSP, 2009). In such situations, serviceability, reliability, functionality and performance are undermined and this affects the useful life duration of the system (Etnier *et al.*, 2005). Sewered infrastructure is also challenged with the same conditions due to service and performance deterioration, blockages and leakages (which result in infiltration and exfiltration), mostly as a result of senescence (deterioration due to age) (Ahmadi, 2015; Ashley & Hopkinson, 2002; Corcoran *et al.*, 2010; Diagne, 2013; El-Sheikh, 2011; Miszta-Kruk, 2016; Qasem *et al.*, 2009; Reynolds & Barrett, 2003; USEPA, 1991a, b). Consequently, EIDT calls attention to the challenges of the adjudged 'improved' facilities with inadequate performance discharging untreated sewage into the receiving environment (Starkl *et al.*, 2013); and in so doing redirects research and development to solutions, especially with non-sewered structures without standardized operation and maintenance criteria.

Functionality and serviceability are broader perspectives in measuring performance and service reliability of sanitation infrastructure guaranteed by certification procedures based on international and national standard certification approval bodies (OFWAT, 2015; Truffer *et al.*, 2013). EIDT seeks to address functional and performance reliability deficiencies inherent in current 'improved' sanitation infrastructures as well as technological gaps in the SSC in order to ensure reliable and resilient existing improved sanitation infrastructure. This dimension is based on the ReGenSan Principles of ApT, HS, NToB, SHI, PaS and PaP (see Section 2.4).

EIDT enhances infrastructure service reliability by ensuring development and adoption of appropriate infrastructure standards, codes of practice, quality control and assurance plans and procedures as well as preventive and proactive maintenance to overcome failure. Reliability implies that the sanitation infrastructure should perform at the specific operational standard requirements, free from failure for a specific period of time under specific conditions (Niku *et al.*, 1979, 1981, 1982; Singhirunnusorn & Stenstrom, 2009; Von Sperling & Oliveira, 2007). Technically, the concept of reliability is the measure of 'probability of success' or 'probability of adequate performance', which is the percentage of the time a facility performs according to statutory regulatory and standard requirements (e.g. the percentage of the time that effluent concentration meets the requirement in the case of WWTPs) (Niku *et al.*, 1981). Existing improved sanitation infrastructure (sewered and non-sewered) is completely reliable if performance response has no failure. This implies that infrastructure could deliver expected services, but with unacceptable risks such as effluent discharge requirement violation (EU, 2012; Starkl *et al.*, 2013).

EIDT proposes that improved sanitation infrastructure is operated and maintained within limits of acceptable measures of risks in order to achieve reliability and safely deliver expected services (Niku *et al.*, 1979, 1981, 1982). The EIDT process is designed to continuously edit out problematic and bottleneck issues that may impact negatively on service reliability in order to ensure optimization, especially with currently aged 'improved' sanitation amenities. This will enable effective and efficient services that respond to higher standards of operational safety, health and environmental protection (El-Sheikh, 2011). Existing improved sanitation provisions (sewered and non-sewered) are challenged by the following factors:

(I) Non-compliance with new regulations on public health, environmental quality and operational standards;
(II) Service expansion due to population growth;
(III) Malfunctions of units and/or equipment in treatment systems;
(IV) Changing outdated equipment and/or technologies due to unavailability of spare parts; and
(V) Retrofitting resource recovery and reuse into existing improved infrastructure (El-Sheikh, 2011).

It seeks to maximize the use of these facilities and minimize rehabilitative cost and risk of failure and its attendant consequences while ensuring service reliability (Diogo *et al.*, 2018; Ward & Savić, 2012). All these should ultimately lead to a crossroad with technology upgrade at the centre as illustrated in the Japanese 'Tandoku-shori Johkasou' (Figures 5.7 and 5.8) (which only treated blackwater), reinforcement in the 'Gappei-shori johkasou' (which treat all categories of wastewater, both blackwater and greywater) from households (ADB, 2016; Mizuochi, 2011; Yang, 2012) (see Section 5.3). To this effect, some strategies that could ensure sustained continuous service improvement and reliability with due regard to environmental quality, public health and safety include:

(I) Functionality and performance assessment to ensure expected reliable and functional services;
(II) Service delivery assessment to ensure appropriate, adequate and acceptable minimum levels of services; and
(III) Integration of risk-based management to ensure that existing improved sanitation infrastructure gives due consideration to 'failure prevention management' and process reliability through continuous monitoring of performance as well as treatment (ASCE, 2017; Marlow *et al.*, 2013).

Therefore, the cardinal role of EIDT is to ensure long-term reliability and avoid the risk of 'slippage', a term coined in India to mean the slipping back of a community from fully covered with improved services to partially or non-covered (Lockwood & Smits, 2011; Rosenqvist *et al.*, 2016; UN-Water & WHO, 2012; Verhagen & Carrasco, 2013).

5.4.1.2 Restorative design technology (ReDT)

ReDT refers to the process of retrofitting and upgrading failed, dysfunctional and abandoned sanitation infrastructure to meet stringent new regulatory and standard requirements, rehabilitate poorly performing unimproved facilities and elongate the lifespan of ageing systems (Bozkurt *et al.*, 2015; Brepols *et al.*, 2008; El-Sheikh, 2011; USEPA, 1974, 2009); with the intention of transforming the 2.4 billion unimproved facilities; meeting the SDG 6 targets (Dixon *et al.*, 2014; Eames, 2011; Giné-Garriga *et al.*, 2017; Pham & Kuyama, 2013; WHO/UNICEF, 2017); and the estimated 1.5 billion sewered facilities without adequate treatment, producing effluents of questionable standards (Baum *et al.*, 2013; Mara, 2003). System dysfunction and failure increase exposure to disease-causing pathogens present in unconfined or poorly disposed-of human excreta and the associated health risks are well established (Boschi-Pinto *et al.*, 2006; Cairncross & Feachem, 1993; Ekane *et al.*, 2014). Therefore, ReDT aims to ensure sustainability service delivery at appropriate levels for quality, quantity, convenience, comfort, continuity, efficiency and reliability (Brikke, 2000). It is based on ReGenSan Principles of ApT, BaB, HS, NToB, SHI, PaS, PaP and RoDF (see Section 2.4) and its key components are rehabilitation, retrofitting, remediation and mitigation (Jenkin & Pedersen, 2009; Mang & Reed, 2012a, b; Mang *et al.*, 2016; Starkl *et al.*, 2013; WSDHWMP, 2002). The main objectives of ReDT include but are not limited to:

(I) Failure management and enhanced predictive maintenance;
(II) Repair and replace unit or equipment malfunctions of treatment facilities, SSC/SVC networks;
(III) Retrofit and optimize resource recovery and reuse into existing sanitation infrastructures;
(IV) Ensure compliance with new regulatory standards with regards to pollution control, particularly effluent discharges, reduction of carbon emissions and energy, etc.;
(V) Repair and replace outdated equipment and/or technologies;
(VI) Clean up and disinfect sites degraded and damaged by bad sanitation practices; and
(VII) Effect infrastructure end-of-life decommissioning.

Considering that 5 billion people are recorded as having 'improved' facilities, 2.4 billion are in the 'unimproved' bracket and 1 billion do not have any facilities at all and are therefore involved in 'open defecation' (WHO/UNICEF, 2017), then achieving the SDG targets by the deadline of 2030 could be made more likely by focusing on upgrading the infrastructure of the 2.4 billion's 'unimproved' (dysfunctional, failed and inadequate) facilities. This is where ReDT comes in to provide rehabilitation, retrofitting, remediation and mitigation strategies in order to increase the percentage of improved and safely managed facilities. In future, the focus will be more on ReDT than building new infrastructure so as to maintain performance quality and continuous improvement and avoid catastrophic failure consequences (Brepols *et al.*, 2008; Marlow *et al.*, 2007).

ReDT seeks to sustain the functionality and performance of sanitation systems that often result in failure, which is a significant threat to infrastructure sustainability (Fenner & Saward, 2004). Failure arises because of structural inadequacies, material degradation and poor construction and workmanship as well as lack of operations and maintenance (Brepols *et al.*, 2008; Brikke, 2000; Cookey *et al.*, 2016). According to the American Bureau of Shipping (ABS) (2004), infrastructure conditions deteriorate over time and before it can be detected (point P in Figure 5.11). As time passes, the infrastructures' conditions worsen and at the end they reach the failure point (point F in Figure 5.11). Figure 5.11 is a hypothetical schematic curve of infrastructure deterioration over time. However, in practice it can vary and can in some cases be very inconsistent (ABS, 2004; Ahmadi, 2015).

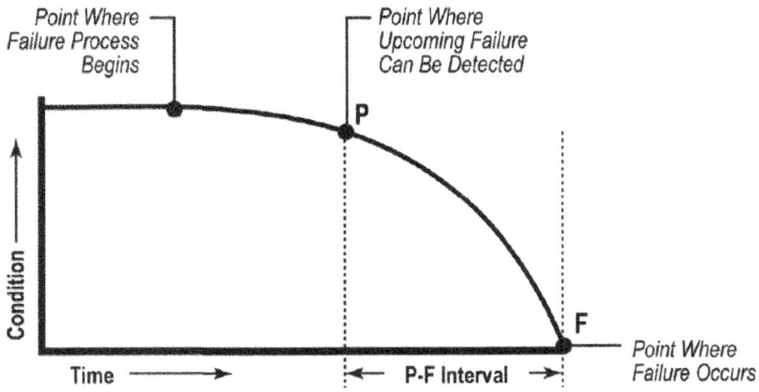

Figure 5.11 Failure process (*Source*: ABS 2004).

On-site sanitation facilities failures are conditions that threaten public health by inadequately treating sewage or by creating a potential for direct or indirect contact between sewage and the public (WSDHWMP, 2002). Failure can be defined as: (i) *failure to dispose,* often referring to a process where sewage is backing up into the house, or surfacing on the ground or entering surface water, and/or (ii) *failure to treat,* which refers to situations where inadequately treated sewage contaminate ground or surface water (CWTRC, 2003; WSDHWMP, 2002). Common non-sewered sanitation infrastructure failures include:

(I) Hydraulic failure – i.e. untreated or partially treated sewage pooling on ground surface, sewage backup in plumbing fixtures and sewage breakouts on hill slopes;
(II) Pollutant contamination of groundwater – i.e. presence of high nitrate levels in drinking water; taste or odour and toxic substances in shallow wells or borehole water caused by untreated, poorly treated or partially treated sewage;
(III) Microbial contamination of ground and surface water – i.e. contamination of drinking water sources such as wells with faecal bacterial or other faecal indicators; and
(IV) Nutrient contamination of surface water – i.e. presence of algae bloom, high aquatic plant productivity, low dissolved oxygen concentrations etc. (Angoli, 1988; CWTRC, 2003; McGauhey & Winneberger, 1964a, b, 1965).

Non-sewered infrastructure failures or dysfunctions, with the exception of problems associated with age, can be attributed to the following factors:

(I) Improper and inadequate facility installations – siting, design, technology application and construction;
(II) Improper operations – this may include change of use at the facility, as in the case of conversion of residential units to commercial or manufacturing purposes and the introduction of incompatible materials that disrupt the biological, chemical or physical processes within the treatment units;
(III) Inadequate maintenance;
(IV) Hydraulic and organic overload; and
(V) Physical, structural and mechanical components failure (Angoli, 1988; CWTRC, 2003; McGauhey & Winneberger, 1964a, b, 1965).

Sewered infrastructure and its associated treatment facilities are affected by many factors such as physical, chemical, biochemical and biological (Al Barqawi, 2006; Al-Barqawi & Zayed, 2006; Best Practice 2003;

Kleiner & Rajani, 2001; Qasem *et al.*, 2009; Urhaman, 2007). These may damage different parts of the infrastructure, which often results in failure and/or collapse (ASCE, 1994; Ward & Savić, 2012). These failures can manifest as hydraulic, structural and/or environmental failures (Abraham & Gillani 1999; Ward & Savić, 2012), as well as treatment plant failure (Brepols *et al.*, 2008; El-Sheikh, 2011):

(I) Structural failure – this type of failure may include mechanical/physical destruction of sewers by erosion, corrosion, deformation and cracking. There may be varying degrees of severity, ranging from minor fabric defects, such as cracking, to complete loss of structural integrity where a full or partial collapse may be observed (Anbari *et al.*, 2017; ASCE, 1994; PDSS, 2009);

(II) Hydraulic failure – consists of wastewater leakages from sewer networks and groundwater infiltration, soil penetration into networks, blockages caused by sedimentation of soil, dust and mud at the bottom of the pipe, fatty deposits (usually at the top of the wastewater flow), sediments due to the effect of groundwater flow, deposits of sulphate and fouling of sewer, penetration and growth of root into sewer and sewer blockage caused by wastes (Anbari *et al.*, 2017; ASCE, 1994);

(III) Environmental failure – refers to severe adverse impacts on public health, environmental and ecosystem services resulting from incidences such as combined sewer overflow (CSO), infiltration and exfiltration, flooding and pollution incidences (Ahmadi, 2015; Anbari *et al.*, 2017; Ashley & Hopkinson, 2002; Corcoran *et al.*, 2010; Diagne, 2013; El-Sheikh, 2011; Miszta-Kruk, 2016; Qasem *et al.*, 2009; Reynolds & Barrett, 2003; USEPA, 1991a, b).

(IV) Wastewater treatment plants (WWPT) failure – refers to inadequate levels of treatment provided by the WWPT, often due to lifespan expiration; WWTP efficiency and effluent quality are reduced. This then causes hydraulic bottlenecks that result in increase of wastewater flow or load and the challenge of higher treatment standards due to change of legal frameworks (Brepols *et al.*, 2008; El-Sheikh, 2011)

ReDT proposes the following actions and strategies to address the challenges of infrastructural dysfunctionality and failure:

(I) Conditions assessment – as a ReDT management tool for existing sanitation infrastructure to ensure suitable and appropriate predictive maintenances and fill specific operations-related information gaps in order to facilitate decision-making. This assessment shall provide information such as: the consequences of sanitation infrastructure failure at local and system levels, cost of infrastructure rehabilitation, retrofitting, remediation and mitigation, alternatives available giving the results of the assessment (partial replacement, non-structural repair, deferment etc.) (Marlow *et al.*, 2007),

(II) Risk-based management – aims to achieve optimum lifecycle management of sanitation infrastructure through consideration of 'risk to service' provision with adequate attention given to 'consequences of failure' and 'failure management' (ASCE, 2017; Marlow *et al.*, 2007). According to the American Bureau of Shipping (ABS, 2004), the main steps of the development of a risk-based approach include (Ahmadi, 2015; Le Gauffre *et al.*, 2009; Tran *et al.*, 2009):

  (a) Determine the risk of potential failure for each infrastructure,
  (b) Identify the degradation mechanisms that will lead to components failures,
  (c) Select most effective inspection techniques,
  (d) Develop and optimize inspection plans, and
  (e) Analyze data from inspections.

These approaches involve the adoption of retrospective and interdisciplinary frameworks to identify sustainable and innovative rehabilitation, retrofitting, remediation and mitigation strategies. The retrospective approach aims to look back to the history of related infrastructure, including the initial planning and design processes, to understand the current failing status. The interdisciplinary approach reflects on the necessity of understanding the influence of PSEP on sanitation technological solutions and management (see Chapter 4) (Cuppens *et al.*, 2013; Tiberghien *et al.*, 2011).

5.4.1.3 Nouveau design technology (NoDT)

NoDT supports the idea that existing sanitation technological infrastructure could be designed better and made more regenerative. As a matter of fact, upgrading today's ageing sewered and non-sewered treatment architecture to new generation resource recovery facilities will require the development and deployment of innovative technologies and processes (Mihelcic *et al.*, 2017) (see Section 5.5.2). A major reason for this is the fact that under the current regulatory frameworks, existing sanitation technologies have a narrow (but very important) focus on treatment of waste to reduce risks to human health and the environment (Mihelcic *et al.*, 2016). Overcoming these barriers requires new-breed technologies that are energy-efficient, recover value-added resources, encourage reuse and are equipped with advanced devices to achieve desirable outcomes (SDE/NSF, 2016; USDE, 2014, 2015). NoDT seeks to inspire the creation of replicable archetypes for sanitation technologies and processes that are systemic, contextual, integrate eco-efficiency with ecological-based approaches and mimic natural systems (see Section 2.4.2) to deliver improved functional components and systems (see Figure 2.2, Section 2.4.6) within the sani-sheds (see Section 2.4.7).

For instance, a septic tank wastewater-processing facility that works well on one soil type or geological structure may not work at all in another. Thus, within a regional system, there might be several different technologies in operation with each responding to local conditions, but playing its part in the larger whole (Lyle, 1994). Tilley *et al.* (2014a) noted that acceptance of any technology or practice is inherently contextual and not necessarily guaranteed. They noted that waterless technologies based on separation and dehydration have ecological benefits, but may not meet cultural expectations in communities that use water for anal cleansing (see Section 4.2.1.1). NoDT should provide solutions at a scale in sani-sheds that reflect the SES and explore ways to develop sanitation technologies and systems where toilets are viewed from within the SSC and extending to the final point of disposal and/or end use of faecal sludge (Dias, 2015; Lyle, 1994; Mang & Reed, 2012a; Spaargaren *et al.*, 2006; Tilley *et al.*, 2014a, b) (see Figures 2.2 and 5.10, Section 4.2). The three principal components of NoDT for technology development are:

(I) Design for products and services (DeSaP),
(II) Design for nature (DeN),
(III) Design for the base-of-the-pyramid (DeBoP).

The special features of NoDT processes include: reasonably fast technology, not necessarily independent of existing infrastructures but working in synergy (see Section 2.4.4); appropriateness (see Section 2.4.1); end users' preference (see Section 2.4.8); no mixing of human faeces with other waste; no-linear-flow; NToB (see Section 2.4.5); incorporation of resource recovery and reuse (see Section 2.4.9); affordable technologies that are resource-reclaimable (Gretchen, 2012; Fan *et al.*, 2017; Sunita, 2012; van Loosdrecht & Brdjanovic, 2014); support food production, prevent environmental pollution and minimize demand on other resources (Fan *et al.*, 2017; Zhou *et al.*, 2010). The 'place and scale' concept ensures that sanitation infrastructure functions in close concert with the natural processes within a range of local environmental conditions (see Section 2.4.7).

The sanitation sector has been risk-averse and slow to adopt new technologies because the path to technological adoption depends on time-consuming, costly and often repetitive cycles of testing and validation, which constitute enormous barriers to investment in novel technologies (Mihelcic *et al.*, 2017). These barriers to investment in NoDT could be technical, regulatory and managerial issues that result in the 'valley of death' syndrome, defined as as situation where a new technology does not advance from demonstration to commercialization (Frank *et al.*, 1996) because a technology developer is unable to obtain financing for scale-up and manufacturing (Frank *et al.*, 1996; Murphy & Edwards, 2003). Details of these barriers are:

(I) Overall aversion to risk that is related to the conservative nature of the environmental regulatory agencies,
(II) Lack of test data on operational performance and cost, and the presence of strict and duplicative requirements;
(III) Technology transfer not seen as important job function for many employed at the sanitation infrastructure management and/or responsible agencies/departments;
(IV) Future environmental regulatory requirements seen as uncertain while regulations currently advance the goal of adequate treatment over goals of sustainability and resource recovery;
(V) Time required for technology development may not fit with the sanitation infrastructure management and/or responsible agencies/departments schedule for capital investment; and
(VI) Failure to transfer new technologies to other sanitation infrastructure management and/or responsible agencies/departments (ELI, 1998; Frank *et al.*, 1996; Greiner & Franza, 2003; Mihelcic *et al.*, 2017).

Current sanitation configurations entail devaluation of human waste, loss of nutrients, waste of water and emission of residual pollutant loads to the environment and no guarantee of sustainability (Cordella *et al.*, 2011; Larsen *et al.*, 2016; Poortvliet *et al.*, 2018). Despite the challenges mentioned above, several new initiatives have been developed that champion new conceptualizations (Hernandez-Leal *et al.*, 2017; Londong, 2013) poised at maximum recovery of energy and resources and minimum use of potable water and reduced emissions (like pharmaceuticals) into the environment (Tuantet *et al.*, 2014); and they operate by integrated holistic systemic approaches rather than the end-of-pipe approach (see Sections 2.3.1, 2.3.2 and 2.4.6, Box 5.6). The drivers for NoDT are hereby attributed to:

(I) Global resource scarcity such as phosphate shortages on the world markets, which are abundant in sewage and critical for agricultural production;
(II) Rising world energy prices that could make large-scale recovery of resources from sanitation very attractive;
(III) Growing awareness of sustainability, liveability and the sanitation economy (Davis, 2008; Poortvliet *et al.*, 2018; Sarkilahti *et al.*, 2017; Swart & Palsma, 2013); and
(IV) Sanitation infrastructure coverage expansion and service improvement (WHO/UNICEF, 2017).

By extension, investing in NoDT could tackle the many dynamic challenges of populations with the existing improved and unimproved infrastructure as well as those without any facility (Mihelcic *et al.*, 2017). Below are some details of the principal components of NoDT.

BOX 5.6 SOME NEW SANITATION TECHNOLOGY CONCEPTS

(I) Alternativen der kommunalen Wasserversorgung und Abwasserentsorgung (Alternative of Municipal water supply and sanitation) AKWA 2100 – is a German project design to identify urban water and wastewater infrastructure concepts intended to reduce the unsustainability of future infrastructure. AKWA 2100 project developed three scenarios with a long-term perspective of 2050 and plans to integrate technological, organizational and institutional innovations into coherent alternative urban water and wastewater systems with improved eco-efficiency (Hiessl *et al.*, 2001, 2003; Schwartz, 1991);

(II) NOVAQUATIS – the Swiss Federal Institute of Aquatic Science and Technology (EAWAG) interdisciplinary project that concentrated on urine source separation technology, resulting in modern urine separating (NoMix) flush toilets (Larsen *et al.*, 2009);

(III) Decentralized sanitation and reuse (DESAR) – Dutch DESAR project aims at radical innovation in sanitation and wastewater infrastructures. The main idea behind DESAR is that water, energy and nutrients are resources in an ecological loop that is broken down in contemporary wastewater infrastructures. Centralized collection and treatment of wastewater means that the useful value of nutrients (phosphorus, potassium, nitrogen) is for a large part lost, while drinking water consumption is unnecessarily high. DESAR technologies aim to close this loop by minimizing water consumption and recovering nutrients from human waste and energy production (using anaerobic digestion) (Larsen *et al.*, 2013; Lens *et al.*, 2001); and

(IV) Reinvent the toilet challenge – a programme of the Bill & Melinda Gates Foundation that calls for implementable ideas at scale aimed at fostering innovative low-cost toilets that sanitize human excreta and recover valuable resources without sewer connections or harmful discharge. The challenge focuses on developing solutions in four specific areas associated with non-networked sanitation for urban areas: (i) hygienic and adequate pit/tank emptying, (ii) recovery of energy from faecal sludge as a means for safe and affordable treatment and disposal, (iii) appropriate sanitation solutions for areas challenged by an abundance of water (e.g. communities that face seasonal flooding, high groundwater tables, riparian or tidal communities, etc.), and (iv) easy to clean, attractive and affordable latrine pans/squatting platform technologies that enhance latrines (Bill & Melinda Gates Foundation, 2011; McConville *et al.*, 2014).

(V) Stimulating Local Innovation on Sanitation for the Urban Poor – a programme of the Bill & Melinda Gates Foundation for developing countries in sub-Saharan Africa, South-East Asia and South America. The project aimed to stimulate local innovation on sanitation for the urban poor and strengthen the sanitation sector in developing countries through education and training. The project was jointly executed by IHE Delft, the Netherlands; Asian Institute of Technology (AIT), Thailand; Institut Teknologi Bandung, Indonesia; University of Cape Town, South Africa; Kwame Nkrumah University of Science and Technology, Ghana; Makerere University, Uganda; International Institute for Water and Environmental Engineering, Burkina Faso; Federal University of Minas Gerais, Brazil and Universidad del Valle, Cali – Colombia (IHE, 2011).

5.4.1.3.1 Design for products and services (DePaS)

DePaS is an innovative component of NoDT that integrates products, services, supporting networks and infrastructure for the design of novel sanitation technologies to jointly satisfy users/customers' needs and ensure NoTB (see Section 2.4.5) as well as recover and reuse resources that were not possible to use with the traditional linear treatment disposal technologies (Ceschin & Gaziulusoy, 2016; Geum & Park,

2011; Goedkoop *et al.*, 1999; Mont, 2002; Tukker & Tischner, 2006). The idea is that new generation sanitation technologies should not only focus on designing and producing 'toilets', but should ensure an integrated offering composed of products, service and infrastructure (rather than only toilet products) by pre-considering and incorporating the entire SSC (CSTEP, 2016; Geum & Park, 2011; Tilley *et al.*, 2014b; Van Dijk, 2012; Wirmer, 2014; Zakari *et al.*, 2015). It requires active engagement by the users/customers, who should have a proper understanding of what alternatives they already have in place and better gauge what they may choose from (for instance, the perspective of user-centred design and consumer research) (Catulli, 2012; Catulli *et al.*, 2013; Rexfelt & Hiort af Ornas, 2009; Vezzoli *et al.*, 2015). It requires appreciating users as part of the innovation group or community from the perspective of PSEP (see Section 4.2). The aim is minimization of material disposal to the ecosystem and maximization of service output, user satisfaction, sustainability and resource efficiency (Mihelcic *et al.*, 2017; Tukker, 2015).

The objective of DePaS is to ensure that technology designers/producers:

(I) Deliver desired functional sanitation products to users/customers;
(II) Integrate service consideration into sanitation components/unit productions, especially after-sales maintenance and other ancillary services;
(III) Close material cycles by ensuring resource recovery and reuse as well as reducing material disposal;
(IV) Provide technological solutions seeking synergy in integrating system elements along with improving resource and functional efficiency of each element;
(V) Improve the total value for users/customers based on increased servicing and better service components, which include activities and schemes that make sanitation infrastructure last longer and extend its function; and
(VI) Improve relationship with users/customers through increased contact and information about preferences (Mont, 2002).

Products are technological infrastructure made up of hard or soft components (Lewis, 2008; Niskanen, 1991 – see Section 5.2.1). The hard parts usually refers to user interface, collection and storage, conveyance, treatment, recovery, reuse and disposal technologies (Mihelcic *et al.*, 2017; Tukker, 2015 – see Section 5.1.1). The soft parts are the social supporting networks, technological testing, validation, verification and certification as well as institutional and organizational structures that ensure adequate interactions and engagement of all actors involved in innovative production and adaptation processes (Geum & Park, 2011; Lewis, 2008; Mihelcic *et al.*, 2017; Mont, 2002; Niskanen, 1991). *Service* is the value and satisfaction products deliver to users/customers as well as product-use-phase activity that the producers render to their customers to ensure maximized benefits (Lewis, 2008; Mihelcic *et al.*, 2017; Mont, 2002). Service also indicates utility, which is the aggregate sum of satisfaction or benefits that accrues to an individual and is derived from the use of sanitation products and its associated activities (Principles of Economics, 2017).

Several studies have recommended measurement criteria for sanitation products' services, such as: functionality, serviceability, reliability, quality, safety, standardization, availability, physical accessibility, affordability and acceptability (Gasiorowski-Denis, 2018; Giné-Garriga *et al.*, 2017; Kaminsky & Javernick-Will, 2015; Kvarnström *et al.*, 2011; Lazarte, 2016; Markard, 2011; NOWRA, 2007; Olsson, 2013; Starkl *et al.*, 2013; Thomas *et al.*, 2018; WHO/UNICEF, 2017). Services could be provided through training, after-sales maintenance and other ancillary supports (Mihelcic *et al.*, 2017; Tukker, 2015) as well as additional resource recovery and reuse from wastewater, faecal sludge and treatment by-products, which could also provide irrigation, fertilizer and energy generation (Geum & Park, 2011; Mihelcic *et al.*, 2017; Mont, 2002; Murray & Buckley, 2009).

DePaS advocates for redesign of sanitation technologies (Mont *et al.*, 2006; Vezzoli *et al.*, 2014; Vezzoli & Manzini, 2008) by innovating and mitigating the negative environmental practices of existing sanitation technologies (Ceschin, 2014). The aim is to incorporate cyclical thinking, cradle-to-cradle and ReGenSan principles and concepts into new generation sanitation technological and infrastructural development (Armstrong *et al.*, 2015).

The main strategy is to shift from just providing 'toilets' to providing functional, reliable, quality, secure and safe sanitation solutions (Cherubini *et al.*, 2015) by offering added value (Barquet *et al.*, 2013) as well as integrated solutions (Manzini & Vezzoli, 2002) to develop ReGenSan compliant infrastructure (Corvellec & Stål, 2017). It also provides incentives for technology designers and producers to ensure that sanitation infrastructures are standardized by appropriate standard bodies (see Section 5.2.2) and used as intensively as possible in order to make them as cost-and material-efficient as possible; and this will shift the goal from being solely that of 'safe disposal' to ensuring that 'resources are safely captured for reuse'. DePaS reinforces the notion that reuse or utilization of wastewater, faecal sludge and its embodied resources are components of complete sanitation in TeS. Hence, the design and development of novel reuse-oriented sanitation infrastructure can leverage the improvement of long-term efficacy of a treatment scheme by providing tangible and quantifiable incentives for sound operation and maintenance that exceed those associated with running disposal facilities (Ceschin & Gaziulusoy, 2016; Geum & Park, 2011; Goedkoop *et al.*, 1999; Mont, 2002; Tukker & Tischner, 2006). DePaS facilitates designing site-specifically based on reuse-oriented facilities (on the PSEP foundation – Section 4.2.1) from the outset of the planning process to ensure standards are met for for agriculture and other applications.

5.4.1.3.2 Design for nature (DeN)

DeN is nature-inspired design with the focus being to develop nature-based sanitation technological solutions (Nesshöver *et al.*, 2017) using 'cradle-to-cradle' strategies (CTC) (Braungart *et al.*, 2007; Ceschin & Gaziulusoy, 2016; McDonough & Braungart, 2002), biomimicry (BM) (Badarnah & Kadri, 2015; Ceschin & Gaziulusoy, 2016; Gamage & Hyde, 2012; Helms *et al.*, 2009; Kennedy & Marting, 2016) and the application of the principles of ecological engineering (Lyle, 1984, 1994; Todd & Josephson, 1996; Todd & Todd, 1980, 1984, 1994) (see Section 2.4.2). DeN involves studying and mimicking nature's models, processes and ecosystems by adopting them to solve sanitation challenges by using ecological standards to judge the rightness of innovations (see Section 2.4.2). It must be noted that mimicry that isolates principles, structures or processes from nature lacks the true application of DeN (Ceschin & Gaziulusoy, 2016; Reap *et al.*, 2005; Volstad & Boks, 2012).

DeN reflects the principle of eco-effectiveness that eliminates or minimizes waste by converting it to resources, and also integrates concepts of low energy consumption and mechanical technology, which are better choices of low-cost treatment facilities for rural areas (Lens *et al.*, 2001; Mahmood *et al.*, 2013). The current nature-based technologies that require innovative technological optimization include:

(I) Wetland systems (WSs), designed to utilize natural processes involving wetlands, vegetation, soils, solar energy, water and associated microbial assemblages to assist in treating faecal matter and domestic wastewater. They are designed to take advantage of many of the same processes that occur in the ecosystems, but do so within a more controlled environment (Brix, 1997; Boonsong *et al.*, 2003; Edwards, 2004, 2005; Edwards *et al.*, 1981; Edwards & Sinchumpasak, 1981; Faulwetter *et al.*, 2009; Mahmood *et al.*, 2013; Polprasert & Koottatep, 2017; Vymazal, 2010; Wang *et al.*, 2010; Wu *et al.*, 2016);

(II) Land treatment systems that use terrestrial biological and physical processes rather than aquatic processes to treat sewage water by spreading over the land (Paranychianakis *et al.*, 2006; Polprasert & Koottatep, 2017); and

(III) Aquatic pond systems, which essentially replicate and intensify the process of the marine environment in organic recycling by simultaneously filtering the water and assimilating the solids into living organisms. They use water as a medium and treat sewage as it comes from the conventional collection infrastructure (Abbas *et al.*, 2006; Lyle, 1994; Mara & Pearson, 1998; Oswald *et al.*, 1959, 1967, 1994; Polprasert & Koottatep, 2017; USEPA, 2011).

DeN's nature-inspired eco-technologies should improve the sanitary and ecological conditions and at the same time introduce a system of green space for recreational purposes in the communities. DeN infrastructure guidelines could be used by landscape architects, planners and developers as the most important and maybe the only possible generator of public green spaces, which are otherwise getting sacrificed to other construction development (Elzein *et al.*, 2016; Stokman, 2008).

DeN as a novel technology will best be utilized in new communities because it is possible to build it into the landscape from the beginning. The treatment facilities can be integrated into subdivision designs to treat wastewater of the communities. The treated water will then irrigate the landscape, consisting of farmland and horticultural land as well as trees that produce fruit and firewood. Water not used by the plants will soak through the land into subsurface storage. Thus, a circular flow will replace the one-way flow by using the capacities of the landscape for assimilation, filtration, productivity and storage. The distribution and redistribution network becomes part of the community's infrastructure. From the aquifer, the community can draw the water up again to supply domestic needs from where the circuit begins (Lyle, 1984, 1994; Oswald *et al.*, 1959, 1967) (see Figure 2.2). DeN innovative sanitation networks could connect land use from the logic of the sani-sheds and/or watershed that provide the potential for shaping the broader human, ecological and aesthetic objectives that have almost been lost (Kuitert, 2008; Stokman, 2008).

5.4.1.3.3 Design for the base-of-the-pyramid (DeBoP)

DeBoP innovative technological designs target the estimated 4 billion people worldwide who live at the base of the income pyramid (BoP) and are resource-constrained, with an annual income below a certain purchasing power parity (PPP) threshold set at USD 2 per day (Ceschin & Gaziulusoy, 2016; Karnani, 2011; Ray & Ray, 2010). DeBoP is a market-based sanitation innovation that combines business and technological development approaches, where the BoP are considered as producers and consumers as well as business partners to be empowered, enabled and involved in the process of business co-invention and co-creation that goes beyond the traditional donor assistance toilet provisions for the poor without any considerations for their needs (Jagtap *et al.*, 2014; Ramani *et al.*, 2012; Simanis & Hart, 2008). DeBoP objectives are to devise the most cost-effective way to provide sophisticated and simple sanitation technologies that are robust, minimal maintenance, zero energy consumption and resource-conservative. Simple technology means any or a combination of simple designs, low part count, ease of assembly and other features; while sophisticated technology entails the use of cutting-edge techniques to make frugal products (Pradel & Adkins, 2014). ReGenSan encourages DeBoP when it integrates PSEP and resource recovery into one technological systemic whole to rejuvenate and revitalize sanitation infrastructure by mimicking and contributing to nature's system (see Chapter 1, 2 and 4).

The 'Reinvent the Toilet' challenge by the Bill & Melinda Gates Foundation developed several sanitation technological prototypes such as the solar-powered toilet that generates hydrogen and electricity, the California Institute of Technology (CalTech); the biological toilet that produces charcoal, minerals, and clean water, Loughborough University, UK; and the toilet that sanitizes faeces and urine and recovers resources and clean water by the University of Toronto, Canada (Bill & Melinda Gates Foundation, 2012). There were also the solar septic tanks of the Asian Institute of Technology (AIT), Thailand (Koottatep *et al.*, 2014, 2015, 2016; Pussayanavin *et al.*, 2015 – see Box 3.3) and a toilet user interface developed by the Swiss Federal Institute of Aquatic Science and Technology (EAWAG) and Austrian design firm EOOS (Bill & Melinda Gates Foundation, 2012). These are classic examples of novel technology designs that target BoPs (see Box 5.6).

DeBoP seeks to develop novel designs that build an equitable and harmonious relationship between supply and demand through development of products and services to meet the expectations and needs of the target populations, and also encourages governments to assume a supportive and promotional role for the development of entrepreneurs, education of citizens and development of financial options that allows the matching of supply and demand (Baskovich, 2010) (see Sections 2.4.3 and 2.4.8). BoP, in this case, refers to the population exposed to open defecation (estimated to be nearly 1 billion people) and the 2.4 billion people using unimproved sanitation facilities (Baum *et al.*, 2013; Brands, 2014; Mara, 2003; WHO/UNICEF, 2017). BOP are the largest but poorest socio-economic groups in the global income pyramid, and are characterized by poor public and private infrastructure (e.g. electricity and roads), poor service availability (e.g. access to health) and/or an overall poor institutional environment (e.g. access to capital or legal advice). Sanitation technologies that overcome these constraints, which are significantly different from those at the top-of-the-pyramid (TOP), are BoP innovations (Ramani *et al.*, 2012; UNDP, 2008; Zeschky *et al.*, 2014a, b).

DeBoP is derived from the concept of 'frugal innovation' often defined as designing products, services, systems and business models to reduce complexities and total lifecycle costs, and enhance functionality, while providing high user-value and affordable solutions for relatively low-income earners (Bhatti, 2012; Hyvärinen *et al.*, 2016; Leliveld & Knorring, 2017; Rao, 2013; Tiwari *et al.*, 2016; Zeschky *et al.*, 2011, 2014a, b). DeBoP provides the opportunity for innovators to respond to limitations in resources (whether financial, material or institutional) and turn these constraints into innovative ideas for practical and good-quality solutions that are accessible and affordable (Open ecosystem network 2017). To deliver sanitation technologies that are ApT (see Section 2.4.1) to serve BoP requires different products and services developed through problem-driven strategies with emphasis on defining the problems and finding solutions within the context of place and scale (see Section 2.4.7). In this case, the designer focuses closely on the problems at hand and not merely the exercise of generating solutions with little time spent on understanding the problems (Christiaans & Restrepo, 2001; Jagtap *et al.*, 2014; Kruger & Cross, 2006).

Therefore, DeBoP works better with identifying critical needs, confirming the appropriateness of the innovation to serve the need, building business delivery models compatible with socio-economic contexts (see Sections 4.2.1 and 4.2.2), confirming the adsorptive capacity of the BoP as consumers, identifying appropriate change agents within the BoP, ensuring that complementary institutions or assets needed for sustained functioning of the innovation are available and ensuring co-creation of value with all stakeholders including user communities, firms, public agents and other forms of existing partnerships (Christensen *et al.*, 2006; Jagtap *et al.*, 2014; Prahalad & Hart, 2002; Ramani *et al.*, 2012).

5.5 REGENSAN APPROACH TO SANITATION TECHNOLOGY

ReGenSan TeS promotes the adoption of practices and strategies that support the achievement of SDG no. 6 by advocating for sanitation technological infrastructures that are standard-driven to deliver functional and safely managed SSC. Therefore, sanitation facilities are not considered regenerative if they are:

(I) One-time use only such as 'dig-use-cover' or 'poo-bag' technologies, except in cases of disaster/emergency,
(II) Infrastructure that combines human and non-human waste together and only for treatment and disposal,
(III) Sewered and non-sewered infrastructure that do not treat all;
(IV) Linear sanitation infrastructure,
(V) Technically, socially and culturally dysfunctional sanitation facilities; and
(VI) Non-standardized sanitation infrastructure.

TeS is integratively and systematically designed to provide three-dimensional strategic solutions to address challenges confronting sanitation technologies through the implementation of the principal dimensions of TeS: EIDT (5.5.1), ReDT (5.5.2) and NoDT (5.5.3). TeS approaches provided in this chapter give new insight and focus to practitioners, policymakers, researchers and entrepreneurs on how to navigate the complexities of sanitation technological solutions and management. TeS dimensions and components may be deployed for diagnostic, prescriptive, normative and analytical solutions and approaches so as to rejuvenate and invigorate complex sanitation infrastructures. 'Diagnostic' tools enhance detailed analytical and assessment processes to determine the status of existing infrastructure in order to develop appropriate interventions that restore and improve existing technological solutions, as well as provide insight for nouveau solutions. TeS 'prescriptive' tools build upon the diagnostics to strategically determine the appropriateness and adequacy of previously applied interventions in order to promote continuous improvement of existing 'safely managed' facilities. 'Normative' approaches support the design, development, adoption and implementation of appropriate sanitation infrastructure standards (terminologies, specifications, test methods and analysis, codes of practice, treatment and management systems standards) that support the applicability, replicability and scalability of sanitation technological solutions (see Section 3.3). In this respect, ReGenSan TeS recommends the following:

(I) Sanitation infrastructure should be adequately organized based on internationally, regionally and nationally recognized and acceptable technical standard specifications for products, services, processes, systems and performance, which will not only act as a place to relieve oneself but also contribute to increased income and nutrition by producing fertilizer for sale or applied directly for subsistent cropping;
(II) They should contribute to livelihood support for sanitation entrepreneurs, good health and environment, in other words, regenerate the community (Kaminsky & Javernick-Will, 2015).

Hence, sanitation technology is not just the physical infrastructure above ground such as the user interface, or the one below ground – the storage/receptacle, which actually holds the human waste. A user interface unconnected to a receptacle for treatment and resource recovery/reuse might better be called a seat and, in the same vein, a sewerage connection without treatment is just a conduit; and it should be noted that these materials by themselves do not constitute sanitation infrastructure. They must be purchased, transported, properly assembled, adequately and appropriately constructed and installed in order to be more than plastics, pipes, wood and concrete. Similarly, sanitation infrastructure needs people to use it and fix it when it breaks down and can only improve health and environmental quality when adopted by entire (geographic) sani-sheds. The World Bank, 2018 report observed that the concept of 'safely managed' directly addresses community safety because of the recognition that sanitation service must be applied on multiple scales to be effective and must strongly support the extension of boundaries beyond household toilets/latrines (household safety) to service chain management in sani-sheds (Thomas *et al.*, 2018).

Simply put, sanitation technologies should gather households, businesses and communities to work together within defined technical and sociotechnical standardized frameworks (Kaminsky and Javernick-Will, 2015). The ReGenSan sani-shed (catchment/community safety) concept is based on this rationale because studies have shown that no comprehensive sanitation treatment and management solutions can take place at household levels only (Zurbrugg & Tilley, 2009). The sani-shed level is a basic management unit where sanitation issues are addressed and resolved with adequate sociotechnical support for optimal performance of sanitation infrastructure that will guarantee household and community safety.

5.6 EXERCISES

(I) What are the differences between sanitation infrastructure and sanitation infrastructure standardization? Give examples.

(II) What are the challenges of sanitation infrastructure globally and why?

(III) Mention some sanitation standards and standard-setting organizations and group them as formal and informal.

(IV) Discuss the challenges of sanitation infrastructure standardization globally, regionally, nationally and locally.

(V) Compare sanitation infrastructure in Japan, Australia, USA, EU, South Africa, China, Thailand, Nigeria, Uganda and Canada (choose any five).

(VI) Briefly discuss ReGenSan proposed elements for sanitation infrastructure standards and give your own insights on each and its potential effectiveness and practicality.

(VII) Give your own insights as to what is happening in Figure 5.10 and compare with the current sanitation situation in your country.

(VIII) Briefly discuss the inter- and transdisciplinary approach to product and service technology design for sanitation infrastructure.

(IX) Briefly discuss the SSC functional groups for describing sanitation technologies.

(X) Design a scenario where sanitation infrastructure at scale runs through the SSC to final use within a particular sani-shed for agriculture, horticulture and aquaculture and finally constructed wetlands. (a) Describe the technologies required at each step based on ReGenSan Principles, (b) Mention and describe the processes of each step, (c) Mention and describe processes and interactions at the point of use, (d) Mention and describe the benefits, risks and mitigation.

(XI) What are the constraints of EIDT and how could they be overcome?

(XII) Critically discuss types of failure in sewered and non-sewered sanitation infrastructure and possible prevention and remediation.

(XIII) What are the factors responsible for failure and dysfunction in sewered and non-sewered sanitation infrastructure that are not related to age? Discuss.

(XIV) What are the barriers and drivers for sanitation infrastructure technological innovations? How can ReGenSan redirect innovation?

(XV) Identify and explain at least five (not included in this chapter) new sanitation technology concepts/projects and indicate how they relate to ReGenSan principles and theory.

(XVI) Briefly discuss the components of NoDT and insights into possible innovation indicators.

(XVII) What are the considerations for DePaS, DeN and DeBoP? Then discuss the interconnection between them.

(XVIII) Assess any two of the Reinventing Toilets Technologies (see Section 5.5.3.1.3) of the Bill & Melinda Gates Foundation, which targeted users in BoP, and determine whether they meet the DeBoP criteria. Give reasons for your claims.

(XIX) Briefly and critically discuss the role of indigenous/local knowledge in ReGenSan implementation.

(XX) What are the opportunities ReGenSan TeS offer for sanitation infrastructure investment?

(XXI) Identify challenges, pitfalls and entryways for sanitation infrastructure/technology investment locally and globally.

(XXII) Discuss the placement and importance of ancillary services within the SSC (e.g. providing waste collection receptacles and disposal for products used for anal cleansing and menstruation and any other such services).

(XXIII) How should acceptability issues be considered in technology development?

5.7 RECOMMENDATIONS FOR FUTURE RESEARCH

(I) Explore context-specific technical specification standards for different sani-technologies (products and services) within sani-sheds at varying levels.

(II) Examine existing standards, their potential impacts and adoption at national, regional and local levels.

(III) Examine the role of standards in sanitation technology design and manufacture as well as investment.

(IV) Do we really need specification and technical standards in sanitation, and how will this impact the SDGs and the NGOs/multilateral organizations involved with sanitation intervention for the vulnerable?

(V) Review the challenges and prospects of standards and regulatory frameworks in sanitation infrastructure management.

(VI) How can the SDG 6 targets of acceptability, quality and safely managed sanitation facilities at local levels be achieved, where possible?

(VII) Explore the use of uniform classifications and typologies for sanitation infrastructure, technology products and services as well as components and resources.

(VIII) Explore ways to ensure applicability, scalability, replicability, standard performance quality and reliability of sanitation technologies irrespective of location.

(IX) Explore ways to ensure return on investment for entrepreneurs and manufacturers of sanitation technologies (products/services) to encourage the uptake of innovative solutions beyond pilot scales.

(X) Conceptualize and design a certification process for any sanitation infrastructural systems at all levels – global, regional, national and local.

(XI) Explore context-specific sanitation infrastructural system designs based on ReGenSan principles.

(XII) Explore the potential of a transdisciplinary approach for developing sanitation technology solutions to rural and urban poor areas as well as densely populated areas in research and practice.

(XIII) Critically examine the ReGenSan TeS process of ecologically sustainable management in sani-sheds i.e. the front-end, middle-end and back-end structures.

(XIV) Explore and develop parameters for reliability and resilience assessment of sewered and non-sewered sanitation infrastructure (serviceability and functionality).

(XV) Develop integrated risk-based assessment and management of EIDT infrastructure (sewered/non-sewered), and take into consideration constraints of urbanization, population growth and technological malfunction.

(XVI) Explore methods failure management in EIDT/ReDT using integrated risk-based strategy for predictive maintenance.

(XVII) How can EIDTs be made to comply with new regulatory standards as regards to pollution control, effluent discharge, carbon emission reduction, etc?

(XVIII) Explore how to rejuvenate abandoned and degraded sites damaged by bad sanitation practices.

(XIX) Explore and develop best approaches and practices for end-of-life decommissioning of irredeemable sanitation infrastructure.

(XX) Investigate using conditions assessment approval for restoration of dysfunctional sanitation technology and infrastructure systems.

(XXI) How can DePaS components be designed into a traditional sanitation infrastructure (e.g. septic tank, VIP latrine and urine-diverting dry (UDDT)?

(XXII) How can the DePaS components be incorporated into the upgrading of existing sewer sanitation infrastructure? (e.g. wastewater treatment plant, conveyance, wetlands etc).

(XXIII) Explore design for context-specific sanitation product technology based on DePaS for densely populated areas and urban centres in developing countries.

(XXIV) Using the principles of BaB (see Section 2.4.2), PaS (see Section 2.4.7) and SES (see Chapter 4), design a context-specific sanitation technology system (e.g. wetlands, land treatment and aquatic system) that mimics the shape, size, structure, processes and functions of particular natural flora and fauna – and reflects the full SSC.

(XXV) Explore and develop design criteria for DeBoP for any sanitation technology and infrastructure.

(XXVI) Explore the role of indigenous/local knowledge in technology development and user acceptability.

(XXVII) Investigate possible options for affordable and quality sanitation technologies (products/services) that fit with the SDG 6 targets of safely managed facilities.

(XXVIII) Explore new models for infrastructure and technology investment along the SSC.

(XXIX) Explore sanitation service infrastructure investment for practical and effective solutions based on ReGenSan principles and TeS dimensions.

(XXX) Explore the design and production of replicable and scalable sanitation technologies (products/services), taking due account of return on investment for uptake by manufacturers.

(XXXI) Explore your university or college campus sanitation infrastructural system by assessing the campus as a sani-shed using ReGenSan principles and TeS dimensions and make recommendations.

REFERENCES

Abbas H., Nasr R. and Seif H. (2006). Study of waste stabilization pond geometry for the wastewater treatment efficiency. *Ecological Engineering*, **28**(1), 25–34.

Abelson H. and Sussman G. J. (1987). Lisp: a language for stratified design. *Byte*, **13**(2), 207–218.

Abiye T. A., Sulieman H. and Ayalew M. (2009). Use of treated wastewater for managed aquifer recharge in highly populated urban centers: a case study in Addis Ababa, Ethiopia. *Environmental Geology*, **58**(1), 55–59.

ABS (2004). American Bureau of Shipping. Guidance Notes on Reliability-centered maintenance, July 2004.

Abraham D. M. and Gillani S. A. (1999). Innovations in materials for sewer system rehabilitation. *Tunnelling and Underground Space Technology* **14**(1), 43–56. https://doi.org/10.1016/S0886-7798(99)00003-6

Abubakar I. R. (2017). Access to sanitation facilities among Nigerian households: determinants and sustainability implications. *Sustainability*, **9**(547), 1–17.

ADB (Asian Development Bank) (2007). Smarter Sanitation, How to Clean Up Your Sanitation and Waste Water Mess. Asian Development Bank, Manila.

ADB (2016). Sanitation and Sustainable Development in Japan. Asian Development Bank, Manila, Publication Stock No. RPT168293-2, e-ISBN: 978-92-9257-588-5.

Adelegan J. and Ojo S. I. A. (1999). Integrated Development for Water Supply and Sanitation: Low-Cost Urban Sanitation in Nigeria. 25th WEDC Conference, Addis Ababa, Ethiopia. http://wedc.lboro.ac.uk/conference/previous.html (accessed 20 October 2005).

AECOM & SANDEC (2010). A Rapid Assessment of Septage Management in Asia: Policies and Practices in India, Indonesia, Malaysia, the Philippines, Sri Lanka, Thailand, and Vietnam. Regional Development Mission for Asia (RDMA) – United States Agency for International Development (USAID), Thailand. http://www.waterlinks.org/library/septage-management/septage-report (accessed 20 May 2018).

AGCA (Associated General Contractors of America) (1982). Our Framework: Why America must Rebuild. Washington D.C, USA.

Ahmadi M., Cherqui F., Aubin J. B. and Le Gauffre P. (2015). Sewer asset management: impact of sample size and its characteristics on the calibration outcomes of a decision-making multivariate model. *Urban Water Journal*, doi: 10.1080/1573062X.2015.1011668

Al-Barqawi H. and Zayed T. (2006). Condition rating model for underground infrastructure sustainable water mains. *ASCE. Journal of Performance of Constructed Facilities*, **20**(2), 126–135.

Al-Barqwi H. A. (2006). Condition rating models for underground infrastructure: sustainable water mains. A Master Thesis, Concordia University, Montreal, Canada. https://spectrum.library.concordia.ca/8720/1/MR14235.pdf

Altaf S. W. (2011). So Much Aid, So Little Development: Stories from Pakistan. Johns Hopkins University Press, Baltimore, USA.

Anand C. K. and Apul D. S. (2014). Composting toilets as a sustainable alternative to urban sanitation – a review. *Waste Management*, **34**, 329–343.

Anbari M. J., Tabesh M. and Roozbahani A. (2017). Risk assessment model to prioritize sewer pipes inspection in wastewater collection networks. *Journal of Environmental Management*, **190**, 91–101.

Angoli T. (1988). National On-site Wastewater Treatment: A National Small Flows Clearinghouse Summary of On-site Systems in the United States, 1993. Proceedings of the 8th National Symposium on Individual and Small Community Sewage Systems, ASAE, St. Joseph, USA, pp. 1–11.

Armstrong C. M., Niinimaki K., Kujala S., Karell E. and Lang C. (2015). Sustainable product-service systems for clothing: exploring consumer perceptions of consumption alternatives in Finland. *Journal of Cleaner Production*, **97**, 30–39.

Arthur W. B. (2009). The Nature of Technology: What It Is and How It Evolves. Penguin Books – Business & Economics, Free Press, New York.

AS/NZS 1546.1:2008. On-site domestic wastewater treatment units (septic tanks). https://www.saiglobal.com/PDFTemp/Previews/OSH/AS/AS1000/1500/1546.1-2008.pdf (accessed 14 March 2018).

AS/NZS 1546.2:2008. On-site domestic wastewater treatment units (Waterless composting toilets). https://www.saiglobal.com/PDFTemp/Previews/OSH/AS/AS1000/1500/1546.2-2008.pdf (accessed 14 March 2018).

AS/NZS 1546.3:2008. On-site domestic wastewater treatment units (aerated wastewater treatment systems). https://www.saiglobal.com/PDFTemp/Previews/OSH/AS/AS1000/1500/1546.3-2008.pdf (accessed 14 March 2018).

AS/NZS 1547:2012. On-site domestic wastewater management. https://infostore.saiglobal.com/en-au/Standards/AS-NZS-1547-2012-1512288/ (accessed 14 March 2018).

ASCE (The American Society of Civil Engineers) (2017). Report Card for America's Infrastructure: A Call to Urgent Attention for Water Infrastructure. https://www.inframark.com/the-american-society-of-civil-engineers-2017-report-card-for-americas-infrastructure-a-call-to-urgent-attention-for-water-infrastructure/ (accessed 26 March 2018).

ASCE/WERF (American Society of Civil Engineers/Water Environmental Research Foundation) (1994). Existing Sewer Evaluation and Rehabilitation, ASCE Manual No.62 or WEF MOP FD-6, 2nd edn. New York, USA.

Ashley R. and Hopkinson P. (2002). Sewer systems and performance indicators—into the 21st century. *Urban Water*, **4**, 123–135. https://doi.org/10.1016/S1462-0758(02)00010-9

ASHRAE (American Society of Heating, Refrigerating and Air Conditioning Engineers) (2005). ASHRAE Handbook – Fundamentals. Self-Published, Atlanta, USA.

Awoke W. and Muche S. (2013). A cross sectional study: latrine coverage and associated factors among rural communities in the District of Bahir Dar Zuria, Ethiopia. *BMC Public Health*, **13**, 99. http://www.biomedcentral.com/1471-2458/13/99 (accessed 6 April 2018).

Badarnah L. and Kadri U. (2015). A methodology for the generation of biomimetic design concepts. *Architectural Science Review*, **58**(2), 120–133, doi: 10.1080/00038628.2014.922458

Barquet A. P. B., de Oliveira M. G., Amigo C. R., Cunha V. P. and Rozenfeld H. (2013). Employing the business model concept to support the adoption of products service systems (PSS). *Industrial Marketing Management*, **42**, 693–704.

Bartram J. and Cairncross S. (2010). Hygiene, sanitation, and water: forgotten foundations of health. *PLoS Medicine*, **7**(11), e1000367, doi: 10.1371/journal.pmed.1000367

Baskovich M. R. (2010). Promoting Sanitation Markets at the Bottom of the Pyramid in Peru: A Win-Win Scenario for Government, the Private Sector, and Communities. IFC Smart Lessons Brief. World Bank, Washington D.C., USA. CC BY-NC-ND 3.0 IGO. https://openknowledge.worldbank.org/handle/10986/10473 (accessed 21 April 2018).

Baum R., Luh J. and Bartram J. (2013). Sanitation: a global estimate of sewerage connections without treatment and the resulting impact on MDG progress. *Advances in Environmental Science and Technology*, **47**(4), 1994–2000, doi: 10.1021/es304284f

Bayo J., Angosto J. M. and Serrano-Aniorte J. (2006). Evaluation of physicochemical parameters influencing bulking episodes in a municipal wastewater treatment plant. In: Water Pollution VIII: Modelling, Monitoring and Management, C.A. Brebbia and J.S. Antunes do Carmo (eds), WIT Press, Southampton, pp. 531–540.

Beal C. D., Gardner E. A. and Menzies N. W. (2005). Process, performance, and pollution potential: a review of septic tank-soil absorption systems. *Australian Journal of Soil Research*, **43**(7), 781–802.

Best Practice (2003). 'Best Practices for Utility-Based Data' – A best practice by National Guide to sustainable Municipal Infrastructure. Issue No 1.0, NRC Infra-guide publication.

Bhagwan J. N., Still D., Buckley C. and Foxon K. (2008). Challenges with up-scaling dry sanitation technologies. *Water Science and Technology*, **58.1**, doi: 10.2166/wst.2008.606

Bhatti Y. (2012). What is frugal? What is innovation? Towards a theory of frugal innovation. SSRN Working Paper Series, Oxford Centre for Entrepreneurship and Innovation.

Bill & Melinda Gates Foundation (2011). Grand challenges: Create the Next Generation of Sanitation Technologies (Round 7). https://gcgh.grandchallenges.org/challenge/create-next-generation-sanitation-technologies-round-7 (accessed 19 April 2018).

Bill & Melinda Gates Foundation (2012). Bill Gates Names Winners of the Reinvent the Toilet Challenge. Press Releases and Statements. https://www.gatesfoundation.org/media-center/press-releases/2012/08/bill-gates-names-winners-of-the-reinvent-the-toilet-challenge (accessed 30 April 2018).

Bisschops I., Seghezzo L., Mels A. and Zeeman G. (2006). Types, Numbers and Performance of Decentralized Sewage Treatment Plants in the EU-15. Final Report, produced by Lettinga Associates Foundation (LeAF), Wageningen, Netherlands. http://www.erasm.org/judownload/0_1520831554_XoHqq4/3.1.2.1.VanGinkel.DeSETPs.Report_KvG.2006.pdf (accessed 12 March 2018).

Blackett I. and Hawkins P. (2015). Institutional Frameworks for Fecal Sludge Management: An Introduction. Faecal Sludge Management (FMS 0.1) Global Faecal Sludge e-Learning Alliance course materials arranged by Martin Mulenga, and Mariska Ronteltap, of UNESCO-IHE Institute for Water Education, Delft, Netherlands. https://ocw.un-ihe.org/mod/resource/view.php?id=3766&forceview=1 (accessed 7 May 2018).

Boonsong K., Piyatiratitivorakul S. and Patanaponpaiboon P. (2003). Potential use of mangrove plantation as constructed wetland for municipal wastewater treatment. *Water Science and Technology*, **48**(5), 257–266.

Bos A. and Gijzen H. (2005). Health benefits versus costs of water supply and sanitation. *Water21*, **October 2005**(7.5), 1–72.

Boschi-Pinto C., Lanata C. F., Mendoza W. and Habte D. (2006). Diarrheal diseases. Disease and mortality in Sub-Saharan Africa. In: D. T. Jamison, R. G. Feachem, M. W. Makgoba, E. R. Bos, F. K. Baingana, K. J. Hofman and K. O. Rogo (eds), World Bank, Washington D.C., USA. http://www.ncbi.nlm.nih.gov/books/NBK2302/

Boujelben I., Samet Y., Messaoud M., Makhlouf M. B. and Maalej S. (2017). Descriptive and multivariate analyses of four Tunisian wastewater treatment plants: a comparison between different treatment processes and their efficiency improvement. *Journal of Environmental Management*, **187**, 63–70. http://dx.doi.org/10.1016/j.jenvman.2016.11.023

Bozkurt H., Gernaey K. and Sin G. (2015). Computer-Aided Framework for Synthesis, Design and Retrofit of Wastewater Treatment Plants. Danmarks Tekniske Universitet (DTU), Kgs. Lyngby. http://orbit.dtu.dk/files/110810194/Hande_Bozkurt_CAP15_08_01_1_.pdf (accessed 30 April 2017).

Bradley B. R., Daigger G. T., Rubin R. and Tchobanoglous G. (2002). Evaluation of onsite wastewater treatment technologies using sustainable development criteria. *Clean Techn Environ Policy*, **4**, 87–99, doi: 10.1007/s10098-001-0130-y, https://link.springer.com/content/pdf/10.1007/s10098-001-0130-y.pdf (accessed 22 March 2018).

Brands E. (2014). Prospects and challenges for sustainable sanitation in developed nations: a critical review. *Environmental Reviews*, **22**, 1–18, doi: dx.doi.org/10.1139/er-2013-0082

Braungart M., McDonough W. and Bollinger A. (2007). Cradle-to-cradle design: creating healthy emissions – a strategy for eco-effective product and system design. *Journal of Cleaner Production*, **15**(13–14), 1337–1348. http://dx.doi.org/10.1016/j.jclepro.2006.08.003

Bremer J. E. and Harter T. (2012). Domestic wells have high probability of pumping septic tank leachate. *Hydrology and Earth System Sciences*, **16**(8), 2453–2467.

Brepols C. H., Dorgeloh E., Frechen F. B., Fuchs W., Haider S., Joss A., de Korte K., Ruiken C. H., Schier W., van der Roest H., Wett M. and Wozniak T. H. (2008). Upgrading and retrofitting of municipal wastewater treatment plants by means of membrane bioreactor (MBR) technology. *Desalination*, **231**, 20–26.

Brikke F. (2000). Operation and Maintenance of Rural Water Supply and Sanitation Systems: A Training Package for Managers and Planners. World Health Organization Geneva, Switzerland. http://apps.who.int/iris/bitstream/10665/66716/1/WHO_SDE_WSH_00.2_p1-124.pdf (accessed 30 April 2017).

Brix H. (1997). Do macrophytes play a role in constructed treatment wetlands? *Water Science and Technology*, **35**(5), 11–17. https://doi.org/10.1016/S0273-1223(97)000047-4

Brown and Caldwell (1984). Utility infrastructure rehabilitation. NTIS No. PB86-N14642, U.S. Department of Housing and Urban Development, Office of Policy Development and Research, Building Technology Division.

BS EN (2016). 12566-4:2016 Small wastewater treatment systems for up to 50 PT. Septic tanks assembled in situ from prefabricated kits. https://shop.bsigroup.com/ProductDetail/?pid=000000000030329426

Cairncross S. and Feachem R. G. (1993). Environmental Health Engineering in the Tropics: An Introductory Text, 2nd edn. Wiley, Chichester, UK. http://eu.wiley.com/WileyCDA/WileyTitle/productCd-0471938858.html

Cardot C. (1999). Les traitements de l'eau -proc_ed_es physico-chimiques et biologiques. (Water treatments – procedures: physicochemical and biological). Ellipses, Paris, France.

Catulli M. (2012). What uncertainty? Further insight into why consumers might be distrustful of product service systems. *Journal of Manufacturing Technology Management*, **23**, 780–793.

Catulli M., Lindley J. K., Reed N. B., Green A., Hyseni H. and Kiri S. (2013). What is mine is not yours: further insight on what access-based consumption says about consumers. *Journal of Consumer Culture*, **15**, 185–208.

CAWST and EAWAG-Sandec (2016). Introduction to Faecal Sludge Management. Trainer Manual. Produced by Centre for Affordable Water and Sanitation Technology (CAWST) the Department of Sanitation, Water and Solid Waste for Development, (Sandec), the Swiss Federal Institute of Aquatic Science and Technology (EAWAG). http://www.eawag.ch/fileadmin/Domain1/Abteilungen/sandec/publikationen/EWM/FSM_workshop/trainer_manual.pdf (accessed 3 March 2018).

Ceschin F. (2014). Sustainable Product-Service Systems: Between Strategic Design and Transition Studies. Springer International Publishing, Cham.

Ceschin F. and Gaziulusoy I. (2016). Evolution of design for sustainability: from product design to design for system innovations and transitions. 0142-694X. *Design Studies*, **47**, 118–163. http://dx.doi.org/10.1016/j.destud.2016.09.002

Chang N.-B., Pires A. and Martinho G. (2011). Empowering systems analysis for solid waste management: challenges, trends, and perspectives. *Critical Reviews in Environmental Science and Technology*, **41**(16), 1449–1530. https://doi.org/10.1080/10643381003608326

Cheng J. J., Schuster-Wallace C. J., Watt S., Newbold B. K. and Mente A. (2012). An ecological quantification of the relationships between water, sanitation and infant, child, and maternal mortality. *Archives of Environmental Health*, **11**(1), 1–8, doi: 10.1186/1476-069X-11-4

Cheng S., Li Z., Mang H.-P. and Huba E.-M. (2013). A review of prefabricated biogas digesters in China. *Renewable & Sustainable Energy Reviews*, **28**, 738–748.

Cheng S., Li Z., Uddin S. M. N., Mang H.-P., Zhou X., Zhang J., Zheng L. and Zhang L. (2017). Toilet revolution in China. *Journal of Environmental Management*, 1–10. http://dx.doi.org/10.1016/j.jenvman.2017.09.043

Cherubini S., Iasevoli G. and Michelini L. (2015). Product-service systems in the electric car industry: critical success factors in marketing. *Journal of Cleaner Production*, **97**, 40–49.

Christensen C. M., Baumann H., Ruggles R. and Sadtler T. M. (2006). Disruptive innovation for social change. *Harvard Business Review*, **84**, 94.

Christiaans H. and Restrepo J. (2001). Information processing in design: a theoretical and empirical perspective. In: Design Research in the Netherlands 2000, H. Achten, B. de Vries and J. Hennessey (eds), Eindhoven University of Technology, Eindhoven, Netherlands.

Cookey P. E. (2013). Public Health Risk Assessment of Decentralized Wastewater Treatment Systems in Port Harcourt City, Nigeria. Master's thesis, Asian Institute of Technology, Bangkok. UNESCO-IHE, Delft, Netherlands.

Cookey P. E., Koottatep T., van der Steen P. and Lens P. N. L. (2016). Public health risk assessment tool: strategy to improve public policy framework for onsite wastewater treatment systems (OWTS). *Journal of Water, Sanitation and Hygiene for Development*, **06.1**, 74–88, doi: 10.2166/washdev.2016.081

E. Corcoran, C. Nellemann, E. Baker, R. Bos, D. Osborn and H. Savelli (2010). Sick Water? The central role of wastewater management in sustainable development. A Rapid Response Assessment. United Nations Environment Programme, UN-HABITAT, GRID-Arendal. https://gridarendal-website-live.s3.amazonaws.com/production/documents/:s_document/208/original/SickWater_screen.pdf?1486721310 (accessed 23 March 2018).

Cordella D., Rosemarin A., Schroder J. J. and Smit A. L. (2011). Towards global phosphorus security: a systems framework for phosphorus recovery and reuse options. *Chemosphere*, **84**, 747–758.

Corvellec H. and Stål H. (2017). Evidencing the waste effect of Product-Service Systems (PSSs). *Journal of Cleaner Production*, **145**, 14–24, doi: 10.1016/j.jclepro.2017.01.033

CPDA (Clay Pipe Development Association Ltd) (2001). The specification, design and construction of drainage and sewerage systems using vitrified clay pipes. http://www.cpda.co.uk/wp-content/uploads/2016/08/CPDA-The-specification-design-and-construction-of-drainage-and-sewerage-systems-using-vitrified-clay-pipes.pdf (accessed 19 March 2018).

CSIR (Council for Scientific and Industrial Research) (2007). Spot check assessments of MIG water and sanitation projects. http://www.dwaf.gov.za/Masibambane/documents/monitoring-evaluation/MIGSpotcheckApr07part1.pdf (accessed 4 June 2013).

CSTEP (Center for Study of Science, Technology and Policy) (2016). Technology Options for the Sanitation Value Chain, Version 1.0, (CSTEP-Report-2016–07). http://cstep.in/uploads/default/files/publications/stuff/CSTEP_Technology_Options_for_the_Sanitation_Value_Chain_Report_2016.pdf (accessed 16 March 2018).

Cuppens A., Smets I. and Wyseur G. (2013). Identifying sustainable rehabilitation strategies for urban wastewater systems: a retrospective and interdisciplinary approach. Case study of Coronel Oviedo, Paraguay. *Journal of Environmental Management*, **114**, 423–432. https://doi.org/10.1016/j.jenvman.2012.10.040

CWTRC (California Wastewater Training & Research Center) (2003). Onsite Wastewater Treatment System Repair of Failure/Malfunction Survey. Technical report prepared for: the California State Water Resources

Control Board, by the California Wastewater Training & Research Center, California State University, Chico. https://www.waterboards.ca.gov/water_issues/programs/owts/docs/onsite_repairsrpt2003.pdf (accessed 5 April 2018).

Davis C. K. (2008). Ethical dilemmas in water recycling. In: Water Reuse e an International Survey of Current Practice, Issues and Needs, B. Jiménez and T. Asano (eds), IWA, London, UK, pp. 250–298.

Diagne N. A. (2013). Evaluation of Sewer Leakage into the Stormwater Drainage System in Singapore. Master's thesis, Massachusetts Institute of Technology, Cambridge, USA. http://hdl.handle.net/1721.1/82810

Dias B. D. (2015). Beyond sustainability - biophilic and regenerative design in architecture. *European Scientific Journal*, 1857–7881. http://eujournal.org/index.php/esj/article/view/5385 (accessed 12 October 2017).

Diogo A. F., Barros L. T., Santos J. and Temido J. S. (2018). An effective and comprehensive model for optimal rehabilitation of separate sanitary sewer systems. *Science of the Total Environment*, **612**, 1042–1057. https://doi.org/10.1016/j.scitotenv.2017.08.315 (accessed 20 May 2018).

Dixon T., Eames M., Britnell J., Watson G. B. and Hunt M. (2014). Urban retrofitting: identifying disruptive and sustaining technologies using performative and foresight techniques. *Technological Forecasting & Social Change*, **89**, 131–144.

Dodane P., Mbegueré M., Sow O. and Strande L. (2012). Capital and operating costs of full-scale fecal sludge management and wastewater treatment systems in Dakar, Senegal. *Advances in Environmental Science and Technology*, **46**, 3705–3711, doi: dx.doi.org/10.1021/es2045234, https://pubs.acs.org/doi/pdf/10.1021/es2045234 (accessed 2 April 2018).

Dubber D. and Gill L. (2014). Application of on-site wastewater treatment in Ireland and perspectives on its sustainability. *Sustainability*, **6**, 1623–1642, doi: 10.3390/su6031623

Dumpert J. W. (2008). Performance Evaluation of VIP Latrines in the Upper West Region of Ghana. Master's thesis, Michigan Technological University, Houghton, USA. http://digitalcommons.mtu.edu/etds/514

Eames M. (2011). Developing urban retrofit scenarios: an outline framework for scenario foresight and appraisal, Retrofit 2050 Working Paper WP 2011/4. www.retrofit2050.org.uk (accessed 1 May 2017).

Edwards A. R. (2005). The Sustainability Revolution: Portrait of a Paradigm Shift. New Society Publishers, Gabriola Island, Canada.

Edwards C. A. (2004). Earthworm Ecology, 2nd edn. CRC Press, Boca Raton, USA.

Edwards P. and Sinchumpasak O. (1981). The harvest of microalgae from the effluent of a sewage fed high rate stabilization pond by Tilapia nilotica. Part I: description of the system and the study of the high rate pond. *Aquaculture*, **23**, 83–105.

Edwards P., Sinchumpasak O., Labhsetwar V. K. and Tabucanon M. (1981). The harvest of microalgae from the effluent of a sewage fed high rate stabilization pond by Tilapia nilotica. Part 3: maize cultivation experiment, bacteriological studies, and economic assessment. *Aquaculture*, **23**, 149–170.

Ekane N., Nykvist B., Kjellén M., Noel S. and Weitz N. (2014). Multi-level sanitation governance: understanding and overcoming the challenges in the sanitation sector in Sub-Saharan Africa. Working Paper no. 2014–4. Stockholm Environment Institute. https://www.sei-international.org/mediamanager/documents/Publications/Water-sanitation/sei-workingpaper-ekane-multi-level-sanitation.pdf (accessed 8 May 2017).

ELI (Environmental Law Institute) (1998). Barriers to Environmental Technology Innovation and Use, ELI Project #960800. https://www.eli.org/sites/default/files/eli-pubs/d8.01.pdf (accessed 24 April 2018).

Ellison T. D. (2017). Standardization of On-Site Wastewater Systems. http://www.aowma.com/2017-convention-proceedings, http://www.aowma.com/files/Standardization-of-On-Site-Wastewater-Systems.pdf (accessed 13 March 2018).

El-Sheikh M. A. (2011). Optimization and upgrading wastewater treatment plants. *Journal of Engineering Sciences, Assiut University*, **39**(4), 697–713. http://www.aun.edu.eg/journal_files/82_J_8366.pdf (accessed 30 April 2017).

Elzein Z., Abdou A. and Abd Elgawad I. (2016). Constructed wetlands as a sustainable wastewater treatmen methods in communities. *Procedia Environmental Sciences*, **34**, 605–617.

Etnier C., Willetts J., Mitchell C. A., Fane S. and Johnstone D. S. (2005). Decentralized Wastewater System Reliability Analysis Handbook. Project No. WU-HT-03-57. Prepared for the National Decentralized Water Resources Capacity

Development Project, Washington University, St. Louis, MO, by Stone Environmental, Inc., Montpelier, USA. http://www.ndwrcdp.org/documents/WU-HT-03-57/WUHT0357.pdf (accessed 4 April 2018).

EU (1991). European Urban Wastewater Treatment Directives (UWWTD: 91/271/EEC), Council Directive. http://ec.europa.eu/environment/water/water-urbanwaste/index_en.html (accessed 12 March 2018).

EU (2012). European Court of Auditors European Union Development Assistance for Drinking Water Supply and Basic Sanitation in Sub-Saharan Countries; Special Report No 13 pursuant to Article 287(4) Treaty on the Functioning of the European Union, ECA, Luxembourg, 2012.

Eurostat (2011). Europe in figures – Eurostat yearbook 2011. http://epp.eurostat.ec.europa.eu/cache/ITY_OFFPUB/KS-CD-11-001/EN/KS-CD-11-001-EN.PDF (accessed 5 November 2013).

EUWATER Network (2005). European Declaration for a New Water Culture, New Water Culture Foundation. Published by Fundacion Nueva Cultura de Aqua, Madrid, Spain, 61 pp.

Exley J. L. R., Liseka B., Cumming O. and Ensink J. H. J. (2014). The sanitation ladder, what constitutes an improved form of sanitation? *Environmental Science & Technology*, **49**, 1086–1094, doi: 10.1021/es503945x

Exley J. L. R., Liseka B., Cumming O. and Ensink J. H. J. (2015). The sanitation ladder, what constitutes an improved form of sanitation? *Advances in Environmental Science and Technology*, **49**(2), 1086–1094, doi: 10.1021/es503945x

Fairbain J. (1996). Application of nitrogen isotope techniques to assess the impact of sewer leakage on shallow urban groundwater quality – the Sherwood Sandstone Aquifer of Nottingham. MSc thesis, University of East Anglia, Norwich, UK.

Fan B., Hu M., Wang H., Xu M., Qu B. and Zhu S. (2017). Get in sanitation 2.0 by opportunity of rural China: scheme, simulating application and life cycle assessment. *Journal of Cleaner Production*, **147**, 86–95.

Faulwetter J. L., Gagnon V., Sundberg C., Chazarenc F., Burr M. D., Brisson J., Camper A. K. and Stein O. R. (2009). Microbial processes influencing performance of treatment wetlands: a review. *Ecological Engineering*, **35**(6), 987–1004. https://doi.org/10.1016/j.ecoleng.2008.12.030

Feng P. (2003). Studying Standardization: A Review of the Literature. The 3rd Conference on Standardization and Innovation in Information Technology, pp. 99–112. 0-7803-8/72-6/03/$17.00 02003 IEEE. http://ieeexplore.ieee.org/stamp/stamp.jsp?tp=&arnumber=1251199 (accessed 23 February 2018).

Fenner R. A. and Saward G. (2004). Towards Assessing Sewer Performance and Serviceability Using Knowledge Based Systems. 9th International Conference on Urban Drainage (9ICUD). https://ascelibrary.org/doi/pdf/10.1061/40644%282002%2982 (accessed 6 April 2018).

Fewtrell L., Kaufmann R. B., Kay D., Enanoria W., Haller L. and Colford J. M. (2005). Water, sanitation, and hygiene interventions to reduce diarrhoea in less developed countries: a systematic review and meta-analysis. *The Lancet Infectious Diseases*, **5**(1), 42–52, doi: 10.1016/S1473-3099(04)01253-8

Fewtrell L., Prüss-Üstün A., Bos R., Gore F. and Bartram J. (2007). Water, Sanitation and Hygiene: Quantifying the Health Impact at National and Local Levels in Countries with Incomplete Water Supply and Sanitation Coverage. World Health Organization, Geneva (WHO Environmental Burden of Disease Series No. 15.).

Fiddes D. and Simmonds N. (1981). Infiltration – do we have to live with it? *Public Health Engineer*, **9**(1), 11.

Finger M., Groenewegen J. and Künneke R. (2005). The quest for coherence between institutions and technologies in infrastructures. *Journal of Network Industry*, **6**(4), 227–259.

Foxon K. M., Pillay S., Lalbahadur T., Rodda N., Holder F. and Buckley C. A. (2004). The Anaerobic Baffled Reactor (ABR): An Appropriate Technology for on-Site Sanitation.

Frank C., Sink C., Mynatt L., Rogers R. and Rappazzo A. (1996). Surviving the 'Valley of Death': a comparative analysis. *Technology Transfer*, **21**, 61–69.

Fulmer J. (2009). What in the world is infrastructure? *PEI Infrastructure Investor*, **July/August**, 30–32.

Gamage A. and Hyde R. (2012). A model based on Biomimicry to enhance ecologically sustainable design. *Architectural Science Review*, **55**(3), 224–235, doi: 10.1080/00038628.2012.709406

Gardner T. and Sharma A. (2013). Development of decentralized systems in Australia. In: Source Separation and Decentralization for Wastewater Management, T. A. Larsen, K. M. Udert and J. Lienert (eds), IWA Publishing, London, UK, pp. 447–454.

Gasiorowski-Denis E. (2018). ISO focus highlights water and sanitation for UN Decade of Action. *ISO focus magazine*. https://www.iso.org/news/ref2256.html (accessed 22 February 2018).

Gaulke L. S. (2006). On-site wastewater treatment and reuses in Japan. *Proceedings of the Institution of Civil Engineers, Waste Management*, **159**(WM2), 103–109. http://www.eawag.ch/fileadmin/Domain1/Abteilungen/sandec/publikationen/EWM/General_FSM/on_site_wastewater_japan.pdf (accessed 12 March 2018).

Geum Y. and Park Y. (2011). Designing the sustainable product-service integration: a product-service blueprint approach. *Journal of Cleaner Production*, **19**, 1601–1614, doi: 10.1016/j.jclepro.2011.05.017

Ghosh B. S. and Elvira M. (2011). Africa's Water and Sanitation Infrastructure: Access, Affordability, and Alternatives. Directions in Development; Infrastructure. World Bank, Washington D.C., USA. http://documents.worldbank.org/curated/en/712211468202191672/Africas-water-and-sanitation-infrastructure-access-affordability-and-alternatives (accessed 24 June 2017).

Gill L. W., O'Súilleabháin C., Misstear B. D. R. and Johnston P. M. (2007). The treatment performance of different subsoils in Ireland receiving on-site wastewater effluent. *Journal of Environmental Quality*, **36**, 1843–1855.

Gine R., Jimenez A. and Perez-Foguet A. (2011). A Closer Look at the Sanitation Ladder: Issues of Monitoring the Sector. Water, Engineering and Development Centre International Conference. 35th WEDC International Conference: The Future of Water, Sanitation and Hygiene in Low-Income Countries: Innovation, Adaptation and Engagement in a Changing World, Loughborough, UK, 2011. http://hdl.handle.net/2117/14199

Giné-Garriga R., Flores-Baquero O., de Palencia A. J. and Pérez-Foguet A. (2017). Monitoring sanitation and hygiene in the 2030 Agenda for Sustainable Development: a review through the lens of human rights. *Science of the Total Environment*, **580**, 1108–1119.

Giovanni L., Ghirardini A. V. and Avezzù F. (2012). To centralise or to decentralise: an overview of the most recent trends in wastewater treatment management. *Journal of Environmental Management*, **94**, 61–68.

Goedkoop M., van Haler C., te Riele H. and Rommers P. (1999). Product Service-Systems, Ecological and Economic Basics Report for Dutch Ministries of Environment (VROM) and Economic Affairs (EZ). http://teclim.ufba.br/jsf/indicadores/holan%20Product%20Service%20Systems%20main%20report.pdf (accessed 20 May 2018).

Grant S. B., Saphores J., Feldman D. L., Hamilton A. J., Fletcher T. D., Cook P. L. M., Stewardson M., Sanders B. F., Levin L. A., Ambrose R. F., Deletic A., Brown R., Jiang S. C., Rosso D., Cooper W. J. and Marusic I. (2012). Taking the 'waste' out of 'wastewater' for human water security and ecosystem sustainability. *Acta Agriculturae Scandinavica Section A Animal Science*, **337**, 681–686.

Greiner M. A. and Franza R. M. (2003). Barriers and bridges for successful environmental technology transfer. *Journal of Technology Transfer*, **28**, 167–177.

Gretchen N. (2012). Finding a new way to go. *Acta Agriculturae Scandinavica Section A Animal Science*, **337**, 673.

Gunady M., Shishkina N., Tan H. and Rodriguez C. (2015). A review of on-site wastewater treatment systems in Western Australia from 1997 to 2011. *Journal of Environmental and Public Health*, Article ID 716957, 12 pages. http://dx.doi.org/10.1155/2015/716957, https://www.hindawi.com/journals/jeph/2015/716957/citations/ (accessed 7 March 2018).

Gutterer B. and Reuter S. (2016). Key Elements for a New Urban Agenda – Integrated Management of Urban Waters and Sanitation. BORDA e.V., Bremen, Germany.

Hatto P. (2010). Standards and Standardization Handbook. Directorate-General for Research Industrial Technologies. European Commission Office CDMA 06/10 B-1049 Brussels, Belgium.

Hawkins P., Blackett I. and Heymans C. (2013). Poor-Inclusive Urban Sanitation: An Overview. Water and Sanitation Program. World Bank, Washington D.C., USA.

Hegger D. L. T., van Vliet J. and van Vliet B. J. M. (2007). Niche management and its contribution to regime change: the case of innovation in sanitation. *Technology Analysis and Strategic Management*, **19**, 729–746.

Helms M., Vattam S. S. and Goel A. K. (2009). Biologically inspired design: process and products. *Design Studies*, **30**(5), 606–622, doi: 10.1016/j.destud.2009.04.003

Hernandez-Leal L. H., Tervahauta T. and Zeeman G. (2017). Resource recovery from source separated domestic wastewater; energy, water, nutrients and organics. In: Innovative Wastewater Treatment & Resource Recovery

Technologies: Impacts on Energy, Economy and Environment, J. M. Lema and S. Suarez Martinez (eds), IWA, London, pp. 61–75.

HES (Health Education Service) (2014). Recommended Standards for Wastewater Facilities. A Report of Committee of the Great Lakes – Upper Mississippi River, Board of State and Provincial Public Health and Environmental Managers, Albany, USA. www.hes.org, http://10statesstandards.com/wastewaterstandards.pdf (accessed 9 March 2018).

Hiessl H., Walz R. and Toussaint D. (2001). Design and Sustainability Assessment of Scenarios of Urban Water Infrastructure Systems. Proceedings of 5th International Conference on Technology Policy and Innovation, Utrecht, Netherlands.

Hiessl H., Toussaint D., Becker M., Dyrbusch A., Geisler S., Herbst H. and Prager J. U. (2003). Alternativen der kommunalen Wasserversorgung und Abwasserentsorgung (Alternatives of Municipal Water Supply and Sanitation), AKWA 2100. Physica-Verlag, Heidelberg, Germany.

Hu M., Fan B., Wang H., Qu B. and Zhu S. (2016). Constructing the ecological sanitation: a review on technology and methods. *Journal of Cleaner Production*, **125**, 1–21. http://dx.doi.org/10.1016/j.jclepro.2016.03.012

Hughes T. P. (1987). The evolution of large technological systems. In: The Social Construction of Technological Systems, W. Bijker, T. P. Hughes and T. Pinch (eds), The MIT Press, Cambridge, USA, pp. 51–82.

Hyvärinen A., Keskinen M. and Varis O. (2016). Potential and pitfalls of frugal innovation in the water sector: insights from Tanzania to global value chains. *Sustainability*, 8(9), 1–16 (888), doi: 10.3390/su8090888

IHE (2011). Stimulating Local Innovation on Sanitation for the Urban Poor – a programme of Bill & Melinda Gates Foundation from developing countries in sub-Saharan Africa, South-East Asia and South America. http://pro-poor-sanitation-innovations.un-ihe.org/sites/pro-poor-sanitation-innovations.un-ihe.org/files/stimulating_local_innovation_in_sanitation.pdf (accessed 30 April 2018).

Ingallinella A. M., Sanguinetti G., Koottatep T., Montangero A. and Strauss M. (2002). The challenge of faecal sludge management in urban areas – strategies, regulations and treatment options. *Water Science and Technology*, **46**(10), 285–294.

IrishEPA (Ireland Environmental Protection Agency) (2009). Code of Practice: Wastewater Treatment and Disposal Systems Serving Single Houses. Environmental Protection Agency, Wexford, Ireland.

IrishEPA (Ireland Environmental Protection Agency) (2012). Focus on Urban Waste Water Discharges in Ireland. Environmental Protection Agency, Wexford, Ireland.

IrishEPA (Ireland Environmental Protection Agency) (2013). A Risk-Based Methodology to Assist in the Regulation of Domestic Waste Water Treatment Systems. Environmental Protection Agency, Wexford, Ireland.

ISEAL (International Social and Environmental Accreditation and Labelling Alliance) (2006). ISEAL Code of Good Practice for Setting Social and Environmental Standards. P005 – Public Version 4. http://orgapet.orgap.org/annexes/annex_A4-4.pdf (accessed 7 May 2018).

ISO (2016). 24521– Activities relating to drinking water and wastewater services – Guidelines for the management of basic on-site domestic wastewater services. https://www.iso.org/standard/64679.html (accessed 12 March 2018).

ISO (2018). Water and Sanitation. ISOfocus January–February 2018. https://www.iso.org/files/live/sites/isoorg/files/news/magazine/ISOfocus%20(2013-NOW)/en/2018/ISOfocus_126/ISOfocus_126_en.pdf (accessed 13 March 2018).

ISO/DIS (2018). 30500 – Non-sewered sanitation systems – Prefabricated integrated treatment units – General safety and performance requirements for design and testing. https://www.iso.org/standard/72523.html (accessed 12 March 2018).

ISO/NP (2018). 23457 – Community scale resource oriented sanitation treatment systems. https://www.iso.org/standard/75633.html (accessed 7 May 2018).

IWA (2006). Sanitation 21: Simple Approaches to Complex Sanitation – A Draft Framework for Analysis. International Water Association, London, UK, p. 39.

IWA (24:2016). Non-sewered sanitation systems – General safety and performance requirements for design and testing. https://www.iso.org/standard/70604.html (accessed 9 May 2017).

Jagtap S., Larsson A., Hiort V., Olander E., Warell A. and Khadilkar P. (2014). How design process for the Base of the Pyramid differs from that for the Top of the Pyramid. *Design Studies*, **35**, 527–558. http://dx.doi.org/10.1016/j.destud.2014.02.007

Jean-Marie I. (2013). Choosing and implementing non-conventional sewers for the provision of sanitation services. Analysis Report – Working Document. © Programme Solidarité Eau. https://www.pseau.org/sites/default/files/fichiers/r_d/non-convetional_sewers_analysis_report.pdf (accessed 7 March 2018).

JECES (Japan Education Center of Environmental Sanitation) (2009). Wastewater treatment with Johkasou systems. http://www.jeces.or.jp/en/technology/structure-treatment.html (accessed 26 March 2018).

Jenkin S. and Pedersen Z. M. (2009). Rethinking Our Built Environments: Towards a Sustainable Future. Ministry for the Environment, Manatu Mo Te Taiao, Wellington, New Zealand. https://www.researchgate.net/publication/261476724_Rethinking_our_built_environments_Towards_a_sustainable_future (accessed 10 February 2018).

Jiang X., Sommer G. S. and Christensen K. V. (2011). A review of the biogas industry in China. *Energy Policy*, **39**, 6073–6081.

Jönsson D. K. (2005). The nature of infrasystem services. *Journal of Infrastructure Systems*, **11**(1), 2–8.

JSWA (Japan Sewage Works Association) (2016). Management of Wastewater in Japan. http://gcus.jp/wp/wp-content/uploads/2017/03/da25d58e311828646b563677c900b90c (accessed 26 March 2018).

Kaijser A. (2004). The Dynamics of Infrasystems. Lessons from History. Proc., 6th. Int. Summer Academy on Technology Studies, Inter-Univ. Research Centre for Technology, Work, and Culture (IFZ), Graz, Austria.

Kaminsky J. and Javernick-Will A. (2015). Theorizing the internal social sustainability of sanitation organizations. *Journal of Water, Sanitation and Hygiene for Development*, **141**(2), 04014071. doi: 10.1061/(ASCE)CO.1943-7862.0000933

Kaneko M., Nambu T. and Tokoro M. (2001). Behaviour of pathogenic E.coli and Salmonella enteritidis in small domestic sewage treatment apparatus ("Johkasou"). *Water Science Technology* **43**(12), 191–193. IWA Publishing, London, UK. https://doi.org/10.2166/wst.2001.0736

Karnani A. (2011). Fighting Poverty Together: Rethinking Strategies for Business, Governments, and Civil Society to Reuce Poverty. Palgrave Macmillan. https://doi.org/10.1057/9780230120235 ISBN 978-1-349-58424-6.

Katukiza A. Y., Ronteltap M., Niwagaba C. B., Foppen J. W. A., Kansiime F. and Lens P. N. L. (2012). Sustainable sanitation technology options for urban slums. *Biotechnology Advances*, **30**, 964–978.

Kemp G. (2000). Sewerage assets, their condition and relationship to performance. Proceedings of IWA/IAHR/CIWEM Joint 1-Day Conference on Performance Indicators for Urban Drainage, University of Hertfordshire, 6 December 2000.

Kennedy E. B. and Marting T. A. (2016). Biomimicry: streamlining the front end of innovation for environmentally sustainable products. *Research-Technology Management*, **59**(4), 40–48. http://dx.doi.org/10.1080/08956308.2016.1185342

Kleiner Y. and Rajani B. (2001). Comprehensive review of structural deterioration of water mains: physically based models. NRCC-43722. *Urban Water*, **3**, 151–164.

Koottatep T., Phuphisith S., Pussayanavin T., Panuvatvanich A. and Polprasert C. (2014). Modeling of pathogen inactivation in thermal septic tanks. *Journal of Water Sanitation and Hygiene for Development*, **4**(1), 81–88.

Koottatep T., Chapagain S. K., Watanatanachart J. and Polprasert C. (2015). Application of nano-silver coated granular activated carbon for inactivation of septic tank effluent. *Journal of Water Sanitation and Hygiene for Development*, **5**(4), 632–637.

Koottatep T., Chapagain S. K., Polprasert C., Panuvatvanich A. and Ahn K.-H. (2016). Sanitation situations in selected Southeast Asian countries and application of innovative technologies. *Environment, Development and Sustainability*, **20**(1), 495–506, doi: 10.1007/s10668-016-9892-6

Korving H., van Noortwijk J. M., van Gelder P. H. A. J. M. and Parkhi R. S. (2003). Coping with uncertainty in sewer system rehabilitation. In: Safety and Reliability, Bedford and van Gelder (eds), Swets & Zeitlinger, Lisse, Netherlands. https://pdfs.semanticscholar.org/3e4f/ceb18d8550fcc35d3c3328680c8105d088eb.pdf (accessed 6 April 2018).

Kruger C. and Cross N. (2006). Solution driven versus problem driven design: strategies and outcomes. *Design Studies*, **27**(5), 527–548.

Kuitert W. (2008). Transforming with water. Proceedings of the 45th World Congress of the International Federation of Landscape Architects IFLA 2008, Blauwdruk/Techne Press, Wageningen, pp. 51–61.

Kvarnström E., McConville J., Bracken P., Johansson M. and Fogde M. (2011). The sanitation ladder – a need for a revamp? *Journal of Water Sanitation and Hygiene for Development*, **1**(1), 3–12.

Kwiakowski V. F. (1986). Infrastructure Assets: An Assessment of Users' Needs and Recommendations for Financial Reporting. PHD thesis, University of Kentucky, USA.

Langergrabera G. and Muelleggera E. (2005). Ecological sanitation – a way to solve global sanitation problems? *Environment International*, **31**(2005), 433–444.

Larsen T. A., Alder A. C., Eggen R. I., Maurer A. and Lienert J. (2009). Source separation: will we see a paradigm shift in wastewater handling? *Environmental Science & Technology*, **43**, 6121–6125.

Larsen T. A., Udert K. and Lienert J. (2013). Source Separation and Decentralization for Wastewater Management. IWA Publishing, London, UK.

Larsen T. A., Hoffmann S., Lüthi C., Truffer B. and Maurer M. (2016). Emerging solutions to the water challenges of an urbanizing world. *Science*, **352**(6288), 928–933.

Lazarte M. (2016). ISO's two-step solution to improving sanitation for 2.4 billion people. *ISO Focus Magazine*. https://www.iso.org/news/2016/11/Ref2145.html (accessed 22 February 2018).

Le Gauffre P. and Cherqui F. (2009). Sewer Rehabilitation Criteria Evaluated by Fusion of Fuzzy Indicators. LESAM 2009 – 3rd IWA Leading Edge Conference on Strategic Asset Management, Miami, USA.

Leliveld A. and Knorring P. (2017). Frugal innovation and development research. *The European Journal of Development Research*, (2018), **30**, 1–16. https://doi.org/10.1057/s41287-017-0121-4

Lens P., Zeeman G. and Lettinga G. (2001). Decentralized Sanitation and Reuse: Concepts, Systems and Implementation. IWA Publishing, London, UK.

Lenton R., Wright A. and Lewis K. (2005). Health, Dignity and Development: What Will It Take? Earthscan, London, UK.

Lettl C. (2007). User involvement competence for radical innovation. *Journal of Engineering and Technology Management*, **24**(1–2), 53–75.

Lewis S. (2008). The Ecology of Infrastructure and the Infrastructure of the Internet, blog Hag Pak Sak. https://hakpaksak.wordpress.com/2008/09/22/the-etymology-of-infrastructure-and-the-infrastructure-of-the-internet/ (accessed 27 February 2018).

Lockwood H. and Smits S. (2011). Supporting Rural Water Supply: Moving towards a Service Delivery Approach. Practical Action Publishing Ltd, Warwickshire, UK.

Londong J. (2013). Practical experience with source separation in Germany. In: Source Separation and Decentralization for Wastewater Management, T. Larsen, K. Udert and J. Lienert (eds), IWA, London, pp. 423–430.

Lüthi C. and Panesar A. (2013). Source separation in middle- and low-income countries In: Source Separation and Decentralization for Wastewater Management, T. A. Larsen, K. M. Udert and J. Lienert (eds), IWA Publishing, London, UK, pp. 455–462.

Lyle J. T. (1984). Designing Human Ecosystems. John Wiley & Sons, Hoboken, USA.

Lyle J. T. (1994). Regenerative Design for Sustainable Development. John Wiley & Sons, Hoboken, USA.

Magara Y. (2003). Status of onsite-treatment of domestic wastewater management in Japan. Proceedings of the Johkassou Session JECES at the 3rd World Water Forum, Kyoto, Japan. https://www.jeces.or.jp/en/downroad/pdf/proceedings.pdf

Magid J., Eilersen A. M., Wrisberg S. and Henze M. (2006). Possibilities and barriers for recirculation of nutrients and organic matter from urban to rural areas: a technical theoretical framework applied to the medium-sized town Hillerød, Denmark. *Ecological Engineering*, **28**, 44–54.

Mahmood Q., Pervez A., Zeb B. S., Zaffar H., Yaqoob H., Waseem M., Zahidullah A. A. and Afsheen S. (2013). Natural treatment systems as sustainable ecotechnologies for the developing countries. *BioMed Research International*, Article ID 796373. http://dx.doi.org/10.1155/2013/796373

Mang P. and Reed B. (2012a). Designing from place: a regenerative framework and methodology. *Building Research & Information*, **40**(1), 23–38.

Mang P. and Reed B. (2012b). Regenerative Development and Design. Regenesis Group and Story of Place, Encyclo, Sustainability Science & Technology.

Mang P., Haggard B. and Regenesis (2016). Regenerative Development and Design: A framework for evolving sustainability. John Wiley & Sons, Inc., Hoboken, USA.

Manzini E. and Vezzoli C. (2002). Product-Service Systems and Sustainability: Opportunities for Sustainable Solutions. United Nations Environment Programme (UNEP), Nairobi, Kenya.

Mara D. D. (2003). Water, sanitation and hygiene for the health of developing nations. *American Journal of Public Health*, **117**, 452–456.

Mara D. D. and Pearson H. W. (1998). Design Manual for Waste Stabilization Ponds in Mediterranean Countries. Lagoon Technology, Leeds, UK.

Mara D. D. and Ryan A. B. (1983). Pit Latrine Ventilation: Field Investigation Methodology. TAG Technical Note Number 4. United Nations Development Programme and the World Bank, Washington D.C., USA. http://documents.worldbank.org/curated/en/883921468016822130/pdf/TAG4000Pit0lat0tigation0methodology.pdf (accessed 4 April 2018).

Markard J. (2011). Transformation of infrastructures: sector characteristics and implications for fundamental change. *Journal of Infrastructure Systems*, **17**(3), 107–117, doi: 10.1061/(ASCE)IS.1943-555X.0000056

Marlow D., Heart S., Burn S., Urquhart A., Gould S., Anderson M., Cook S., Ambrose M., Madin M. and Fitzgerald A. (2007). Condition assessment strategies and protocols for water and wastewater utility assets. Report of Water Environmental Research Foundation (WERF) sponsored workshop. http://simple.werf.org/simple/media/documents/1_03-CTS-20CO_MWH_Final%20WERF%20Report_thru%20Appndx%20E.pdf (accessed 26 March 2018).

Marlow D. R., Moglia M., Cook S. and Beale D. J. (2013). Towards sustainable urban water management: a critical reassessment. *Water Research*, **47**, 7150–7161.

Maurer M., Bufardi A., Tilley E., Zurbrügg C. and Truffer B. (2012). A compatibility-based procedure designed to generate potential sanitation system alternatives. *Journal of Environmental Management*, **104**, 51–61.

McConville J., Kunzle R., Messmer U., Udert K. M. and Larsen T. A. (2014). Decision support for redesigning wastewater treatment technologies. *Environmental Science & Technology*, **48**(20), 12238–12246.

McDonough W. and Braungart M. (2002). Cradle to Cradle: Remaking the Way We Make Things. North Point Press, New York, USA.

McGauhey P. H. and Winneberger J. H. T. (1964a). Studies of the failure of septic tank percolation systems. *Journal of the Water Pollution Control Federation*, **36**(5), 593–606.

McGauhey P. H. and Winneberger J. H. T. (1964b). Causes and Prevention of Failure of Septic Tank Percolation Systems, FHA No. 533. Federal Housing Administration, Washington D.C., USA.

McGauhey P. H. and Winneberger J. H. T. (1965). Final Report on a Study of Methods of Preventing Failure of Septic Tank Percolation Systems. SERL Report No. 65–17, Sanitary Engineering Research Laboratory, University of California, Berkeley, USA.

Medland L. S., Scott R. E. and Cotton A. P. (2016). Achieving sustainable sanitation chains through better informed and more systematic improvements: lessons from multi-city research in Sub-Saharan Africa. *Environmental Science: Water Research & Technology*, **2**, 492–501, doi: 10.1039/C5EW00255A, http://pubs.rsc.org/en/content/articlepdf/2016/EW/C5EW00255A (accessed 22 March 2018).

Melo J. C. (2005). The Experience of Condominial Water and Sewerage Systems in Brazil: Case Studies from Brasilia, Salvador and Parauebas. Water and Sanitation Program. World Bank, Washington, D.C., USA. http://documents.worldbank.org/curated/en/505601468226737476/The-experience-of-condominial-water-and-seweragesystems-in-Brazil-case-studies-from-Brasilia-Salvador-and-Parauebas (accessed 13 November 2017).

Mesthene E. G. (1970). Technological Change: Its Impact on Man and Society. Harvard Studies in Technology and Society. Harvard University Press, Cambridge, USA.

Mihelcic J. R., Ren Z. J., Cornejo P. K., Fisher A., Simon A. J., Snyder S. W., Zhang Q., Rosso D., Huggins T. M., Cooper W., Moeller J., Rose B., Schottel B. L. and Turgeon J. (2017). Accelerating innovation that enhances resource recovery in the wastewater sector: advancing a national testbed network. *Environmental Science & Technology*, **51**(14), 7749–7758, doi: 10.1021/acs.est.6b05917

Miszta-Kruk K. (2016). Reliability and failure rate analysis of pressure, vacuum and gravity sewer systems based on operating data. *Engineering Failure Analysis*, **61**, 37–45. http://dx.doi.org/10.1016/j.engfailanal.2015.07.034

Mizuochi M. (2011). Small-Scale Domestic Wastewater Treatment Technology in Japan, and the Possibility of Technological Transfer, Asian Environment Research Group, National Institute for Environmental Studies, Japan. http://www.wepa-db.net/pdf/0712forum/paper18.pdf (accessed 26 February 2018).

MOA (Ministry of Agriculture) (2009). Technology Specification of Biogas Digester for Domestic Sewage Treatment. NY/T 1702-2009 (in Chinese). www.bzfxw.com

MOA (Ministry of Agriculture) (2014a). Collection of Standard Design Drawings of Biogas Digester for Domestic Sewage Treatment. NY/T 2597-2014 (in Chinese). www.bzfxw.com

MOA (Ministry of Agriculture) (2014b). Collection of Standard Design Drawings of Biogas Digester for Domestic Sewage Treatment. NY/T 2597-2014 (in Chinese). www.bzfxw.com

MOA (Ministry of Agriculture) (2014c). Construction Regulations of Biogas Digester for Domestic Sewage Treatment. NY/T 2601/2014 (in Chinese). www.bzfxw.com

MOA (Ministry of Agriculture) (2014d). Operation Maintenance Specification of Biogas Digester for Domestic Sewage Treatment. NY/T 2602/2014 (in Chinese). www.bzfxw.com

Mont O. (2002). Clarifying the concept of product service systems. *Journal of Cleaner Production*, **10**, 237–245.

Mont O., Dalhammar C. and Jacobsson N. (2006). A new business model for baby prams based on leasing and product remanufacturing. *Journal of Cleaner Production*, **14**, 1509–1518.

Morita A. (2015). Advantages of Japanese On-Site Treatment System: The History of Night Soil Treatment in Japan. Japan Sanitation consortium/Japan Environmental Sanitation Center. https://www.jica.go.jp/information/seminar/2015/ku57pq00001zht6b-att/20150910_01_03.pdf (accessed 28 March 2018).

Murphy H. M., McBean E. A. and Farahbakhsh K. (2009). Appropriate technology – a comprehensive approach for water and sanitation in the developing world. *Technology in Society*, **31**, 158–167, doi: 10.1016/j.techsoc.2009.03.010

Murphy L. M. and Edwards P. L. (2003). Bridging the Valley of Death: Transitioning from Public to Private Sector Financing, NREL/MP-720-34036; 2003.

Murray A. and Buckley C. (2009). Designing reuse-oriented sanitation infrastructure: the design for service planning approach. In: Wastewater Irrigation and Health Assessing and Mitigating Risk in Low-income Countries, P Drechsel, C. A. Scott, L. Raschid-Sally, M. Redwood and A. Bahri (eds), Earthscan, London, UK.

Murray A. and Drechsel P. (2011). Positive deviance in the sanitation sector in Ghana: why do some wastewater Treatment facilities work when the majority fail? *Waterlines*, **30**(2), 135–149. https://jstor.org/stable/24686701

Nakagiri A., Niwagaba C. B., Nyenje P. M., Kulabako R. N., Tumuhairwe J. B. and Kansiime F. (2016). Are pit latrines in urban areas of SubSaharan Africa performing? A review of usage, filling, insects and odour nuisances. *BMC Public Health*, **16**, 120, doi: 10.1186/s12889-016-2772-z

Nakajima J., Fujimura Y. and Inamori Y. (1999). Performance evaluation of on-site treatment facilities for wastewater from households, hotels and restaurants. *Water Science and Technology*, **39**(8), 85–92.

NDHS (Nigeria Demographic and Health Survey) (2008). Demographic and Health Survey. National Population Commission, Federal Republic of Nigeria, Abuja, Nigeria.

Nelson K. L. and Murray A. (2008). Sanitation for unserved populations: technologies, implementation challenges, and opportunities. *Annual Review of Environment and Resources*, **33**, 119–151.

Nesshöver C., Assmuth T., Irvine K. N., Rusch G. M., Waylen K. A., Delbaere B., Haase D., Jones-Walters L., Keune H., Kovacs E., Krauze K., Külvik M., Rey F., van Dijk J., Vistad O. I., Wilkinson M. E. and Wittmer H. (2017). The science, policy and practice of nature-based solutions: an interdisciplinary perspective. *Science of the Total Environment*, **579**, 1215–1227.

NHFPC (2015). China Health and Family Planning Statistical Yearbook. Peking Union Medical College Press, Beijing (in Chinese).

NHFPC (2016). China Health and Family Planning Statistical Yearbook. Peking Union Medical College Press, Beijing (in Chinese).

Niku S., Schroeder E. D. and Samaniego F. J. (1979). Performance of activated sludge process and reliability-based design. *Journal of Water Pollution Control Association*, **51**(12), 2841–2857.

Niku S., Schroeder E. D., Tchobanoglous G. and Samaniego F. J. (1981). Performance of activated sludge process: reliability, stability and variability. Environmental Protection Agency, EPA Grant no. R805097-01, pp. 1–124.

Niku S., Schroeder E. D. and Haugh R. S. (1982). Reliability and stability of trickling filter processes. *Journal of Water Pollution Control Association*, **54**(2), 129–134.

Nilsson M. and Olsson L. (2014). Sanitation in an informal settlement – a user-oriented study on sanitation solutions and factors influencing decisions for implementation in Cochabamba, Bolivia. Master's thesis, Chalmers University of Technology, Gothenburg, Sweden. http://publications.lib.chalmers.se/records/fulltext/207030/207030.pdf

NISER (Nigerian Institute of Social, Economic Research) (2010). Socio-Economic Impact of Climate Change in Nigeria. NISER, Ibadan, Nigeria.

Niskanen W. A. (1991). The soft infrastructure of a market economy. Wayback Machine. https://web.archive.org/web/20110328213000/http://www.cato.org/pubs/journal/cj11n2/cj11n2-4.pdf (accessed 27 February 2018).

NOWRA (National Onsite Wastewater Recycling Association) (2007). Model Code Framework for the Decentralized Wastewater Infrastructure. Volume I: Workbook for Writing the Code/ Volume II: Code Design Philosophy and Guidance. http://082511a.membershipsoftware.org/files/About%20NOWRA/Model%20Code/VOLUMEone-WorkbookfortheCode.pdf (accessed 16 March 2018).

NSF International (2012). NSF's Advance Onsite Wastewater Treatment Certification Program. https://www.nsf.org/newsroom_pdf/WW_Onsite_Treatment_Insert_LT_EN_LWW11110712.pdf (accessed 14 March 2018).

NSF/ANSI (2010a). NSF/ANSI Standard 40: Residential wastewater treatment systems, NSF International.

NSF/ANSI (2010b). NSF/ANSI Standard 360: Field Performance Verification. NSF International.

NSF/ANSI (2011). Standard 350 and 350-1: Onsite residential and commercial greywater treatment systems for subsurface discharge. NSF International.

NSF/ANSI (2016). NSF/ANSI Standard 41: Non-Liquid Saturated Treatment Systems. NSF International.

NSF/ANSI (2017). NSF/ANSI Standard 46: Water Treatment System Components. NSF International.

OFWAT (Office of Water Services) (2000). Serviceability of the water main and sewerage networks in England and Wales. Information Note 35A. OFWAT, Birmingham, UK, March 2000.

OFWAT (Office of Water Services) (2015). Annex 1: Our process for assessing serviceability. https://www.ofwat.gov.uk/wp-content/uploads/2015/11/ltr_rd1506_annexes.doc (accessed 26 March 2018).

Oladoja N. A. (2017). Appropriate technology for domestic wastewater management in under-resourced regions of the world. *Applied Water Science*, **7**, 3391–3406, doi: 10.1007/s13201-016-0495-z

Oliveira S. C. and Von Sperling M. (2008). Reliability analysis of wastewater treatment plants. *Australian Journal of Marine and Freshwater Research*, **42**, 1182–1194.

Olsson G. (2013). The potential of control and monitoring. In: Source Separation and Decentralization for Wastewater Management, T. A. Larsen, K. M. Udert and J. Lienert (eds), IWA Publishing, London, UK, pp. 179–192.

Open ecosystem network (2017). Frugal Innovation for Sustainable Growth and Development. https://open-ecosystem.org/articles/frugal-innovation-sustainable-growth-and-development (accessed 20 May 2018).

Orenco System (2017). Effluent Sewer Design Manual: A design aid for Septic Tank Effluent Gravity (STEG)/ Septic Tank Effluent Pumping (STEP). NDA-EFS-1, Rev. 3.0, pp. 1–56. Orenco Systems Inc., USA. https://odl.orenco.com/documents/NDA-EFS-1.pdf (accessed 12 April 2018).

Ormiston A. W. and Floyd R. E. (2004). On-Site Wastewater Systems: Design and Management Manual, 3rd edn. ARC Technical Publication. Auckland Regional Council Technical Publication No. 58 (TP 58). http://www.aucklandcity.govt.nz/council/documents/technicalpublications/TP58%20On-site%20wastewater%20systems%202004%20Chapters%201-8.pdf (accessed 7 March 2018).

Orner K. D. and Mihelcic J. R. (2018). A review of sanitation technologies to achieve multiple sustainable development goals that promote resource recovery. *Environmental Science: Water Research & Technology*, **4**, 16.

Oswald W. J., Golueke C. G. and Gee H. K. (1959). Waste Water Reclamation through Production of Algae. Contribution 22, Water Resources Center, University of California, Berkley, USA.

Oswald W. J., Golueke C. G. and Tyler R. W. (1967). Integrated pond systems for subdivisions. *Journal Water Pollution Control Federation*, **38**, 1289.

Oswald W. J., Bailey-Green F. and Lundquist T. J. (1994). Performance of methane fermentation pits in advanced integrated wastewater pond systems. *Water Science and Technology*, **30**(12), 287–295.

Palaniappan M., Lang M., Gleick P. H. and Pacific I. (2008). A Review of Decision-Making Support Tools in the Water, Sanitation, and Hygiene Sector. Pacific Institute, Oakland, USA.

Paranychianakis N. V., Angelakis A. N., Leverenz H. and Tchobanoglous G. (2006). Treatment of wastewater with slow rate systems: a review of treatment processes and plant functions. *Critical Reviews in Environmental Science and Technology*, **36**(3), 187–259.

Parkman M., Patchett J. and Odongo O. (2008). Kisumu Water Supply and Sanitation Project, Long Term Action Plan: Water Design Report, Lake Victoria South Water Services Board, Kisumu, Kenya.

PDSS (President Deputy of Strategic Supervision) (2009). Operation and Maintenance of Wastewater Collection Networks Manual. Code No. 520. President Deputy for Strategic Supervision, Tehran, Iran.

Peal A., Evans B. E., Blackett I., Hawkins P. and Heymans C. (2014a). Fecal Sludge Management: analytical tools for assessing FSM in cities. *Journal of Water, Sanitation and Hygiene for Development*, **4**(3), 371–383. https://doi.org/10.2166/washdev.2014.139

Peal A., Evans B. E., Blackett I., Hawkins P. and Heymans C. (2014b). Fecal sludge management: a comparative assessment of 12 cities. *Journal of Water, Sanitation and Hygiene for Development*, **4**(4), 563–575.

Pham N. B. and Kuyama T. (2013). Decentralized domestic wastewater management in Asia – challenges and opportunities. Policy Brief – Series 1, Water Environment Partnership in Asia (WEPA), Institute for Global Environmental Strategies [IGES]. http://www.wepa-db.net/pdf/1403policy_brief/WEPA_PB1_2013.pdf (accessed 30 April 2017).

Polprasert C. and Koottatep T. (2017). Organic Waste Recycling – Technology, Management and Sustainability, 4th edn. IWA Publishing, the Hague, Netherlands.

Poortvliet P. M., Sanders L., Weijma J. and De Vries J. R. (2018). Acceptance of new sanitation: the role of end-users' pro-environmental personal norms and risk and benefit perceptions. *Australian Journal of Marine and Freshwater Research*, **131**, 90–99. https://doi.org/10.1016/j.watres.2017.12.032

Potter A., Klutse A., Snehalatha M., Batchelor C., Uandela A., Naafs A., Fonseca C. and Moriarty P. (2011). Assessing sanitation service levels. IRC International Water and Sanitation Centre, Working Paper 3 2nd edn. https://www.ircwash.org/sites/default/files/Potter-2011-Assessing.pdf (accessed 18 May 2017).

Powell K. L., Barrett M. H., Pedley S., Tellam J. H., Stagg K. A., Gresswell R. B. and Rivett M. O. (2000). Enteric virus detection in groundwater using a glass wool trap. Groundwater: Past Achievements and Future Challenges. (Proc. of XXX Int. Assoc. of Hydrologists' Congress on Groundwater, Cape Town, South Africa. November 2000), pp. 813.

Powell K. L., Cronin A. A., Barrett M. H., Sellwood J. and Pedley S. (2001). Distribution of microbiological contaminants in Triassic Sandstone urban aquifers. Environment Agency R&D Technical Report P2–255/TR, 2001.

Pradel P. and Adkins D. (2014). Towards a design for frugal: review of implications for product design. Emerging Practices: Design Research and Education Conference (EPC), Shanghai, China, October 10–17. https://dspace.lboro.ac.uk/dspace-jspui/bitstream/2134/26324/3/26324.pdf (accessed 21 April 2018).

Prahalad C. K. and Hart S. L. (2002). The fortune at the Bottom of the Pyramid. Strategy þ Business. https://www.strategy-business.com/article/11518?gko=9a4ba (accessed 21 April 2018).

Principles of Economics (2017, August 2). Wikibooks. The Free Textbook Project. https://en.wikibooks.org/w/index.php?title=Principles_of_Economics&oldid=3253900 (accessed 25 April 2018).

Programme Solidarité Eau (2018). The Sustainable Development Goals for Water and Sanitation Services. Interpreting the Targets and Indicators. pS-Eau, Paris. www.pseau.org/en/agenda-2030 (accessed 19/03/2018).

Pruss A., Kay D., Fewtrell L. and Bartram J. (2002). Estimating the burden of disease from water, sanitation, and hygiene at a global level. *Environmental Health Perspectives*, **110**(5), 537–542.

Prüss-Üstün A., Bos R., Gore F. and Bartram J. (2008). Safer Water, Better Health: Costs, Benefits and Sustainability of Interventions to Protect and Promote Health. Geneva. http://www.who.int/water_sanitation_health/publications/safer_water/en/ (accessed 20 May 2018).

Pussayanavin T., Koottatep T., Eamrat R. and Polprasert C. (2015). Enhanced sludge reduction in septic tanks by increasing temperature. *Journal of Environment Science and Health Part A*, **50**(1), 81–89.

Qasem A., Zayed T. and Chen Z. (2009). A condition rating system for wastewater treatment plants infrastructures. *World Academy of Science, Engineering and Technology International Journal of Environmental and Ecological Engineering*, **3**(4). https://waset.org/publications/15241/a-condition-rating-system-for-wastewater-treatment-plants-infrastructures (accessed 23 March 2018).

Ramani S. V., SadreGhazi S. and Duysters G. (2012). On the diffusion of toilets as bottom of the pyramid innovation: lessons from sanitation entrepreneurs. *Technological Forecasting & Social Change*, **79**, 676–687, doi: 10.1016/j.techfore.2011.06.007

Ramoa A., Matos J. and Luthi C. (2015). System-based decision trees for the selection of sanitation technologies. 38th WEDC International Conference, Loughborough University, UK. http://wedc.lboro.ac.uk/resources/conference/38/Ramoa-2250.pdf (accessed 26 June 2017).

Rao B. C. (2013). How disruptive is frugal? *Technology in Society*, **35**(1), 65-73. https://doi.org/10.1016/j.techsoc.2013.03.003Rao K. C., Kvarnström E., Di Mario L. and Drechsel P. (2016). Business Models for Fecal Sludge Management. CGIAR Research Program on Water, Land and Ecosystems (WLE). International Water Management Institute (IWMI), Colombo, Sri Lanka. 80p. (Resource Recovery and Reuse Series 6). doi:10.5337/2016.21

Ratanatamskul C., Chiemchaisri C. and Yamamoto K. (1995). The use of a zeolite–iron column for residual ammonia and phosphorus removal in the effluent from a membrane process as an on-site small-scale domestic wastewater treatment. *Water Science and Technology*, **31**, 145–152.

Ray P. K. and Ray S. (2010). Resource-constrained innovation for emerging economies: the case of the Indian telecommunications industry. *IEEE Transactions on Engineering Management*, **57**(1), 144–156.

Reap J., Baumeister D. and Bras B. (2005). Holism, Biomimicry and Sustainable Engineering. ASME 2005 International Mechanical Engineering Congress and Exposition (IMECE2005) ASME, pp. 423–431.

Reed R. and Vines M. (1992). Reduced Cost Sewerage in Orangi, Karachi. WEDC, Loughborough.

Rexfelt O. and Hiort af Ornas V. (2009). Consumer acceptance of product-service systems: designing for relative advantages and uncertainty reductions. *Journal of Manufacturing Technology Management*, **20**, 674–699.

Reynolds J. H. and Barrett M. H. (2003). A review of the effects of sewer leakage on groundwater quality. *Water and Environment Journal*, **17**(11), 34–39. https://doi.org/10.1111/j.1747-6593.2003.tb00428.x

Rodgers A. F., Ajono L. A., Gyapong J. O., Hagan M. and Emerson P. M. (2007). Characteristics of latrine promotion participants and non-participants; inspection of latrines; and perceptions of household latrines in northern Ghana. *Tropical Medicine & International Health*, **12**(6), 772–782.

Rosencqvist T., Mitchell C. and Willetts J. (2016). A short history of how we think and talk about sanitation services and why it matters. *Journal of Water, Sanitation and Hygiene for Development*, **06.2**, doi: 10.2166/washdev.2016.118

Russell A. L. (2005). Standardization in history: a review essay with an eye to the future. In: The Standards Edge: Future generations, S. Bolin (ed.), Sheridan Press, Ann Arbor, USA, pp. 247–260. http://arussell.org/papers/futuregeneration-russell.pdf (accessed 23 February 2018).

Sarkilahti M., Kinnunen V., Kettunen R., Jokinen A. and Rintala J. (2017). Replacing centralized waste and sanitation infrastructure with local treatment and nutrient recycling: expert opinions in the context of urban planning. *Technological Forecasting and Social Change*, **118**, 195–204.

Schertenleib R. (2005). From conventional to advanced environmental sanitation. *Water Science and Technology*, **51** (10), 7–14.

Schwartz P. (1991). The Art of the Long View. Curency Doubleday, New York, USA.

Seetharam K. (2015). Challenges and opportunities for sanitation in developing countries. *Journal of Science Policy & Governance*, **7**(1), 1–12. http://www.sciencepolicyjournal.org/uploads/5/4/3/4/5434385/seetharam-jspg-issue7.pdf

Simanis E. and Hart S. (2008). Base of the Pyramid Protocol, 2nd edn. Cornell University, Ithaca, USA.

Singhirunnusorn W. and Stenstrom M. K. (2009). Appropriate wastewater treatment systems for developing countries: criteria and indictor assessment in Thailand. *Water Science and Technology*, **59.9**, 1884–1873, doi: 10.2166/wst.2009.215

Smith A., Vob J. and Grin J. (2010). Innovation studies and sustainability transitions: the allure of the multi-level perspective and its challenges. *Research Policy*, **39**(4), 435–448.

Spaargaren G., Mommaas H., Van Den Burg S., Maas L., Drissen E. and Dagevos H. (2007). More sustainable lifestyles and consumption patterns: a theoretical perspective for the analysis of transition processes within consumption domains. Contrast Research Report, TMP project. Wageningen: Environmental Policy Group, Wageningen University; Telos: Tilburg University; RIVM; LEI, Wageningen, Netherlands.

Spaargaren G., Oosterveer P., Van Buuren J. and Mol A. P. J. (2006). Mixed Modernities: Towards Viable Urban Environmental Infrastructure Development in East Africa. Wageningen University, Wageningen, Netherlands.

Sparkman D. (2012). More than just counting toilets: the complexities of monitoring for sustainability in sanitation. *Waterlines*, **31**, 260–271.

Springer-Heinzer A. (2017). ValueLinks 2.0 – Manual on Sustainable Value Chain Development. Value Chain Analysis, Strategy and Implementation. Draft volume 1, GTZ. http://valuelinks.org/wp-content/uploads/2015/09/ValueLinks-Manual-2.0-Vol-1-final-draft-July-2017.pdf (accessed 20 May 2018).

Sridhar M. K. C. (1995). Sullage/Wastewater in Nigeria: Problems and Scope for Utilization for Gardening. A report published by UNICEF, Lagos, Nigeria.

Starkl M., Brunner N. and Stenstrom T. (2013). Why do water and sanitation systems for the poor still fail? Policy analysis in economically advanced developing countries. *Advances in Environmental Science and Technology*, **47**(12), 6102–6110, doi: 10.1021/es3048416

Starkl M., Brunner N., Feil M. and Hauser A. (2015). Ensuring Sustainability of Non-Networked Sanitation Technologies: An Approach to Standardization. *Advances in Environmental Science and Technology*, **49**, 6411–6418, doi: 10.1021/acs.est.5b00887

Statistics Canada (2008). The age of public infrastructure. http://www.statcan.gc.ca/daily-quotidien/080213/dq080213aeng.htm (accessed 5 April 2018).

Statistics Canada (2015). Remaining useful service life of infrastructure assets. https://www.statcan.gc.ca/daily-quotidien/180119/dq180119e-eng.htm (accessed 5 April 2018).

Still D. and Foxon K. (2012a). Tackling the Challenges of Full it Latrines. Water Research Commission (WRC) Report no. 1745/1/12, Pretoria, South Africa, Vol. **1**.

Still D. and Foxon K. (2012b). Tackling the Challenges of Full it Latrines. Water Research Commission (WRC) Report no. 1745/2/12, Pretoria, South Africa, Vol. **2**.

Still D. and Foxon K. (2012c). Tackling the Challenges of Full it Latrines. Water Research Commission (WRC) Report no. 1745/3/12, Pretoria, South Africa, Vol. **3**.

Still D. A., Salisbury R. H., Foxon K. M., Buckley C. A. and Bhagwan J. N. (2010). The challenges of dealing with full pit latrines. Proceedings WISA Biennial Conference and Exhibition, Durban ICC, South Africa. Water Research Commission (WRC) Report no. TT 357/08. ISBN 978-1-77005-718-0

Stokman A. (2008). Water purification landscapes - constructed ecologies and contemporary urbanism. In: K. Wybe (ed.) Transforming With Water. Proceedings of the 45th World Congress of the International Federation of Landscape Architects (IFLA). Blauwdruk/Techne Press, Waginingen, Netherlands, pp. 51–61. https://www.freiraum.uni-hannover.de/uploads/tx_tkpublikationen/Water_Purificative_Landscapes_-_Constructed_Ecologies_and_Contemporary_Urbanism_01.pdf (accessed May 17 2017)

L. Strande, M. Ronteltap and D. Brdjanovic (2014). Faecal Sludge Management: Systems Approach for Implementation and Operation. IWA Publishing, London, UK.

Sunita N. (2012). Sanitation for all. *Bionature*, **486**, 185.

Swart D. B. and Palsma A. J. B. (2013). The Netherlands: 'Nieuwe Sanitatie'. In: Source Separation and Decentralization for Wastewater Management, T. A. Larsen, K. M. Udert and J. Lienert (eds), IWA Publishing, London, UK, pp. 431–438.

Tchobanoglous G. and Leverenz H. (2013). The rationale for decentralization of wastewater infrastructure. In: Source Separation and Decentralization for Wastewater Management, T. A. Larsen, K. M. Udert and J. Lienert (eds), IWA Publishing, London, UK, pp. 101–115.

Templeton M. R. (2015). Pitfalls and progress: a perspective on achieving sustainable sanitation for all. *Environmental Science: Water Research & Technology*, **1**, 17–21, doi: 10.1039/C4EW00087K

Thomas E., Andrés L. A., Borja-Vega C. and Sturzenegger G. (2018). Innovations in WASH Impact Measures: Water and Sanitation Measurement Technologies and Practices to Inform the Sustainable Development Goals. Directions in Development – Infrastructure. World Bank, Washington D.C., USA, License: CC BY 3.0 IGO. https://openknowledge.worldbank.org/handle/10986/29099

Tiberghien J. E., Robbins P. T. and Tyrrel S. F. (2011). Reflexive assessment of practical and holistic sanitation development tools using the rural and peri-urban case of Mexico. *Journal of Environmental Management*, **92**, 457–471, doi: 10.1016/j.jenvman.2010.08.032

Tilley E. (2013). Conceptualizing sanitation systems to accounts for new complexities in processing and management. In: Source Separation and Decentralization for Wastewater Management, T. A. Larsen, K. M. Udert and J. Lienert (eds), IWA Publishing, London, UK, pp. 227–239.

Tilley E., Strande L., Luthi C., Mosler H., Udert K. M., Gebauer H. and Hering J. G. (2014a). Looking beyond technology: an integrated apparoach to water, sanitation and hygiene in low income countries. *Advances in Environmental Science and Technology*, **48**(17), 9965–9970, doi: 10.1021/es501645d

Tilley E., Ulrich L., Luthi C., Reymond P. H. and Zurbrugg C. (2014b). Compendium of Sanitation Systems and Technologies, 2nd rev. edn. Swiss Federal Institute of Aquatic Science and Technology (EAWAG), Dübendorf, Switzerland. https://www.eawag.ch/fileadmin/Domain1/Abteilungen/sandec/schwerpunkte/sesp/CLUES/Compendium_2nd_pdfs/Compendium_2nd_Ed_Lowres_1p.pdf (accessed 11 September 2018).

Tiwari R., Fischer L. and Kalogerakis K. (2016). Frugal Innovation in Scholarly and Social Discourse: An Assessment of Trends and Potential Societal Implications. Working Paper. Project Potenziale Frugaler Innovationen. https://tubdok.tub.tuhh.de/bitstream/11420/1291/1/Tiwari_et_al_2016_Frugal_Innovation_BMBF_ITA.pdf (accessed 10 October 2017).

Tobias R., O'Keefe M., Künzle R., Gebauer H., Gründl H., Morgenroth E., Pronk W. and Larsen T. A. (2017). Early testing of new sanitation technology for urban slums: the case of the blue diversion toilet. *Science of the Total Environment*, **576**, 264–272.

Todd J. and Josephson B. (1996). The design of living technologies for waste treatment. *Ecological Engineering*, **6**, 109–136. https://doi.org/10.1016/0925-8574(95)00054-2

Todd J. and Todd N. J. (1980). Tomorrow is Our Permanent Address: The Search for an Ecological Science of Design. Harper and Row, New York, USA.

Todd N. J. and Todd J. (1984). Bioshelters, Ocean Arks, City Farming: Ecology as the Basis of Design. Sierra Club Books, San Francisco, USA.

Todd N. J. and Todd J. (1994). From Eco-Cities to Living Machines: Principles of Ecological Design. North Atlantic Books, Berkeley, USA.

Tran D. H., Perera B. J. C. and Ng A. W. M. (2009). Comparison of structural deterioration models for stormwater drainage pipes. *Computer-Aided Civil and Infrastructure Engineering*, **24**(2), 145–156.

Truffer B., Binz C., Gebauer H. and Störmer E. (2013). Conceptualizing sanitation systems to accounts for new complexities in processing and management. In: Source Separation and Decentralization for Wastewater Management, T. A. Larsen, K. M. Udert and J. Lienert (eds), IWA Publishing, London, UK, pp. 210–223.

Tuantet K., Temmink H., Zeeman G., Janssen M., Wijffels R. and Buisman C. (2014). Nutrient removal and microalgal biomass production on urine in a short lightpath photobioreactor. *Water Research*, **55**, 162–174.

Tukker A. (2015). Product services for a resource-efficient and circular economy – a review. *Journal of Cleaner Production*, **97**, 76–91. http://dx.doi.org/10.1016/j.jclepro.2013.11.049

Tukker A. and Tischner U. (2006). New Business for Old Europe: Product-Service Development, Competitiveness and Sustainability. Greenleaf Publishers, Sheffield, UK.

Uddin W., Hudson R. W. and Haas R. (2013). Public Infrastructure Asset Management, 2nd edn. McGraw-Hill Education, New York, USA.

Ulrich A., Reuter S., Gutterer B., Sasse L., Panzerbieter T. and Reckerzügel T. (2009). Decentralised Wastewater Treatment Systems (DEWATS) and Sanitation in Developing Countries. WEDC, Loughborough, UK. https://www.wbdg.org/FFC/ARMYCOE/PWTB/pwtb_420_46_9.pdf (accessed 4 April 2018).

UNDP (2006). Human Development Report 2006: Beyond Scarcity: Power, Poverty and the Global Water Crisis. Palgrave Macmillan, Basingstoke, New York. http://www.undp.org/content/dam/undp/library/corporate/HDR/200620Global20HDR/HDR-2006-Beyond%20scarcity-Power-poverty-and-the-global-water-crisis.pdf (accessed 5 August 2018).

UNDP (2008). Creating Value for All: Strategies for Doing Business with the Poor. United Nations Development Programme, New York, USA. http://www.growinginclusivemarkets.org/reports (accessed 20 May 2018).

UN-Water and WHO (2012). GLAAS 2012 Report. UN-Water Global Analysis and Assessment of Sanitation and Drinking-Water. The Challenge of Extending and Sustaining Services. WHO, Geneva, Switzerland. http://www.un.org/waterforlifedecade/pdf/glaas_report_2012_eng.pdf (accessed 28 June 2017).

UN-Water (2015). Wastewater Management A UN-Water Analytical Brief. http://www.unwater.org/publications/wastewater-management-un-water-analytical-brief/ (accessed 2 May 2018).

Urhaman F. (2007). Integrated condition assessment models for sustainable sewer pipelines. M.A.Sci. thesis, Concordia University, Montreal, Canada.

USDE (United State Department of Energy) (2014). The Water-Energy Nexus: Challenges and Opportunities, DOE/EPSA-0002; U.S. Department of Energy. https://www.energy.gov/under-secretary-science-and-energy/downloads/waterenergy-nexus-challenges-and-opportunities (accessed 24 April 2018).

USDE (United State Department of Energy) (2015). Waste-to-Energy Workshop Summary. U.S. Department of Energy, Office of Energy Efficiency and Renewable Energy, Bioenergy Technologies Office: Arlington, USA. http://www.energy.gov/sites/prod/files/2015/08/f25/beto_wte_workshop_report.pdf (accessed 24 April 2018).

USDE/NSF (United State Department of Energy/National Science Foundation) (2016). Energy-Positive Water Resource Recovery Workshop Report. National Science Foundation, U.S. Department of Energy, U.S. Environmental Protection Agency: Arlington, USA. http://www.energy.gov/sites/prod/files/2016/01/f28/epwrr_workshop_report.pdf (accessed 24 April 2018).

USEPA (1974). Process design manual for upgrading existing wastewater treatment plants. U.S. Environmental Protection Agency, Technology Transfer. http://www.un.org/waterforlifedecade/pdf/glaas_report_2012_eng.pdf (accessed 28 June 2017).

USEPA (1991a). Handbook of Sewer System Infrastructure Analysis and Rehabilitation. Report Number EPA 625/6-91/030. U.S. Environmental Protection Agency, Office of Research and Development, Center for Environmental Research Information, Cincinnati, USA. October 1991.

USEPA (U.S. Environmental Protection Agency) (1991b). Sewer System Infrastructure Analysis and Rehabilitation. EPA 625/6-91/030. U.S. Environmental Protection Agency, Office of Research and Development, Cincinnati, USA.

USEPA (1997). Response to congress on use of decentralized wastewater treatment systems. EPA 832-R-97-00 lb, U S Environmental Protection Agency Office of Water and Office of Wastewater Management Washington D.C., USA. https://nepis.epa.gov/Exe/ZyPDF.cgi/200047VF.PDF?Dockey=200047VF.PDF (accessed 24 February 2018).

USEPA (U.S. Environmental Protection Agency) (2008a). Review of Sewer Design Criteria and RDII Prediction Methods. Office of Research Environmental Protection and Development. Washington D.C., USA. EPA/600/R-08/010. https://nepis.epa.gov/Exe/ZyPDF.cgi/P1008BP3.PDF?Dockey=P1008BP3.PDF (accessed 09 April 2018).

USEPA (U.S. Environmental Protection Agency) (2008b). Clean Watersheds Needs Survey (CWNS) – 2008 Report and Data. https://www.epa.gov/cwns/clean-watersheds-needs-survey-cwns-2008-report-and-data (accessed 9 April 2018).

USEPA (2009). Rehabilitation of Wastewater Collection and Water Distribution Systems – White Paper, U.S. Environmental Protection Agency, Office of Research and Development – National Risk Management Research Laboratory, Cincinnati, USA. EPA/600/R-09/048.

USEPA (U.S. Environmental Protection Agency) (2010). State of technology for rehabilitation of wastewater collection systems. EPA/600/R-10/078. http://nepis.epa.gov/Adobe/PDF/P1008C45.pdf (accessed 6 November 2013).

USEPA (U.S. Environmental Protection Agency) (2011). Principles of Design and Operations of Wastewater Treatment Pond Systems For Plant Operators. Engineers and Managers, U.S. Environmental Protection Agency, Cincinnati, USA.

Van Dijk M. P. (2012). Sanitation in developing countries: innovative solutions in a value chain framework. In: Management of Technological Innovation in Developing and Developed Countries, H. Sun (ed.), Intech, Rijkeka, Netherlands, pp. 65–83.

van Loosdrecht M. C. and Brdjanovic D. (2014). Anticipating the next century of wastewater treatment. *Acta Agriculturae Scandinavica Section A Animal Science*, **344**, 1452–1453.

van Vliet B. J. M., Spaargaren G. and Oosterveer P. (2011). Sanitation under challenge: contributions from the social sciences. *Water Policy*, **13**, 797–809.

Vanier D. J. and Rahman S. (2004). Survey on Municipal Infrastructure Assets. Client Report B5123.2, National Research Council Canada, Ottawa. irc.nrc-cnrc.gc.ca/pubs/fulltext/b5123.2/ (accessed 5 April 2018).

Von Sperling M. and Oliveira S. M. A. C. (2007). Reliability analysis of stabilization ponds systems. *Water Science and Technology*, **55**(11), 127–134.

Verbong G. P. J. and Geels F. W. (2010). Exploring sustainability transitions in the electricity sector with socio-technical pathways. *Technological Forecasting and Social Change*, **77**(8), 1214–1221.

Verhagen J. and Carrasco M. (2013). Full-Chain Sanitation Services that Last: Non-Sewered Sanitation Services. IRC International Water and Sanitation Centre, the Hague, Netherlands. https://www.ircwash.org/sites/default/files/201306_sanitation_framework_final.pdf (accessed 17 June 2017).

Verhagen J. and Ryan P. (2008). Sanitation services for the urban poor: symposium background paper. IRC Symposium: Sanitation for the Urban Poor – Partnership and Governance, Delft, Netherlands.

Vermaas P., Kroes P., van de Poel I., Franssen M. and Houkes W. (2011). A Philosophy of Technology: From Technical Artefacts to Sociotechnical Systems. Synthesis Lectures on Engineers, Technology, and Society. https://doi.org/10.2200/S00321ED1V01Y201012ETS014 (Accessed 4 April 2018).

Vezzoli C. and Manzini E. (2008). Design for Environmental Sustainability. Springer, Berlin, Germany.

Vezzoli C., Kohtala C., Srinivasan A., Xin L., Fusakul M., Sateesh D. and Diehl J. C. (2014). Product-Service System Design for Sustainability. Greenleaf Publishing Limited, Sheffield, UK.

Vezzoli C., Ceschin F., Diehl J. C. and Kohtala C. (2015). New design challenges to widely implement 'Sustainable Product – Service Systems'. *Journal of Cleaner Production*, **97**, 1–12.

Viessman W. (1964). Ventilation Control of Odors. Annals of the New York Academy of Sciences, pp. 630–637.

Vinnerås B. and Jönsson H. (2013). The Swedish experience with source separation. In: Source Separation and Decentralization for Wastewater Management, T. A. Larsen, K. M. Udert and J. Lienert (eds), IWA Publishing, London, UK, pp. 415–422.

Volstad N. L. and Boks C. (2012). On the use of biomimicry as a useful tool for the industrial designer. *Sustainable Development*, **20**(3), 189–199. http://dx.doi.org/10.1002/sd.1535

Vymazal J. (2010). Constructed wetlands for wastewater treatment. *Water* **2**, 530–549. MDPI (www.mdpi.com/journal/water). https://doi.org/10.33390/w2030530

Wang R., Korboulewsky N., Prudnt P., Domeizel M., Rolando C. and Bonin G. (2010). Feasibility of using an organic substrate in a wetland system treating sewage sludge: impact of plant species. *Bioresource Technology* **101**(1), 51–57. https://doi.org/10.1016/j.biortech.2009.07.080

Wang H., Wang T., Zhang B., Li F., Toure B., Omosa I. B., Chiramba T., Abdel-Monem C. and Pradhan M. (2014). Water and wastewater treatment in Africa – current practices and challenges. *Clean – Soil, Air, Water*, **42**(8), 1029–1035, doi: 10.1002/clen.201300208

Ward B. and Savić D. A. (2012). A multi-objective optimisation model for sewer rehabilitation considering critical risk of failure. *Water Science and Technology*, **66**(11), 2410–2417.

WEDC (Water, Engineering and Development Centre) (2005). Comparing National Sanitation Policy Content. An Initial Review of Nine Country Profiles. Edited by WEDC, United Kingdom.

WHO/UNICEF (World Health Organization/United Nations Children's Fund Joint Monitoring Programme for Water and Sanitation) (2015a). JMP Green Paper: Global Monitoring of Water, Sanitation and Hygiene Post-2015. World Health Organization, Geneva, Switzerland.

WHO/UNICEF (World Health Organization/United Nations Children's Fund Joint Monitoring Programme for Water and Sanitation) (2015b). Progress on Sanitation and Drinking Water – 2015 Update and MDG Assessment. World Health Organization, Geneva, Switzerland.

WHO/UNICEF (2017). Progress on Drinking Water, Sanitation and Hygiene: 2017 Update and SDG Baselines. World Health Organization (WHO) and the United Nations Children's Fund (UNICEF), Geneva, Licence: CC BY-NC-SA 3.0 IGO.

Wilderer P. A. and Schreff D. (2000). Decentralized and centralized wastewater management: a challenge for technology developers. *Water Science and Technology*, **41**(1), 1–8.

Wirmer B. (2014). A functional approach to guide sustainable innovations in the sanitation chain. MSc thesis, Innovation Sciences Master of Science Eindhoven University of Technology, Utrecht, Netherlands.

Wood S., Sawyer R. and Simpson-Herbert M. (1998). PHAST Step-by-Step Guide: A Participatory Approach for the Control of Diarrhoeal Disease. World Health Organization, Geneva, Switzerland (unpublished documents WHO/EOS/98.3).

WSDHWMP (Washington State Department of Health Wastewater Management Program) (2002). Failing Systems – Rule Development Committee Issue Research Report. https://www.doh.wa.gov/portals/1/Documents/Pubs/337-098.pdf (accessed 4 April 2018).

WSP (Water and Sanitation Program) (2007). Community-Led Total Sanitation in Rural Areas. World Bank, New Delhi, India.

WSP (Water and Sanitation Program) (2011). The Political Economy of Sanitation: How can we increase investment and improve service for the poor? https://openknowledge.worldbank.org/bitstream/handle/10986/17276/599330V20WSP0R10Box358311B01PUBLIC1.pdf?sequence=1&isAllowed=y (accessed 8 January 2018).

WSP (Water and Sanitation Program) (2014). The Missing Link in Sanitation Service Delivery: A Review of Fecal Sludge Management in 12 Cities. Water and Sanitation Program: Research Brief. https://www.wsp.org/sites/wsp.org/files/publications/WSP-Fecal-Sludge-12-City-Review-Research-Brief.pdf (accessed 20 June 2017).

WSP (Water and Sanitation Program) (2015). Improving On-site Sanitation and Connections to Sewers in Southeast Asia Insights from Indonesia and Vietnam. The Global Water Practice and Urban Sanitation, Water and Sanitation Program: Research Brief. http://wsp.org/sites/wsp.org/files/publications/WSP-Improving-On-site-Sanitation-Connections-to-Sewers-Southeast-Asia.pdf (accessed 26 February 2018).

Wu S., Caralho P. N., Muller J. A., Valsa R. M. and Dong R. (2016). Sanitation in constructed wetlands: a review on the removal of human pathogens and fecal indicators. *Science of Total Environment*, **541**, 8–22. https://doi.org/10.106/j.scitotenv.2015.09.047

Yang N. (2004). Disease prevention, social mobilization and spatial politics: the anti-germ warfare incident of 1952 and the 'patriotic health campaign'. *The Chinese Historical Review*, **11**, 155–182.

Yang X. (2012). Japan's Experience in Decentralized Sanitation (Johkasou). Presentation at the first consultation with stakeholders workshop in Best Practices for Septage Management, organized by Asian Development Bank, February, 2012. Aizawl, INDIA. http://www.jsanic.org/news/Awareness_Workshop_NERUDP_India_28Feb2012/Japan's_Experience_in_Decentralized_Sanitation.pdf (accessed 26 February 2018).

Yang X. M., Yahashi T., Kuniyasu K. and Ohmori H. (2001). On-site systems for domestic wastewater treatment (johkasous) in Japan. In: Decentralised Sanitation and Reuse – Concepts, Systems and Implementation, P. Lens, G. Zeeman and G. Lettinga (eds), IWA, London, UK, pp. 256–280.

Yardley S. (2010). Joining the dots: Why better water, sanitation and hygiene are necessary for progress on maternal, newborn and child health. http://tilz.tearfund.org/~/media/files/tilz/research/j20066%20wash%20report%20web.pdf (accessed 20 May 2018).

Zaharia C. (2017). Decentralized wastewater treatment systems: efficiency and its estimated impact against onsite natural water pollution status. A Romanian case study. *Process Safety and Environmental Protection*, **108**, 74–88. https://doi.org/10.1016/j.psep.2017.02.004

Zakaria F., Garcia H. A., Hooijmans C. M. and Brdjanovic D. (2015). Decision support system for the provision of emergency sanitation. *Science of the Total Environment*, **512–513**, 645–658. http://dx.doi.org/10.1016/j.scitotenv.2015.01.051

Zeschky M., Widenmayer B. and Gassmann O. (2011). Frugal innovation in emerging markets. *Research-Technology Management*, **54**(4), 38–45.

Zeschky M. B., Winterhalter S. and Gassmann O. (2014a). From cost to frugal and reverse innovation: mapping the field and implications for global competitiveness. *Research-Technology Management*, **57**(4), 20–27.

Zeschky M. B., Winterhalter S. and Gassmann O. (2014b). Resource-Constrained Innovation: Classification and Implications for Multinational Firms. This Paper was Presented at the XXV ISPIM Conference – Innovation for Sustainable Economy & Society, Dublin, Ireland, 8–11 June 2014, The publication is available to ISPIM members at. www.ispim.org

Zhou C., Liu J., Wang R., Yang W. and Jin J. (2010). Ecological-economic assessment of ecological sanitation development in the cities of Chinese Loess Plateau. *Ecological Complexity*, **7**, 162–169.

Zurbrügg C. and Tilley E. (2007). Evaluation of existing low cost conventional as well as innovative sanitation system and technologies – NETSSAF deliverable D22&23. Workpackage 3 – Assessment of Sanitation Systems and Technologies Swiss Federal Institute of Aquatic Science and Technology (EAWAG). Dübendorf, Switzerland. https://www.sswm.info/sites/default/files/reference_attachments/NETSSAF_Evaluation%20of%20existing%20low-cost%20conventional%20as%20well%20as%20innovative%20sanitation%20system%20and%20technologies.pdf (accessed 15 March 2018).

Zurbrugg C. and Tilley E. (2009). A system perspective in sanitation – Human waste from cradle to grave and reincarnation. *Desalination*, **248**, 410–417.

Chapter 6

Resource system

'humans will need to collect, sort, recover and recycle to complete the material cycle as well as turn waste into a resource'

Veolia Environmental Services

6.1 INTRODUCTION

The ReGenSan resource system (ReS) reinforces the resource-based paradigm (rather than waste-based) on the premise that sanitation matter should be managed as a resource within a closed loop or cyclical flow from collection to reuse (Park & Chertow, 2014). Sanitation provision in ReS is then regarded as the powerhouse of the sanitation economy (within a circular economy) with a robust marketplace for products and services, renewable resource flows and data/information vault that could transform cities, communities and businesses with vast opportunities for economic growth, resilience, collective advantage, urban regeneration and universal access to safely managed sanitation facilities (TBC, 2017). In the circular economy, it is expected that natural resource utilization will follow a process of material flow, regeneration and bio-geo-chemical circulation that reroutes them into appropriate value chains (VCs) so as to enhance efficient resource usage (Winker *et al.*, 2009), whereby substances are neither sources nor sinks and links are considered both receivers and donors (Brunner & Rechberger, 2017; Yan *et al.*, 1993a, b, 1998). This supports the sanitation economy as the regeneration and circulation of sanitation materials are fundamental to ensuring adequate integration of sanitation systems within SES (Chapter 4). It is universally expected that any product in the world will inevitably turn into waste eventually (Yan *et al.*, 1993a, b, 1998). This circulation of products and waste is essential for the ReGenSen-ReS, which advocates that sanitation management should follow the biochemical circulation of materials route (see Section 4.2.3) by ensuring that infrastructural and technological solutions extend beyond 'access and safely managed' to include 'resource recovery and reuse' (RRR) from onset of planning (Esrey, 2001; Langergrabera & Muelleggera, 2005; Masi *et al.*, 2017; Simha *et al.*, 2017; Werner *et al.*, 2009).

© IWA Publishing 2019. Regenerative Sanitation: A New Paradigm for Sanitation 4.0
Editor(s): Thammarat Koottatep, Peter Emmanuel Cookey and Chongrak Polprasert
doi: 10.2166/9781780409689_0209

It follows the concept that 'waste equals food' and the continuous flow of resources (Benetto *et al.*, 2009; Langergrabera & Muelleggera, 2005; Otterpohl, 2002; Werner *et al.*, 2003) because in nature, nothing is useless (Hodges, 2006; McDonough & Michael, 2001a, b). ReS, therefore, holds to the process of 'sanitize and recycle', which is based on pollution prevention in order to meet the needs of local communities (Haq & Cambridge, 2012; Kujawa-Roeleveld & Zeeman, 2006; Lens *et al.*, 2001; Maurer *et al.*, 2012; Simha *et al.*, 2016; Wielemaker *et al.*, 2016; Zeeman, 2012); that is, sanitation management should be a complete circle e.g. materials recycling and circular economy.

Human excreta have traditionally been used for crop fertilization in many countries. In Japan, urine and faeces recycling was introduced in the 12th century and in China human and animal excreta have been composted for thousands of years (Esrey *et al.*, 1998; Höglund, 2001). Swedish cities organized collection and transportation of latrine products to farmers in the 18th century and these products were also mixed with lime to produce limed ammonium nitrate and ammonium sulphate (Höglund, 2001; Tingsten, 1911) as well as to treat wounds, or drink as treatment therapy (Frode-Kristensen, 1966; Höglund, 2001), and in the tanning of hides and production of gunpowder (Höglund, 2001; Stenström, 1996). In the 19th century in Denmark, urine was stored and used as detergent for washing clothes (Hansen, 1928; in Drangert, 1998; Höglund, 2001). In other words, sanitation stock has been transformed into a resource since ancient times. Accordingly, sustainability demands that we acknowledge by-products of human digestion as renewable resources from which water (Daigger, 2008), materials e.g. fertilizers (Larsen, *et al.* 2009), bioplastics (Kleerebezem & van Loosdrecht, 2007) and energy (Daigger, 2008) can be recovered instead of merely removed (Guest *et al.*, 2009).

To this effect, ReS incorporates design for reuse, from the initial phase of planning, that is tailored to the system of the place with considerations for local institutions, market demand and supply chains that will be necessary for them to thrive (Bahri, 1999; Jenkins & Sugden, 2006; Lazarova *et al.*, 2001). This facilitates a shift from the 'design-for-disposal' to 'design-for-reuse' paradigm (Table 6.1) because sanitation resources are increasingly being considered valuable and right for certain specific purposes (Asano & Levine, 1996; Curtis, 2010; Lu *et al.*, 2015; Mihelcic *et al.*, 2011, 2017; McCarty *et al.*, 2011; Mo & Zhang, 2013; Nouri *et al.*, 2007; Ren & Umble, 2016).

Table 6.1 Resource-Based Paradigm versus Waste-Based Paradigm.

	Waste-Based Paradigm	**Resource-Based Paradigm**
Underlying thoughts	Human waste can cause harm to public health and safety until shown otherwise	Human waste is a potential resource until shown otherwise
Main strategy	Safe disposal and containment	Environmentally sound processing and reuse
Language and taxonomy	Waste	Secondary materials
Operationalized metric for waste	None	Reuse potential

(Reprinted from Park J. Y. and Chertow M. R. (2014). Establishing and testing the 'reuse potential' indicator for managing wastes as resources. *Journal of Environmental Management*, 137, 45–53, with permission from Elsevier).

This perspective effectively shifts the goal of sanitation from being solely the safe disposal of waste to maximizing the extent to which embodied resources are safely captured and allocated to deliver public and environmental health benefits (Balkema *et al.*, 2002). The resource-based paradigm could potentially contribute productively to the local economy and livelihood (Murray & Buckley, 2009) by closing the nutrient loop and connecting it with agriculture, recovering or processing organic matter into energy, maintaining soil quality and recycling/recharging water entering human settlements (Brands, 2014; Langergrabera & Muelleggera, 2005; Larsen *et al.*, 2009; Wang *et al.*, 2006; Werner *et al.*, 2009). Therefore, to move towards the goal of efficiency in sanitation matter utilization, a change that blends the definitions of 'wastes' and 'resources' is required (Ganesapillai *et al.*, 2015). Waste could then be defined as the 'opposite of value' (Gille, 2007) from 'waste treatment system' to 'resource recovery facilities' (Mayer *et al.*, 2016) with high resource-reuse potential often expressed as a real value between 0 and 1, where 0 represents 'waste' and 1 represents potential 'resource'; the movement along this line represents how biomass from sanitation systems transform into reusable sanitation-derived products (SDPs) (Park & Chertow, 2014 – Figure 6.1) and represents a change in emphasis from 'treatment and disposal' to 'cradle-to-cradle' (Braungart *et al.*, 2007; Hodges, 2006; Masi *et al.*, 2017; McDonough & Michael, 2001a, b).

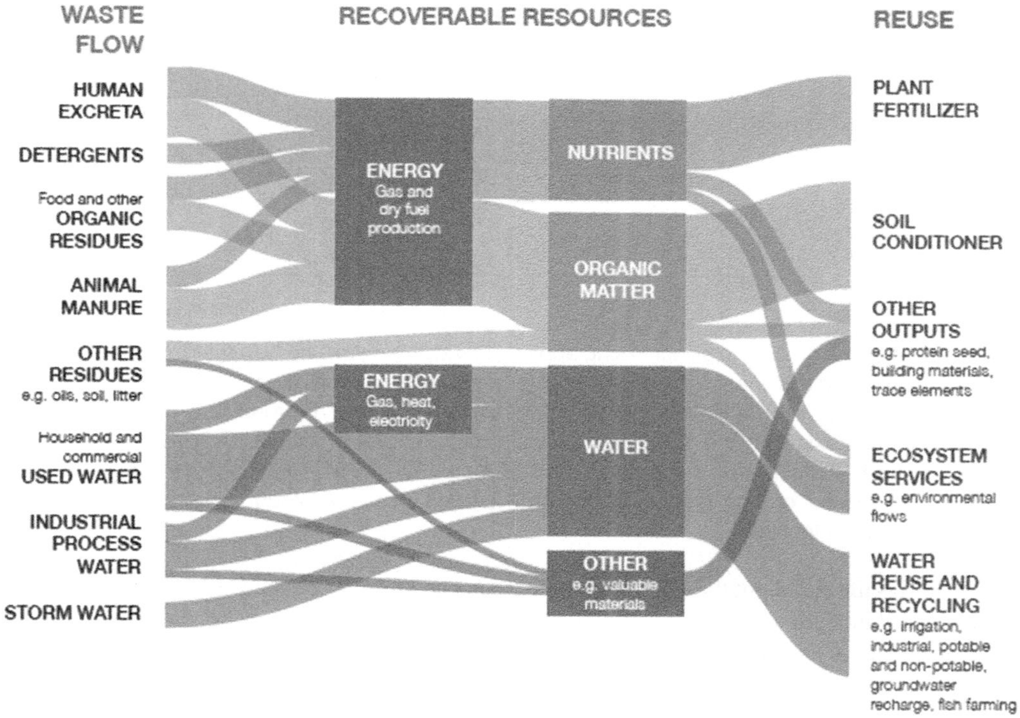

Figure 6.1 Overview of waste stream and resources with potentials for recovery and reuse of nutrients, organic matter, water and energy (*Source*: Andersson *et al.*, 2016a).

For instance, huge potential exists for nutrient and energy recovery as well as safe reuse of sanitation materials in agriculture (Mihelcic *et al.*, 2011; Munamati *et al.*, 2016; Verbyla *et al.*, 2013). Sanitation will particularly have to cooperate with agriculture in order to avoid emissions and allow for reuse of water and nutrients (Otterpohl, 2000). It is estimated that by the year 2050, sub-Saharan Africa (SSA) will have the potential to recover over a million tonnes of phosphorus per year from faeces and urine (Mihelcic *et al.*, 2011), which provides an incentive to immediately implement low-cost innovative sanitation technologies to unserved populations in the region so as to enable recovery and reuse of nutrients and water (Munamati *et al.*, 2016). Studies have shown that crop fertilization with sanitation-derived nutrients (SDNs) increases yield due to limiting growth factors of nutrients such as nitrogen, phosphorus and potassium. Consequently, it is important that SDNs are used efficiently and in differing ways, which will depend on the amount of available nutrients in relation to available space and fertilizer requirement per area unit (Jönsson *et al.*, 2004). When the area is not a limiting factor, the full fertilizing effect can easily be gained from SDNs such as urine, even if the urine is applied at different dosages in different places, as long as the dosage in all places is below rate A (Figure 6.2).

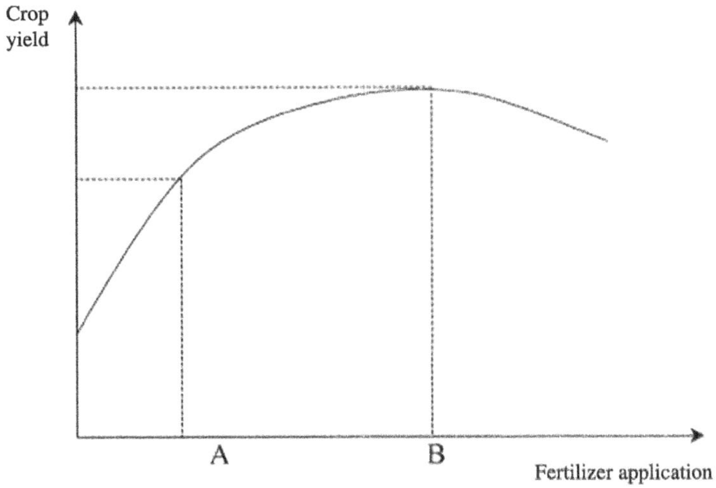

Figure 6.2 The effect on crop yield of increasing application rates of available N, e.g. in the form of urine. Up to rate A, the increase in yield is linear to the addition of urine. Between rate A and rate B, the yield still increases in response to the increased fertilizer application, but at a slower rate. Beyond rate B, additional fertilizer application becomes toxic and the yield decreases if the application rate is increased (Reprinted from Jönsson H., Richert Stintzing A., Vinnerås B. and Salomon E. (2004). EcoSanRes Programme Report 2004–2, 137, Guidelines on the Use of Urine and Faeces in Crop Production, with permission from Stockholm Environment Institute).

Effective ReS is the paradigm that provides products and services across the entire sanitation value chain (SVC), interlinks with agriculture and other sectors to generate benefits to all economies, actors and entities in the VC and creates connectivity of resources among physical and biological systems, ensuring circularity in the process from start to finish (Otoo & Drechsel, 2017). It should also open up opportunities for VC creation within the sanitation system in general and the service chain (SC) in particular that would have been lost to cost savings and/or cost recovery in the sanitation sector (Otoo & Drechsel, 2017). SC refers to the sequence of actions and infrastructure that is applied to long-term provision of sanitation services for all and at scale and also describes the ways in which sanitation services are provided (Zurbrügg & Tilley, 2007 – see Section 5.2.1). The VC includes all producing and marketing enterprises operating in the entire sanitation system creating and delivering services in the sanitation market. They include product design and development and product transformation enterprises that use SSC raw materials for safe recovery of water, nutrients, organic matter and energy, which are finally delivered to customers. These enterprises interact constantly with the activities of buying and selling products and services, exchanging information and cooperating to pursue shared interests (Springer-Heinzer, 2017). This offers immense and scalable opportunities for entrepreneurs within the VC through transforming sanitation matter into low-carbon assets for use in agriculture and other sectors. And with appropriate implementation, it could bridge technological gaps in conventional sanitation management that optimize the recovery of beneficial materials, which could be used to support related economic opportunities (Tilley *et al.*, 2014a). Along the lines of the ReGenSan framework, ReS approaches should not be technology-centred or toilet-focused, but should target resources and their management in a manner that emphasizes continuous protection of the environment and human health (Peccia & Westerhoff, 2015; Scheierling *et al.*, 2010).

However, to avoid past failures of countless similar projects, determining the cost-effectiveness of different measures in specific contexts and under local conditions requires an economic analysis (WHO, 2006a), while improvement will require significant investments in market research, bankable business models for cost recovery, stakeholder buy-in and innovative partnerships (Otoo & Drechsel, 2017). To address this challenge, several landmark activities by national governments and international non-governmental organizations have been established to improve resource recovery in the sanitation sector. For example, the strategic research action plan of the U.S. Environmental Protection Agency (USEPA) Office of Research and Development's guiding objectives towards recovery of energy, nutrients, water and other valuable substances embedded in wastewater (Mihelcic *et al.*, 2017; USEPA, 2015); U.S. Department of Energy (DOE) Institute; research opportunities in the area of waste conversion technologies (Mihelcic *et al.*, 2016; USDOE, 2016 – Box 6.1); the Water Environment and Reuse Foundation (WE&RF)'s projects on the identification of recoverable products from wastewater streams as a key strategy to bridging the knowledge gap in nutrient recovery (Mihelcic *et al.*, 2016; WE&RF, 2011); and the International Water Association (IWA)'s global initiatives for recovery of water, energy and other valuable materials found in wastewater (IWA, 2016; Mihelcic *et al.*, 2016; USDOE, 2014). Also, the United Nations Sustainable Development Goals have provided specific targets relating to resource recovery and reuse (RRR), from sanitation materials (SDG 6), harvesting renewable energy resources from materials derived from sanitation systems (SDG 7) and supporting sustainable production and consumption by recovery and reuse of materials from sanitation to support agriculture, aquaculture and other industrial processes that deplete non-renewable natural resources (SDG 12), using SDPs to help restore degraded soils (SDG 15), utilizing resources from sanitation to improve food security through the practice of resilient agriculture that help maintain ecosystems (SDG 2) and designing sanitation infrastructure that supports urban regeneration so as to help build resilient cities (SDG 11) (Otoo & Drechsel, 2017; UNDP, 2018). Therefore, the key questions that need to be answered are: (i) what resources are available in the sanitation streams? (ii) what demand might there be for them? and (iii) how could they be recovered? (Andersson *et al.*, 2016a, b).

BOX 6.1 GLOBAL EXAMPLES OF LARGE-SCALE REUSE OF TREATED WASTEWATER FOR IRRIGATION, AQUACULTURE AND INDUSTRY (MURRAY & RAY, 2010)

Irrigation in Tunisia

Tunisia uses about 30 to 43% of the country's treated wastewater for agricultural and landscape irrigation, an option that is always considered at the planning stage of any treatment plant (Bahri, 2009). Reuse is motivated by interests in protecting coastal waters and mitigating water scarcity. By 2020, plans in Tunisia include irrigating 20,000 to 30,000 hectares with reclaimed wastewater (Bahri, 2009).

Aquaculture in Kolkata, India

Kolkata's decades-old wastewater-fed aquaculture system is the largest in the world. The system was built in the 1930s and consists of a series of waste stabilization ponds that feed into fish ponds, which span nearly 4000 hectares (Jana, 1998; WHO, 2006b). The combination of treatment pond with aquaculture is credited with simultaneously providing a low-cost means of wastewater treatment, 10 to 20% of the fish consumed in Kolkata and employment for more than 25,000 local residents (WHO, 2006b).

Industrial Reuse in Durban, South Africa

In Durban, industries reuse municipal wastewater from the Southern Wastewater Treatment Works. The reclamation facility comprises tertiary treatment and has a capacity of 47,00 cubic metres per day, equivalent to 7% of the city's potable/industrial water demand. The effluent is used primarily by Mondi Paper as well as oil refineries (Gisclon *et al.*, 2002; USEPA, 2004).

6.2 SANITATION VALUE CHAIN (SVC)

SVC is a market-based approach where value is added at different stages of the SC activities from design, production and marketing to deliver products and services, especially in RRR within the SSC (Sievers & Kelly, 2016). In general, VC depicts how customers' value accumulates along a chain of activities that lead to an end product or service (Figure 6.3). The concept of VC has several dimensions, such as:

(i) The input-output structure, which describes a chain as a set of products and services linked together in a sequence of value-adding economic activities;
(ii) Geographic spread. Some chains are truly global (with activities taking place in many countries and on different continents) while others are more limited (involving only a few locations in different parts of the world); and

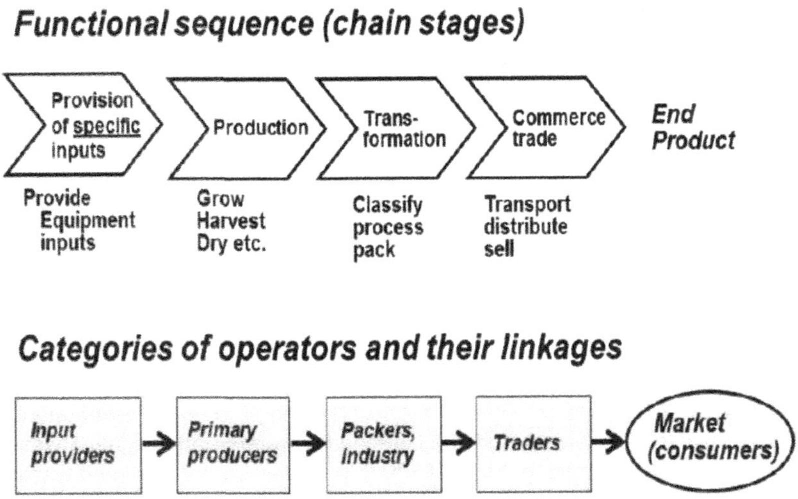

Figure 6.3 Generic elements of a basic value chain map (*Source*: Springer-Heinzer, 2017).

(iii) The control that different actors can exert upon what makes up the chain. The actors in a chain are directly controlling their own activities or indirectly controlled by other actors (McCormick & Schmitz, 2001). Porter (1985) describes the VC as processes or activities by enterprises 'to design, produce, market, deliver and support its product' (IMA, 1996; Springer-Heinzer, 2017).

In other words, VC conceptualizes activities needed to provide products or services to customers and depicts the way a product gains value (and costs) as it moves along the path of design, production, marketing, delivery and service to customers (Ensign, 2001; Springer-Heinzer, 2017). But in sanitation the VC is often focused within the SSC actors and local businesses involved in FSM activities (Blackett *et al.*, 2014 – Figure 6.4); however, the SSC (see Section 5.2.1) does not really present the whole picture of the SVC; this makes it more restrictive.

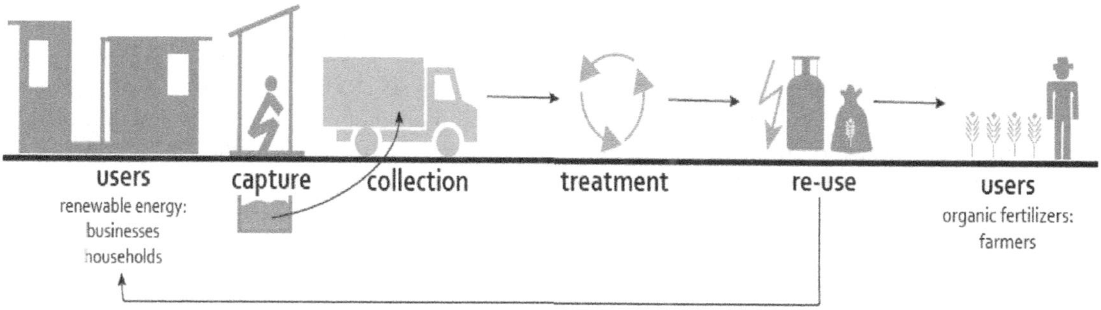

Figure 6.4 Sanitation value chain (SVC) used to illustrate sanitation service chain (SSC) (*Source*: van Welie & Romijn, 2018).

The SVC could be used to identify the technical steps (such as production, transformation, etc) within the sanitation system and the value-added activities within the SSC through the collaboration and interaction of enterprises and their partners in the delivery of products and services to customers and users. It could also be useful in analyzing how sanitation entrepreneurs operate in faecal sludge management (FSM) markets in specific cities and how such markets could be improved (Blackett *et al.*, 2014; SuSanA, 2017). However, the current structure of the SVC only addresses the value-added activities at the SSC stage of the sanitation system; while some argue that value is added at every phase of the SSC (Blackett *et al.*, 2014), others suggest that the value that could generate additional financial flow in the sanitation system is at the back end of the SSC (Murray & Ray, 2010). This implies that tapping into the customer segment interested in resources recovery from sanitation-derived products (SDPs) (Diener *et al.*, 2014) could produce additional revenue stream through sales of such products and this could alter the financial flow of the SSC, which could also result in the off-setting of disposal costs (Gebauer & Reynoso, 2013; London & Hart, 2010).

However, as impressive and positive as this is, it focuses the VC activities within the SSC, which is probably the reason for misidentifying the SSC and the SVC. This limits the efficiency and practicality of the SVC and breaks the circularity from initial production to reuse (i.e. weakens the force behind the sanitation economy). The complete sanitation system begins from the point of product and service development and design (e.g. toilets, faucets, tanks, pipes, treatment plants etc) up to the point of reuse and continues through feedback to the design stage. Therefore, the SVC should not be restricted to the SSC. There is, then, a need to differentiate between the SSC and SVC so that the SVC captures all the value-added activities and enterprises within the entire sanitation system. Van Dijk (2012) recognized this necessity and proposed three stages of SVC before products' raw materials could be 'harvested and manufactured:

(1) Designing and building toilets as well as going for individual or collective solutions;
(2) Operation and maintenance outsourced to small enterprises and emptying and transportation done by small private operators; and
(3) Recycling services provided by separate actors, preferably local enterprises.

ReGenSan presumes that solving the sanitation problems will require innovative solutions in infrastructure, technology, production, service delivery and cost recovery and so the SVC should be analyzed throughout the value-adding activities of the entire sanitation system. A lack of profitable and financially viable options for managing the complete sanitation system and its associated enterprises and partners constitute a major barrier to development. Finding innovative ways to create viable businesses in the sanitation sector could greatly enhance the ReS and improve access and delivery (Diener *et al.*, 2014). This could be achieved by distilling the motivations and incentives at work on each link across the SVC (Kennedy-Walker *et al.*, 2014). Several series of studies on the sanitation value chain analysis (SVCA) have been conducted in an attempt to understand the SVC, its components and various interventions that improve access to SDPs based on the needs of actors or their place within the chain (Peal *et al.*, 2010; Pedi, 2012; Tayler *et al.*, 2000; Willetts *et al.*, 2017). SVCA is valuable in that it helps in comprehending the workings of the whole chain, the functions of each link along the chain and the influence of external parties (supporters and enablers). It also aids mapping of the SVC and examines costs, output and the physical flow of commodities along the chain (Willetts *et al.*, 2017). In addition, it could unravel issues of power, inclusion/exclusion, inequality and vulnerability (Bolwig *et al.*, 2010) as well as desirability, feasibility and viability (Pedi, 2012).

6.3 INTEGRATED FUNCTIONAL SANITATION VALUE CHAIN (IFSVC)

Based on the concept of tracking the SVC from the initial stage of product/service design/development and following the ReGenSan principles, an integrated version of the SVC is proposed – integrated functional sanitation value chain (IFSVC). The IFSVC addresses the operational functions within sanitation systems in combination with sanitation enterprise operators and external actors such as supporters that regulate the sector (professional associations) and support SVC growth, and enablers that govern and regulate (governments) SVC activities (Table 6.2) (Springer-Heinzer, 2017).

Table 6.2 IFSVC Combination Structure.

Operational Functions	Enterprise Operators	Supporters	Enablers
Primary production	Designers	Public agencies	Policies
Transformation	Producers	Regulators	Legislation
Marketing	Integrators	Financing institutions	Regulations
Final sale	Processors	NGOs/CBOs	Standards
	Traders		Cultures
	Distributors		Environments
	Suppliers		
	Fabricators		

IFSVC is made up of seven stages and links in a closed loop and represents the 'enterprisation' of the entire sanitation system, which includes the SSC (Figure 6.4, Table 6.1). This implies that the SVC covers more than the SSC, but explores all enterprises, ventures and activities within the whole system of a sani-shed or sani-district (sanitation district). An attempt is made to expand the considerations for sanitation activities beyond just the SSC because, until sanitation management is addressed comprehensively and solutions are integrated systematically, the wins will cancel themselves across range and the sanitation economy will be weakened. The sanitation economy is dependent on the value-added activities across all stages of ReS and so when some aspects are disregarded or left out, the gains from those enterprises are not counted and the SVC is incomplete. To this effect, the IFSVC tries to identify the functional linkages within a systemic loop that captures the various functions at different stages related to sanitation management from conceptualization to the final market as well as the enterprises and ventures within each stage. This concept implies that value-added activities within each stage and the way the stages link to each other through enterprise interactions depicts the integrated value chain system. This integrated VC system assumes that the SVC (and sanitation economy) cannot be effective unless all stages are working connectively. Figure 6.5 illustrates the seven stages of the proposed IFSVC:

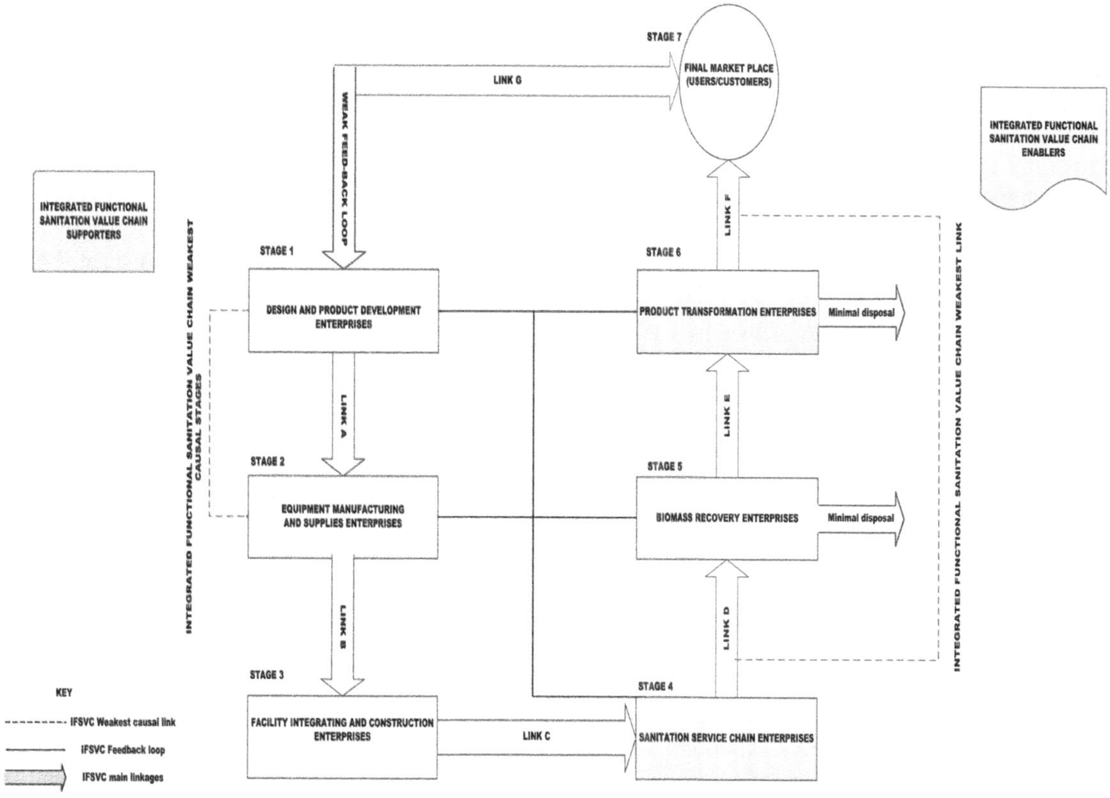

Figure 6.5 Illustration of integrated functional sanitation value chain (IFSVC).

(I) Design and development i.e. the initial stage of planning and conceptualization; this stage is directly linked to the next (link A);
(II) Equipment manufacturing and supplies, i.e. the stage of primary production of user products, supply and distribution to the market and service delivery such as toilet combo, faucets, pipes etc. which then links (link B) to stage III directly;
(III) Facility integration and construction, i.e. the stage of installation, construction, connections, operations and fabrication of sanitation facilities at point of use (such as septic tanks, public toilets, treatment facilities, pipe-laying etc), and directly connect to the next stage (link C);
(IV) Sanitation service chain, i.e. the stage of activities after the user interface from collection to disposal and/or treatment (such as emptying, transportation, disposal etc); there is then a direct link (link D) to recovery stage;
(V) Biomass recovery, i.e. the recovery of organic nutrients, biogas, manure through various processes; this then moved on to transformation (link E);
(VI) IV Product transformation, i.e. the utilization of changed products for other purposes such as agriculture, aquaculture etc. and is where connection to the final market for resource reuse is effected (link F);
(VII) VII End users/customers' markets, i.e. the final market for the sanitation resources obtained from the recovery and transformation processes (e.g. farmers). Subsequently, the market provides feedback to the designers and developers at the conceptualization stage for appropriate, acceptable and marketable products/services (link G).

All these stages are anchored (all through the chain) by supporting and enabling structures and mechanisms. In essence, design and conceptualization of sanitation products and services take place at stage 1, then moves on to stage 2, the equipment manufacturers that produce end-to-end equipment and technologies as well as distributors. Stage 3 refers to enterprises involved in integrating, installing, fabricating and operating sanitation infrastructure using the products and services from stage 2, leading to the enterprises and partners operating the processes of the SSC in stage 4 (Figure 6.5). Subsequently, it is expected that resources would be recovered after treatment at stage 5 and transformed for utilization at stage 6, where it should be transmitted to the final marketplace at stage 7. Feedback from stage 7 back to the stages from stage 1 will relate users'/customers' satisfaction, expectations, requirements, preferences and acceptability, which will provide fodder for innovation, continuous improvement and market comprehension.

The reason for starting from the stage of conceptualization (stage 1) is based on the observation that most of the time, sanitation products available in the marketplace (stage 2) were not originally designed to enhance resource capturing and facilitate reuse, while inadequate supporting/enabling governance, management and financial environment in most developing countries further hinder the growth of the sanitation sector. This is probably responsible for the low patronage experienced from stage 5 and 6 as there are very few enterprises invested in this stage of the value chain; and even though stage 4 (SSC) has seen a lot of activities over the years, there is still minimal standard business participation with the exception of sewered systems in developed countries. Therefore, the weakest links in this IFSVC are links D, E, F and G; this is directly related to the fact that there are almost no direct or indirect interactions between stages 1 and 2, with stages that could help enterprises to design and produce technologies that fit from the beginning (which could create a more integrated value chain and effective and robust sanitation economy). Although sewered sanitation in developed countries (see Section 5.2.5) is suitably established, a review of enterprises on the IFSVC indicated that the weakest links are links D, E and F; however, this is strengthened through infrastructural standardization (see Section 5.3). In addition, consistent investment, innovation, research and development often lead to technological and regulatory improvements, which also aid the market. Furthermore, the lifespan of sewered sanitation components is well defined and so the products are manufactured to acceptable durability specifications that give players and partners within the existing SVC enough time to adjust and adapt to market forces and associated changes (PWC, 2012). For instance, in the USA, a shower head is replaced every 12.5 years, a lavatory faucet is replaced every 15 years and a toilet is replaced every 30 years (GMP Research Inc., 2015); thus, innovation has approximately 12 years to mature, which gives enough room for investments, user preparedness and acceptance in the marketplace (Table 6.3).

Table 6.3 Enterprises and Businesses Operating in the Integrated Functional Sanitation Value Chain (IFSVC).

Stages	Enterprises/Businesses
Design and product development enterprises	Research and development, education and training institutes, wastewater treatment plant designers, home, commercial and toilets/accessories designers, architectural and interior designers, sanitary engineering consultants, sanitary wares and accessories designers, recovery, recycling and reuse system designers, town and urban planners, public and environmental health consultants etc.
Equipment manufacturing and supplies enterprises	Sanitary wares and accessories, treatment plant/accessories, toilets/accessories, recovery, recycling and reuse accessories, cement manufacturers, importers, suppliers, retailers, distributors, plumbing materials/accessories, cleaning and hygiene products, metal, cements and wood work, suppliers/distributors etc.
Facility integration and construction enterprises	Fabrication and installation, logistics and transportation, architecture and real estate, sanitary engineering, town and urban planners, wastewater treatment plant installation, faecal and sewage sludge treatment plant installation, recovery, recycling and reuse, home and commercial building toilet installation, plumbing and cement works, public and mobile toilet installation, microenterprises services, sales, installation and construction, local mason, metal, woodwork and concrete building contractors, testing and quality control laboratories etc.
Sanitation service chain enterprises	Logistics and transportation, wastewater treatment plant operations and maintenance, faecal and sewage sludge collection, emptying and treatment, public and private utilities, environment and public health consultants, operations and maintenance, sanitary engineering, town and urban planners, cleaning and hygiene, testing and quality control laboratories, public, home, commercial building, mobile, toilet maintenance, plumbers, local artisan and masons, education, training and capacity development etc.
Biomass recovery enterprises	Resource recycling and recovery, composting and organic matter recovery, wastewater resource recovery, faecal and sewage sludge treatment recovery, public and private utilities, certification and verification, health, safety, quality control and assurance, training/capacity building etc.

(*Continued*)

Table 6.3 Enterprises and Businesses Operating in the Integrated Functional Sanitation Value Chain (IFSVC) (*Continued*).

Stages	Enterprises/Businesses
Product transformation enterprises	Organic fertilizers and manure, recycled water suppliers, biogas and energy producers and suppliers, aquaculture, horticulture, animal feeds producers, parks and gardens management, farmers' cooperative organizations, aquaculture cooperative organizations, certification and verification, health, safety, quality control and assurance etc.
Final marketplace (users/customers)	Microenterprises, institutions e.g. schools, prisons, horticulturists, farmers, aquaculturists, households, biogas users, producers of organic fertilizers etc.
IFSVC enablers	Policy, legislators, regulators, guidelines and standards developers, land-use planning, sustainable financing, investors and banks, policymakers etc.
IFSVC supporters	Ministries/departments/agencies responsible for environment, health, water resources, economic planning and cooperatives, natural resources, agriculture, fisheries and aquaculture, trade, commerce, industry, gender and development, education, information and communication, financial and insurance institutions, public and private investors, marketing and advertising, multilateral organizations, research community, international, regional, national and local non-governmental organizations, community based organizations etc.

Non-sewered sanitation in developing countries is not so well established. The challenges of standardization (see Section 5.3), short lifespan of products and high rate of dysfunctionality impact negatively on the effectiveness and efficiency of existing SVCs, especially since these products and their components were not designed for RRR. Unsurprisingly, the enterprisation of non-sewered sanitation in the IFSVC is still in its infancy at stages 5 and 6 and the weakest links are D, E, F and G. This could be related to the fact that most products used to configure non-sewered sanitation infrastructure were not originally designed for its processes, but are actually borrowed from sewered structures and so have different requirements and specifications. For example, water closet toilets used with septic tanks that require minimal water and/or other improvised products and services provided by local artisans, particularly in developing countries. These products are hardly standardized and often unattractive to users/customers (there are no certifications for these artisans either) and so often create an impression that there are no markets for sanitation products and services in developing countries (as demand is based on known necessity). Recent studies indicate that household demands go beyond value propositions for actual relevant benefits or package of benefits (OECD, 2011; Sy *et al.*, 2014). Accordingly, the proposed IFSVC entrepreneurs, investors and marketers require more qualitative and quantitative insights in order to design products and services that people truly want and adequate investment in deep understanding of customers' needs, desires and expectations could fill up this loop (BoP Innovation Centre, Aqua for All and Sanitation Window, 2014) (see Box 6.2). In addition, researchers and interventionists need to interact directly with the enterprises at all stages so as to share knowledge/information on users' expectations, innovations and discoveries.

BOX 6.2 SANERGY FRESH LIFE TOILETS, NAIROBI, KENYA (SANERGY, 2018)

The development of a classic case of an integrated SVC is aptly illustrated by Sanergy, a social enterprise based in Nairobi, Kenya, with a mission to make hygienic sanitation affordable, and accessible for everyone. Sanergy has a local brand, Fresh Life, under which the FLT (Fresh Life Toilets) are franchised to entrepreneurs in informal settlements. The Fresh Life Operators (FLT owners) maintain the FLT, keeping it clea,n and make a profit by charging per use of the FLT. The FLT is a prefabricated concrete structure that uses a urine-diverting squat plate. The urine-diverting squat plate is a dry toilet design that eliminates the need for water after use. Members of the Sanergy staff collect the waste from the FLTs every day and bring it to a central processing site where it is converted into fertilizer (http://www.chefcooke.com/sanergy/ – Figure 6.6).

Figure 6.6 Sanergy integrated sanitation value chain (http://www.chefcooke.com/sanergy/).

Strengthening the ReS will require direct interaction between stages A and D, E and F (see Figure 6.5) so that product designers and developers can incorporate key requirements of the SSC, resource capturing and transformation into the initial conceptualization of products and services. Other upscaling strategies include, but are not limited to (BoP Innovation Centre, Aqua for All and Sanitation Window, 2014):

(i) Establish and/or scale up local businesses and industries to manufacture, distribute and install products that meet customer demand as well as offering choice, flexibility, upgrading potential etc.;
(ii) Stimulate innovation of sanitation business models and solutions to address all areas of needs covering households, institutions, public and communal;
(iii) Bundle microenterprises and/or attract larger companies to be able to invest in new value propositions, reduce costs and develop demand through marketing;
(iv) Facilitate the participation of industry actors, especially in non-sewered sanitation, and provide the needed support to overcome its barriers;
(v) Design for hygiene and high quality standards by ensuring that products and services offered by companies adhere to demands and desires;
(vi) Encourage the development of new partnerships in both developed and developing markets by attracting larger private-sector actors to collaborate with pioneers in sewered and non-sewered sanitation to stimulate hybrid financing mechanisms and/or results-based financing mechanisms;
(vii) Encourage investment in standardization in order to create conditions for further scaling that could reduce costs;
(viii) Explore/focus on potential revenue streams for RRR;
(ix) Facilitate the creation of new industries and markets for SDPs, or enter current industries that may have created barriers for new-entrants;
(x) Develop innovative strategies to stimulate the demand for SDPs and provide incentives for further investments in the sanitation sector; and
(xi) Facilitate the continued pioneering of new products and services through a range of start-up funding facilities for intermediaries to facilitate inclusive innovations to reach the market.

Currently regulations, financial incentives and innovation instruments are lacking, and these result in huge barriers to new entrants, especially for non-sewered sanitation; meanwhile poor performance of the SDG targets for sanitation in most developing countries is not encouraging. Therefore, industry facilitation that lowers barriers for companies in energy, organic agriculture inputs, fish feeds, biogas etc. (without distortion of market systems) could be instrumental to scaling sanitation up to enhance the delivery of improved sanitation services (BoP Innovation Centre, Aqua for All and Sanitation Window, 2014).

6.4 REGENSAN RESOURCE SYSTEM

So what is the ReGenSan ReS? It is nested in the SES, which provides the catchment area for harnessing sanitation resources (see Chapter 4). It is particularly based on ReGenSan principles (see Section 2.4) interlinked with agriculture and other sectors to generate benefits to all economies, actors and entities in the IFSVC while also creating resource connectivity among physical and biological systems (Otoo & Drechsel, 2017). This subsystem of the ReGenSan framework is made up of three dimensions and 15 components. The three major dimensions are: (i) design for recovery and reuse (DeRaR), which provides technological and infrastructural support for the operationalization of ReS; (ii) sanitation service chain raw materials (SSC-RaMs), which provides the input materials for ReS; and (iii) sanitation-derived products (SDPs), which refers to recoverable and reusable products from sanitation systems (Box 6.3, Figure 6.7). ReS highlights re-materialization i.e. the innovative recovery and reuse of sanitation matter to create entirely new value-added products (Simha & Ganesapillai, 2017). To this effect, the circular economy could be accomplished if agriculture (food security) is introduced into the sanitation-water-health equation (Rose *et al.*, 2015).

Globally, RRR of human waste have been studied and treated wastewater used for various crops in agriculture {from vegetables cultivated for human consumption (Cifuentes, 1998) to willow coppice used as biofuels (Carlander *et al.*, 2000) and for irrigation as well as domestic potable and non-potable purposes (Asano *et al.*, 1992; Höglund *et al.*, 2002; Okun, 2000). The underlying philosophy behind ReS is that economic incentives can result in more efficient and productive services as values are added at each stage and they cut across key stages and links of the IFSVC, to stimulate new integrated solutions that generate profits or at least recover operational costs (O'Keefe *et al.*, 2015).

BOX 6.3 RES DIMENSIONS AND COMPONENTS

(I) *Design for Recovery and Reuse (DeRaR)* advocates for designing site-specific and recovery/reuse-oriented sanitation infrastructural and technological solutions from the onset of the ReGenSan planning process. DeRaR comprises four components: (i) recovery and reuse from existing facilities, (ii) alternative designs for recovery and reuse, (iii) demand based-designs and (iv) health and safety.

(II) *Sanitation Service Chain Raw Materials (SSC-RaMs)* are the input materials in ReS. SSC-RaMs consists of six components: (i) faeces, (ii) urine, (iii) flushwater, (iv) greywater, (v) anal cleansing water and other materials and (vi) sludge (faecal and sewage)

(III) *Sanitation-derived Products (SDPs)* focus on the recovery and reuse of sanitation-derived nutrients (SDNs), bioenergy and used water products from the SSC-RaMs. It consists of five components: (i) used water, (ii) sanitation-derived fertilizers(SDFs) (iii) bioenergy, (iv) soil conditioners and (v) other output materials, e.g. protein feed, building materials, trace elements.

Therefore, ReS views the sanitation economy as an interaction between different actors including designers, producers, consumers, suppliers, organizations, labour force and the government (Fligstein & Dauter, 2007). These actors and the markets are seen to be in constant co-evolution within different institutional settings (Beckert, 2009). A poor understanding of how markets are structured, operated and, indeed, performed leads to an underestimation of the complexity of the sanitation system that often results in unstandardized products and service delivery (O'Keefe *et al.*, 2015).

The principal drivers for ReS are largely:

(I) Increasing resource scarcity and stress with diminishing global nutrient reserves (e.g. depletion of phosphorus ores);

(II) Increasing fertilizer prices and stricter regulations for safeguarding the environment from pollution;

(III) Expanding population with increasingly limited resources under progressing climate change; and

(IV) Recognizing the resource value of wastewater, excreta and greywater (Otoo & Drechsel, 2017; WHO, 2006a, b).

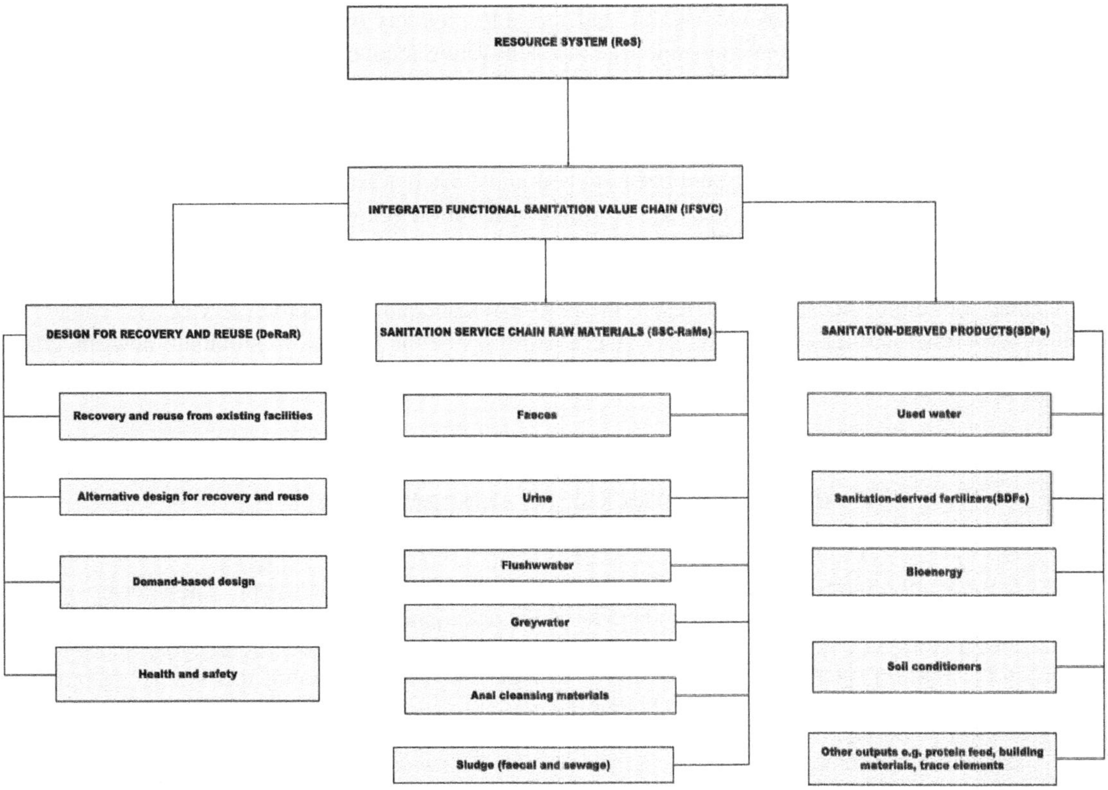

Figure 6.7 Linkage between the ReS dimensions and the integrated functional sanitation value chain.

6.4.1 ReGenSan resource system dimensions

There are three dimensions to the ReGenSan ReS: (i) Design for Recovery and Reuse (DeRaR) recommends site-specific recovery and reuse sanitation infrastructure and technology design solutions from the onset of sanitation planning; (ii) Sanitation Service Chain Raw Materials (SSC-RaMs) requires that input materials recovered from the SSC are appropriately captured in the right quality and quantity required for necessary bio-physical-chemical value-added transformation processes that deliver the expected products and services to satisfy customers' needs; and (iii) Sanitation-Derived Products (SDPs) requires that products derived from sanitation are obtained with the highest standards of technological, best applicable and appropriate processes at that point in time and such products should meet the general uniform global and national standard characteristics and quality for the industry category and the satisfaction of the consumers. Details of the three dimensions are presented below.

6.4.1.1 Design for Recovery and Reuse (DeRaR)

DeRaR incorporates the philosophy of recovery and reuse from the conceptual stage of sanitation initiatives to provide tangible and quantifiable incentives to deliver products and services with focus on the provision of safely managed sanitation infrastructure such as toilets and treatment plants, to facilitate the ease of product recovery. The aim of DeRaR is to facilitate a culture of designing site-specific and RRR-oriented sanitation technological infrastructure from the onset of the planning process rather than as an afterthought and at the same time protect public health and environment (Bahri, 1999; Murray & Buckley, 2009). This strategy is born out of global acknowledgement of the inherent resource potentials of human waste materials rather than resource scarcity. Therefore, integrating DeRaR into the design of the 'sanitation of the future' creates windows of opportunity that could provide a large number of new sanitation infrastructures (sewered and non-sewered) around the world and meet the need of the estimated 2.4 billion people currently without 'safely managed' sanitation facilities, and in addition cover the shortfall arising from the ever-increasing global population, estimated to be over 9 billion by 2050 (Orner & Mihelcic, 2018). This reiterates the urgent need for new-generation sanitation technologies that embody resource recovery goals (see Section 5.4.1.3). However, practitioners could be challenged, especially in the area of technological and infrastructure redesigns and development, while users would be faced with pyscho-social-cultural issues of 'faecophobia' and 'faecophilia' (see Section 4.2.1.1).

For instance, Murray and Buckley (2009) observed that when designing a treatment scheme for reuse in agriculture, it is desirable to maintain the embodied nutrients in the water, a factor that can significantly reduce capital and operational costs to the aquatic environment. Therefore, the goal of DeRaR is to support the development of sanitation technological infrastructure that is able to recover and reuse inherent resources from sanitation systems, deliver public health and environmental quality benefits associated with 'safely managed' sanitation and make productive contributions to the local economy and livelihood (Murray & Buckley 2009). DeRaR core components are: (a) recovery and reuse from existing facilities, (b) alternative designs for recovery and reuse, (c) demand based-designs and (d) health and safety.

6.4.1.1.1 Recovery and reuse from existing facilities

Existing sanitation technologies have varying capacities to recover resources that could enhance food security, reduce pollution from untreated waste and reduce waterborne diseases (Heinonen-Tanski & van Wijk-Sijbesma 2005). However, resource recovery from existing sanitation systems is constrained by available technologies, quality and concentration of materials for recovery as well as space and capital for infrastructure modification (Brands, 2014). Recovering resources from existing conventional technologies will require additional biophysical, mechanical and chemical processes as well as procedures (Mulder, 2003; Stark, 2004; WERF, 2011). Safely managed facilities that allow treatment and disposal to occur on site (non-sewered) can also collect, transport and manage faecal sludge (FS) off site or convey it through networks of sewers, and these would be good options for resource recovery (Orner & Mihelcic, 2018). Also, safe collection of urine from urine-diversion composting toilets (UDCT) would allow for recovery and reuse of sanitation-derived nutrients (SDNs) (e.g. N, P and K). Technologies that incorporate stored faeces, such as latrines and septic tanks, can incorporate faecal sludge management (FSM), which is defined as the effective collection, transportation, treatment and/or disposal of faecal sludge that has not travelled to a sewer (ISO/IWA, 28:2018; Singh *et al.*, 2017; Strande *et al.*, 2014) Figure 6.8.

FSM can provide cost-effective sanitation solutions with potential beneficial use of SDPs for soil amendment or energy production (Kennedy-Walker *et al.*, 2016). For instance, FS can be composted in a latrine and perhaps even also collected from latrines, then transported and treated to recover resources (Orner & Mihelcic, 2018) Figure 6.9.

Consequently, resources could also be recovered from existing conventional sanitation infrastructure. There is, of course, the exception of the traditional 'dig-use-cover' pit latrines that return the majority of the nutrients to the soil (but not N, which is primarily lost to the atmosphere) and do not have the potential to strategically recover nutrients because of the large number of distributed disposal sites associated with this technology (Orner & Mihelcic, 2018). Existing technologies such as septic tanks, VIPs, pour flush, etc. have moderate potential for resource recovery, particularly with integration of

FSM or centralized resource recovery facilities (ISO/IWA 28:2018; Orner & Mihelcic, 2018; Singh *et al.*, 2017; Strande *et al.*, 2014). The double-vault composting and the urine-diverting composting latrines have shown the greatest potential for resource recovery among existing non-sewered sanitation facilities, if properly operated (Orner & Mihelcic, 2018).

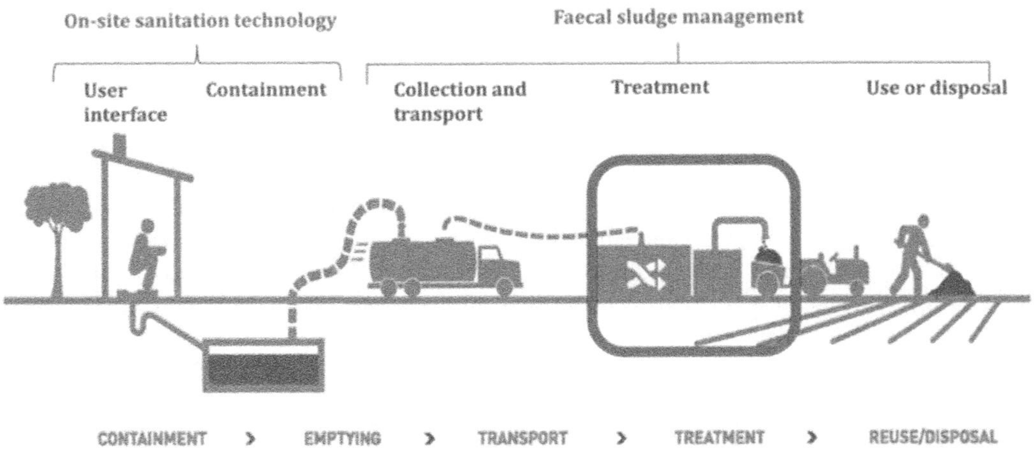

Figure 6.8 Non-sewered sanitation faecal sludge management (FSM) chain (*Source*: ISO/IWA 28:2018).

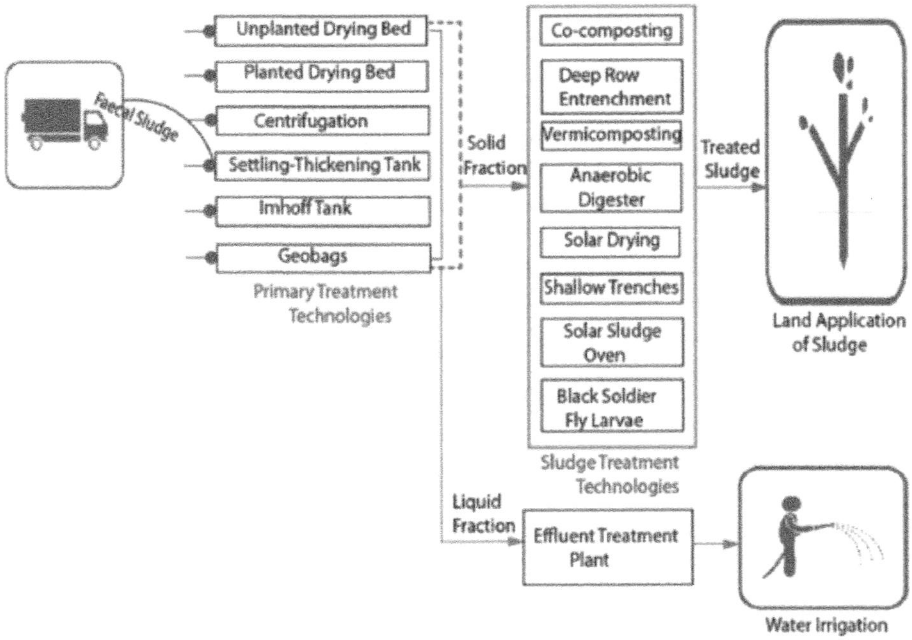

Figure 6.9 Overview of technology options for faecal sludge treatment (Reprinted from Singh S., Mohan R. R., Rathi S. and Raju N. J. (2017). Technology options for faecal sludge management in developing countries: benefits and revenue from reuse. *Environmental Technology & Innovation*, 7, 203–218, with permission from Elsevier).

Centralized treatment facilities provide enormous opportunities for resource recovery such as used water, bioenergy and SDFs. Several current options of centralized resource recovery infrastructure existing options are anaerobic digestion, bioenergy from biogas and struvite precipitation for nutrients (Asano & Levine, 1996; Orner & Mihelcic, 2018; WERF, 2011). For instance, there are four potential aspects of traditional sewage treatment plants that could be utilized for P recovery (Cornel & Schaum, 2009; Ye *et al.*, 2017). Figure 6.10 illustrates that the liquid phase for P recovery is sludge liquor (A) while the dewatered sewage sludge (1) and ash (2) are considered to constitute the sludge phase for P recovery as well as the ash from the incineration process (Bourioug *et al.*, 2015; Egle *et al.*, 2016; Li *et al.*, 2016; Ye *et al.*, 2017).

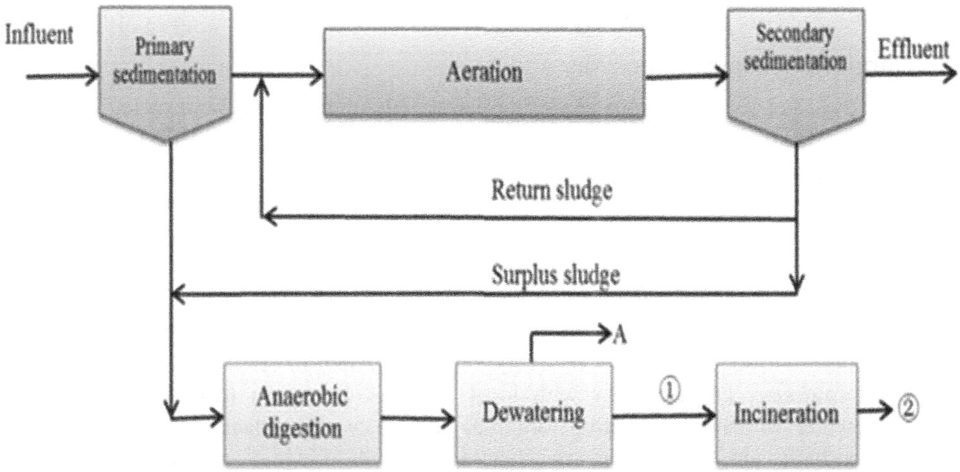

Figure 6.10 Different potential locations for phosphate recovery in conventional municipal wastewater treatment (Reprinted from Ye Y., Ngo H. H., Guo W., Liu Y., Li J., Li Y., Zhang X. and Jia H. Insight into chemical phosphate recovery from municipal wastewater. *Science of the Total Environment*, 576, 159–171, with permission from Elsevier).

In addition, the sludge liquor is returned to the influent in the municipal wastewater treatment so that P recovery from the sludge liquor can decrease the load of P by up to 20% (Evans, 2007). Some strategies for recovery and reuse from existing sanitation facilities are:

(I) VIP latrines that recover sludge through manual or mechanical excavation. Mechanical removal of sludge into slurry tankers may not be available or not well advertised and studies have shown that the majority of users of VIP in Blantyre (Malawi) were unaware that mechanical removal was an option (Grimason *et al.*, 2000; Orner & Mihelcic, 2018),

(II) Composting latrines produce compost, a soil conditioner (Fry *et al.*, 2008; Orner & Mihelcic, 2018),

(III) Urine-separated composting latrines additionally produce struvite, a nutrient-rich fertilizer (Orner & Mihelcic, 2018),

(IV) Septic tanks produce sludge that can be recovered through faecal sludge management (Orner & Mihelcic, 2018; Strande *et al.*, 2014),

(V) Sewered toilets have several options for resource recovery. Although sewer effluents are often reused in agriculture without treatment, the pathogens present may make this an unsafe option. With pathogen treatment, nutrients in the wastewater can be reused and more safely applied to crops (Guest *et al.*, 2009; Orner *et al.*, 2017, Orner & Mihelcic, 2018; Verbyla *et al.*, 2013).

Thus, achieving optimum resource recovery with existing sanitation infrastructure will require rehabilitation and upgrading of existing facilities and balancing several interrelated factors, such as:

(I) Nutrient removal or recovery efficiency,
(II) Energy requirements,
(III) Cost effectiveness,
(IV) Emissions, and
(V) Chemical consumption (e.g. magnesium used for struvite precipitation).

Land areas required to complete removal, recovery and purity of the final product(s) (e.g. presence of metals or toxics) are also important considerations (Brands, 2014).

6.4.1.1.2 Alternative designs for recovery and reuse

This component advocates for continuous exploration and development of new ideas and concepts for the optimization of RRR from sanitation system taking into account the principles of ReGenSan (see Section 2.4). Alternative designs are also referred to as 'new sanitation' by most scholars and indicate a new paradigm for the collection, transportation, treatment and recovery of resources from sanitation systems such as urine-deviated vacuum toilets, anaerobic digesters and struvite precipitation with the aim to recover resources (i.e. water, nutrients, organic matter), increase efficiency, reduce energy costs and/or offer solutions to sanitation management (Brands, 2014; Fan *et al.*, 2017; Larsen *et al.*, 2009; Wielemaker *et al.*, 2016). These technologies minimize transport and are, therefore, locally oriented systems that may be described as resource recovery and reuse when in close proximity and the technical design serves this purpose. Designs vary with local contexts, but often include source separation of waste and wastewater streams, collecting organic kitchen waste, blackwater (urine and faeces), greywater, (shower/bath, sink, laundry, dishwasher) and/or yellow-water (urine) separately (Fan *et al.*, 2017; Thibodeau *et al.*, 2014; Wielemaker *et al.*, 2016). In essence, alternative designs seek to provide safely managed infrastructure that sanitizes urine and faeces, recovers nutrients for production and prevents environmental pollution, which minimizes the demand on other resources (Fan *et al.*, 2017; Zhou *et al.*, 2010).

Several alternative resource recovery concepts from sanitation have been proposed since the early 2000s, such as EcoSan, resource-oriented sanitation, decentralized sanitation and reuse (DESAR) and source separation and decentralization (SSD); SSD are the latest addition (Esrey, 2002; Brands, 2014; Grant *et al.*, 2012; Larsen *et al.*, 2013; Lens *et al.*, 2001 and Box 5.6 on new sanitation technology concepts). The basic premise of SSD is that valuable resources in human excreta can be best recovered by keeping waste streams separate and minimizing or eliminating the use of water as a waste transport vehicle (Larsen *et al.*, 2009). In so doing, industrial effluents and domestic excreta are not mixed and domestic blackwater, kitchen waste and greywater are, to varying degrees, also stored and treated separately. Nutrient cycles are closed and reconnected with agriculture, organic matter is recovered or processed into energy and soil quality is maintained; there is no connection between waste or residue water and surface water bodies, and most of the water entering human settlements is recycled/recharged (Figures 1.4 and 2.2) (Brands, 2014; Langergrabera & Muelleggera, 2005; Larsen *et al.*, 2009; Wang *et al.*, 2006; Werner *et al.*, 2009). SSD alternative designs have been expressed in many forms of technologies, as presented below:

(1) *Urine separating (Nomix) flush toilet* Urine source separation (NoMix technology) has the potential to become a cost-efficient (Maurer *et al.*, 2005) and effective technology for nutrient recycling

(Larsen *et al.*, 2010). In this concept, yellow-water is collected with low or better still, no dilution and can be used directly on farms. Brownwater is converted to small volumes by two-chamber composting tanks with a filtration system, where each chamber is used for a year and left without further charge the other year (Otterpohl, 2000). The concept of NoMix toilets is very appealing to the general public, despite design-related drawbacks of existing NoMix toilets (Lienert & Larsen, 2010). Larsen *et al.* (2010) observed that considerable development and technical improvements are needed to make the NoMix technology viable and economically attractive to users/customers. NoMix flush toilets are still mainly at pilot stages, although some Swedish municipalities subsidize broader installation for environmental reasons (Kvarnström *et al.*, 2006; Larsen *et al.*, 2009 – Figure 6.11). However, modern versions of the urine-faeces separating toilets have been available since at least the 1970s (Winblad, 1994) and in China nearly 700,000 of such toilets were in use by 2003, especially in the arid and water-deficient areas (Kvarnström *et al.*, 2006) (see Table 5.2).

Figure 6.11 Roediger NoMix toilet. Roediger Vacuum (www.roevac.com) (Reprinted from Larsen T. A., Alder A. C., Eggen R. I., Maurer A. and Lienert J. Source separation: will we see a paradigm shift in wastewater handling? *Environmental Science & Technology*, 43(16), 6121–6125, with permission from American Chemical Society).

The NOVAQUATIS project of EAWAG, the NoMix toilet is designed to work with the sewered system, unlike other types that work with non-sewered facilities. In EAWAG's NoMix toilets, urine stored in-house is released into sewers at controlled times and then transported to the wastewater treatment plant, where it can be diverted for separate treatment (Larsen *et al.*, 2001a, b). A urine-separating NoMix toilet saves 80% of the water used for toilet flushing, accounting for 30% of the average western European's direct daily use and 10% of the total freshwater use in Switzerland. Since urine accounts for a large fraction of the wastewater nutrient load, this approach can also reduce emissions from fertilizer production and halt the contamination of agricultural soils by heavy metals found in raw materials used to produce synthetic fertilizer (Larsen & Gujer, 1996; Larsen *et al.*, 2001a, b – Figure 6.12).

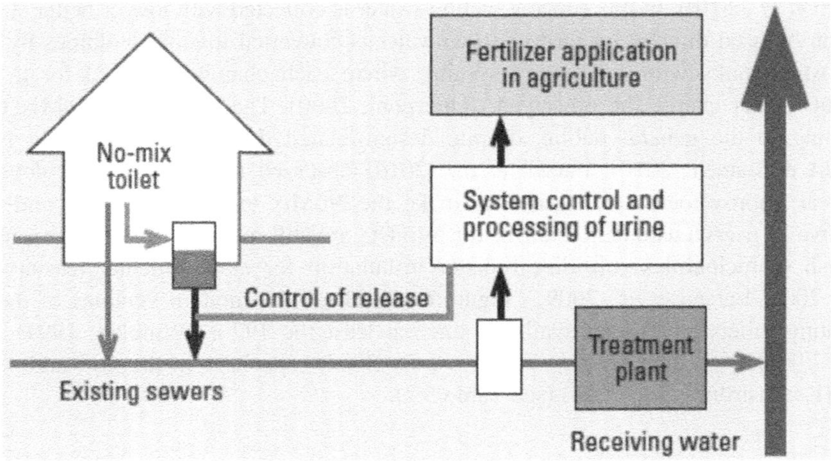

Figure 6.12 Urine source separation technology (Reprinted from Larsen T. A., Peters I., Alder A. C., Eggen R. I., Maurer M. and Muncke J. (2001). Re-engineering the toilet for sustainable wastewater management. *Environmental Science & Technology*, May 1, 194A, with permission from American Chemical Society).

(2) *Vacuum source-separation* is a thoroughly studied resource-oriented technology (Remy, 2010; Todt, 2015; Zhang *et al.*, 2014). It uses the vacuum system to separately collect and transport toilet wastes (slurry), which are then transformed into biogas for household cooking and organic fertilizers for nutrient utilization in rural communities' farmlands (Fan *et al.*, 2017). Vacuum toilets are widely deployed on aeroplanes, ships and trains; and vacuum sewer is also a common municipal wastewater collection technology, mostly adopted where gravity sewers are inapplicable (Zeeman *et al.*, 2007). Vacuum source separated sanitation technologies have been tested in many locations around the world, especially in new settlements such as Flintenbreite within the city of Lubeck (Baltic Sea, Germany) and Hahuageng Settlement, Guli Town of Changshu, China (for illustration purposes see Fan *et al.*, 2017; Otterpohl, 2000).

The new settlement of Flintenbreite within the city of Lubeck is an area totalling 3.5 ha. and inhabited by about 350 inhabitants. It is not connected to the central sewerage system, but serviced by vacuum toilets and vacuum-transported to a biogas plant (Otterpohl, 2000) unit. This system consists mainly of: (a) vacuum closet (VC) with collection and anaerobic treatment and co-treatment of organic household waste in a semi-centralized biogas-plant, recycling of digested anaerobic sludge to agriculture with further storage for growth periods, (b) use of biogas in a heat and power generator (heat for houses and digester plus electricity) in addition to natural gas, (c) decentralized treatment of greywater in vertical-flow constructed wetlands with interval feeding (very energy efficient) and (d) rainwater retention and infiltration in a swale system (Otterpohl, 2000). The heat for the settlement is produced by a combined heat- and power-generating engine that switches to biogas when the storage is filled; heat is also used for the biogas plant. In addition, there is a passive solar system to support heating of the houses and an active solar system for warm water production. Figure 6.13 is not meant to show all the features, but to give an idea of the concept with collection and treatment of faeces.

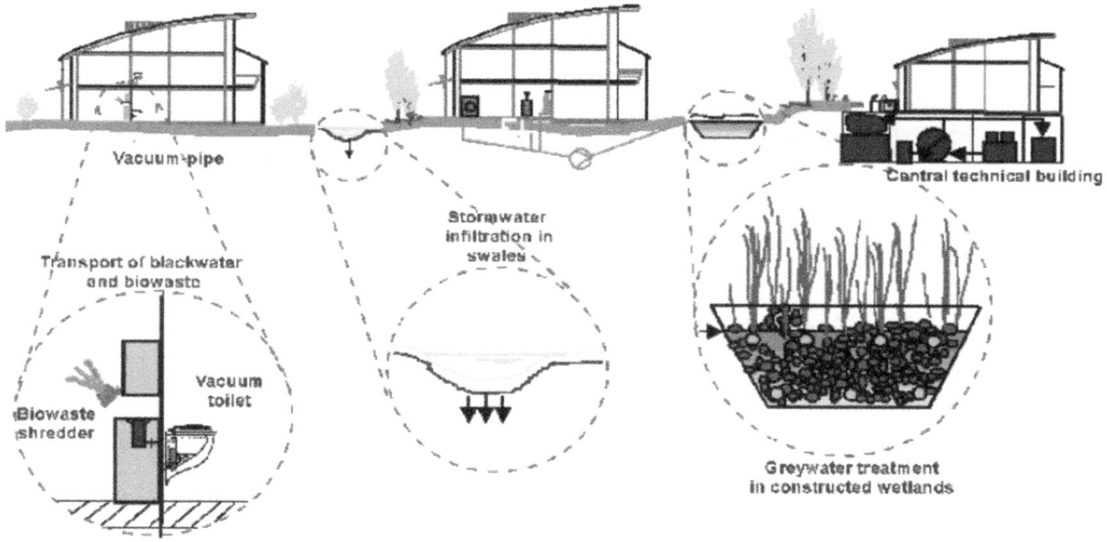

Figure 6.13 Flintenbreite vacuum toilet and vacuum-transport to a biogas plant located in the central technical building (*Source*: Otterpohl, 2000).

In China, a rural sanitation unit using vacuum source-separation was built in 2011 at Hahuageng Settlement, Guli Town of Changshu; it has been operating for the past six years and has gained high recognition (Fan *et al.*, 2017; Hu *et al.*, 2016). The most outstanding merit of vacuum source-separation is its use of minimal water to immediately transfer wastes from living-places. Figure 6.14 illustrates Guli Town's vacuum source-separation sanitation technology, a resource recovery facility that serves 200 residents in 50 households. The system is made up of four major compartments: collection, treatment, disposal and resource recovery process. The collection section covers the facilities and processes of treatment and reclamation. The disposal and resource recovery process section cover all the processes after treatment, including recovery, reuse and disposal (Fan *et al.*, 2017).

(3) *Decentralized circular system* is a conceptual redesigning and transformation of the centralized sanitation system to localize decentralized sanitation treatment and nutrient recycling as well as integrate source-separation and urine-diverting-low-water toilets. An example is a small scale biogas plant that treats faeces or blackwater, biowaste, energy crops and plant residues proposed by Särkilahti *et al.* (2017) for the city of Tampre, Finland, with its population of 226,000 (Figure 6.15). In this concept, blackwater from toilets is collected separately from other domestic wastewater and the low-water (dry or vacuum) toilets enable concentrations of blackwater to ensure direct treatment in an anaerobic digester. Urine is diverted from blackwater using a urine-diverting toilet. Nutrients recovered from urine and anaerobic digestion feedstock are used in local scenery fields to cultivate energy crops (such as corn or sugar cane) and/or in nearby greenhouse cultivation. After upgrading, biogas can be used locally as in household gas cookers, vehicle fuel, or injected into a gas grid. Greywater is treated either on site or directed to centralized treatment; it can also be reused, e.g. in greenhouse irrigation or as flush-water, if it fulfils quality criteria (Särkilahti *et al.*, 2017).

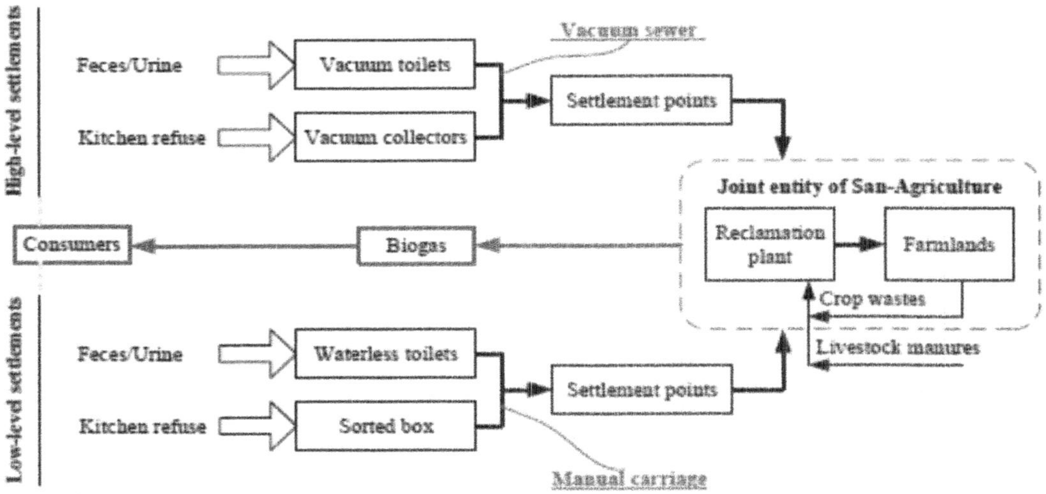

Figure 6.14 Guli Town vacuum source-separation sanitation technology. *Note: Greywater drains together with rainwater, or is treated or reclaimed; other wastes are sorted and collected, and then treated or disposed of* (Reprinted from Fan B., Hu M., Wang H., Xu M., Qu B. and Zhu S. (2017). Get in sanitation 2.0 by opportunity of rural China: scheme, simulating application and life cycle assessment. *Journal of Cleaner Production*, 147, 86–95, with permission from Elsevier).

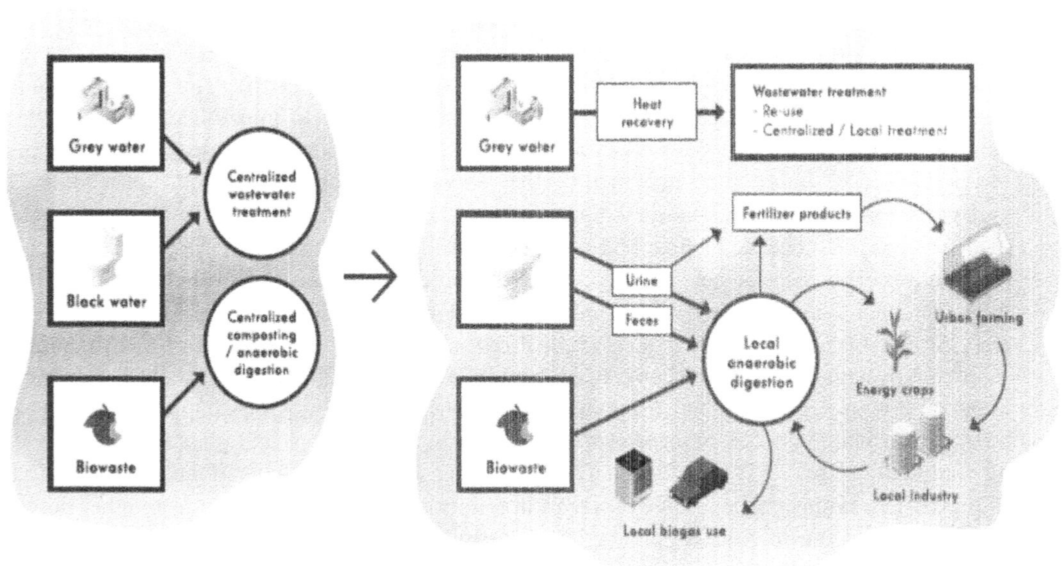

Figure 6.15 Decentralized circular system. Note: *The transition from a conventional sanitation system to a decentralized circular system, which consists of source-separating toilets; a small-scale AD plant to treat local waste streams; and local utilization of end products* (Reprinted from Särkilahti M., Kinnunen V., Kettunen R., Jokinen A. and Rintala J. (2017). Replacing centralised waste and sanitation infrastructure with local treatment and nutrient recycling: expert opinions in the context of urban planning. *Technological Forecasting & Social Change*, 118, 195–204, with permission from Elsevier).

(4) *Up-concentration for zero-wastewater* Verstraete and Vlaeminck (2011) proposed a zero-wastewater concept with the up-concentration process, followed by organic anaerobic digestion to achieve maximum energy recovery in domestic used water (Gong *et al.*, 2017). This approach aims to recycle energy and valuable materials, while adequately abating pathogens, heavy metals and trace organics. It is designed to ensure that potential energy present in wastewater is captured for reuse and requires less treatment, thus making wastewater treatment a net energy producer (McCarty *et al.*, 2011; Mezohegyi *et al.*, 2012). Therefore, the overall goal is to convert the wasteful linear urban wastewater chain into a smaller, local scale in which the resources present in wastewater are short-cycled (Cardoen, 2011; Wett *et al.*, 2007). Thus wastewater is made into a valuable resource to society by linking sanitation to food production, a connection that existed long before the advent of waterborne sanitation risks (Cordell, 2010, 2011; Gumbo, 2005; Verstraete & Vlaeminck, 2011).

Up-concentration is a process layout that enables maximal recovery of water, energy and inorganic and organic fertilizers from domestic discharges. The domestic 'used water' is initially pre-treated by mechanical means (screening, grit removal, microstrainer) to remove large particles and sand. To up-concentrate the water, dynamic sand filtration (DSF), dissolved air flotation (DAF), membrane filtration, biological sorption or a combination thereof produce an effluent with low levels of suspended and colloidal solids. Simultaneously, removal of soluble organic matter by appropriate usage of poly-electrolytes during DSF or DAF could be necessary. The key point is that quality of the effluent must be suitable for ultrafiltration/reverse osmosis (UF/RO). Thus, freshwater can be recovered as a first product (Verstraete *et al.*, 2009). The generated high-strength concentrated stream of valuable resources of residual biosolids is further treated by the anaerobic digestion (AD) process, which converts organic compounds to biogas, a mixture of methane, carbon-dioxide (CO_2) and some trace gases such as ammonia and hydrogen sulphide following the enzymes-mediated process of hydrolysis, acidogenesis, acetogenesis and methanogenesis, particularly good for pathogen removal (Appels *et al.*, 2008; Sonakya *et al.*, 2001; Verstraete, 2011 – Figure 6.16).

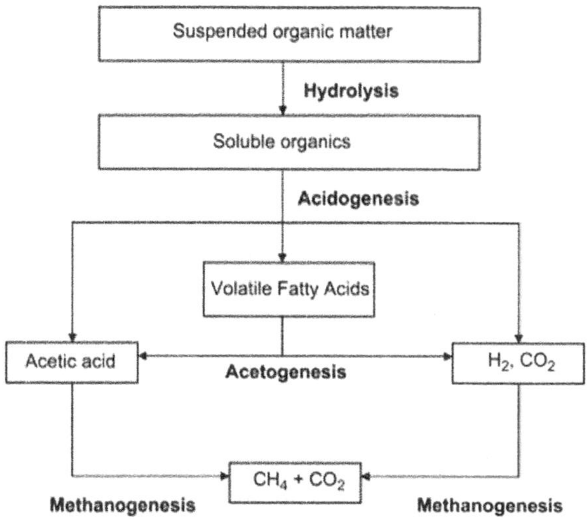

Figure 6.16 Subsequent steps in the anaerobic digestion process (Reprinted from Appels L., Baeyens J., Degreve J., and Dewil R. (2008). Principles and potential of the anaerobic digestion of waste-activated sludge. *Progress in Energy and Combustion Science*, 34, 755–781, with permission from Elsevier).

The biogas after purification is combusted in a combined heat and power generation (CHP) unit to produce electricity at a typical efficiency of 30–40% and heat, which can be used to warm the anaerobic reactor and for post-treatment of the anaerobic effluent such as sludge drying (Verstraete *et al.*, 2009). The CO_2 produced in the combustion could ultimately even be used to culture algae in closed photo-bioreactors, which could also serve as a co-substrate in the AD (Zamalloa *et al.*, 2011).

Furthermore, the anaerobic effluent can be lime conditioned to precipitate P and subsequently dewatered by a filter press or centrifuge. This process produces a cake that is then pyrolyzed to produce biochar (Verstraete *et al.*, 2009). The ammonium-rich liquor stream created during dewatering can either be physico-chemically treated to recover the nitrified ammonium that has been up-concentrated by RO and turned into natural stable fertilizers (Cardoen, 2011 – Figure 6.17).

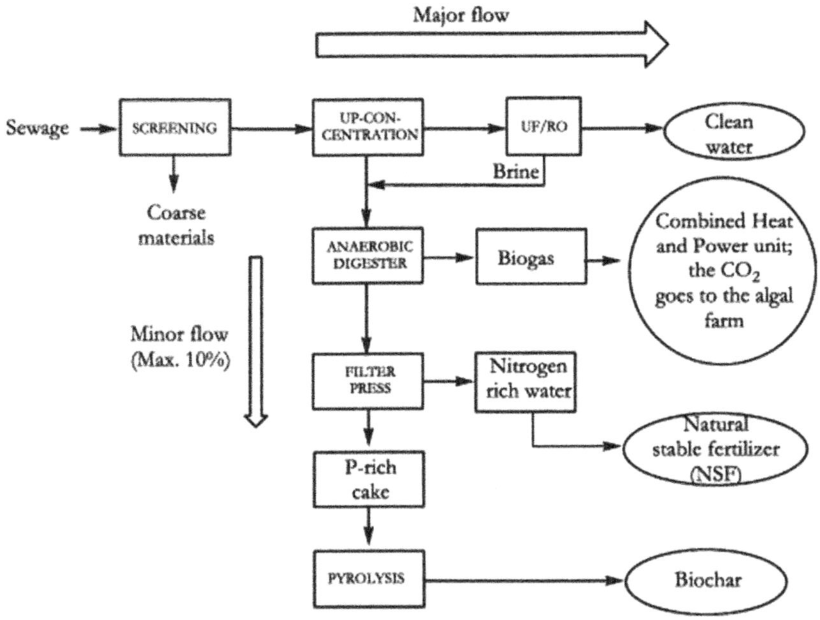

Figure 6.17 Process scheme based on up-concentration of wastewater at a centralised plant, thus allowing maximal recovery of resources from domestic wastewater (Reprinted from Verstraete W., Van de Caveye P. and Diamantis V. (2009). Maximum use of resources present in domestic 'used water'. *Bioresource Technology*, 100, 5537–5545, with permission from Elsevier.)

(5) *Nature-based recovery and reuse* proposes continuous replacement (through its own functional processes) of energy and materials. The energy is produced by solar and internally generated by its own processes such as AD, while materials are replaced by cycling and reuse (Lyle, 1994) (see Sections 1.4, 2.3.2, 4.2.1.3, and 5.4.1.3.1). This concept incorporates nature's multiple targets into the sanitation system, such as water saving, water use maximization, rainwater retention, the most complete reuse possible of all involved substances, crop production maximization, the reduction of non-renewable energy usage and mitigation of urban heat island effects (Masi *et al.*, 2017) (see Section 5.4.1.3.2). For instance, the significance of nature-based resource recovery and reuse is highlighted in the use of an integrated sewage treatment

aquaculture-wetland ecosystem (AWE) for food production and inorganic nitrogen recycling at the Lyle Centre for Regenerative Studies, located on 16 acres (6.5 ha.) and accommodating more than 1600 community members within the university campus of Pomona, California, USA (Figure 6.18).

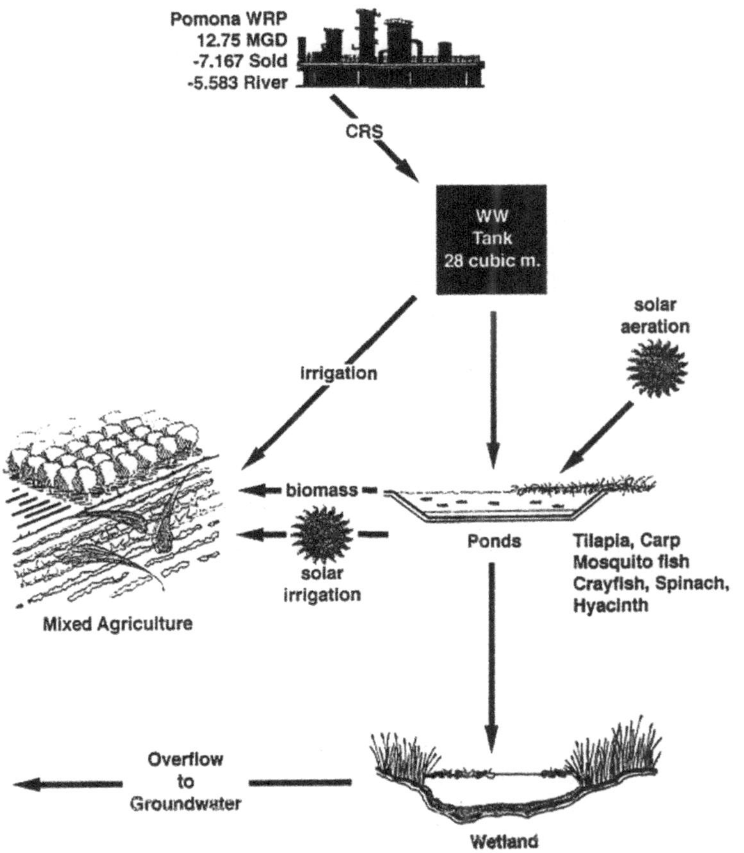

Figure 6.18 Lyle Centre for Regenerative Studies aquaculture-wetland ecosystem (AWE). *Note: 12.75MGD = 48,264 m^3/d; 7.167 MGD = 27,130 m^3/d; 5.583 MGD = 21,134 m^3/d*; (Reprinted from Costa-Pierce B. A. (1998). Preliminary investigation of an integrated aquaculture–wetland ecosystem using tertiary-treated municipal wastewater in Los Angeles County, California. *Ecological Engineering*, 10, 341–354, with permission from Elsevier).

Thus, the problem of sewage was converted into a community asset, serving the deprived people of a problematic neighborhood at a cost considerably less than that of treating the sewage (Costa-Pierce, 1998; Lyle, 1994). The Lyle Centre for Regenerative Studies' aquaculture-wetland ecosystem (AWE) is unique and this Pomona tertiary treatment plant produced approximately 12.8 million gallons per day (MGD) (48,453 m^3/d) of wastewater. It sold 7.2 MGD (28,391 m^3/d) and discharged 5.6 MGD (21,198 m^3/d) out of it to local waterways. The AWE used approximately 28 m^3 per week of Pomona's wastewater (Costa-Pierce, 1998). Thus, nature-based recovery and reuse systems do not only provide solutions for some major sanitation management issues, but also contribute to regulating the local climate, preventing floods, providing interesting recreational space and facilities, increasing biodiversity and providing nutrients for crop production (Masi *et al.*, 2017). Therefore, redesign and realignment of constructed wetlands technological systems should begin to focus on the following new goals as stated by Masi *et al.* (2017):

(I) Water reuse: (a) greywater treatment (outdoor, indoor) for local reuse and recreational purposes, possibly as only liquid treatment, while excreta are collected and processed separately (Masi *et al.*, 2010, 2016); (b) rainwater (including first flush) treatment and storage (Nolde, 2007); (c) treating persistent organic molecules in low concentrations for water reuse (Matamoros *et al.*, 2017; Verlicchi & Zambello, 2014); (d) reducing pathogens to acceptable levels as part of a multi-barrier approach; and (e) polishing of secondary treated wastewaters, as long as these still exist, for reuse (Ayaz, 2008; Rousseau *et al.*, 2008).

(II) Nutrient recovery: (a) CWs as pre-treatment for fertilization (disease vector and hazardous substance reduction, separation of liquid and solid phase); and (b) biomass production from secondary sludge (as long as such sludge is still produced), digestate or primary sludge.

(III) Energy production: (a) anaerobic reactor (biogas) + CW as polishing stage; (b) CWs as biomass production plots; and (c) greywater collection and heat recovery + CW for greywater treatment.

(IV) Ecosystem services: (a) multipurpose CWs for rainwater storage, recreation and wetland ecosystems following the sustainable urban drainage concepts (Fletcher *et al.*, 2015); and (b) re-adaptation of ornamental green areas in terms of ecosystem services (green roofs, green walls, indoor green areas, parks, permaculture productive areas) comprising organic food production in integrated habitats.

These options are open for improvement, while other more innovative, practical, place- and context-specific solutions should be investigated/explored.

6.4.1.1.3 Demand based-designs (DBD)

DBDs ensure that the sanitation products and services respond to what users/customers want rather than what sanitary engineers and scientists believe they should have (WSP, 2004). The aim is to link technology development with sanitation marketing as a strategy to uptake ReS integration into sanitation for ease of user buy-in. Sanitation marketing is targeted at users/customers so as to create sustained and effective demand for sanitation products and services (WSP, 2011b). Research has shown that lack of DBD is responsible for the high numbers of abandoned, unused and unimproved sanitation infrastructures; coupled with low levels of customer satisfaction with current solutions as well as unsustainable inefficient public and private investments required to promote the sanitation market for current innovative technologies (WSP, 2011a).

Borsuk *et al.* (2008) observed that while conceptualizing may be simple, the NoMiX (modern urine separating flush toilets) technology concept (see Box 5.6) has proven significantly more difficult to implement (Berndtsson, 2006) because successful adoption depends not only on the priorities of users, but also on those of developers and suppliers in the IFSVC (see Section 6.2.1). Surveys have shown household users to be receptive to the NoMiX concept as long as it does not deviate from the current standards of bathroom convenience and hygiene and no additional costs are incurred (Borsuk *et al.*, 2008; Lienert & Larsen, 2006; Pahl-Wostl *et al.*, 2003). As a result, while toilet manufacturers believe they can develop NoMiX units that meet these requirements, they do not see an immediate market for such product (Larsen & Lienert, 2003) (see Section 6.4.1.2i). It is essential that designers incorporate users'/customers' preferences so that solutions are satisfactory and usable, as well as seeking input from potential consumers about which products they would like to pay for. Also, when developing prototypes, designers should consult manufacturers to understand what the cost of the technology is and how it will be made, transported and installed (Chapin & Pedi, 2013). This information must be communicated to the users/customers so they can make an informed decision about whether the product will work in their homes.

A viable and sustainable DBD relies on an IFSVC that stretches from raw materials to manufacturing and distribution. Market research is a critical first step in understanding the landscape for the users'/customers' preferences and what they are willing to pay (Sievers & Kelly 2016). The objectives of DBD include ensuring:

(I) Users/customers choose to receive what they want and are willing to pay for;
(II) Financial sustainability by assuring return on technological investment;
(III) Cost-effectiveness and ease of scale-up of technological innovation; and
(IV) Technological designs address users'/customers' psycho-social-cultural issues (see Section 4.2.1.1).

Hence, ensuring effective demands for new generation resource-oriented sanitation technologies will require more exploration and comprehension of customers interested in resources recovered (SDPs) (Larsen & Lienert, 2003).

6.4.1.1.4 Health and safety

A fundamental function of all sanitation infrastructural technological solutions is to prevent human contact with hazardous pathogens and chemicals, even when the main aim is recovery and reuse. The use of SDPs poses possible risks of transmission of microbial and non-microbial contaminants to grazing cattle, vectors, surface and groundwater and through standing crops to humans (Höglund *et al.*, 1999; Stenström *et al.*, 2011). Thus, a well-designed recoverable and reusable infrastructure should not only protect health, but also promote it by contributing to food security (Andersson *et al.*, 2016a, b). For a long time, pathogens (disease-causing micro-organisms), organic pollutants and some heavy metals have been well-known as major constraints when using SDPs (Winker *et al.*, 2009). Heavy metal concentrations are low in all products of domestic origins as long as no industrial wastewater is included, although it may not be the case with municipal wastewater treatment plants that receive industrial wastewater (Simons *et al.*, 2005). Vinnerås *et al.* (2006) observed that the use of heavy metal in households can result in a large impact on sanitation-derived fertilizers (SDFs) (Winker *et al.*, 2009).

All SDPs can contain organic micropollutants such as pharmaceuticals used for human medication and substances originating from cleansing agents, personal care products etc. in their original sanitation waste stream (Eriksson *et al.*, 2003; Lienert *et al.*, 2007; Polprasert & Koottatep, 2017; Ronteltap *et al.*, 2007; Tettenborn *et al.*, 2007). Therefore, appropriate treatment is required in order to eliminate these unwanted and potentially harmful substances before the product can be used (Gajurel, 2007; Maurer *et al.*, 2006; Polprasert & Koottatep, 2017; Tettenborn *et al.*, 2007; Winker *et al.*, 2009). According to Morsing (1994), there are four approaches to minimize the risk for contamination through the use of SDFs: (i) reduction of pathogens, (ii) reduction of vector attraction (rats, flies, etc.), (iii) soil treatment and (iv) restrictions on cultivation methods and crops. The use of SDFs may consequently reduce the risks of pathogen exposure, if treatment and other barriers against exposure are accounted for in the process of resource recovery and reuse (Höglund *et al.*, 1999). This should, then, be aimed at from the conceptualization stage of design.

The four major groups of micro-organisms that can transmit infectious diseases from sanitation systems are: bacteria, protozoa, viruses, helminths (worms that infect humans) and fungi (Dixon & Fromtling, 1991; Feachem *et al.*, 1983; Haug, 1993; Höglund, 2001). These pathogens are divided into six categories depending on their infectious dose, transmission route (e.g. the need for an intermediate host), latency, persistence and multiplication (Cairncross & Feachem, 1993; Feachem *et al.*, 1983; Höglund, 2001); and infect the gastrointestinal tract, causing diarrhoea and many other clinical manifestations with severe signs and symptoms (Feachem *et al.*, 1983; Havelaar *et al.*, 2000; WCED, 1987). These intestinal micro-organisms, particularly those of the coliform group (Walker *et al.*, 1982), are used as indicator organisms of faecal contamination and possible risks related to water (WHO, 2017). They are, however, not suitable as sole indicators of the hygienic quality of SDPs since other groups of microorganisms are considered more resistant (Cooper & Olivieri, 1998; Höglund, 2001).

Meanwhile, transmission of infections can either be direct, through means of person-to-person contact including airborne short-distance; or indirect (secondary), which includes vehicle-borne (food, water, formites), vector-borne, airborne long-distance and parenteral (injections with contaminated syringes) transmissions (Beaglehole *et al.*, 1993; Höglund, 2001). The secondary transmission routes for pathogens emanating from faeces and urine are illustrated in Figure 6.19.

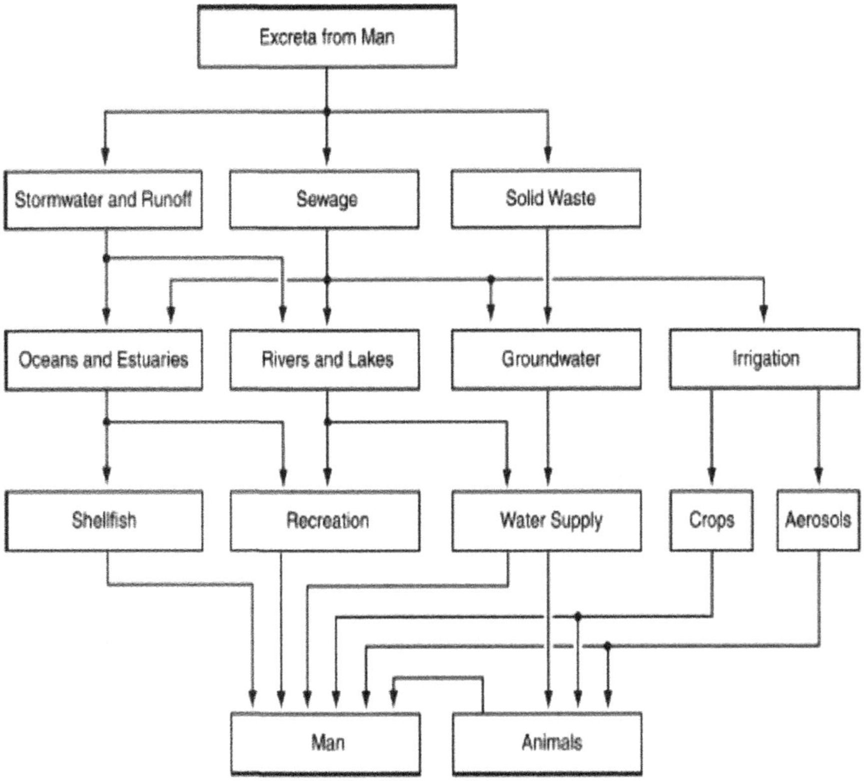

Figure 6.19 Secondary transmission routes for pathogens originating from human urine and faeces (*Source*: Höglund, 2001).

These routes can largely be defined as *faecal-oral (or urine-oral)*, i.e. pathogens are excreted in faeces (or urine) and infect another person by ingestion or contaminate other articles and environmental media. The pathogens may be transmitted via hands or through food or water and other fluids (Höglund, 2001; Reed & Scott, 2014 – Figure 6.20). Exposure to microbial hazards can

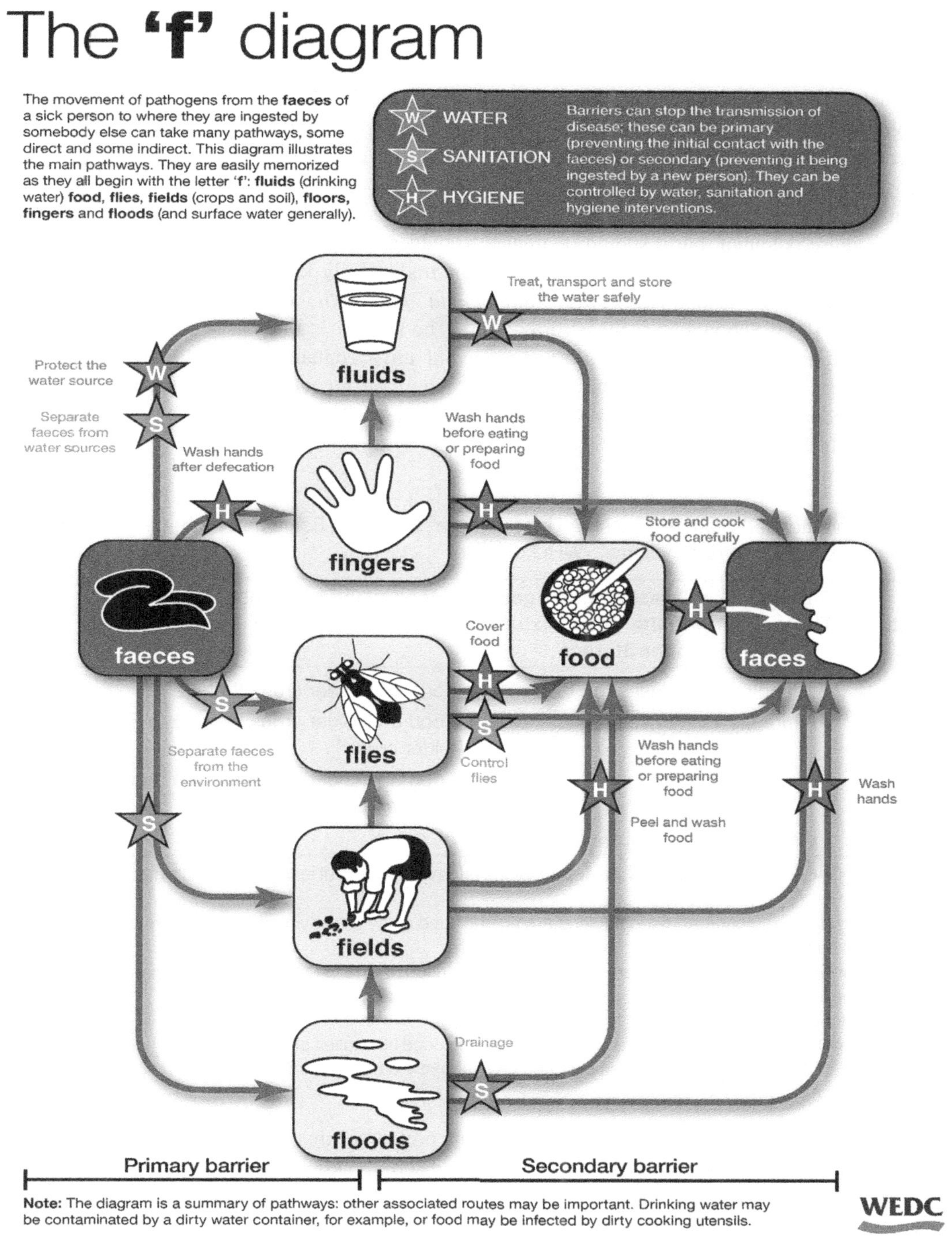

Figure 6.20 Faecal-oral route of disease transmission from contaminated sanitation-derived products (Reprinted from Reed B. and Scott R. (2014). Preventing the Transmission of Faecal-Oral Diseases, 1–8, with permission from Practical Action Publishing for WEDC).

happen at different points in the sanitation system. It may occur during normal operations, e.g. due to improper use and operation, lack of maintenance); during partial or full system failure (e.g. power failure, equipment breakdown, faulty infrastructure, system overloading); or seasonally due to climatic factors (e.g. flooding). The population at risk usually includes users, sanitation workers responsible for operation and maintenance of the system, population living nearby, farmworkers using recovered resources (e.g. sludge and water) and people consuming agricultural products grown with recovered resources (Andersson *et al.*, 2016a, b).

Human urine does not usually contain pathogens and industrial chemical contaminants that may hamper their reuse potential, but it should be treated to reduce human pathogens to a safe level (Jönsson *et al.*, 2004; Schönning & Stenström, 2004). Pathogens traditionally known to be excreted in urine are *Leptospira interrogans, Salmonella typhi, Salmonella paratyphi* and *Schistosoma haematobium* (Feachem *et al.*, 1983; Schönning & Stenström, 2004). The main risk, though, is contamination of urine by some faecal matter and so it is important to avoid cross-contamination (Höglund *et al.*, 2002; Schönning *et al.*, 2002). Some health-based targets for nutrient recovery from urine based on storage time and temperature have been proposed (Jönsson *et al.*, 2000; Höglund, 2001; Schönning & Stenström, 2004; WHO, 2006b), Table 6.4.

Table 6.4 Recommended Storage Times for Urine Mixture[a] Based on Estimated Pathogen Content[b] and Recommended Crop for Larger Systems[c] (WHO, 2006a).

Storage Temperature	Storage Time	Possible Pathogens in the Urine Mixture after Storage	Recommended Crops
4°C	≥1 month	Viruses, protozoa	Food and folder crops that are to be processed
4°C	≥6 months	Viruses	Food and folder crops that are to be processed folder crops[d]
20°C	≥1 month	Viruses	Food and folder crops that are to be processed folder crops[d]
20°C	≥6 months	Probably none	All crops[e]

[a]Urine or urine and water. When diluted, it is assumed that the urine mixture has at least pH 8.8 and a nitrogen concentration of at least 1 g/l.
[b]Gram-positive bacteria and spore-forming bacteria are not included in the underlying risk assessments, but are not normally recognized as causing any of the infections of concern.
[c]A larger system in this case is a system where the urine mixture is used to fertilize crops that will be consumed by individuals other than members of the household from which the urine was collected.
[d]Not grasslands for production of fodder.
[e]For food crops that are consumed raw, it is recommended that the urine be applied at least one month before harvesting and that it be incorporated into the ground if the edible parts grow above the soil surface.

Health-based targets, defined as that level of health protection that is relevant for each hazard, are often measured as a standard metric of disease, such as 'disability adjusted life year' or DALY (i.e. 10^{-6} DALY), or they could be based on an appropriate health outcome, such as prevention of exposure to pathogens in urine. To achieve a health-based target, health protection measures are developed (WHO, 2006b). During storage, the urine should be contained in a sealed tank or container. This prevents humans and animals coming into contact with the urine and hinders evaporation of ammonia, thus decreasing the risk of odour and loss of plant-available nitrogen (Schönning & Stenström, 2004).

Meanwhile, faeces are the major source of transmission for most prevalent enteric infections by bacteria, viruses, parasitic protozoa and helminths. Bacterial infections have traditionally been considered the leading group of organisms causing gastrointestinal illness in developing countries and are often transmitted through the faecal-oral route (Figure 6.20), while viruses are considered to cause the majority of gastrointestinal infections in developed countries (Reed & Scott, 2014; Schönning & Stenström, 2004; Svensson, 2000). Therefore, health risks may be enhanced due to improper handling, treatment and use of excreta and urine (Jönsson *et al.*, 2004; Schönning & Stenström, 2004). Higher incidences of enteric infections have been recorded in epidemiological investigations in areas where wastewater was used on crops (Bouhoum & Amahmid, 2000; Cifuentes, 1998; Katzenelson *et al.*, 1976); while in some studies no risk of infection was recorded (Devaux *et al.*, 2000; Ward *et al.*, 1989). Foodborne outbreaks caused by wastewater irrigation of vegetables and fruits have also been documented (Höglund, 2001; Yates & Gerba, 1998). Health-based targets for the use of excreta-fertilizer products (EFP) can be achieved through different treatment barriers or health promotion measures. Barriers relate to verification monitoring mainly in large-scale systems (Table 6.5) for excreta and greywater, but do not apply to urine.

Table 6.5 Guideline Values for Verification Monitoring in Large-Scale Treatment Systems of Greywater, Excreta and Faecal Sludge foe Use in Agriculture (WHO, 2006a).

	Helminth Eggs (Number per Gram Total Solids or per Litre)	E.coli (Number per 100 ml)
Treated faeces and faecal sludge	<1/g total soilds	<1000 g/total solids
Greywater for use in: Restricted irrigation	<1/litre	$<10^{5a}$ Relaxed to $<10^6$ when exposure is limited or regrowth is likely $<10^3$
Unrestricted irrigation of crops eaten raw		Relaxed to $<10^4$ for high-growing leaf crops or drip irrigation

[a]These values are acceptable due to the regrowth potential of E.coli and other faecal coliforms in greywater.

Health-based targets may also relate to operational monitoring, such as storage as an on-site treatment measure or further treatment off site after collection. This is exemplified in faeces collection from small-scale systems (Table 6.6).

Hence, increased materials recovery and reuse from sanitation systems requires strengthened approaches that identify and manage health risks and safety issues (WHO, 2016). In addition, special attention should be given to issues of environmental health by ensuring full compliance with safety and public health protection, critical control points and mitigation measures (Otoo & Drechsel, 2017). Public health policies for interventions should ensure that the most cost-effective measures are applied in specific contexts. Measures from a range of categories may be applied at different points during the lifecycle, and they are normally used in combination to reach desired goals:

(I) Treatment of wastewater, excreta and greywater should prevent the contaminants from entering the environment;
(II) Crops and produce should be processed or cooked before they are eaten in order to minimize health risks to product consumers;

(III) Waste application techniques should aim to reduce contamination of the products by allowing sufficient time for pathogen die-off in the environment prior to harvest;
(IV) Protective equipment and good hygiene should be implemented to prevent environmental contamination from reaching exposed groups;
(V) Produce washing/rinsing/disinfection and cooking should be implemented to reduce exposures for product consumers;
(VI) Vector control should be implemented to reduce exposures of workers and local communities; and
(VII) Immunization should be used to prevent illness for likely exposed persons and treat those who are ill as well as reduce future pathogen input into the wastewater, excreta or greywater (WHO, 2006a).

Table 6.6 Recommendations for Storage Treatment of Dry Excreta and Faecal Sludge before Use at the Household and Municipal Levels[a] (WHO, 2006a).

Treatment	Criteria	Comments
Storage: ambient temperature 2–20°C	2–5 years	Will eliminate bacterial pathogens; regrowth of E.coli and salmonella may need to be considered if rewetted; will reduce viruses and parasitic protozoa below risk levels. Some soil-borne ova may persist in low numbers.
Storage: ambient temperature >20–35°C	>1 year	Substantial to total inactivation of viruses, bacterial and protozoa; inactivation of schistosome eggs (<1 month); inactivation of nematode (roundworm) eggs, e.g. hookworm (*Ancylostoma Necator*) and whipworm (*Trichuris*); survival of a certain percentage (10–30%) of *Ascaris* eggs (≥4 months), whereas a more or less complete inactivation of *Ascaris* eggs will occur within 1 year.
Alkaline treatment	pH >9 over >6 months	If temperature >35°C and moisture <25%, lower pH and/or wetter material will prolong the time for absolute elimination.

[a]No addition of new material.

6.4.1.2 Sanitation service chain raw materials (SSC-RaMs)

SSC-RaMs is another ReS dimension that identifies sanitation input materials as resources such as faeces, urine, flushwater, greywater, anal cleansing water, sludge (faecal and sewage) and other materials for SDPs' production. Some of the SSC-RaMs are generated directly (e.g. urine and faeces), others are required in the functioning of technologies (e.g. flushwater to move excreta in water-based sanitation systems), some are generated as a function of storage (faecal sludge) or treatment (sewage sludge), some are generated from washing food, clothes and dishware as well as from bathing (greywater) and some are generated from cleaning oneself after defecating and/or urinating; that is, water and other dry materials such as papers, leaves etc. (anal cleansing water and dry materials). Faecal sludge comes from non-sewered sanitation technologies, which means it has not been transported through sewers, while sewage sludge (wastewater sludge) originates from sewer-based technologies and semi centralized treatment processes (Tilley *et al.*, 2014b).

SSC-RaMs require a holistic management approach aimed at pollution control, RRR and water conservation (Maurya, 2012). Human excreta consisting of urine and faeces not mixed with any flush water is small in volume, but concentrated in both nutrients and pathogens (Tilley *et al.*, 2014b). Everyday a huge amount of human excreta is produced by humans (Maurya, 2012) and the average weight produced daily varies greatly. Literature indicates that the amount of faeces produced at minimum is 69 g/capita/day in China (Snell, 1943) and the highest up to 520 g/capita/day in Kenya (Pieper, 1987). Table 6.7 indicates that generally, developed countries produce less faeces in comparison to developing countries. Similarly, generation of urine also varied from 600 to 1500 mL/capita/day (Maurya, 2012). Thus, the amount of human excreta produced varies from country to country and the variations may be attributed to different dietary habits due to economical and climate conditions of the region. Below are the types and characterizations of SSC-RaMs:

Table 6.7 Human Excreta Generation in Different Countries (Maurya, 2012).

Country	Human Excreta per Capita per Day		References
	Faeces (g)	Urine (mL)	
Africa	400	1200	Mann (1976)
USA	86	1055	Snell (1943)
China	69	845	Snell (1943)
	350	–	Gao *et al.* (2002)
Europe and America	100–200	1200	Feachem *et al.* (1983)
	100–200	–	Edwards (1992)
Developing countries	130–520	–	Edwards (1992)
Vietnam	1370 g (faeces + urine)		Nimpuno (1983)
	130–140	820–1200	Polprasert *et al.* (1981)
Thailand	1000 g (faeces + urine)		Stoll and Parameswaran (1996)
	120–400	600–1200	Schouw *et al.* (2002)
Japan and China	116–200	–	Polprasert *et al.* (1981)
Denmark	1700 g (faeces + urine)		Hansen and Tjell (1979)
Sweden	140	1500	Vinnerås *et al.* (2006)
Kenya	520		Pieper (1987)

(a) *Faeces* consist of a solid material that leaves the body through the anus usually defined as excrement discharged from the intestine. Faeces consist of undigested foods, mixed with materials extracted from the bloodstream or shed from glands and the intestines (Niwagaba, 2007). Faeces contain mainly water, food residues, nutrients and about 15–30% dry solids. They also contain large concentrations of pathogenic micro-organisms such as viruses, cysts of protozoa and eggs of helminths (Feachem *et al.*, 1983). Carbon is the main constituent in dry solids of faeces and contributes the major fraction of solids (≈44–55%), followed by nitrogen (≈5–7%), phosphorous (≈3–5.4% as P_2O_5) and potash (≈1–2.5% as K_2O) (Gijzen, 2001; Maurya, 2012; Polprasert, 1996; Polprasert & Koottatep, 2017 – Figure 6.9).

Also, faeces contain fewer nutrients than urine. However, the humus produced from faeces contains significant concentrations of phosphorus and potassium. Thus, the application of humus produced from faeces in agricultural fields may increase the fraction of the organic carbon in soils. More organic matter in soils is especially important to improve the soil structure and makes the soil more resistant to erosion, drought and salinity (Engel *et al.*, 2001; Heinonen-Tanski & van Wijk-Sijbesma, 2005; Maurya, 2012; Smith *et al.*, 2001). In addition, biogas (methane) could also be produced from anaerobic degradation of human excreta (Rahman *et al.*, 1996). Faeces also contain other elements such as calcium, magnesium, zinc, copper, nickel, cadmium, lead, mercury and barium (Maurya, 2012; Schouw *et al.*, 2002). However, before applying faeces-based products as soil conditioners and fertilizers, complete destruction of pathogens is necessary (Maurya, 2012).

(b) *Urine* is a liquid waste product of the human body secreted by the kidneys through a process of filtration from the blood called urination and excreted through the urethra. As urine is produced after filtration of the blood in the kidneys, it contains low-molecular-weight compounds as proteins that are not filtered (Karak & Bhattacharyya, 2011). It is relatively much less infectious than faeces, but also contains some of the major valuable resources available in human excreta (Niwagaba, 2007) (about 90% nitrogen, 50–65% phosphorus and 50–80% potassium) (Heinonen-Tanski & van Wijk-Sijbesma, 2005; Maurya, 2012). The amount of plant nutrients excreted via urine per person/per year has been estimated at 2.5–4.3 kg nitrogen, 0.4–1.0 kg phosphorus and 0.9–1.0 kg potassium (Guyton, 1992; Vinnerås *et al.*, 2006). The concentration of heavy metals in human urine is negligible (Palmquist & Jönsson, 2004). All plant nutrients consumed are excreted over time, mainly via urine and faeces (Guyton, 1992).

Also, a range of pollutants can occur in urine, including pharmaceutical compounds, natural and artificial hormones and pathogens (Pronk & Kone, 2008). However, since most nutrients in sanitation materials come from urine, with about 80% N and 50% P (Larsen & Gujer, 1996), with P being a depletable raw material, its recovery and reuse contributes to environmental sustainability. N is not a depletable raw material, but it is present in urine in the form of ammonia or urea, which is a valuable material for use as fertilizer (Maurer *et al.*, 2003). Recovered urine products could then replace chemical fertilizers (which requires high energy and auxiliary materials) as inherent nutrients in urine reflect the components necessary for plant growth, thereby making these products suitable as sanitation-derived fertilizers (SDFs) in agricultural application (Heinonen-Tanski & van Wijk-Sijbesma, 2005; Kirchmann & Pettersson, 1995).

(c) *Flushwater* is the water discharged through the user interface from toilets for transporting the contents and/or clean water. It can be freshwater, rainwater, recycled greywater or any combination of the three (Tilley *et al.*, 2014b). Toilet flushing is the single highest use of water in the average home, so it also presents a prime opportunity for water conservation. With the average person flushing five times a day, toilets make up about 31% of overall household water consumption. Toilets made from the early 1980s to 1992 typically use about 13.2 litres per flush (lpf) or more, while toilets made prior to 1980 typically use 18.9 lpf to 26.5 lpf; the oldest toilets can use more than 30 lpf. New high-efficiency toilets (HETs) use five lpf, and a family of four will use approximately 34 m^3 per year in total toilet water use (AWF, 2017).

(d) *Greywater* is a voluminous stream characterized by lower concentrations (or even absence) of some components in comparison with blackwater (Kujawa-Roeleveld & Zeeman, 2006). It includes wastewater from baths, showers, hand basins, washing machines and dishwashers, laundries and kitchen sinks (Jefferson *et al.*, 2004). The most commonly described application for greywater reuse is toilet/urinal flushing, which reduces water demand within dwellings by up to 30% (Karpiscak *et al.*, 1990). Untreated greywater, although less contaminated than other wastewater sources, does contain pathogens, salts, solid particles, fats, oils and chemicals (Morel & Diener, 2006). However, greywater has been considered for many other applications including irrigation of lawns at cemeteries, golf courses and college campuses (Okun, 1997), vehicle washing, fire protection, boiler feed water, concrete production (Santala *et al.*, 1998) and preservation of wetlands (Jefferson *et al.*, 2004; Otterpohl *et al.*, 1999).

Water quality requirements for each application are geospecific, but normally include criteria based on organic, solids and microbiological content of the water (Morel & Diener, 2006). Greywater also contains a major fraction of heavy metals, but a minor proportion of nutrients. The heavy metals in greywater originate from dust (wiped out during house-cleaning activities) and chemicals used in the household (Cd in phosphorus-containing detergents and zinc in anti-fungal shampoo). The nutrients in greywater are mainly inorganic, but K and P are used in detergents and their concentrations will mainly reflect the usage rate of these products (Vinnera, 2002). (Blackwater or yellow-water may include menstrual blood in homes with post-puberty females).

(e) *Anal cleansing materials* are often overlooked in the design of sanitation facilities (McMahon *et al.*, 2011; Zomerplaag & Mooijman, 2005). It may be water or dry materials used for cleaning oneself after defecation and/or urinating. If it is water, the volume used per cleaning typically ranges from 0.5 L to 3 L; while solid materials used include paper, leaves, corncobs, rags or stones (commonly used in rural areas) (Tilley *et al.*, 2014b) and this could include menstrual hygiene products, especially in homes that are predominantly female.

(f) *Sludge (faecal and sewage)* is a mixture of solids and liquids, containing mostly excreta and water, in combination with sand, grit, metals, trash and/or various chemical compounds (Tilley *et al.*, 2014b). Faecal sludge comes from non-sewered sanitation technologies made up of variable consistencies (Aalbers, 1999; Heinss *et al.*, 1998) and can be raw or partially digested, a slurry or semi-solid, and results from the collection and storage/treatment of excreta or blackwater, with or without greywater. Sewage sludge originates from sewer-based wastewater treatment collection and semi-centralized treatment processes (Strande *et al.*, 2014; Tilley *et al.*, 2014b). Most existing treatment technologies for faecal and sewage sludge do not support treatment processes aimed at resource recovery and reuse (Aalbers, 1999; Heinss *et al.*, 1998).

6.4.1.3 Sanitation-derived products (SDPs)

SDPs are high on the development agenda because sustainability requires keeping resources in circulation and making productive use of them at every stage (Andersson *et al.*, 2016b; Otoo & Drechsel, 2017). A prime example of SDPs is the SDN P, an essential plant nutrient also found in human by-products of digestion, which is inefficiently managed as a vital sanitation resource. Only 20% of P mined for production systems ends up in food consumed (Schroder *et al.*, 2010). Much of the remainder is lost to rivers and coastal waters, where it can cause eutrophication (Andersson *et al.*, 2016b). SDPs 'close the loop', turning linear sanitation resource management cyclical and thereby enriching the 'sanitation economy'. The two most important 'loops' are (i) linking sanitation with food production using nutrients and organic matter; and (ii) linking wastewater not only to agriculture, but also ecosystem flows and a variety of other human uses, including industrial purposes (Andersson *et al.*, 2016b; Langergrabera & Muelleggera, 2005; UN, 1992 – see Sections 1.4.2 and 1.4.5).

SDPs also take into account the possible linkages between sanitation and urban economic development because sanitation materials contain natural and recyclable resources with high ecological and economic value, and so processing units should be designed to operate as sustainable value chain businesses. Integrating SDPs' potential into the design of sanitation infrastructure can establish strong links to economic development needs and agenda (Kone, 2007; Pronk & Kone, 2009). They close the loop in sanitation systems to meet livelihood needs, provide food and energy security, generate economic benefits and protect public health and environmental quality as well as potentially attract investments (Jewitt, 2011). These products are generated from SSC-RaMs and are different from the classical sanitation processes focused on transport and treatment of a single waste stream. Enhancing SDPs will mean designing sanitation infrastructure that addresses several parallel processes for the collection and transformation of SSC-RaMs in proper form, with each carried out at different scales and stages and managed by different operators within the IFSVC (Tilley, 2013).

The SSC infrastructural technologies should be designed to fit the intended purpose of the raw materials collected and transformed. The specific characteristics of these raw materials present resource recovery opportunities (see Section 6.4.2) (Andersson *et al.*, 2016b; Langergrabera & Muelleggera, 2005), and as these materials have inherent nutrients, energetic potential and water, they cannot be referred to as 'waste' but connote 'resource' and constitute the 'input' materials for transformation processes that generate 'output' products. For example, as shown in Figure 6.21, urine is an SSC-RaM from a urine-diverting user interface needed for the production of struvite in a urine-based struvite reactor (Tilley, 2013).

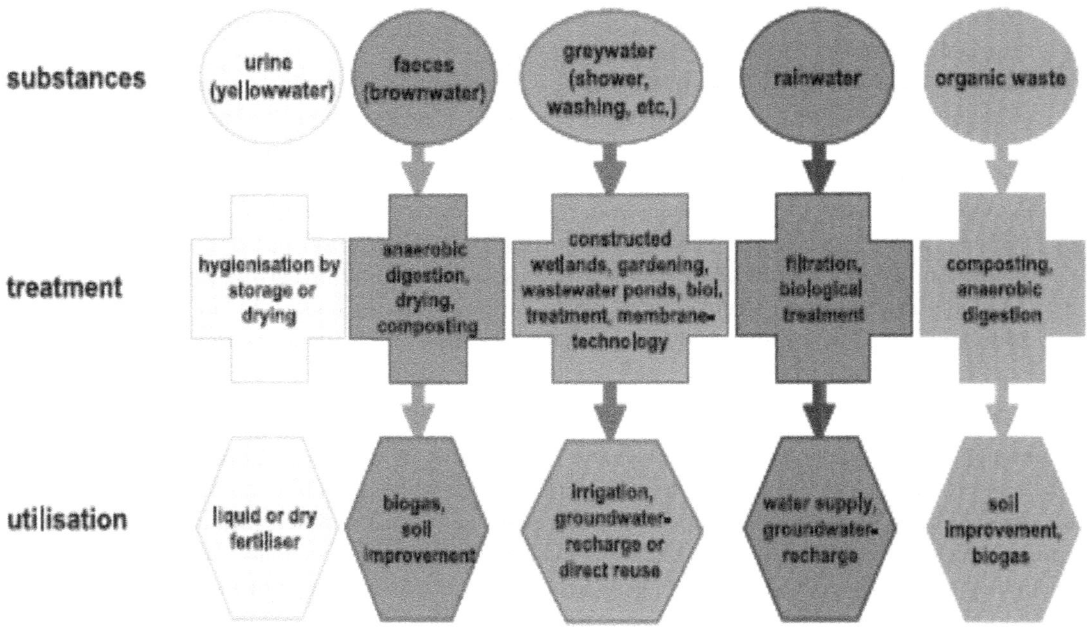

Figure 6.21 Treatment and utilization options for the separated streams of sanitation raw materials (GTZ, 2002).

SDPs generated from SSC-RaMs include but are not limited to:

(I) Used water reuse (Burn, 2014; Futran, 2013; Guest *et al.*, 2009; Kone, 2007; Lahnsteiner & Lempert, 2007; Mayer *et al.*, 2016; Otoo & Drechsel, 2017; Pronk & Kone, 2009; United Nations, 2009);

(II) SDFs, such as direct use of urine, struvite, hydroxyapatite, ammonium sulphate (Andersson *et al.*, 2016a, b; de-Bashan & Bashan, 2004; Guest *et al.*, 2009; Mayer *et al.*, 2016; SEI, 2006; Otoo & Drechsel, 2017);

(III) Bioenergy, such as biogas and hydrogen gas (Funamizu *et al.*, 2001; Guest *et al.*, 2009; Logan *et al.*, 2006; Mayer *et al.*, 2016; Otoo & Drechsel, 2017);

(IV) Organic soil conditioners, such as compost manure (Andersson *et al.*, 2016a, b; Otoo & Drechsel, 2017); and

(V) Other output products, such as protein feed, building materials, trace elements etc. (Andersson *et al.*, 2016a; Mihelcic *et al.*, 2011; Munamati *et al.*, 2016; Verbyla *et al.*, 2013).

***(a)** Sanitation-derived fertilizers (SDFs)* are organic plant fertilizers that originate from the SSC-RaMs through the process of nutrient recovery like phosphorus, nitrogen and potassium from wastewater, faecal and sewage sludge to produce organic fertilizers and soil conditioners. This recovery process decreases their discharge to receiving waters, potential eutrophication and excess micro-biological growth (Guest *et al.*, 2009; Mayer *et al.*, 2016). There have also been recent developments in struvite harvesting (magnesium ammonium phosphate, $MgNH_4PO_4.6H_2O$) from solids treatment processes (de-Bashan & Bashan, 2004) as well as recovery of nutrients from source-separated urine (Andersson *et al.*, 2016a, b; Guest *et al.*, 2009; SEI, 2006). Experience with full-scale wastewater treatment plants in Japan shows the possibility of struvite recovery from wastewater (Mayer *et al.*, 2016). Examples of SDFs are described below:

***(a1)** Direct use of urine* Urine is rich in plant nutrients and contains most of the nutrients present in human food that were not used for new cell growth or energy consumption (Heinonen-Tanski *et al.*, 2007). Urine contributes 81% of the nitrogen and 50% of the phosphorus in purely domestic wastewater (DWA, 2008; Larsen *et al.*, 2009). Recent advances have focused more on recovering nitrogen and phosphorus from urine as value-added products than on removal during treatment processes as nitrogen gas via nitrification-denitrification processes (Larsen *et al.*, 2009; Tarpeh *et al.*, 2018; Wielemaker *et al.*, 2016).

In the presence of urease, the urea is quickly degraded to ammonium and carbon (Equation 6.1) and the hydroxide ions produced normally increases the pH to 9–9.3. Normally urease accumulates within the urine system and therefore this transformation is very swift; usually within hours (Jönsson *et al.*, 2000; Vinnerås *et al.*, 1999) as shown in Equations 6.1 to 6.5.

urea water urease ammonium hydroxide carbonate

$$CO(NH_2)_2 + 3H_2O \rightarrow 2NH_4^+ + OH^- + HCO_3^- \tag{6.1}$$

Autotrophic ammonia oxidation

$$2NH_4^+ + 3O_2 \rightarrow NO_2^- + 2H_2O + 4H^+ \tag{6.2}$$

Autotrophic nitrite oxidation

$$2NO_2^- + O_2 \rightarrow 2NO_3^- \tag{6.3}$$

Heterotrophic denitrification (see Equation 4.14)

$$4NO_3^- + 5CH_2O + 4H^+ \rightarrow 2N_2 + 5CO_2 + 7H_2O \tag{6.4}$$

Chemical phosphorus removal

$$PO_4^{3-} + Fe^{3+} \rightarrow FePO_{4(s)} \tag{6.5}$$

SDPs recovered and reused from urine reduce both external input of virgin resources and output of waste (Wielemaker *et al.*, 2016). Several studies have shown that human urine has a high content of readily available SDNs (N, P & K) and fertilizing effects are reported to be similar to that of nitrogen-rich chemical fertilizers (Heinonen-Tanski *et al.*, 2007; Jönsson *et al.*, 2004; Kirchmann & Pettersson, 1995; Pronk & Kone, 2009; Richert Stintzing *et al.*, 2002 – Figure 6.22). Urine has been used to cultivate several crops and if the amounts of nutrients are correct, urine could totally replace commercial fertilizers, as equal cereal yields are obtained with urine as with 'normal' fertilizer methods (Figure 6.22) (Heinonen-Tanski *et al.*, 2007; Richert Stintzing *et al.*, 2002).

Figure 6.22 New business: urine collection service for reuse, Burkina Faso. Photo: Linus Dagerskog (Andersson *et al.*, 2016a).

Storage of urine is a crucial part of the cycle before its application to agricultural fields. Losses of nitrogen during storage can be minimized by reducing temperature and avoiding aeration above the liquid surface in storage tanks. High temperature, pH and nutrients concentration as well as long storage periods are favourable for hygiene reasons (Höglund *et al.*, 1998). For unrestricted use with respect to pathogens Vinnerås *et al.* (2008) and Karak and Bhattacharyya (2011) recommended that the storage period for human urine be six months at 2°C or higher.

The goals and requirement of urine collection and treatment can be summarized as follows:

(I) Recovery of nutrients,
(II) Removal of micropollutants,
(III) Increase the concentration of nutrients,
(IV) Prevent ammonia volatilization during storage and fertilizer application,
(V) Hygienisation, and
(VI) Economic viability/market-driven implementation (Pronk & Kone, 2009).

(a2) Ammonium sulphate recovery Urine is an ideal wastewater component for N removal and recovery because it is only 1% of wastewater volume but contributes to 80% of wastewater N (Moerman, 2012; Tarpeh *et al.*, 2018). N is currently recovered with P beyond that recovered in biosolids as struvite (Mayer *et al.*, 2016). One main approach for N recovery is ammonia (NH_4), which could easily be removed from the liquid phase of urine through air stripling (Basakcilardan-Kabakci *et al.*, 2007) and subsequently be absorbed in sulphuric acid (H_2SO_4) to form ammonium sulphate, a liquid fertilizer

(Karak & Bhattacharyya, 2011), as shown in Equations 6.6.

$$H_2SO_4 + 2NH_3 \rightarrow (NH_4)_2SO_4 \downarrow \quad (6.6)$$

(a3) *Phosphate recovery* P from municipal wastewater has been widely used due to its high stability (Moerman *et al.*, 2009; Simavi, 2015; Verstraete *et al.*, 2009). Beyond land applications of sewage sludge (SS) and faecal sludge (FS) or other organic soil amendments, P recovery can be achieved through separation of P from the liquid and sludge phase by relying on precipitation/crystallization to recover P minerals that can be used by the fertilizer industry (Mayer *et al.*, 2016). The current technologies of chemical phosphate recovery from municipal wastewater are summarized in Table 6.8.

Table 6.8 Overview of Main Technologies for Chemical Phosphate Recovery from Municipal Wastewater.

Technology	Recovery spot	Main Chemical	P Recovery	Lab-Scale	Pilot-Scale	Plant-Scale
Chemical precipitation	Liquid phase	Mg/Ca materials	Over 80%	√	√	√
Adsorption	Liquid phase	Metal-based adsorbents	Over 90%	√	√	√
Wet-chemical process	Sludge phase	Acid/alkaline solution	Over 60%	√	√	√
Thermochemical process	Sludge phase	Chlorides	N/A	√	√	√

(Reprinted from Ye Y., Ngo H. H., Guo W., Liu Y., Li J., Li Y., Zhang X. and Jia H. (2017). Insight into chemical phosphate recovery from municipal wastewater. *Science of the Total Environment*, 576, 159–171, with permission from Elsevier.)

However, P recovery in conventional sanitation technological infrastructure is limited by the design of existing treatment systems since additional steps are required beyond nutrient removal. The focus has been on nutrient control measures over several decades even though processes exist for nutrient recovery and reuse at all stages in sanitation management and technological solutions (Brands, 2014). Figure 6.23 illustrates how struvite precipitation processes (using a struvite crystallizer reactor (SCR)) operate (Bird, 2015; Hanzen & Sawyer, 2010).

SCRs operate in reverse gravitational flows by pumping reject waters from anaerobic digesters and solids-dewatering machinery into the bottom of a SCR (Bird, 2015; Britton *et al.*, 2005; Cullen *et al.*, 2013; Fattah *et al.*, 2010; Rahman *et al.*, 2014). The upward flow through increasingly large reactor chambers allows struvite crystals to separate by density and size. The smallest particles (prills) remain in the uppermost reactor until enough minerals accumulate, increasing the size and density to drop the prill down into the smaller reactor chambers below. From the top down, each reactor zone has a reduced liquid detention time. The largest diameter zone at the top of the structure has the smallest struvite particles with the longest detention time to allow for crystal growth (Bird, 2015; Cullen *et al.*, 2013; Westerman *et al.*, 2009). The largest prills present in the lowest and smallest reactor chambers are removed, dried and bagged. Reject waters provide adequate amounts of ammonium and phosphate, but are low in magnesium concentrations. Therefore, magnesium additions are necessary for struvite crystal formations (Acelas *et al.*, 2014; Bergmans *et al.*, 2014; Bird, 2015; Chanan *et al.*, 2013; Munch & Barr, 2001; Xavier *et al.*, 2014): see Equations 6.7 and 6.8.

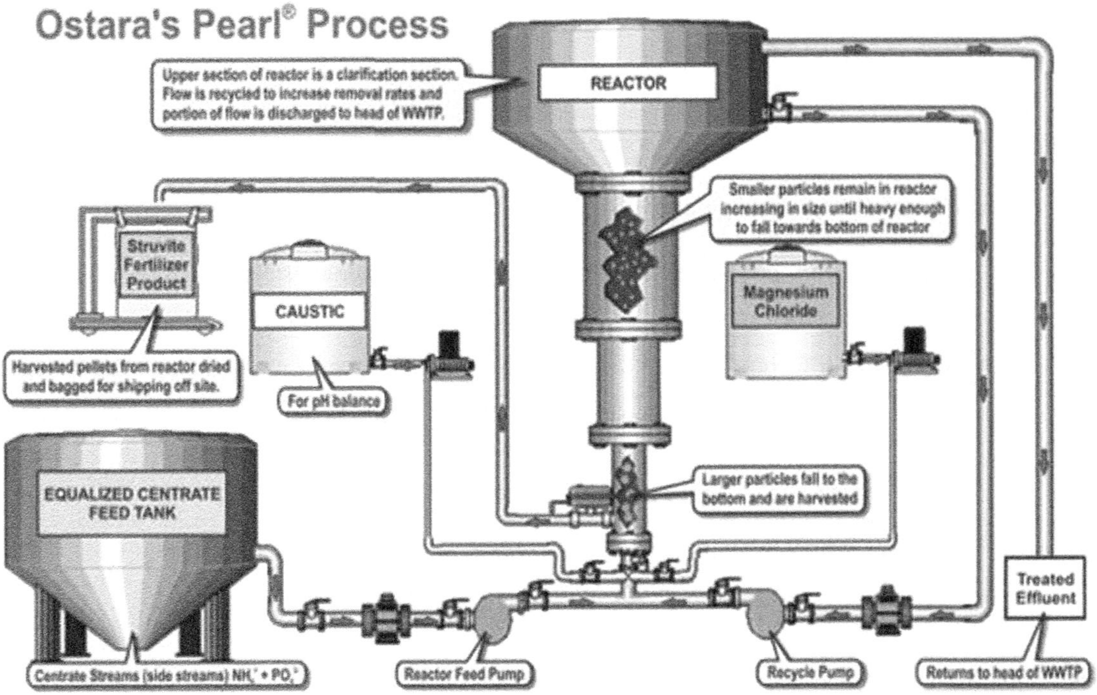

Figure 6.23 Schematic of Ostara's Pearl® struvite crystallizer process (*Source*: Hanzen & Sawyer, 2010; https://www.hazenandsawyer.com/news/new-process-makes-asset-of-phosphorus-and-nitrogen-from-wastewater/).

Cordell *et al.* (2011) reported on two additional commercial-scale methods of recovering P from wastewater and by-products:

(i) The Ostara fluidized bed reactors (Pearl® Process), implemented in Portland, Oregon, USA (and elsewhere), recovered approximately 90% of P from the wastewater stream, and the product is Crystal Green® struvite, magnesium ammonium phosphate ($MgNH_4PO_4{\cdot}6H_2O$) (Wilsenach *et al.*, 2007).

(ii) The partnership between Dutch company N.V. Sliberwerking Nord-Brabant (SNB) and Thermophos International, in which P was recovered from sewage sludge ash. Detailed chemical P recovery from liquid and sludge phases are presented in Table 6.9.

(a3i) *P recovery from liquid phase* In a liquid phase, chemical and adsorption are mainly utilized to recover P (Chen *et al.*, 2009; Jones *et al.*, 2015; Spears *et al.*, 2013). The chemical precipitation requires choosing the appropriate chemical precipitator, which could be added before, after or during the conventional biological treatment of the acquired raw materials. P recovered by this process could be easily dewatered and subsequently, potentially reused as fertilizer (Chen *et al.*, 2009). Calcium and magnesium are commonly employed as precipitators (Desmidt *et al.*, 2015) as shown in Equations 6.7 and 6.8.

Table 6.9 Overview of Applications of Chemical Phosphate Recovery at Plant Scale.

Plant-Scale Process	Recovery Spot	Chemicals	Recovered Product	P Recovery	Type of Reactor
NuReSys®	Liquid phase	$MgCl_2$, NaOH	Struvite	80–85%	Continuous stirred tank reactor
Crystalactor®	Liquid phase	$Ca(OH)_2$, NaOH, H_2SO_4, Sand	HAP	70–80%	Fluidized bed
AirPrex®	Liquid phase	$MgCl_2$, Flocculent	Struvite	80–90%	Continuous stirred tank reactor
PHOSPAQ®	Liquid phase	MgO	Struvite	70–95%	Continuous stirred tank reactor
Seaborne®	Sludge phase	H_2SO_4, Na_2S, NaOH, MgO, flocculent	Struvite	>90%	Continuous stirred tank reactor
AshDec®	Sludge phase	$MgCl_2$, $CaCl_2$	Multi-nutrient fertilizers including phosphorus, nitrogen and potassium	N/A	Furnace

(Reprinted from Ye Y., Ngo H. H., Guo W., Liu Y., Li J., Li Y., Zhang X. and Jia H. (2017). Insight into chemical phosphate recovery from municipal wastewater. *Science of the Total Environment*, 576, 159–171, with permission from Elsevier.)

Calcium reacts to P to form $Ca_5(OH)(PO_4)_3$ (hydroxyapatite = HAP).

$$5Ca^{2+} + 3PO_4^{2-} + OH^- \rightarrow Ca_5(OH)(PO_4)_3 \downarrow \tag{6.7}$$

Magnesium reacts to P to form $MgNH_4PO_4{\cdot}6H_2O$ (struvite = MAP).

$$Mg^{2+} + PO_4^{3-} + NH_4^+ + 6H_2O \rightarrow MgNH_4PO_4 + 6H_2O \downarrow \tag{6.8}$$

P in the form of struvite can be applied directly as soil fertilizer, while HAP is recycled by the fertilizer industry. Iron (Fe) and aluminium (Al) ions are not recommended precipitators for P recovery as they are not suitable for application as fertilizer because P is strongly bounded in the recovery process, thus generating little P release (Egle *et al.*, 2016; Petzet & Cornel, 2012; Ye *et al.*, 2017).

(a3ii) *P recovery from sludge phase* P fixed in the sludge phase as sewage sludge (SS) and sewage sludge ash (SSA) uses wet-chemical and thermochemical treatments as the main recovery technologies (Appels *et al.*, 2010; Kahiluoto *et al.*, 2015; Qin *et al.*, 2015). P bounded in the SS/SSA is released by adding strong acids (e.g. HCl and H_2SO4) or alkalis (e.g. NaOH) to the liquid phase, which is then followed by chemical precipitation and adsorption processes to recover P from the supernatant. This process reduces the content of heavy metals, pathogens and other toxic substances in the supernatant. Usually, extracting P from the SS/SSA by using acids is more effective than alkalis (Egle *et al.*, 2016; Ye *et al.*, 2017).

(i) Acid leaching released over 80% of P from SSA/SS with a low pH (<2.0) and strong acids of about 0.3–0.68 kg/SSA yielding dissolved P of about 66.5–99.4% (Petzet *et al.*, 2012). H_2SO_4 is usually

prioritized for use in the wet-chemical process due to its low cost and ease of accessibility (Cohen, 2009; Donatello & Cheeseman, 2013; Donatello *et al.*, 2010). When P-rich leachate is processed by struvite precipitation for P recovery, sulphuric acid (H_2SO_4) can react with Ca^{2+} to reduce the disturbances of the process. Also, phosphoric acid (~52% H_3PO_4) of industrial grade can be used to dissolve phosphate bound in SSA in a rotary kiln. This scenario could enhance the enrichment of P and create a product that is similar to triple-superphosphate (Weigand *et al.*, 2013) as shown in Equation 6.9.

$$Ca_4Mg_5(PO_4)_6 + 12H_3PO_4 + 2H_2O \rightarrow 4Ca(H_2PO_4)_2 + 5Mg(H_2PO_4)_2 + 12H_2O \quad (6.9)$$

Ion exchange (Donatello *et al.*, 2010) and nano-filtration (Niewersch *et al.*, 2009) are currently applied to remove heavy metals from P-rich leachate to make them less restrictive to P recovery (Ye *et al.*, 2017).

(ii) *Alkali leaching* could directly separate P from heavy metals in SS/SSA because most heavy metals cannot dissolve in alkaline solutions. However, this reaction causes precipitation of some metal ions such as Mg^{2+} and Ca^{2+}. The release efficiency of P increases at high pH because the sludge cells can be disintegrated only at pH 11 and then release P to the liquid phase (Becerra *et al.*, 2010; Li *et al.*, 2013a, b). The wastewater treatment plant in Gifu, Japan, utilizes alkali leaching for P recovery, achieving about 60–70% of total P recovered from SSA (Takaoka *et al.*, 2010). In addition, NaOH is the preferred alkali leachate because in most cases it produces more dissolved P (Torres & Lloréns, 2008; Ye *et al.*, 2017).

(iii) *Thermochemical treatment* uses chloride additives such as NaCl, KCl, $MgCl_2$ and $CaCl_2$ to mix with SS/SSA at high temperatures (800–1000°C) (Adam *et al.*, 2009). The reaction removes most of the heavy metals due to the formation of volatile heavy metal chlorides captured in the flue gas by a further cleaning process (Herzel *et al.*, 2016). For instance, humid reaction atmosphere may easily result in the formation of either HCl or Cl_2 and then volatile Zn can react with them to form $ZnCl_2$, which is easily removed with the emission of flue gas due to its low evaporation temperature as shown in Equations 6.10 and 6.11.

$$2HCl + ZnO \rightarrow H_2O + ZnCl_2 \quad (6.10)$$

$$2Cl_2 + 2ZnO \rightarrow O_2 + 2ZnCl_2 \quad (6.11)$$

Most heavy metals can be removed from the sludge phase via the thermochemical process, but some heavy metals (e.g. As and Ni) are still retained in the sludge phase (Nowak *et al.*, 2013). The heavy metals in the flue gas could be treated by filter while $MgCO_3$ and $NaHCO_3$ are utilized to adsorb chloride and remove SO_2, respectively. In addition, Na_2SO_4 could replace the chlorides as an additive to treat SS/SSA for the formation of $NaCaPO_4$ (Vogel *et al.*, 2016; Ye *et al.*, 2017) (Figure 6.24).

(b) *Used water* Water reclamation and reuse is arguably the most important product recovery process because of the ability of sanitation systems to provide additional water resources in water-stressed areas (Burn, 2014; Guest *et al.*, 2009; Mayer *et al.*, 2016). Water reuse and recycling is the application of technologies for wastewater treatment to meet water quality requirements for the intended use based on the concept of 'fit-for-purpose'; and it emphasizes the efficiencies realized by designing reuse for specific applications (Table 6.4) (USEPA, 2012). The positive benefits of used water include:

(I) Improved agricultural production;
(II) Irrigating landscapes and agricultural fields,

Figure 6.24 Struvite crystal produced from lagoon wastewater (*Source*: Rahman *et al.*, 2014).

(III) Reduced energy consumption associated with production, treatment and distribution of water; and
(IV) Significant environmental benefits, such as recharging groundwater aquifers, providing critical stream flows, reduced nutrient loads to receiving waters (Burn, 2014; USEPA, 2012 – Table 6.10).

Many nations recycle water to some extent, but most currently recycle less than 10% of total wastewater flows (Mayer *et al.*, 2016). Successful examples include:

(I) The Orange County Water District, California, USA and the Upper Occoquan Sewage Authority, Virginia, USA, which have been achieving indirect potable reuse for over 30 years;
(II) Singapore's indirect potable and direct non-potable reuse of reclaimed water (NEWater) that contributes 30% of the country's drinking water; and
(III) Namibia's direct potable reuse found in Windhoek, where water resources are particularly scarce (Andersson *et al.*, 2016a, b; Guest *et al.*, 2009; Lahnsteiner & Lempert, 2007).

Furthermore, Israel is increasingly leveraging substantial fractions of municipal wastewater to augment water resources with used water (Futran, 2013; United Nations, 2009). It should be noted that most recycled wastewater is used for irrigation, which provides all of the nutrients needed by crops and adds value, through cost savings, to farmers (Mayer *et al.*, 2016).

(c) Bioenergy recovery The conversion of organic-rich materials from sanitation system into bioenergy has a long history, especially through anaerobic digestion (AD) (McCarty *et al.*, 2011; Puyol *et al.*, 2017). Several technologies are under current development to convert organic matter to bioenergy such as biohydrogen, biodiesel, bioethanol and microbial cell fuels, etc. (Puyol *et al.*, 2017). The most common and efficient method for energy recovery from sanitation materials is generating biogas derived from the AD process (Batstone & Virdis, 2014; Jewitt, 2011; Puyol *et al.*, 2017). For instance, biogas digesters are widely used for energy production in many parts of the world with China leading in the field since the late nineteenth century (Mae-Wan, 2008; Jewitt, 2011).

Meanwhile in developed countries, biogas methane generated from municipal wastewater treatment plants is widely used for cooking and heating, electricity, vehicle fuel and 'mains' gas when processed

Table 6.10 Categories of Water Reuse Applications.

Category of Reuse		Description
Urban Reuse	Unrestricted	The use of reclaimed water for non-potable applications in municipal settings where public access is not restricted.
	Restricted	The use of reclaimed water for non-potable applications in municipal settings where public access is controlled or restricted by physical or institutional barriers, such as fencing, advisory signage, or temporal access restriction.
Agricultural Reuse	Food Crops	The use of reclaimed water to irrigate food crops that are intended for human consumption.
	Processed Food Crops and Non-food Crops	The use of reclaimed water to irrigate crops that are either processed before human consumption or not consumed by humans.
Impoundments	Unrestricted	The use of reclaimed water in an impoundment in which no limitations are imposed on body-contact water recreation activities.
	Restricted	The use of reclaimed water in an impoundment where body contact is restricted.
Environmental Reuse		The use of reclaimed water to create, enhance, sustain or augment water bodies including wetlands, aquatic habitats or stream flow.
Industrial Reuse		The use of reclaimed water in industrial applications and facilities, power production and extraction of fossil fuels.
Groundwater Recharge – Non-potable Reuse		The use of reclaimed water to recharge aquifers that are not used as a potable water source.
Potable Reuse	Indirect Potable Reuse (IPR)	Augmentation of a drinking water source (surface or groundwater) with reclaimed water followed by an environmental buffer that precedes normal drinking water treatment.
	Direct Potable Reuse (DPR)	The introduction of reclaimed water (with or without retention in an engineered storage buffer) directly into a water treatment plant, either collocated or remote from the advanced wastewater treatment system

(*Source*: USEPA, 2012).

and introduced into natural gas pipelines (Defra, 2010; Dutta *et al.*, 1997; Energy Saving Trust, 2008; EREC, 2009; Guardian, 2008; Jewitt, 2011). Even greater benefits can be achieved where biogas units are directly linked to farm, sewage and municipal waste streams as part of a broader waste management strategy (Defra, 2009a, b, 2010; Gasworld, 2010; Guardian, 2008; Taglia, 2010; Jewitt, 2011; Thames Water, 2009) (Figure 6.25).

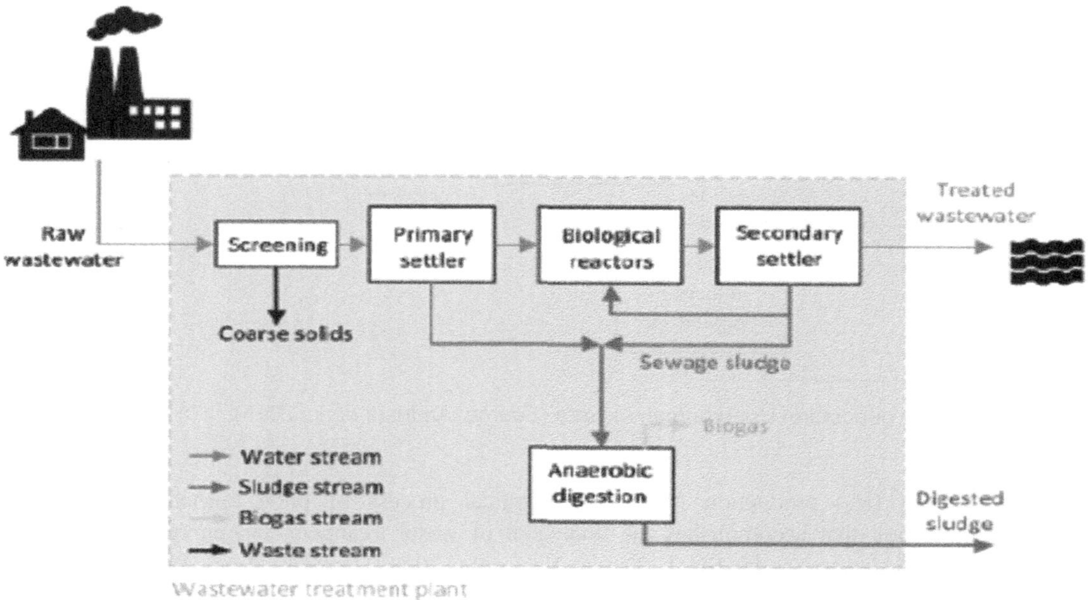

Figure 6.25 Schematic illustration of a typical wastewater treatment plant with AD (*Source*: Bachmann, 2015).

In Stockholm, Sweden, the Bromma wastewater plant produces around 10,000 tonnes of sewage sludge annually with anaerobic treatment conditions to produce biogas for use in vehicles. The remaining biogas sludge is used in agriculture as a soil conditioner (Jewitt, 2011; Welsh School of Architecture, 2008). Biogas produced at Heidelberg's municipal sewage treatment works is used to generate electricity and German engineers have developed a new processing system that generates biogas within five days. A third of the electricity generated from this biogas is used to power the plant while the rest is sold to the national grid (EDIE, 2002; Jewitt, 2011).

But in developing countries, biogas projects are small-scale rural and semi-urban digesters mainly used as fuel for cooking and lighting (the most convenient use of biogas) (Andriani *et al.*, 2015). A biogas burner that burns with blue flames without odour is available in a wide-ranging capacity from 14 m^3 to 170 m^3 biogas consumption per hour. The biogas lamp consumes 3.40–5.10 m^3 per hour, having an illumination capacity equivalent to 40 W electric bulbs at 220 volt (Andriani *et al.*, 2015; Pathak, 2014). For example, 200 biogas plants using human excreta and animal dung as feedstock have been implemented in Ghana (Diener *et al.*, 2014); there is also the Sulabh biogas toilet system in India and biogas in Kibera, Kenya (Katukiza *et al.*, 2012) and Indonesia (Andriani *et al.*, 2015), to mention but a few (Figure 6.26). Other forms of bioenergy recovery include:

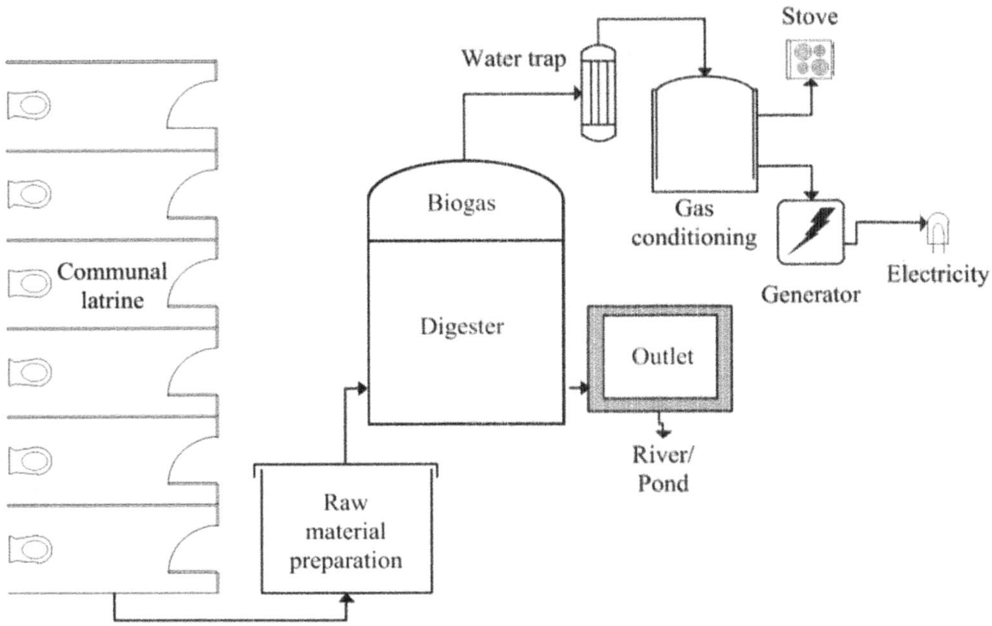

Figure 6.26 Biogas generation using human excreta (*Source*: Andriani *et al.*, 2015).

(I) Hydrogen (H_2) production from the biological process of photo-fermentation and dark fermentation that accomplishes the dual goal of waste treatment and energy recovery as it utilizes wastewater as feedstock (Das & Veziroglu, 2008; Dincer & Acar, 2015; Han & Shin, 2004; Puyol *et al.*, 2017);

(II) Biofuel (biodiesel, bioethanol and microbial cell fuels) production from the conversion of organic-rich wastewater stream by microalgae (Funamizu *et al.*, 2001; Logan *et al.*, 2006; McCarty *et al.* 2011; Polprasert & Koottatep, 2017; Rittmann, 2013); and

(III) Biochar or hydrochar production from the conversion of biomass into coal through the process of pyrolysis called hydrothermal carbonization (HTC) used in cooking briquette production and solid fuel for industrial application (Fakkaew *et al.*, 2015a, b, 2018; Koottatep *et al.*, 2016, 2017; Titirici *et al.*, 2007).

(d) Soil conditioners These are the end use of recovered and recycled organic matter from FS and SS as organic manure-cum-soil conditioner – an effect not shared by chemical fertilizers applied on agricultural land (Nikiema *et al.*, 2014). FS comes from on-site sanitation technologies (non-sewer) (Strande *et al.*, 2014), while SS originates from sewered sanitation technologies that yield 'biosolids' with further treatment. They are nutrient-rich organic materials that can be recycled and applied as fertilizer to improve and maintain productive soils and stimulate plant growth (USEPA, 1994; WEF, 2010). Sludge is a renewable source of plant nutrients, such as nitrogen, phosphorus, potassium and organic matter as well as some trace elements such as zinc, important for plant growth (Cofie *et al.*, 2005; Nartey *et al.*, 2017). This organic matter has the potential to be anything, from use in agriculture to bagged compost products sold for use at household level (Diener *et al.*, 2014; Strande *et al.*, 2014). There are processes presented through different treatment technologies and each technology results in end products based on the intended recovery product of choice. The main objectives for the treatment of sludge are to eliminate potential health hazards, stabilize organic matter and recover materials for reintegration into the material cycle (Otterpohl & Buzie, 2013; USEPA, 2003). For example, sludge from sanitation systems could be used on agricultural land as soil conditioners in the form of dried or partially dried sludge, compost, pellets, leachates, biosolids etc.

Also, they could be co-composted together with other waste streams such as municipal solid waste and the application of vermicomposting (Andersson *et al.*, 2016a, b; Peccia & Westerhoff, 2015; Polprasert, 1996; Polprasert & Koottatep, 2017; Strande *et al.*, 2014). Koottatep *et al.* (2005) showed that sunflowers irrigated with leachate from planted drying beds improved, with increased seed and oil yields, in a six-year study in Thailand. Therefore, sludge used as soil conditioners increases soil organic matter (SOM), supports soil functions such as retaining N and other nutrients, retains water to protect roots from diseases and parasites and makes retained nutrients available to plants (Bot & Benites, 2005). The organic matter itself also contains nutrients that will release gradually as it is broken down through natural processes (Andersson *et al.*, 2016a).

6.5 WHERE DO WE GO FROM HERE?

The use of SDPs is gaining attention due to the many opportunities to close the bio-geo-chemical cycle. The importance of sanitation as a resource is quite clear, the challenge is where to focus on. There is a wide variety of possible alternatives as the array of technical options grow (Gates-Foundation, 2010; van der Hoek *et al.*, 2016). Some recovery processes and resources are well established, such as used water, bioenergy and SDFs (Daigger, 2008, 2009; Doyle & Parsons, 2002; Garcia-Belinchón *et al.*, 2013; Lee *et al.*, 2013; McCarty *et al.*, 2011; Puchongkawarin *et al.*, 2015; Sutton *et al.*, 2011), while the emerging ones are: recovery of cellulose fibres (Ruiken *et al.*, 2013), biopolymers (Tamis *et al.*, 2014), bioplastics (Kleerebezem & Van Loosdrecht, 2007) and proteins (Matassa *et al.*, 2015). Despite these advances, there are still limited studies on integrated energy-nutrient-water recovery that explore the combined benefits and potential trade-offs of these methods under different scales (McCarty *et al.*, 2011; Mo & Zhang, 2013; Slater, 2009; Verstraete *et al.*, 2009). Other challenges of ReS include:

(I) Uncertainties and limitations in energy recovery, also impacted by capital costs of infrastructure and lack of reliable and specific requirements for climate and local conditions as well as lack of integration and trade-offs in on-site energy generation, especially in large municipal infrastructure (Mo & Zhang, 2013);

(II) Nutrient-recycling technologies have not been widely implemented beyond biosolids and FS land application. The challenges of nutrient recycling are more complicated than on-site bioenergy generation technologies and include land and financial resource constraints, integration into existing infrastructure, safety and technology imperfections. Mo and Zhang (2013) observed that the application of urine separation needs cooperation from local communities and governments; and changing the existing infrastructure incurs high construction costs that could hinder application. Furthermore, more studies are needed to address issues of PSEP (4.2); and

(III) Water recovery and reuse are not adequately linked to treatment technology and certainly not at the desired quality standards associated with certain applications because most of the time researchers, industries and other users strive for better treated-water quality, but neglect the fact that the over-treated wastewater brings no additional benefits, but does bring higher energy consumption and costs. Also, health and safety concerns are still major barriers to broader application and reclamation (Mo & Zhang, 2013).

Moving forward will require studies that identify challenges and opportunities as well as develop sani-shed and scale research and practices on integrated resource recovery; they have to be encouraged through funding, policy instruments and regulations (Mo & Zhang, 2013; Rittmann & McCarty, 2001; Verstraete *et al.*, 2009). Studies have also shown that the major challenges of ReS exceed the absence of technology; they include a lack of social-technological planning and design methodologies to identify and deploy the most sustainable solutions in a given geographic and cultural context (Guest *et al.*, 2009). According to Li *et al.* (2015), uncertainties about which techniques are most useful and how to combine them stand in the way of creating SDPs (van der Hoek *et al.*, 2016) and/or a nutrients-energy-water-environment-land nexus (NEWEL) (Papa *et al.*, 2017). Systematic planning approaches are needed to guide practitioners through the process of designing ReS. Beyond being technically feasible, a recovery and reuse scheme should be socially, economically and institutionally robust for sustainability and should also incentivize appropriate operations and maintenance of the ReGenSan systems (Murray & Buckley, 2009).

Murray and Buckley (2009)'s design for service framework (DSF) approach used for planning and the van der Hoek *et al.* (2016) options could both serve as viable and coherent strategies to scale up ReS in the ReGenSan framework. These could include:

(I) Generate list of all potential SDPs with focus on material flow in the sanitation chain. Material Flow Analysis (MFA) tool can be deployed to quantify the flow of materials through a defined system (Chevre *et al.*, 2013; Cooper & Carliell-Marquet, 2013; Yuan *et al.*, 2011);

(II) Assess the environmental and cost characteristics of technology options available for RRR. Identify and characterize measures (measure here means plans or courses of action that change resource flow and/or recovery) (Murray & Buckley, 2009; van der Hoek *et al.*, 2016);

(III) Assess the demand for SDPs in communities and local production/use through market analysis and optimization modeling (Murray & Buckley, 2009; van der Hoek *et al.*, 2016);

(IV) Assess performance of 'business as usual' provisions for SDPs. This entails gap analysis between existing performance using optimization modelling, ecosystems impact assessment, institutional stability analysis, user satisfaction and perception survey (Murray & Buckley, 2009; van der Hoek *et al.*, 2016);

(V) Design appropriate sanitation technology/measures for the provision of SDPs with lowest performance. Some tools to use include lifecycle assessment (LCA), Material Flow Analysis (MFA), mathematical modelling and economic cost-benefit analysis; determine realistic and necessary institutional and user change. Some critical questions for consideration are:
   (a) How many of these technologies/measures influence MFA?
   (b) How much of these resources are recovered by the technology/measures? (Murray & Buckley, 2009; van der Hoek *et al.*, 2016); and

(VI) Develop strategic measures necessary for possible recovery and reuse and what changes and commitments are required for recovery and when they can be implemented (Murray & Buckley, 2009; van der Hoek *et al.*, 2016). Since strategies imply actions, plans, measures and choices towards a vision (Rampersad, 2002), it is vital to know which measure and/or SDP is preferred in order to determine the most appropriate strategy.

Ultimately, ReGenSan proposes that the ReS in sanitation be given as much attention as TeS and built within the boundaries of the specific SES in which the system is domiciled. The sanitation economy is strongly dependent on the ReS and so its growth and sustainability is equally dependent on a practical, comprehensive, holistic and effective IFSVC. The SVC cannot be restricted to the SSC as there are other businesses within the sanitation sector that enable the SSC to function. All parties that contribute to and benefit from the entire sanitation system of any sani-shed within a given SES should be included in the SVC to capture the real picture of the sanitation economy. As it is now, the sanitation economy and a vibrant/robust IFSVC need to be thoroughly investigated in order to increase investment, sustainable financing and upscaling.

ReS within SES could also benefit from integrated resource recovery transformation whereby different facilities that capture separate resources are contained in one location to collect across sani-sheds. SDPs could then be made available directly or indirectly to the market/users in whatever form they are needed from this same location. This could reduce overhead burdens through shared costs, human resource and markets. In addition, agriculture and aquaculture farms could have integrated facilities in-house with cross-supply of SDPs relevant to either cultivators (crops or aquatic) or other purposes. These are areas open to research and innovations in technological, ecological, cultural, social, economic and other fields. Researchers and innovators need to include product users/customers, designers/developers, manufacturers, service providers, regulators/policymakers, investors/financiers, suppliers/distributors, microenterprises, sanitation workers/professionals, professional/occupational associations as well as local government officials at the planning and conceptualization stage in order to design and develop appropriately and get a buy-in even before going to the drawing board. This partnership and collaboration could also fund and keep such projects within focus.

6.6 EXERCISES

(I) ReS considers sanitation as a production factory. Briefly explain your understanding and position on this statement.

(II) Briefly trail the historical use of human digestive by-products. Use a timeline to illustrate.

(III) ‘Waste equals food’ and ‘in nature nothing is useless’. Briefly explain with practical examples related to sanitation and illustrations.

(IV) What is ReS and how can it be effective? And what are the benefits from an effective ReS?

(V) From Figure 6.1, explain what you think is happening in the diagram and how it fits into the IFSVC and the ReS.

(VI) Describe with the aid of an illustration the interactions between sanitation and agriculture, aquaculture, horticulture and urban regeneration.

(VII) From your perspective, describe the sanitation value chain (SVC) and its advantages and disadvantages to the communities of your country.

(VIII) How does the BGC (i.e. bio-geo-chemical cycle) contribute to resource recovery and transformation of sanitation raw materials?

(IX) Why should RRR be introduced from the planning and conceptualization stage?

(X) What is the role of standards and governance instruments (legislation, policy and regulations) in effective and practical sanitation ReS?

(XI) How does the current SVC differ from the ReGenSan IFSVC?

(XII) Briefly and critically describe some of the enterprises and activities within the IFSVC (generally) and outline the systemic interactions (use illustrations where possible).

(XIII) Briefly and critically describe the IFSVC within a specific SES or san-shed and outline the systemic interactions (whether they exist or not).

(XIV) Identify the value-added activities and resources within each of the IFSVC.

(XV) What are the challenges for upscaling and commercialization of SDPs?

(XVI) What are some critical challenges of PSEP within a given ReS?

(XVII) Identify opportunities for micro-enterprise bundling in sanitation management and how it could attract large investments.

(XVIII) What financing mechanism could be adopted for non-sewered sanitation in developing countries?

(XIX) What are some standards that should concern RRR in both sewered and non-sewered sanitation products and service provision?

(XX) What are some options for RRR from existing non-sewered sanitation systems?

(XXI) Briefly and critically discuss some innovations of recovery and reuse of SDPs from existing sewered and non-sewered sanitation infrastructure.

(XXII) What are some necessary considerations for RRR in existing sanitation facilities.

(XXIII) Briefly and critically describe some alternative designs for RRR in sanitation and suggest some possible innovations.

(XXIV) Briefly and critically describe decentralized circular systems and up-concentration for zero wastewater; and show how they could be integrated.

(XXV) What are nature-based design options in sanitation management and how significant are their overall contributions?

(XXVI) Identify some constructed wetlands options for sanitation and their possible functions and suggest innovations.

(XXVII) How can demand-based-designs reduce abandoned, dysfunctional and unimproved sanitation facilities/devices and improve innovative provisions' ability to meet customers/users' needs and expectations as well as service the market appropriately?

(XXVIII) Briefly consider what kind of facilities and innovations customers/users would be willing to accept and pay for in your own country.

(XXIX) Critically discuss the customer segment at the final market for SDPs.

(XXX) How does an effective and practical ReS contribute to agriculture, aquaculture, horticulture and livelihoods?

(XXXI) Critically and briefly describe the ReS dimension and its components.

(XXXII) What are some options for designing to recover and reuse SDPs and how?

(XXXIII) What are the safety and health hazards inherent in the production and transformations process of SDPs and their reuse? How can they be averted?

(XXXIV) Identify the four major pathogens that transmit sanitation-related infection in your community and how they are categorized.

(XXXV) With the aid of an illustration, briefly and critically describe the route of transmission from SDPs productions, transformation and utilization.

(XXXVI) Briefly and critically discuss the storage time and temperature required to treat different sanitation matters.

(XXXVII) Critically describe with an illustration the differences between sanitation-related transmitted pathogens in developed and developing countries (use examples if possible).

(XXXVIII) What are sanitation-raw materials? How are they generated? Critically and briefly discuss their reactions, interactions and transformation.

(XXXIX) Identify some SDPs, their productions and transformation, processes, possible reuse, challenges, (barriers, health and safety, demand, market etc.) technology using examples.

(XL) Identify some ways that SDPs could contribute to rural development and livelihood support.

(XLI) What are the significance, benefits and challenges of an integrated recovery facility for a nutrient-energy-water-environment-land (NEWEL) combination?

6.7 RECOMMENDATIONS FOR FUTURE RESEARCH

(I) Critically review and analyze contemporary and traditional use of sanitation matter for agriculture, aquaculture and horticulture to identify trends, failures, advantages/disadvantages and the possible ways forward.

(II) Explore the current assumptions of the SVC and how it fits into the whole sanitation system.

(III) Using the ReGenSan ReS, develop a scenario in a particular sani-shed for a sanitation resource management programme.

(IV) Explore the role of BGC (i.e. bio-geo-chemical cycle) in RRR of sanitation matter from storage to transformation.

(V) How could RRR concept be introduced at preliminary conceptualization (i.e. design and development) stage of sanitation devices and infrastructure? What are some risks and benefits?

(VI) How could standardization and governance instruments (such as policies, legislation, and regulations) be used to improve ReS?

(VII) Explore the ReS of a particular sani-shed within a specific SES from an inter (trans) disciplinary perspective to determine the possible most appropriate ReS programme (including technology selection/uptake) for the given area.

(VIII) Explore users'/customers' needs, desires, expectations and preferences as regards sanitation provisions and how this could affect design, development and marketability of sanitation devices and facilities.

(IX) Explore the existing SVC within a specific SES/sani-shed using the ReGenSan IFSVC to determine the actors, activities, enterprises, investors/financiers, governance structure/provisions, facilities and infrastructure of the IFSVC and how these play out in the marketplace, and what lessons could there be for researchers, innovators, investors, manufacturers, government and interventionists.

(X) Explore the value chain within each stage of the IFSVC to determine the value they add to the whole system.

(XI) How can we address the principle of frugal designs to known RRR processes for SDPs to address the need of those at the Base of the Pyramid (BoP)?

(XII) How can upscaling and commercialization of SDPs be achieved and what are the challenges involved?

(XIII) How can the PSEP challenges within the ReS be properly handled to overcome cultural and personal preferences, acceptability, willingness to buy and use as well as shared facility/resources?

(XIV) Explore place-based and context-specific options for RRR from existing non-sewered sanitation systems and final markets.

(XV) How can RRR be embedded into innovative and new designs for non-sewered sanitation, particularly in dense urban populations of developing countries?

(XVI) Explore opportunities to establish and scale up local businesses and industries that could manufacture, distribute and install recovered products and sanitation facilities particularly for non-sewered.

(XVII) What kind of innovative sanitation business solutions could serve the needs of ReS in households, communities, institutions, large gatherings, etc.? How can such businesses be stimulated within SES?

(XVIII) How can shared-cost and shared-facilities assist micro-enterprises in sanitation products/service provisions through merging or cooperatives?

(XIX) Examine any of the available alternative designs for SDP recovery, transformation and reuse (or more than one) to specifically address sanitation in place, scale and particular context and if it is possible to combine options for improved results.

(XX) Critically examine the challenges, barriers and gains of current RRR options from existing sanitation infrastructure (sewered and non-sewered) and how their process, management, technologies and final products match up with required standards, regulations (policies and legislation-global, national and local), health and safety considerations.

(XXI) How could large investors and manufacturing companies be attracted to sanitation device/facility designs that include recovery, transformation and reuse options particularly in non-sewered?

(XXII) How could investment be drawn down for conglomeration of micro-enterprises that deliver sanitation services/products to customers, especially of non-sewered systems?

(XXIII) How can hybrid financing mechanism be achieved in non-sewered sanitation of developing countries?

(XXIV) Explore other possible uses for SDPs that could meet consumer needs and attract investment and industry actors.

(XXV) How can standardization improve and strengthen the ReS and ensure/enhance health and safety of SDP users and collection/recovery/transformation employees for sewered, but particularly non-sewered, sanitation?

(XXVI) Explore the contributions of ReS to food security and urban regeneration and identify key areas of interactions/markets and possible financial materials gains.

(XXVII) How can some current innovations of recovery and reuse of SDPs from existing sewered and non-sewered sanitation infrastructure be improved and made cost-effective and upscale?

(XXVIII) Examine the significant contributions (if any) of constructed wetlands to sanitation management and consider some innovative (integrated) options.

(XXIX) Examine the potential of demand-based design (DBD) to turn the trend of unimproved, dysfunctional and abandoned sanitation facilities in developing countries and its possible impact on innovations satisfaction and market expectations.

(XXX) Explore customers'/users' demands, expectation and willingness to accept, utilize and pay for existing and new sanitation provisions in a place-based and context-specific mixed method study.

(XXXI) What kind of customers/users will be interested in SDPs and how could DBD be incorporated to meet their needs and expectations?

(XXXII) How can inherent hazards in the production, transformation and reuse of SDPs be minimized and controlled for all the populations at risk?

(XXXIII) Explore precautions and barriers to ensure occupational health and safety of employees in the production, transformation and reuse of SDPs.

(XXXIV) Explore considerations for preventing pathogens from RRR processes entering the environment, particularly surface/groundwater, soil and air.

(XXXV) Explore technology and operational options that prevent pathogenic infections during collection, production, transformation and reduction (i.e. limit human contact) – considering the various populations at risk.

(XXXVI) What are possible issues that could arise from pharmaceutical residues and female bodily fluid inside sanitation raw materials? How can they be managed?

(XXXVII) Explore the linkages between ReGenSan and food security, ecosystem flows, industrial and human purposes, with recommendations.

(XXXVIII) Explore the connections between ReGenSan, SDPs and urban regeneration, with recommendations for urban development.

(XXXIX) Explore the possible contributions of ReGenSan and SDPs to rural development and livelihood support, with recommendations.

(XL) Explore the use and market of sanitation-derived cellulose fibres, biopolymers, bioplastics and proteins towards attracting the buy-in of investors and manufacturers.

(XLI) Investigate the recovery of cellulose fibres, biopolymers, bioplastics and proteins from sanitation matter (BGC, engineering, productions, etc) for relevance, practicality, efficacy, cost-effectiveness and potential significance to the sanitation economy.

(XLII) What kind of technologies and systemic processes could deliver an integrated recovery facility for a nutrient-energy-water-environment-land (NEWEL) combination? Is this possible?

(XLIII) How can agriculture and aquaculture firms combine to recover and reuse sanitation resources to benefit both systems and urban/rural regeneration?

(XLIV) Explore constructed wetlands options for sanitation management in coastal areas (context-specific) and, if possible, specify a place. Identify possible SSC-RaMs, SDPs, challenges, contributions to the community and market.

(XLV) Develop a possible ReS for arid or water-scarce regions (context-specific) using a case area and investigate possible recovery and reuse options as well as impact on the region.

(XLVI) Explore the possible combined benefits, potential trade-offs, processes and shared points from integrated energy-nutrient-water recovery facilities (existing and innovative options).

(XLVII) Explore the uncertainties and limitations in energy SDPs considering frugal options, finance mechanisms, demand/use, tradeoffs, standards and regulations, benefits, technological, health and safety risks and other challenges, particularly in large infrastructure, dense communities and accommodations and rural communities.

(XLVIII) Examine nutrient-recycling technologies for efficiency, practicality, financial viability, available capital resources, health and safety, environmental protection, acceptability, government support and integration into existing sanitation infrastructure.

REFERENCES

Aalbers H. (1999). Resource Recovery from Faecal Sludge using Constructed Wetlands: A survey of the literature. UWEP Working Document. https://www.pseau.org/outils/ouvrages/waste_resource_recovery_from_faecal_sludge_using_constructed_wetlands_a_survey_of_the_literature_1999.pdf (accessed 7 June 2018).

Acelas N. Y., Flórez E. and López D. (2014). Phosphorus Recovery through Struvite Precipitation from Wastewater: Effect of the Competitive Ions. *Desalination and Water Treatment*, 1–12, Taylor & Francis, doi: 10.1080/19443994.2014.902337

Adam C., Peplinski B., Michaelis M., Kley G. and Simon F.-G. (2009). Thermochemical treatment of sewage sludge ashes for phosphorus recovery. *Waste Management*, **29**(3), 1122–1128.

Andersson K., Rosemarin A., Lamizana B., Kvarnström E., McConville J., Seidu R., Dickin S. and Trimmer C. (2016a). Sanitation, Wastewater Management and Sustainability: From Waste Disposal to Resource Recovery. UNEP/GPA/SEI, Stockholm, Sweden.

Andersson K., Dickin S. and Rosemarin A. (2016b). Towards 'sustainable' sanitation: challenges and opportunities in urban areas. *Sustainability*, **8**, 1289.

Andriani D., Wresta A., Saepudin A. and Prawara B. (2015). A review of recycling of human excreta to energy through biogas generation: Indonesia case. *Energy Procedia*, **68**, 219–225, doi: 10.1016/j.egypro.2015.03.250

Appels L., Baeyens J., Degreve J. and Dewil R. (2008). Principles and potential of the anaerobic digestion of waste-activated sludge. *Progress in Energy and Combustion Science*, **34**(6), 755–781.

Appels L., Degrève J., Van der Bruggen B., Van Impe J. and Dewil R. (2010). Influence of low temperature thermal pre-treatment on sludge solubilisation, heavy-metal release and anaerobic digestion. *Bioresource Technology*, **101**(15), 5743–5748.

Asano T. and Levine A. D. (1996). Wastewater reclamation, recycling and reuse: past, present, and future. *Water Science and Technology*, **33**(10–11), 1–14.

Asano T., Leong L. Y. C., Rigby M. G. and Sakaji R. H. (1992). Evaluation of the California wastewater reclamation criteria using enteric virus monitoring data. *Water Science and Technology*, **26**(7–8), 1513–1524.

AWF (Alliance for Water Efficiency) (2017). Frugal Flushing. https://www.home-water-works.org/indoor-use/toilets (accessed 7 June 2018).

Ayaz S. Ç. (2008). Post-treatment and reuse of tertiary treated wastewater by constructed wetlands. *Desalination*, **226**(1e3), 249–255.

Bachmann N. (2015). Sustainable Biogas Production in Municipal Wastewater Treatment Plants. Technical Brochure (eBook electronic edition), International Energy Agency (IEA). http://task37.ieabioenergy.com/files/daten-redaktion/download/Technical%20Brochures/Wastewater_biogas_grey_web-1.pdf (accessed 20 June 2018).

Bahri A. (1999). Agricultural reuse of wastewater and global water management. *Water Science and Technology*, **40**(4), 339–346.

Bahri A. (2009). Managing the Other Side of the Water Cycle: Making Wastewater an Asset. TEC Background Papers no. 13, Global Water Partnership, Stockholm, Sweden.

Balkema A. J., Preisig H. A., Otterpohl R. and Lambert F. J. D. (2002). Indicators for the sustainability assessment of wastewater treatment systems. *Urban Water*, **4**, 153–161.

Basakcilardan-Kabakci S., Ipekoglu A. N. and Talini I. (2007). Recovery of ammonia from human urine by stripping and absorption. *Environmental Engineering Science*, **24**(5), 615–624.

Batstone D. J. and Virdis B. (2014). The role of anaerobic digestion in the emerging energy economy. *Current Opinion in Biotechnology*, **27**, 142–149, doi: 10.1016/j.copbio.2014.01.013

Beaglehole R., Bonita R. and Kjellström T. (1993). Basic Epidemiology. World Health Organization, Geneva, Switzerland.

Becerra F. Y. G., Acosta E. J. and Allen D. G. (2010). Alkaline extraction of wastewater activated sludge biosolids. *Bioresource Technology*, **101**(18), 6972–6980.

Beckert J. (2009). The social order of markets. *Theory and Society*, **38**(3), 245–269.

Benetto E., Nguyen D., Lohmann T., Schmitt B. and Schosseler P. (2009). Life cycle assessment of ecological sanitation system for small-scale wastewater treatment. *Science of the Total Environment*, 1506–1516, doi: 10.1016/j.scitotenv.2008.11.016

Bergmans B. J. C., Veltman A. M., van Loosdrecht M. C. M., van Lier J. B. and Rietveld L. C. (2014). 'Struvite formation for enhanced dewaterability of digested wastewater sludge'. *Environmental Technology*, **35**(5), 549–555, doi: 10.1080/09593330.2013.837081

Berndtsson J. C. (2006). Experiences from the implementation of a urine separation system: goals, planning, reality. *Building and Environment*, **41**, 427–437.

Bird A. R. (2015). Evaluation of the feasibility of struvite precipitation from domestic wastewater as an alternative phosphorus fertilizer resource. *Master's Projects and Capstones*, **141**, https://repository.usfca.edu/capstone/141

Blackett I., Hawkins P. and Heymans C. (2014). The Missing Link in Sanitation Service Delivery: A Review of Fecal Sludge Management in 12 Cities. Water and Sanitation Program, International Bank for Reconstruction and Development/The World Bank, Washington D.C., USA.

Bolwig S., Ponte S., du Toit A., Riisgaard L. and Halberg N. (2010). Integrating poverty and environmental concerns into value-chain analysis: a conceptual framework. *Development Policy Review*, **28**, 173–194.

BoP Innovation Centre, Aqua for All and Sanitation Window (2014). Ready for Funding: Innovative sanitation businesses. https://aquaforall.org/wp-content/uploads/2014/12/Ready-for-Funding_Innovative-Sanitation-Businesses.pdf (accessed 5 June 2018).

Borsuk M. E., Maurer M., Lienert J. and Larsen T. A. (2008). Charting a path for innovative toilet technology using multicriteria decision analysis. *Advances in Environmental Science and Technology*, **42**(6), 1855–1862, doi: 10.1021/es702184p

Bot A. and Benites J. (2005). The importance of soil organic matter: key to drought-resistant soil and sustained food production. *FAO Soils Bulletin*, **80**. UN Food and Agriculture Organization, Rome, Italy.

Bouhoum K. and Amahmid O. (2000). Health effect of wastewater reuse in agriculture. In: Water, Sanitation and Health. Proceedings of the International Conference, Bad Elster, Germany, 24–28 November 1998, I. Chorus, G. Ringelband, G. Schlag and O. Schmoll (eds), IWA Publishing, London, UK, pp. 241–247.

Bourioug M., Gimbert F., Alaoui-Sehmer L., Benbrahim M., Aleya L. and Alaoui-Sossé B. (2015). Sewage sludge application in a plantation: effects on trace metal transfer in soil – plant – snail continuum. *Science of the Total Environment*, **502**, 309–314.

Brands E. (2014). Prospects and challenges for sustainable sanitation in developed nations: a critical review. *Environmental Reviews*, **22**, 1–18, doi: 10.1139/er-2013-0082

Braungart M., McDonough W. and Bollinger A. (2007). Cradle-to-cradle design: creating healthy emissions e a strategy for eco-effective product and system design. *Journal of Cleaner Production*, **15**(13–14), 1337–1348.

Britton A., Koch F. A., Adnan A., Oldham W. K. and Mavinic D. S. (2005). Pilot-scale struvite recovery from anaerobic digester supernatant at an enhanced biological phosphorus removal wastewater treatment plant. *Journal of Environmental Engineering and Science*, **4**(4), 265–277, doi: 10.1139/s04-059

Brunner P. H. and Rechberger H. (2017). Handbook of Material Flow Analysis: For Environmental, Resource, and Waste Engineers. CRC Press, Taylor & Francis, Boca Raton, USA, p. 435.

Burn S. (2014). Resource Recovery from Wastewater: A Research Agenda for Australia. IWA, the Hague, Netherlands.

Cairncross S. and Feachem R. (1993). Environmental Health Engineering in the Tropics – An Introductory Text, 2nd edn. John Wiley and Sons Ltd, Chichester, UK.

Cardoen D. (2011). Up-Concentration Techniques for Zero-Waste Water Treatment. Masters of Environmental Sanitation Thesis Universiteit, Gent, Belgium. https://www.researchgate.net/profile/Dennis_Cardoen/publication/287748090_Up-concentration_techniques_for_zero-waste_water_treatment/links/567917e808ae0ad265c99b61/Up-concentration-techniques-for-zero-waste-water-treatment.pdf

Carlander A., Aronsson P., Allestam G., Stenström T. A. and Perttu K. (2000). Transport and retention of bacteriophages in two types of willow-cropped lysimeters. *Journal of Environmental Science and Health*, **A35** (8), 1477–1492.

Chanan A. P., Vigneswaran S., Kandasamy J. and Johir M. A. (2013). Need for a fresh look at phosphorus management in wastewater treatment: trash to treasure. *Sustainable Environment Research*, **23**(1), 23–31.

Chapin J. and Pedi D. (2013). Getting the Product and Service Right: How do we design affordable, desirable latrines that businesses can profitably produce and sell? UNICEF Sanitation Marketing Learning Series: Guidance Note 5. NICEF, New York, USA.

Chen X., Kong H., Deyi W. U. and Wang X. (2009). Phosphate removal and recovery through crystallization of hydroxyapatite using xonotlite as seed crystal. *Journal of Environmental Sciences*, **21**(5), 575–580.

Chevre N., Coutu S., Margot J., Kyi Wynn H., Bader H., Scheidegger R. and Rossi L. (2013). Substance flow analysis as a tool for mitigating the impact ofpharmaceuticals on the aquatic system. *Water Research*, **47**, 2995–3005.

Cifuentes E. (1998). The epidemiology of enteric infections in agricultural communities exposed to wastewater irrigation: perspectives for risk control. *International Journal of Environmental Health Research*, **8**, 203–213.

Cofie O. O., Kranjac-Berisavljevic G. and Drechsel P. (2005). The use of human waste for peri-urban agriculture in Northern Ghana. *Renewable Agriculture and Food Systems*, **20**(2), 73.

Cohen Y. (2009). Phosphorus dissolution from ash of incinerated sewage sludge and animal carcasses using sulphuric acid. *Environmental Technology*, **30**(11), 1215–1226.

Cooper J. and Carliell-Marquet C. (2013). A substance flow analysis of phosphorus in the UK food production and consumption system. *Resources Conservation and Recycling*, **74**, 82–100.

Cooper R. C. and Olivieri A. W. (1998). Infectious disease concerns in wastewater reuse. In: Wastewater Reclamation and Reuse, T. Asano (ed.), Technomic Publishing Company, Inc., Lancaster, USA, pp. 489–520.

Cordell D. (2010). The Story of Phosphorus Sustainability Implications of Global Phosphorus Scarcity for Food Security. PhD dissertation, Linkoping Studies in Arts and Science No. 509, Linkoping University. Linkoping University Press. ISBN 978-91-7393-440-4.

Cordell D., Rosemarin A., Schröder J. J. and Smit A. L. (2011). Towards global phosphorus security: a systems framework for phosphorus recovery and reuse options. *Chemosphere*, **84**(6), 747–758, doi: 10.1016/j.chemosphere.2011.02.032, PMID:21414650

Cornel P. and Schaum C. (2009). Phosphorus recovery from wastewater: needs, technologies and costs. *Water Science and Technology*, **59**(6).

Costa-Pierce B. A. (1998). Preliminary investigation of an integrated aquaculture–wetland ecosystem using tertiary-treated municipal wastewater in Los Angeles County, California. *Ecological Engineering*, **10**, 341–354.

Cullen N., Baur R. and Schauer P. (2013). Three years of operation of North America's first nutrient recovery facility. *Water Science and Technology*, **68**(July 2011), 763–768, doi: 10.2166/wst.2013.260

Curtis T. P. (2010). Low-energy wastewater treatment: strategies and technologies. In: Environmental Microbiology, R. Mitchell and J. D. Gu (eds), 2nd edn, John Wiley & Sons, Inc., Hoboken, USA.

Daigger G. T. (2008). New approaches and technologies for wastewater management. *The Bridge*, **38**(3), 38–45.

Daigger G. T. (2009). Evolving urban water and residuals management paradigms: water, reclamation and reuse, decentralization, and resource recovery. *Water Environment Research*, **81**(9), 809–823.

Das D. and Veziroglu N. (2008). Advances in biological hydrogen production processes. *International Journal Hydrogen Energy*, **33**(21), 6046–6057. https://doi.org/10.1016/j.ijhydene.2008.07.098.

de-Bashan L. E. and Bashan Y. (2004). Recent advances in removing phosphorus from wastewater and its future use as a fertilizer (1997–2003). *Water Research*, **38**(19), 4222–4246.

Defra (2009a). Anaerobic digestion. http://www.defra.gov.uk/farm/environment/climate-change/anaerobic-workshop/0907.htm (accessed 2 March 2009).

Defra (2009b). Farm waste. http://www.defra.gov.uk/farm/waste/sludge/index.htm (accessed 2 March 2009).

Defra (2010). Anaerobic digestion. http://www.biogas-info.co.uk/index.php/localgovernment (accessed 30 July 2010).

Desmidt E., Ghyselbrecht K., Zhang Y., Pinoy L., Van der Bruggen B., Verstraete W., Rabaey K. and Meesschaert B. (2015). Global phosphorus scarcity and full-scale P-recovery techniques: a review. *Critical Reviews in Environmental Science and Technology*, **45**(4), 336–384.

Devaux I., Gerbaud L., Planchon C., Bontoux J. and Glanddier P. Y. (2000). Infectious risk study associated with wastewater reuse: an epidemiological approach applied to the case of Clermont-Ferrand, France. Conference Preprint, Book 7, Health-Related Water Microbiology, 1st World Water Congress of the International Water Association (IWA), Paris, France, 3–7 July 2000, pp. 41–42.

Diener S., Semiyaga S., Niwagaba C. B., Muspratt A. M., Gning J. B., Mbéguéré M., Ennin J. E., Zurbrugg C. and Strande L. (2014). A value proposition: resource recovery from faecal sludge – can it be the driver for improved sanitation? *Resources, Conservation and Recycling*, **88**, 32–38, http://dx.doi.org/10.1016/j.resconrec.2014.04.005

Dincer I. and Canan A. (2015). Review and evaluation of hydrogen production methods for better sustainability. *International Journal of Hydrogen Energy*, **40**, 11094–11111. https://dx.doi.org/10.1016/j.ijhydene.2014.12.035

Dixon D. M. and Fromtling R. A. (1991). Morphology, taxonomy, and classification of the fungi. In: Manual of Clinical Microbilology, A. Balows, W. J. Hausler, Jr, H. D. Isenberg and H. J. Shadomy (eds), 5th edn, American Society for Microbiology, Washington D.C., USA, pp. 579–587.

Donatello S., Tong D. and Cheeseman C. R. (2010). Production of technical grade phosphoric acid from incinerator sewage sludge ash (ISSA). *Waste Management*, **30**(8–9), 1634–1642. https://doi.org/10.1016/j.wasman.2010.04.009.

Donatello S. and Cheeseman C. R. (2013). Recycling and recovery routes for incinerated sewage sludge ash (ISSA): a review. *Waste Management*, **33**(11), 2328–2340.

Doyle J. D. and Parsons S. A. (2002). Struvite formation, control and recovery. *Water Research*, **36**, 3925–3940.

Drangert J.-O. (1998). Fighting the urine blindness to provide more sanitation options. *Water SA*, **24**(2), 157–164.

Dutta S., Rehman I. H., Malhotra P. and Venkata Ramana P. (1997). Biogas. The Indian NGO Experience. Tata Energy Research Institute, New Delhi, India.

DWA (German Association for Water, Wastewater and Waste) (2008). Neuartige Sanitärsysteme (New sanitary systems) (in German); DWA-Themen KA1. www.dwa.de

Edwards P. (1992). Reuse of Human Wastes in Aquaculture – A Technical Review. UNDP-World Bank Water and Sanitation Program, Washington D.C., USA.

Egle L., Rechberger H., Krampe J. and Zessner M. (2016). Phosphorus recovery from municipal wastewater: an integrated comparative technological, environmental and economic assessment of P recovery technologies. *Science of the Total Environment*, **571**, 522–542, http://dx.doi.org/10.1016/j.scitotenv.2016.07.019

Energy Saving Trust (2008). Biogas. http://www.energysavingtrust.org.uk/fleet/technology/alternativefuels/biogas/ (accessed 15 August 2008).

Engel A., Kunz A. and Blanke M. (2001). Effects of compost and wood chippings on soil nutrients and on vegetative and reproductive growth and fruit quality of apple to overcome the replant problem. *Erwerbsobstbau*, **43**, 153–160.

Ensign P. (2001). Value chain analysis and competitive advantage. *Journal of General Management*, **27**(1), 18–42, https://www.researchgate.net/publication/265181192_Value_Chain_Analysis_and_Competitive_Advantage (accessed 3 June 2018).

Environmental Data Interactive Exchange (EDIE) (2002). Germany boasts biogas from newsewage treatment plant 06/09/2002. EDIE Newsroom. http://www.edie.net/news/news_story.asp?id¼45936 (accessed 15 August 2008).

EREC (2009). Renewable energy policy review. Sweden. www.erec.org/fileadmin/erec./SWEDEN_RES_Policy_Review_Final.pdf Support for Biofuels (accessed 30 July 2010).

Eriksson E., Auffahrt K., Eilersen A., Henze M. and Ledin A. (2003). Household chemicals and personal care products as sources for xenobiotic organiccompounds in grey wastewater. *Water SA*, **29**, 135–146.

Esrey S. A. (2001). Towards a recycling society ecological sanitation – closing the loop to food security. *Water Science & Technology*, **43**(4), 177–187.

Esrey S. A. (2002). Philosophical, ecological and technical challenges for expanding ecological sanitation into urban areas. *Water Science and Technology*, **45**(8), 25–228, PMID:12019825.

Esrey S. A., Gough J., Rapaport D., Sawyer R., Simpson-Hébert M., Vargas J. and Winblad U. (1998). Ecological Sanitation. Swedish International Development Cooperation Agency, Stockholm, Sweden.

Evans T. D. (2007). Reovering ammonium and struvite fertilisers from digested sludge dewatering liquors. Proceedings of IWA Specialist Conference:Moving Forward-Wastewater Biosolids Sustainability, Moncton, Canada. http://www.bvsde.paho.org/bvsaar/cdlodos/pdf/recoveringammonium713.pdf (accessed 18 February 2018).

Fakkaew K., Koottatep T. and Polprasert C. (2015a). Effects of hydrolysis and carbonization reactions on hydrochar production. *Bioresource Technology*, **192**, 328–334.

Fakkaew K., Koottatep T., Puassayanavin T. and Polprasert C. (2015b). Hydrochar production by hydrothermal carbonization of faecal sludge. *Journal of Water, Sanitation and Hygiene for Development*, **5**(3), 439–447.

Fakkaew K., Koottatep T. and Polprasert C. (2018). Faecal sludge treatment and utilization by hydrothermal carbonization. *Journal of Environmental Management*, **216**, 421–426, http://dx.doi.org/10.1016/j.jenvman.2017.09.031

Fan B., Hu M., Wang H., Xu M., Qu B. and Zhu S. (2017). Get in sanitation 2.0 by opportunity of rural China: scheme, simulating application and life cycle assessment. *Journal of Cleaner Production*, **147**, 86–95.

Fattah K. P., Zhang Y., Mavinic D. S. and Koch F. A. (2010). Use of carbon dioxide stripping for struvite crystallization to save caustic dosage: performance at pilotscale operation. Paper submitted to the journal of environmental engineering and science. *Canadian Journal of Civil Engineering*, **37**(9), 1271–1275, doi: 10.1139/L10–055

Feachem R. G., Bradley D. J., Garelick H. and Mara D. D. (1983). Sanitation and Disease: Health Aspects of Excreta and Wastewater Management. John Wiley and Sons, Chichester, UK.

Fletcher T. D., Shuster W., Hunt W. F., Ashley R., Butler D., Arthur S., Trowsdale S., Barrand S., Semadeni-Daview A., Betrand-Krajewski J., Mikkelson P. S., Rivard G., Uhl M., Dagenais D. and Viklander M. (2015). SUDS, LID, BMPS, WSUD and more – The evolution and application of terminology surrounding urban drainage. *Urban Water Journal*, **12**(7), 525–542. https://doi.org/10.1080/1573062X.2014.916314

Fligstein N. and Dauter L. (2007). The sociology of markets. *Annual Review of Sociology*, **33**(1), 105–128.

Frode-Kristensen S. (1966). By the well at the turn of the century. (Vid brunnen – En kulturbild från sekelskiftet) (At the well – A cultural image from the turn of the century). Collected works from the archives of life of the people in Lund no. 8, Lund, Sweden (In Swedish).

Fry L. M., Mihelcic J. R. and Watkins D. W. (2008). Water and nonwater-related challenges of achieving global sanitation coverage. *Environmental Science & Technology*, **42**, 4298–4304.

Funamizu N., Iida M., Sakakura Y. and Takakuwa T. (2001). Reuse of heat energy in wastewater: implementation examples in Japan. *Water Science and Technology*, **43**(10), 277–285.

Futran V. (2013). Tackling Water Scarcity: Israel's Wastewater Recycling As a Model for the World'S Arid Lands? GWF Discussion Paper 1311; Global Water Forum: Canberra, Australia.

Gajurel D. (2007). Untersuchung zur Elimination ausgewählter Pharmaka bei der Lagerung der UV-Bestrahlung sowie der Ozonisierung von Urin (Investigation on the elimination of selected drugs during the storage of UV radiation and the ozonization of urine). Report No. GA 1063/2–1, Institute of Wastewater Management and Water Protection,Hamburg University of Technology, Hamburg, Germany. http://www.tu-harburg.de/aww/forschung/pdf/endbericht_dfg01.pdf (accessed 17 February 2009).

Ganesapillai M., Simha P. and Zabaniotou A. (2015). Closed-loop fertility cycle: realizing sustainability in sanitation and agricultural production through the design and implementation of nutrient recovery systems for human urine. *Sustainable Production and Consumption*, **4**, 36–46, https://doi.org/10.1016/j.spc.2015.08.004

Gao X. Z., Shen T., Zheng Y., Sun X., Huang S., Ren Q., Zhang X., Tian Y. and Luan G. (2002). Practical Manure Handbook. Chinese Agricultural Publishing House, Beijing, China.

Garcia-Belinchón C., Rieck T., Bouchy L., Galí A., Rougé P. and Fàbregas C. (2013). Struvite recovery: pilot-scale results and economic assessment of different scenarios. *Water Practice & Technology*, **8**(1), 119–130.

Gasworld (2010). Chesterfield biogas to supply UK's first biogas upgrade plant to Centrica. http://www.gasworld.com/news.php?a¼4929 (accessed 30 July 2010).

Gates-Foundation (2010). Water, Sanitation & Hygiene Fact Sheet. Bill & Melinda Gates Foundation, Seattle (accessed). https://docs.gatesfoundation.org/documents/water-sanitation-hygiene-fact-sheet-2010.pdf (accessed 04 December 2017).

Gebauer H. and Reynoso J. (2013). An agenda for service research at the base of the pyramid. *Journal of Service Management*, **24**(5), 482–502.

Gijzen H. J. (2001). Low cost wastewater treatment and potentials for re-use – a cleaner production approach to wastewater management. International Symposium on Low-Cost Wastewater Treatment and Re-Use, Cairo, Egypt, 3–4 February.

Gille Z. (2007). Toward a social theory of waste. In: From the Cult of Waste to the Trash Heap of History: The Politics of Waste in Socialist and Postsocialist Hungary, Z. Gille (ed.), Indiana University Press, Indianapolis, USA, pp. 11–40.

Gisclon A., McCarley S. and McNally K. (2002). The Durban Water Recycling Project – The Vision Becomes Reality. Biennial Conference of the Water Institute of Southern Africa, Durban, South Africa.

GMP Research Inc (2015). US market penetration of watersense shower heads, lavatory faucets and toilets. A GMP Research Industry Report commissioned by Plumbing Manufacturers International (PMI). https://www.safeplumbing.org/files/safeplumbing.org/documents/press_release_downloads/9-15-15-WaterSense-market-penetration-study.pdf (accessed 5 June 2018).

Gong H., Wang Z., Zhang X., Jin Z., Wang C., Zhang L. and Wang K. (2017). Organics and nitrogen recovery from sewage via membrane-based pre-concentration combined with ion exchange process. *Chemical Engineering Journal*, **311**, 13–19, http://dx.doi.org/10.1016/j.cej.2016.11.068

Grant S. B., Saphores J. D., Feldman D. L., Hamilton A. J., Fletcher T. D., Cook P. L., Stewardson M., Sanders B. F., Levin L. A., Ambrose R. F., Deletic A., Brown R., Jiang S. C., Rosso D., Cooper W. J. and Marusic I. (2012). Taking the 'waste' out of 'wastewater' for human water security and ecosystem sustainability. *Acta Agriculturae Scandinavica Section A Animal Science*, **337**, 681–686.

Grimason A. M., Tembo K. C., Jabu G. C. and Jackson M. H. (2000). Problems associated with the use of pit latrines in Blantyre, Republic of Malawi. *Journal of the Royal Society for the Promotion of Health*, **120**, 175–182, PMID:11077806.

GTZ (Deutsche Gesellschaft für Technische Zusammenarbeit) (2002). Ecosan – recycling beats disposal. Eschborn (Germany) 7 GTZ. http://www.waterfund.go.ke/watersource/Downloads/002.%20ecosan-projectinfo-2002.pdf (accessed 9 September 2018).

Guardian (2008). Digesting the problem. http://www.guardian.co.uk/environment/2008/jul/30/biofuels.waste?gusrc¼rss&feed¼environmentDigestingtheproblem (accessed 7 July 2009).

Guest J. S., Skerlos S. J., Barnard J. L., Beck M. B., Daigger G. T., Hilger H., Jackson S. J., Karvazy K., Kelly L., Macperson L., Mihelcic J. R., Pramanik J. B., Raskin L., Van Loosdrecht M. C. M., Yeh D. and Love N. G. (2009). A new planning and design paradigm to achieve sustainable resource recovery from wastewater. *Advances in Environmental Science and Technology*, **43**, 6126–6130, https://pubs.acs.org/doi/pdf/10.1021/es9010515

Gumbo B. (2005). Short-Cutting the Phosphorus Cycle in the Urban Ecosystems. PhD dissertation, UNESCO-IHE Institute of Water Education, Delft, Netherlands. Taylor and Francis Inc., London, UK. ISBN 041 5384842.

Guyton A. C. (1992). Human Physiology and Mechanisms of Disease. W.B. Saunders Co, Philadelphia, USA.

Han S. and Shin H. (2004). Biohydrogen production by anaerobic fermentation of food waste. *International Journal of Hydrogen Energy*, **29**(6), 569–577. https://doi.org/10.1016/j.ijhydne.203.09.001

Hansen J. A. and Tjell J. C. (1979). Human Excretion of Heavy Metals and Other Elements. Working Document, Tehnical University of Denmark, The Laboratory for Technical Hygiene.

Hansen H. P. (1928). About hygiene in the old days. (Om renlighed i gamla dage). Nordisk Forlag, Copenhagen, Denmark. (In Danish). In: Drangert J.-O. (1998). Fighting the urine blindness to provide more sanitation options. *Water SA*, **24**(2), 157–164.

Hanzen and Sawyer (2010). New Process Makes Asset of Phosphorus and Nitrogen from Wastewater. Hampton Roads, Virginia. http://www.hazenandsawyer.com/news/new-process-makes-asset-of-phosphorusand-nitrogen-from-wastewater/ (accessed 20 June 2018).

Haq G. and Cambridge H. (2012). Exploiting the co-benefits of ecological sanitation. *Current Opinion in Environmental Sustainability*, **4**(4), 431–435.

Haug R. T. (1993). The Practical Handbook of Compost Engineering. Lewis Publishers, Boca Raton, USA.

Havelaar A. H., de Wit M. A. S. and van Koningsveld R. (2000). Health burden in the Netherlands (1900–1995) due to infections with thermophilic Campylobacter species. Report 284550 004, National Institute of Public Health and the Environment, Bilthoven, Netherlands.

Heinonen-Tanski H. and van Wijk-Sijbesma C. (2005). Human excreta for plant production. *Bioresource Technology*, **96**, 403–411.

Heinonen-Tanski H., Sjöblom A., Fabritius H. and Karinen P. (2007). Pure human urine is a good fertiliser for cucumbers. *Bioresource Technology*, **98**, 214–217, doi: 10.1016/j.biortech.2005.11.024

Heinss U., Larnie S. A. and Strauss M. (1998). Solids Separation and Pond Systems for the Treatment of Septage and Public Toilet Sludges in Tropical Climate – Lessons Learnt and Recommendations for Preliminary Design. EAWAG/SANDEC Report no. 05/98. https://www.ircwash.org/sites/default/files/342=98SO-1423.pdf (accessed 8 February 2018).

Herzel H., Krüger O., Hermann L. and Adam C. (2016). Sewage sludge ash – a promising secondary phosphorus source for fertilizer production. *Science of the Total Environment*, **542**, 1136–1143.

Hodges N. (2006). Regenerative Design Theory and Practice: A Demonstration of the Integrated Framework in a Resort Development at Mountain Lake, VA. Master's thesis, Virginia Polytechnic and State University, USA. https://theses.lib.vt.edu/theses/available/etd-05082006-101306/

Höglund C. (2001). Evaluation of microbial health risks associated with the reuse of source-separated human urine. Doctoral thesis, Royal Institute of Technology (KTH) and Department of Water and Environmental Microbiology, Applied Microbiology and Swedish Institute for Infectious Disease Control (SMI), Stockholm, Sweden. https://pdfs.semanticscholar.org/498b/9a04c9c2c3725cb1245841f7197c3daf6fa1.pdf

Höglund C., Stenström T., Jönsson H. and Sundin A. (1998). Evaluation of faeal contamination and microbial die-off in urine separating sewage systems. *Water Science Technology*, **38**(6), 17–25. https://doi.org/10.1016/S0272-1223(98)00563-0

Höglund C., Stenström T. A., Sundin A., Jönsson H. and Vinnerås B. (1999). Hygienic quality of source separated urine based on the analysis of faecal indicator bacteria. Proceedings from 8th International Conference on Management Strategies for Organic Waste Use in Agriculture (Ramiran 98), FAO and Cemagref, France, pp. 517–528.

Höglund C., Stenström T. A. and Ashbolt N. (2002). Microbial risk assessment of source-separated urine used in agriculture. *Waste Management & Research*, **20**(3), 150–161.

Hu M., Fan B., Wang H. L., Qu B. and Zhu S. K. (2016). Constructing the ecological sanitation: a review on technology and methods. *Journal of Cleaner Production*, **125**, 1–21.

IMA (Institute of Management Accountants) (1996). Value Chain Analysis for Assessing Competitive Advantage. https://www.imanet.org.cn/uploads/resource/2015-11/1447061044-16209.pdf (accessed 3 June 2018).

ISO/IWA (International Organization for Standardization)/(International Water Association) (28:2018). Faecal sludge treatment units – Energy independent, prefabricated, community-scale resource-recovery units – Safety and performance. https://libnor.isolutions.iso.org/obp/ui#iso:std:iso:iwa:28:ed-1:v1:en (accessed 16 June 2018).

IWA (International Water Association) (2016). The Principles of Water Wise Cities. International Water Association, London, UK. http://www.iwa-network.org/projects/water-wise-cities/

Jana B. (1998). Sewage-fed aquaculture: the Calcutta model. *Ecological Engineering*, **11**(1–4), 73–85. https:/doi.org/10.1016/S0925-8574(98)00024-x.

Jefferson B., Palmer A., Jeffrey P., Stuetz R. and Judd S. (2004). Greywater characterisation and its impact on the selection and operation of technologies for urban reuse. *Water Science & Technology*, **50**(2), 157–644.

Jenkins M. and Sugden S. (2006). Rethinking sanitation: Lessons and innovation for sustainability and success in the new millennium. in Human Development Report 2006, United Nations Development Programme, New York, p. 37.

Jewitt S. (2011). Poo gurus? Researching the threats and opportunities presented by human waste. *Applied Geography*, **31**, 761–769, doi: 10.1016/j.apgeog.2010.08.003

Jones J., Chang N.-B. and Wanielista M. P. (2015). Reliability analysis of nutrient removal from stormwater runoff with green sorption media under varying influent conditions. *Science of the Total Environment*, **502**, 434–447.

Jönsson H., Vinnerås B., Höglund C., Stenström T. A., Dalhammar G. and Kirchmann H. (2000). Källsorterad humanurin i kretslopp (Recycling source separated human urine) (In Swedish, English summary). VA-FORSK Report 2000•1. VA-FORSK/VAV. Stockholm, Sweden.

Jönsson H., Richert Stintzing A., Vinnerås B. and Salomon E. (2004). Guidelines on the Use of Urine and Faeces in Crop Production. EcoSanRes Programme. Report 2004–2. Stockholm Environment Institute, Sweden. http://www.ecosanres.org/pdf_files/ESR_Publications_2004/ESR2web.pdf (accessed 2 June 2018).

Kahiluoto H., Kuisma M., Ketoja E., Salo T. and Heikkinen J. (2015). Phosphorus in manure and sewage sludge more recyclable than in soluble inorganic fertilizer. *Advances in Environmental Science and Technology*, **49**(4), 2115–2122.

Karak T. and Bhattacharyya P. (2011). Human urine as a source of alternative natural fertilizer in agriculture: a flight of fancy or an achievable reality. *Resources, Conservation and Recycling*, **55**, 400–408, doi: 10.1016/j.resconrec.2010.12.008

Karpiscak M. M., Foster K. E. and Schmidt N. (1990). Residential water conservation. *Australian Journal of Marine and Freshwater Research*, **26**, 939–948.

Katukiza A. Y., Ronteltap M., Niwagaba C. B., Foppen J. W. A., Kansiime F. and Lens P. N. L. (2012). Sustainable sanitation technology options for urban slums. *Biotechnology Advances*, **30**, 964–978.

Katzenelson E., Buium I. and Shuval H. I. (1976). Risk of communicable disease infection associated with wastewater irrigation in agricultural settlements. *Acta Agriculturae Scandinavica Section A Animal Science*, **26**, 944–946.

Kennedy-Walker R., Evans B., Amezaga J. and Paterson C. (2014). Challenges for the future of urban sanitation planning: critical analysis of John Kalbermatten's influence. *Journal of Water, Sanitation and Hygiene for Development*, **4**, 1–14.

Kennedy-Walker R., Holderness T., Alderson D., Amezaga J. M. and Paterson C. A. (2016). Optimisation and costing of faecal sludge management options for Lusaka's informal settlements. *Environmental Science: Water Research & Technology*, **2**, 97–106, doi: 10.1039/c5ew00179j

Kirchmann H. and Pettersson S. (1995). Human urine – chemical-composition and fertilizer use efficiency. *Fertilizer Research*, **40**(2), 149–154.

Kleerebezem R. and van Loosdrecht M. C. M. (2007). Mixed culture biotechnology for bioenergy production. *Current Opinion in Biotechnology*, **18**(3), 207–212.

Kone D. (2007). Excreta and wastewater management contributing to Cities' economic development. A paradigm shift. *Sandec News*, **8**, 8–9.

Koottatep T., Surinkul N., Polprasert C., Kamal A. S. M., Kone D., Montangero A., Heinss U. and Strauss M. (2005). Treatment of septage in constructed wetlands in tropical climate: lessons learnt from seven years of operation. *Water Science and Technology*, **51**(9), 119–126.

Koottatep T., Fakkaew K., Tajai N., Pradeep S. V. and Polprasert C. (2016). Sludge stabilization and energy recovery by hydrothermal carbonization process. *Renewable Energy*, **99**, 978–985.

Koottatep T., Fakkaew K., Tajai N. and Polprasert C. (2017). Isotherm models and kinetics of copper adsorption by using hydrochar produced from hydrothermal carbonization of faecal sludge. *Journal of Water, Sanitation and Hygiene for Development*, **7**(1), 102–110.

Kujawa-Roeleveld and K. and Zeeman G. (2006). Anaerobic treatment in decentralized and source-separation-based sanitation concepts. *Reviews in Environmental Science and Bio/Technology*, **5**, 115–139.

Kvarnström E., Emilsson K., Stintzing A. R., Johansson M., Jönsson H., af Petersens E., Schonning C., Christensen J., Hellstrom D., Qvarnstrom L., Ridderstolpe P. and Drangert J.-O. (2006). Urine Diversion: One Step Towards Sustainable Sanitation, EcoSanRes Publication Series; SEI Stockholm Environment Institute. http://www.ecosanres.org

Lahnsteiner J. and Lempert G. (2007). Water management in Windhoek, Namibia. *Water Science and Technology*, **55** (1–2), 441–448.

Langergrabera G. and Muelleggera E. (2005). Ecological Sanitation – a way to solve global sanitation problems? *Environment International*, **31**(2005), 433–444.

Larsen T. A. and Gujer W. (1996). Separate management of anthropogenic nutrient solutions (human urine), WaterSci. *Technol.*, **34**(3–4), 87–94.

Larsen T. A. and Lienert J. (2003). Societal implications of re-engineering the toilet. In Water Intelligence Online, UNIQUEID: 200303006. www.iwaponline.com/wio/2003/03/default001.htm

Larsen T. A., Peters I., Alder A. C., Eggen R. I., Maurer M. and Muncke J. (2001a). Peer Reviewed: Re-engineering the toilet for sustainable wastewater management. *Environmental Science & Technology*, **May 1**, American Chemical Society. https://pubs.acs.org/doi/pdf/10.1021/es012328d

Larsen T. A., Peters I., Alder A., Eggen R., Maurer M. and Muncke J. (2001b). Re-engineering the Toilet for sustainable wastewater management: Could one answer be a new toilet that separates urine from wastewater streams? *Advances in Environmental Science and Technology*, **35**(9), 192A–197A, doi: 10.1021/es012328d

Larsen T. A., Alder A. C., Eggen R. I., Maurer A. and Lienert J. (2009). Source separation: will we see a paradigm shift in wastewater handling. *Environmental Science & Technology*, **43**(16), 6121–6125, doi: 10.1021/es803001r

Larsen T. A., Maurer M., Eggen R. I. L., Pronk W. and Lienert J. (2010). Decision support in urban water management based on generic scenarios: the example of NoMix technology. *Journal of Environmental Management*, **91**, 2676–2687, doi: 10.1016/j.jenvman.2010.07.032

Larsen T. A., Udert K. and Lienert J. (2013). Source Separation and Decentralization for Wastewater Management. IWA Publishing, London, UK.

Lazarova V., Levine B., Sack J., Cirelli G., Jeffrey P., Muntau H., Salgot M. and Brissaud F. (2001). Role of water reuse for enhancing integrated water management in Europe and Mediterranean countries. *Water Science and Technology*, **43**(10), 25–33.

Lee E. J., Criddle C. S., Bobel P. and Freyberg D. L. (2013). Assessing the scale of resource recovery for centralized and satellite wastewater treatment. *Environmental Science & Technology*, **47**, 10762–10770.

Lens P., Zeeman G. and Lettinga G. (2001). Decentralized Sanitation and Reuse: Concepts, Systems and Implementation. IWA Publishing, London, UK.

Li H., Zou S., Li C. and Jin Y. (2013a). Alkaline post-treatment for improved sludge anaerobic digestion. *Bioresource Technology*, **140**, 187–191.

Li W., Feng X., Yan Y., Sparks D. L. and Phillips B. L. (2013b). Solid-state NMR spectroscopic study of phosphate sorption mechanisms on aluminum (hydr)oxides. *Advances in Environmental Science and Technology*, **47**(15), 8308–8315.

Li W., Yu H. and Rittmann B. E. (2015). Reuse water pollutants. *Nature*, **528**(7580), 29–31. https:/www.nature.com/news/chemistry-reuse-water-pollutants-1.18899 (accessed 14 March 2018).

Li M., Liu J., Xu Y. and Qian G. (2016). Phosphate adsorption on metal oxides and metal hydroxides: a comparative review. *Environmental Review*, **24**, 319–332. https://doi.org/10.1139/er-2015-0080.

Lienert J. and Larsen T. A. (2006). Considering user attitude in early development of environmentally friendly technology: a case study of NoMix toilets. *Advances in Environmental Science and Technology*, **40**(16), 4838–4844.

Lienert J., Bürki T. and Escher B. (2007). Reducing micropollutants with source control: substance flow analysis of 212 pharmaceuticals in faeces and urine. *Water Science & Technology*, **56**, 87–96.

Lienert J. and Larsen T. A. (2010). High acceptance of urine source separation in seven European countries: a review. *Environmental Scienc and Technoogy*, **44**(2), 555–566. https://doi.org/10.1021/es9028765.

Logan B. E., Hamelers B., Rozendal R., Schröder U., Keller J., Freguia S., Aelterman P., Verstraete W. and Rabaey K. (2006). Microbial fuel cells: methodology and technology. *Advances in Environmental Science and Technology*, **40** (17), 5181–5192.

London T. and Hart S. L. (2010). Next Generation Business Strategies for the Base of the Pyramid: New Approaches for Building Mutual Value. Pearson Education Australia PTY, Limited Prentice Hall: FT. http://ptgmedia.pearsoncmg.com/images/9780137047895/samplepages/9780137047895.pd (accessed 17 June 2018).

Lu L., Huang Z., Rau G. and Ren Z. J. (2015). Microbial electrolytic carbon capture for carbon negative and energy positive wastewater treatment. *Advances in Environmental Science and Technology*, **49**(13), 8193–8201.

Lyle J. T. (1994). Regenerative Design for Sustainable Development. John Wiley & Sons, Hoboken, USA.

Mae-Wan H. (2008). Biogas bonanza for third world development. Institute of Science in Society. http://www.i-sis.org.uk/BiogasBonanza.php (accessed 20 January 2008).

Mann H. T. (1976). Sanitation without sewers. Overseas Building Notes no. 168, Garston, England.

Masi F., El Hamouri B., Abdel Shafi H., Baban A., Ghrabi A. and Regelsberger M. (2010). Segregated black/grey domestic wastewater treatment by Constructed Wetlands in the Mediterranean basin: the Zer0-m experience. *Water Science and Technology*, **61**(1), 97–105, https://doi.org/10.2166/wst.2010.780

Masi F., Bresciani R., Rizzo A., Edathoot A., Patwardhan N., Panse D. and Langergraber G. (2016). Green walls for greywater treatment and recycling in dense urban areas: a case-study in Pune. *Journal of Water, Sanitation and Hygiene for Development*, **6**(2), 342–347, https://doi.org/10.2166/washdev.2016.019

Masi F., Rizzo A. and Regelsberger M. (2017). The role of constructed wetlands in a new circular economy, resource oriented, and ecosystem services paradigm. *Journal of Environmental Management*, **216**, 275–284, https://doi.org/10.1016/j.jenvman.2017.11.086

Matamoros V., Rodríguez Y. and Bayona J. M. (2017). Mitigation of emerging contaminants by full-scale horizontal flow constructed wetlands fed with secondary treated wastewater. *Ecological Engineering*, **99**, 222–227.

Matassa S., Batstone D. J., Hülsen T., Schnoor J. and Verstraete W. (2015). Can direct conversion of used nitrogen to new feed and protein help feed the world? *Advances in Environmental Science and Technology*, **49**, 5247–5254.

Maurer M., Schwegler P. and Larsen T. A. (2003). Nutrients in urine: energetic aspects of removal and recovery. *Water Science and Technology*, **48**(1), 37–46.

Maurer M., Rothenberger D. and Larsen T. (2005). Decentralised wastewater treatment technologies from a national perspective: at what cost are they competitive? *Water Science Technology: Water Supply*, **5**(6). https://doi.org/10.2166/ws.2005.0059

Maurer M., Pronk W. and Larson T. (2006). Treatment processes for source-separated urine. *Water Research*, **40**, 3151–3166.

Maurer M., Bufardi A., Tilley E., Zurbrügg C. and Truffer B. (2012). A compatibility-based procedure designed to generate potential sanitation system alternatives. *Journal of Environmental Management*, **104**, 51–61.

Maurya N. S. (2012). Is human excreta a waste? *International Journal of Environmental Technology and Management*, **15**(3/4/5/6), 325–332.

Mayer B. K., Baker L. A., Boyer T. H., Drechsel P., Gifford M., Hanjra M. A., Parameswaran P., Stoltzfus J., Westerhoff P. and Rittmann B. E. (2016). Total value of phosphorus recovery. *Advances in Environmental Science and Technology*, **50**, 6606–6620, doi: 10.1021/acs.est.6b01239

McCarty P. L., Bae J. and Kim J. (2011). Domestic wastewater treatment as a net energy producer – can this be achieved? *Advances in Environmental Science and Technology*, **45**(17), 7100–7106.

McCormick D. and Schmitz H. (2001). Manual for Value Chain Research on Homeworkers in the Garment Industry, Institute for Development Studies. https://www.ids.ac.uk/ids/global/pdfs/WiegoManualendNov01.pdf (accessed 6 June 2018).

McDonough W. and Michael B. (2001a). The Next Industrial Revolution, William McDonough, Michael Braungart and the Birth of the Sustainable Economy. Earthome Productions, Stevenson, USA.

McDonough W. and Michael B. (2001b). Cradle to Cradle: Remaking the Way We Make Things. Northpoint Press, New York, USA.

McMahon S., Caruso B. A., Obure A., Okumu F. and Rheingans R. D. (2011). Anal cleansing practices and faecal contamination: a preliminary investigation of behaviours and conditions in schools in rural Nyanza Province, Kenya. *Tropical Medicine and International Health*, **16**(12), 1536–1540.

Mezohegyi G., Bilad M. R. and Vankelecom I. F. J. (2012). Direct sewage up-concentration by submerged aerated and vibrated membranes. *Bioresource Technology*, **118**, 1–7, http://dx.doi.org/10.1016/j.biortech.2012.05.022

Mihelcic J. R., Fry L. M. and Shaw R. (2011). Global potential of phosphorus recovery from human urine and feces. *Chemosphere*, **84**(6), 832–839.

Mihelcic J. R., Ren Z. J., Cornejo P. K., Fisher A., Simon A. J., Snyder S. W., Zhang Q., Rosso D., Huggins T. M., Cooper W., Moeller J., Rose B., Schottel B. L. and Turgeon J. (2017). Accelerating innovation that enhances resource recovery in the wastewater sector: advancing a national testbed network. *Environmental Science & Technology*, **51**(14), 7749–7758, doi: 10.1021/acs.est.6b05917

Mo W. and Zhang Q. (2013). Energy–nutrients–water nexus: integrated resource recovery in municipal wastewater treatment plants. *Journal of Environmental Management*, **13**(127), 255–267.

Moerman W. (2012). Method for purifying wastewater with ammonium removal EP2431336A1, Google Patents. https://patents.google.com/patent/EP2431336A1/en (accessed 18 June 2018).

Moerman W., Carballa M., Vandekerckhove A., Derycke D. and Verstraete W. (2009). Phosphate removal in agro-industry: pilot-and full-scale operational considerations of truvite crystallization. *Water Research*, **43**(7), 1887–1892.

Morel A. and Diener S. (2006). Greywater Management in Low and Middle-Income Countries, Review of Different Treatment Systems for Households or Neighbourhoods. Swiss Federal Institute of Aquatic Science and Technology (Eawag), Dübendorf, Switzerland. https://www.sswm.info/sites/default/files/reference_attachments/MOREL%20and%20DIENER%202006%20Greywater%20Management.pdf (accessed 7 June 2018).

Morsing M. (1994). The Use of Sludge in Forestry and Agriculture. A Comparison of the Legislation in Different Countries. Forskningsserien no. 5-1994, Danish Forest and Landscape Research Institute, Lyngby, Denmark.

Mulder A. (2003). The quest for sustainable nitrogen removal technologies. *Water Science and Technology*, **48**(1), 67–75, PMID:12926622.

Munamati M., Nhapi I. and Misi S. (2016). Exploring the determinants of sanitation success in Sub-Saharan Africa. *Australian Journal of Marine and Freshwater Research*, **103**, 435–443, http://dx.doi.org/10.1016/j.watres.2016.07.030

Munch E. V. and Barr K. (2001). Controlled struvite crystallisation for removing phosphorus from anaerobic digester side Streams. *Australian Journal of Marine and Freshwater Research*, **35**(1), 151–159, doi: 10.1016/S0043-1354(00)00236-0

Murray A. and Buckley C. (2009). Designing reuse-oriented sanitation infrastructure: the design for service planning approach. In: Wastewater Irrigation and Health Assessing and Mitigating Risk in Low-income Countries, P. Drechsel, C. A. Scott, L. Raschid-Sally, M. Redwood and A. Bahri (eds), Earthscan, London, UK.

Murray A. and Ray I. (2010). Commentary: back-end users: the unrecognized stakeholders in demand-driven sanitation. *Journal of Planning Education and Research*, OnlineFirst, published on May 6, 2010, doi: 10.1177/0739456X10369800

Nartey E. G., Amoah P., Ofosu-Budu G. K., Muspratt A. and Pradhan S. K. (2017). Effects of co-composting of faecal sludge and agricultural wastes on tomato transplant and growth. *International Journal of Recycling of Organic Waste in Agriculture*, **6**, 23–36, doi: 10.1007/s40093-016-0149-z

Niewersch C., Abels C., Li R., Wintgens T. and Melin T. (2009). Mass transport modelling to estimate the efficiency of nanofiltration application for the recovery of phosphorus from sewage sludge. *Desalination and Water Treatment*, **6**(1–3), 86–93.

Nikiema J., Cofie O. and Impraim R. (2014). Technological options for safe resource recovery from fecal sludge. CGIAR Research Program on Water, Land and Ecosystems (WLE). 47p. (Resource Recovery and Reuse Series 2). International Water Management Institute (IWMI), Colombo, Sri Lanka. doi: 10.5337/2014.228

Nimpuno K. (1983). Sanitation Options. Institute for Housing Studies (IHS), Rotterdam, Netherlands.

Niwagaba C. (2007). Human Excreta Treatment, Technologies – Prerequisites, Constraints and Performance. Licentiate thesis 005, Department of Biometry and Engineering, Swedish University of Agriculture Sciences (SLU), Uppsala, Sweden. https://www.researchgate.net/publication/253275382_Human_Excreta_Treatment_Technologies_-_prerequisites_constraints_and_performance (accessed 18 June 2018).

Nolde E. (2007). Possibilities of rainwater utilisation in densely populated areas including precipitation runoffs from traffic surfaces. *Desalination*, **215**, 1e11.

Nouri J., Naddafi K., Nabizadeh R. and Jafarinia M. (2007). Energy recovery from wastewater treatment plant. *Asian Journal of Water, Environment and Pollution*, **4**(1), 145–149.

Nowak B., Aschenbrenner P. and Winter F. (2013). Heavy metal removal from sewage sludge ash and municipal solid waste fly ash – a comparison. *Fuel Processing Technology*, **105**, 195–201.

O'Keefe M., Luthi C., Tumwebaze I. K. and Tobias R. (2015). Opportunities and limits to market-driven sanitation services: evidence from urban informal settlements in East Africa. *Environment & Urbanization*, **27**(2), 421–440, doi: 10.1177/0956247815581758

OECD (2011). Benefits of investing in water and sanitation: an OECD perspective. http://www.oecd.org/berlin/47630231.pdf (accessed 5 June 2018).

Okun D. A. (1997). Distributing reclaimed water through dual systems. *American Water Works Association Journal*, **89**(11), 153–160.

Okun D. A. (2000). Water reclamation and unrestricted non-potable reuse: a new tool in urban water management. *Annual Review of Public Health*, **21**, 223–245.

Orner K. D. and Mihelcic J. R. (2018). A review of sanitation technologies to achieve multiple sustainable development goals that promote resource recovery. *Environmental Science: Water Research*, **4**, 16.

Orner K. D., Ozcan O. Y., Saetta D., Boyer T. H., Yeh D. H., Anderson D. and Cunningham J. A. (2017). House of quality planning matrix for evaluating wastewater nutrient management technologies at three scales within a sewershed. *Environmental Engineering Science*, doi: 10.1089/ees.2017.0016

Otoo M. and Drechsel P. (2017). Resource Recovery from Waste Business Models for Energy, Nutrient and Water Reuse in Low- and Middle-Income Countries. Routledge, London. Page 890, ISBN 978-1-138-01655-2.

Otterpohl R. (2000). Design of highly efficient Source Control Sanitation and practical Experiences. EURO-Summer School DESAR, Wageningen June 18-23, Netherlands. https://cgi.tu-harburg.de/~awwweb/susan/downloads/desar.pdf (accessed 1 June 2018).

Otterpohl R. (2002). Options for alternative types of sewerage and treatment systems directed to improvement of the overall performance. *Water Science and Technology*, **45**(3), 149–158.

Otterphol R. and Buzie C. (2013). Treatment of the solid fraction. In: Source Separation and Decentralization for Wastewater Management, T. A. Larsen, K. M. Udert and J. Lienert (eds), IWA Publishing, London, UK, pp. 260–273.

Otterpohl R., Albold A. and Olgenburg M. (1999). Sources control in urban sanitation and waste management: ten systems with reuse of resources. *Water Science & Technology*, **39**(5), 153–160.

Pahl-Wostl C., Schonborn A., Willi N., Muncke J. and Larsen T. A. (2003). Investigating consumer attitudes towards the new technology of urine separation. *Water Science and Technology*, **48**(1), 57–65.

Palmquist H. and Jönsson H. (2004). Urine, faeces, greywater and biodegradable solid waste as potential fertilisers. 2nd International Symposium on Ecological Sanitation, Incorporating the 1st IWA Specialist Group Conference on Sustainable Sanitation, Lübeck, Germany, 7–11 April.

Papa M., Foladori P., Guglielmi L. and Bertanza G. (2017). How far are we from closing the loop of sewage resource recovery? A real picture of municipal wastewater treatment plants in Italy. *Resources, Conservation and Recycling*, **113**, 53–64.

Park J. Y. and Chertow M. R. (2014). Establishing and testing the 'reuse potential' indicator for managing wastes as resources. *Journal of Environmental Management*, **137**, 45–53, http://dx.doi.org/10.1016/j.jenvman.2013.11.053

Pathak B. (2014). Biogas Technology. Community Toilet Linked Biogas Plant, Published 4 August. http://www.sulabhinternational.org/content/biogas-technology (accessed 2 June 2018).

Peal A., Evans B. and Voorden C. v. d. (2010). Hygiene and Sanitation Software: An Overview of Approaches. Water Supply & Sanitation Collaborative Council, Geneva, Switzerland. https://www.wsscc.org/wp-content/uploads/2016/06/Hygiene-and-Sanitation-Software-An-overview-of-approaches-WSSCC.pdf (accessed 18 June 2018).

Peccia J. and Westerhoff P. (2015). We should expect more out of our sewage sludge. *Advances in Environmental Science and Technology*, **49**, 8271–8276, doi: 10.1021/acs.est.5b01931

Pedi D. (2012). Sanitation Supply Chain Analysis Pemagetshel Dzongkhag PHED-SNV Rural Sanitation and Hygiene Programme, SNV, Thimphu, Bhutan. https://docplayer.net/56114774-Sanitation-supply-chain-analysis-pemagetshel-dzongkhag-phed-snv-rural-sanitation-and-hygiene-programme.html (accessed 20 June 2018).

Petzet S. and Cornel P. (2012). Prevention of struvite scaling in digesters combined with phosphorus removal and recovery-the FIX-Phos process. *Water Environment Research*, **84**(3), 220–226.

Petzel S., Peplinski B. and Cornel P. (2012). On wet chemical phosphorus recovery from sewage sludge ash by acidic or alkaline leaching and an optimized combination of both. *Water Research*, **46**(12), 3769–3780. https://doi.org/10.1016/j.watres.2012.03.068.

Pieper W. (1987). Das Scheiss-Buch – Entstehung, Nutzung, Entsorgung menschlicher Fäkalien (The Shit Book – Genesis, Use, Disposal of Human Feces), Der Grüne Zweig 123, Werner Pieper and Grüne Kraft, Germany.

Polprasert C. (1996). Organic Waste Recycling. John Wiley & Sons Ltd, New York (as cited in Gijzen, 2001).

Polprasert C. and Koottatep T. (2017). Organic Waste Recycling – Technology, Management and Sustainability, 4th edn. IWA Publishing, Netherlands.

Polprasert C., Lohani B. N. and Chan C. B. (1981). Human Faeces, Urine and their Utilization. Asian Institute of Technology, Bangkok, Thailand (as cited in Schouw, 2002).

Porter M. E. (1985). Competitive Advantage. Free Press, New York, USA.

Pronk W. and Kone D. (2009). Options for urine treatment in developing countries. *Desalination*, **248**(2009), 360–368, doi: 10.1016/j.desal.2008.05.076

Puchongkawarin C., Gomez-Mont C., Stuckey D. C. and Chachuat B. (2015). Optimization-based methodology for the development of wastewater facilities for energy and nutrient recovery. *Chemosphere*, **140**, 150–158.

Puyol D., Batstone D., Hülsen T., Astals S., Peces M. and Krömer J. O. (2017). Resource recovery from wastewater by biological technologies: opportunities, challenges, and prospects. *Frontiers in Microbiology*, **7**, 1–23, doi: 10.3389/fmicb.2016.02106

PWC (PricewaterhouseCoopers) (2012). Water: challenges, drivers and solutions. https://www.pwc.com/gx/en/sustainability/publications/assets/pwc-water-challenges-drivers-and-solutions.pdf (accessed 18 June 2018).

Qin C., Liu H., Liu L., Smith S., Sedlak D. L. and Gu A. Z. (2015). Bioavailability and characterization of dissolved organic nitrogen and dissolved organic phosphorus inwastewater effluents. *Science of the Total Environment*, **511**, 47–53.

Rahman M. H., Mottalin M. A. and Bhuiyan M. H. (1996). A study on biogas technology in Bangladesh. 22nd WEDC Conference, New Delhi, India. https://wedc-knowledge.lboro.ac.uk/resources/conference/22/Rahman.pdf (accessed 18 June 2018).

Rahman M. M., Salleh M. A. M., Rashid U., Ahsan A., Hossain M. M. and Ra C. S. (2014). Production of slow release crystal fertilizer from wastewaters through struvite crystallization – a review. *Arabian Journal of Chemistry*, **7**, 139–155, http://dx.doi.org/10.1016/j.arabjc.2013.10.007

Rampersad H. (2002). Total Performance Scorecard, NUGI684–Management. Scriptum, Schiedam, Netherlands (in Dutch).

Reed B. and Scott R. (2014). Preventing the transmission of faecal-oral diseases. WEDC guides on sanitation, Water, Engineering and Development Centre (WEDC), school of Civil and Building Engineering, Loughborough University, UK. https://wedc-knowledge.lboro.ac.uk/resources/booklets/G020-Preventing-transmission-of-disease-online.pdf (accessed 20 June 2018).

Remy C. (2010). Life Cycle Assessment of Conventional and Source-Separation Systems for Urban Wastewater Management. Technische Universitaet Berlin, Berlin, Germany.

Ren Z. J. and Umble A. K. (2016). Water treatment: recover wastewater resources locally. *Bionature*, **529**, 25.

Richert Stintzing A., Rodhe L., Åkerhielm H. and Steineck S. (2002). Human urine as a fertiliser – plant nutrients, application technique and environmental effects. In: Proceedings of the 10th International Conference on Ramiran 2002 Network, J. Venglovsky and G. Gréserová (eds), FAO European System of Cooperative Research Network, pp. 161–162.

Rittmann B. E. (2013). The energy issue in urban water management. In: Source Separation and Decentralization for Wastewater Management, T. A. Larsen, K. M. Udert and J. Lienert (eds), IWA Publishing, London, UK, pp. 13–27.

Rittmann B. E. and McCarty P. E. (2001). Environmental Biotechnology: Principles and Applications. McGraw-Hill, Boston, USA. ISBN 9780072345537.

Ronteltap M., Mauer M. and Gujer W. (2007). The behaviour of pharmaceuticals and heavy metals during struvite precipitation in urine. *Water Research*, **41**, 1859–1868.

Rose C., Parker A., Jefferson B. and Cartmell E. (2015). The characterisation of feces and urine: a review of the literature to inform advanced treatment technology. *Critical Reviews in Environmental Science and Technology*, **5**(45), 1827–1879.

Rousseau D. P. L., Lesage E., Story A., Vanro H. and De Pauw N. (2008). Constructed Wetlands for water reclamation. *Desalination*, **218**, 181–189. https://doi.org/10.1016/j.desal.2006.09.034.

Ruiken C. J., Breurer G., Klaversma E., Santiago T. and Van Loosdrecht M. C. M. (2013). Sieving wastewater cellulose recovery, economic and energy evaluation. *Water Research*, **47**, 43–48.

Sanergy (2018). Sanergy integrated sanitation value chain. http://www.chefcooke.com/sanergy/ (accessed 5 June 2018).

Santala E., Uotila J., Zaitsev G., Alasiurua R., Tikka R. and Tnegvall J. (1998). Microbiological greywater treatment and recycling in an apartment building. IN: AWT98 – Advanced Wastewater Treatment, Recycling and Reuse, Milan, 14–16 September 1998, pp. 319–324.

Särkilahti M., Kinnunen V., Kettunen R., Jokinen A. and Rintala J. (2017). Replacing centralised waste and sanitation infrastructure with local treatment and nutrient recycling: expert opinions in the context of urban planning. *Technological Forecasting & Social Change*, **118**, 195–204, http://dx.doi.org/10.1016/j.techfore.2017.02.020

Scheierling S. M., Bartone C., Mara D. D. and Drechsel P. (2010). Improving wastewater use in agriculture: an emerging priority (English). Policy Research working paper; no. WPS 5412. World Bank, Washington D.C., USA. http://documents.worldbank.org/curated/en/710411468176653818/Improving-wastewater-use-in-agriculture-an-emerging-priority (accessed 13 June 2018).

Schonning C., Leeming R. and Stenstrom T. (2002). Faecal contamination of source-separated human urine based on the content of faecal sterols. *Water Research*, **36**(8), 1965–1972. https://doi.org/10.1016/S0043-1354(01)00427-4.

Schönning C. and Stenström T. A. (2004). Guidelines on the Safe Use of Urine and Faeces in Ecological Sanitation Systems. Report 2004-1. Stockholm Environment Institute. http://www.ecosanres.org/pdf_files/ESR_Publications_2004/ESR1web.pdf (accessed 2 June 2018).

Schouw N. L., Danteravanich S., Mosbaek H. and Tjella J. C. (2002). Composition of human excreta – a case study from Southern Thailand. *The Science of the Total Environment*, **286**, 155–166.

Schröder J. J., Cordell D., Smit A. L. and Rosemarin A. (2010). Sustainbale Use of Phosphorus. European Union Contract ENV.B.1/ETU/2009/0025. Report 357, Plant Research International, Wageningen University and Research Centre, Wageningen, The Netherlands. https://library.wur.nl/WebQuery/wurpubs/fulltext/163942.

SEI (Stockholm Environment Institute) (2006). Urine Diversion: One Step Towards Sustainable Sanitation; EcoSanRes Publication Series; Stockholm Environment Institute: Stockholm, Sweden. www.ecosanres.org/pdf_files/Urine_Diversion_2006-1.pdf (accessed 20 June 2018).

Sievers D. and Kelly G. (2016). Evidence Series: Market-based Approaches to Sanitation. Population Services International Washington D.C., USA. https://www.psi.org/wp-content/uploads/2016/06/Sanitation_EvidenceSeries_Jun2016.pdf (accessed 3 June 2018).

Simavi (2015) Factsheet Sustainable Sanitation Service Chain. Simavi, Haarlem https://simavi.org/wp-content/uploads/2015/03/150317_7300037_Factsheet_Simavi_Sustainable-Sanitation.pdf. **7**, 203–218. http://dx.doi.org/10.1016/j.eti.2017.02.004 (accessed 20 June 2018).

Simha P. and Ganesapillai M. (2017). Ecological Sanitation and nutrient recovery from human urine: how far have we come? A review. *Sustainable Environment Research*, **27**, 107–116, http://dx.doi.org/10.1016/j.serj.2016.12.001

Simha P., Mathew M., Jaind P. and Ganesapillai M. (2016). Resource recovery and recycling in sanitation is key to health, water and food security. *Procedia Technology*, **25**, 201–207.

Simha P., Lalander C., Vinnerås and Ganesapillai M. (2017) Farmer attitudes and perceptions to the re–use of fertiliser products from resource–oriented sanitation systems – the case of Vellore, South India. *Science of the Total Environment*, **581–582**, 885–896.

Simons J., Clemens J. and Kaub J. (2005). Biologische Behandlung der Fakalien. In: *Nahrstoffrennung und – verwertung in der Abwassertechnik am Beispiel der "Lambertsmuhle", Bonner Agrikulturchemische Reihe*, **Vol. 21**. INRES, Bonn, Germany.

Singh S., Mohan R. R., Rathi S. and Raju N. J. (2017). Technology options for faecal sludge management in developing countries: benefits and revenue from reuse. *Environmental Technology & Innovation*, **7**, 203–218, http://dx.doi.org/10.1016/j.eti.2017.02.004

Slater S. (2009). Resources from Waste: A Guide to Integrated Resource Recovery. Ministry of Community Development, Victoria, Canada. https://www2.gov.bc.ca/assets/gov/british-columbians-our-governments/local-governments/planning-land-use/resources_from_waste_irr_guide.pdf (accessed 20 June 2018).

Smith D. C., Beharee V. and Hughes J. C. (2001). The effects of composts produced by a simple composting procedure on the yields of Swiss chard (Beta vulgaris L. var. flavescens) and common bean (Phaseolis vulgaris L. var. nanus). *Scientia Horticulture*, **91**, 394–406.

Snell J. R., III (1943) Anaerobic digestion of undiluted human excreta. *Sewage Works Journal*, **15**, 679–701.

Sonakya V., Raizada N. and Kalia V. C. (2001). Microbial and enzymatic improvement of anaerobic digestion of waste biomass. *Biotechnology Letters*, **23**(18), 1463–1466.

Spears B. M., Meis S., Anderson A. and Kellou M. (2013). Comparison of phosphorus (P) removal properties of materials proposed for the control of sediment p release in UK lakes. *Science of the Total Environment*, **442**, 103–110.

Springer-Heinzer A. (2017). ValueLinks 2.0 – Manual on Sustainable Value Chain Development. Value Chain Analysis, Strategy and Implementation. Draft volume 1, pp 349, GTZ. http://valuelinks.org/wp-content/uploads/2015/09/ValueLinks-Manual-2.0-Vol-1-final-draft-July-2017.pdf

Stark K. (2004). Phosphorus recovery – experience from European countries. In: Proceedings of Polish-Swedish seminars. Integration and optimization of urban sanitation systems, E. Plaza, E. Levlin and B. Hultman (eds), Stockholm, Sweden, pp. 19–30. http://www2.lwr.kth.se/forskningsprojekt/Polishproject/rep12/StarkSthlm19.pdf (accessed 6 November 2013).

Stenström T. A. (1996). Pathogenic microrganisms in wastewater systems – Risk assessment of traditional and alternative wastewater systems. (Sjukdomsframkallande mikroorganismer i avloppssystem – Riskvärdering av traditionella och alternativa avloppslösningar). Report 4683, Swedish Environmental Protection Agency and National Board of Health and Welfare, Stockholm, Sweden (In Swedish).

Stenström T. A., Seidu R., Ekane N. and Zurbrügg C. (2011). Microbial Exposure and Health Assessments in Sanitation Technologies and Systems, EcoSanRes Series, 2011-1. Stockholm Environment Institute, Stockholm. http://www.susana.org/_resources/documents/default/2-1236-microbialexposurehealthassessmentsinsanitationtechnologiessystems.pdf (accessed 3 June 2018).

Stoll U. and Parameswaran K. (1996). Treatment and disposal of domestic sewage sludge and nightsoil sludge for Bangkok. *Water Science and Technology*, **34**, 209–217.

L. Strande, M. Ronteltap and D. Brdjanovic (eds) (2014). Faecal Sludge Management: Systems Approach for Implementation and Operation. IWA Publishing, London, UK.

SuSanA (2017). Definition of sanitation value chain. https://forum.susana.org/166-definitions-glossaries-dictionaries-mapping-tools/21715-definition-of-sanitation-value-chain.

Sutton P. M., Melcer H., Schraa O. J. and Togna A. P. (2011). Treating municipal wastewater with the goal of resource recovery. *Water Science and Technology*, **63**(1), 25–31.

Svensson L. (2000). Diagnosis of foodborne viral infections in patients. *International Journal of Food Microbiology*, **59**(1–2), 117–126.

Sy J., Warner R. and Jamieson J. (2014) Tapping the Markets: Opportunities for Domestic Investments in Water and Sanitation for the Poor. Directions in Development. World Bank, Washington D.C., USA. License: Creative Commons Attribution CC BY 3.0. doi:10.1596/978-1-4648-0134-1

Taglia P. J. (2010). Emerging policy opportunities for the biogas sector. 5th Agstar National Conference, 28 April 2010. www.epa.gov/agstar/pdf/conf10/Taglia-BiogasAgSTAR.pdf (accessed 30 July 2010).

Takaoka M., Oshita K., Cui X., Matsukawa K. and Fujiwara T. (2010) Phosphorus material flow and its recovery from wastewater and solid waste. UNEP-DTIE-IETC, Regional Workshop on Waste Agricultural Biomass, 2–5 March 2010 at Global Environment Centre Foundation in Osaka. Japan Institute of Wastewater Engineering Technology (JIWET).

Tamis J., Marang L., Jiang Y., Van Loosdrecht M. C. M. and Kleerebezem R. (2014). Modeling PHA-producing microbial enrchment cultures – towards a generalized model with predictive power. *New Biotechnology*, **31**(4), 324–334.

Tarpeh W. A., Barazesh J. M., Cath T. Y. and Nelson K. L. (2018). Electrochemical stripping to recover nitrogen from source-separated urine. *Advances in Environmental Science and Technology*, **52**(3), 1453–1460, doi: 10.1021/acs.est.7b05488

Tayler K., Colin J. and Parkinson J. (2000). Strategic Planning for Municipal Sanitation: A Guide. GHK Research and Training, London.

TBC (Toilet Board Coalition) (2017). The sanitation economy in India: Market estimates and insights. http://www.toiletboard.org/media/38-The_Sanitation_Economy_in_India.pdf (accessed 5 June 2018).

Tettenborn F., Behrendt J. and Otterpohl R. (2007). Resource Recovery and Removal of Pharmaceutical Residues. Treatment of Separate Collected Urine within the EUFundedSCST-Project. Institute of Wastewater Management and Water Protection, Hamburg University of Technology, Hamburg, Germany. http://www.tu-harburg.de/aww/publikationen/pdf/endbericht_tutech151009.pdf (accessed 16 February 2009).

Thames Water (2009). Thames Water Asset Standard – Wastewater Process - WWP 1.6-Chemical Dosing Systems, s.I. Thames Water Utilities Ltd, Reading, UK.

Thibodeau C., Monette F., Bulle C. and Glaus M. (2014). Comparison of blackwater source-separation and conventional sanitation systems using life cycle assessment. *Journal of Cleaner Production*, **67**, 45–57, http://dx.doi.org/10.1016/j.jclepro.2013.12.012

Tilley E. (2013). Conceptualizing sanitation systems to accounts for new complexities in processing and management. In: Source Separation and Decentralization for Wastewater Management, T. A. Larsen, K. M. Udert and J. Lienert (eds), IWA Publishing, London, UK, pp. 227–239.

Tilley E., Strande L., Luthi C., Mosler H., Udert K. M., Gebauer H. and Hering J. G. (2014a). Looking beyond technology: an integrated apparoach to water, sanitation and hygiene in low income countries. *Advances in Environmental Science and Technology*, **48**(17), 9965–9970, doi: 10.1021/es501645d

Tilley E., Ulrich L., Luthi C., Reymond P. H. and Zurbrugg C. (2014b). Compendium of Sanitation Systems and Technologies, 2nd rev. edn. Swiss Federal Institute of Aquatic Science and Technology (EAWAG), Dübendorf, Switzerland.

Tingsten K. (1911). The Cleaning of Stockholm from the Oldest Times Until Our Days. (Stockholms renhållningsväsende från äldsta tider till våra dagar). P.A. Norstedt & Söners Förlag, Stockholm, Sweden (In Swedish).

Titirici M. M., Thomas A. and Antonietti M. (2007). Back in the black: hydrothermal carbonization of plant material as an efficient chemical process to treat the CO2 problem? *New Journal of Chemistry*, **31**, 787–789.

Todt D. (2015) Source Separating Sanitary Systemse Energy Efficient Treatment of Blackwater and Minimizing Greenhouse Gas Emissions. *PhD thesis*, Faculty of Environmental Science and Technology, Department of Environmental Sciences, Norwegian University of Life Sciences, Akershus, Norway. https://brage.bibsys.no/xmlui/handle/11250/2448556 (accessed 13 June 2018).

Torres M. L. and Lloréns M. d. C. E. (2008). Effect of alkaline pretreatment on anaerobic digestion of solid wastes. *Waste Management*, **28**(11), 2229–2234.

UN (United Nations) (1992) The United Nation Program of Action from Rio-Agenda. United Nations, New York, USA, vol. **21**.

UN (United Nations) (2009). Water in a Changing World. World Water Development Report 3. http://unesdoc.unesco.org/images/0018/001819/181993e.pdf (accessed 2 June 2018).

UNDP (2018). Sustainable development goals. http://www.undp.org/content/undp/en/home/sustainable-development-goals.html (accessed 31 May 2018).

USDOE (United States Department of Energy) (2014). The Water-Energy Nexus: Challenges and Opportunities, DOE/EPSA-0002. https://www.energy.gov/under-secretary-science-and-energy/downloads/waterenergy-nexus-challenges-and-opportunities (accessed 18 June 2018).

USDOE (United States Department of Energy) (2016). Hydrogen, Hydrocarbons, and Bioproduct Precursors from Wastewaters Workshop; United States Department of Energy; Office of Energy Efficiency and Renewable Energy Bioenergy Technologies Office/Fuel Cell Technologies Office: Washington D.C., USA. http://energy.gov/eere/ fuelcells/hydrogen-hydrocarbons-and-bioproductprecursors-wastewaters-workshop (accessed 15 June 2018).

USEPA (U.S. Environmental Protection Agency) (1994). EPA, Biosolids: Frequently Asked Questions. see also, EPA, Biosolids Recycling: Beneficial Technology for a Better Environment (1994). http://www.epa.gov/owm/mtb/biosolids/genqa.htm (accessed 11 June 2018).

USEPA (U.S. Environmental Protection Agency) (2003). Control of Pathogens and Vector Attraction in Sewage Sludge. United State Environmental Protection Agency, National Risk Management Research Laboratory, Cincinnati, USA.

USEPA (U.S. Environmental Protection Agency) (2004). Guidelines for Water Reuse. Municipal Support Division, Office of Wastewater Management, EPA, Washington D.C., USA.

USEPA (U.S. Environmental Protection Agency) (2012). Guidelines for Water Reuse. EPA/600/R-12/618, AR-1530. https://www3.epa.gov/region1/npdes/merrimackstation/pdfs/ar/AR-1530.pdf (accessed 11 June 2018).

USEPA (U.S. Environmental Protection Agency) (2015). Safe and Sustainable Water Resources Strategic Research Action Plan 2016-2019. EPA 601/K-15/004. https://www.epa.gov/research/safe-andsustainable-water-resources-strategic-research-action-plan-2016-2019

van der Hoek J. P., de Fooij H. and Struker A. (2016). Wastewater as a resource: strategies to recover resources from Amsterdam's wastewater. *Resources, Conservation and Recycling*, **113**, 53–64, http://dx.doi.org/10.1016/j.resconrec.2016.05.012

van Dijk M. P. (2012). Sanitation in developing countries: Innovative solutions in a value chain framework. In Management of technological innovation in developing and developed countries., ed. Hongyi SunIntech. http://www.intechopen.com/books/management-of-technological-innovation-in-developing-and-developed-countries/sanitation-in-developing-countries-innovative-solutions-in-a-value-chain-framework (accessed 4 June 2018).

van Welie M. J. and Romijn H. A. (2018). NGOs fostering transitions towards sustainable urban sanitation in low-income countries: insights from transition management and development studies. *Environmental Science & Policy*, **84**, 250–260, http://dx.doi.org/10.1016/j.envsci.2017.08.011

Verbyla M. E., Oakley S. M. and Mihelcic J. R. (2013). Wastewater infrastructure for small cities in an urbanizing world: integrating *protection of human health and the environment with resource recovery and food security. Advances in Environmental Science and Technology*, **47**, 3598–3605, http://dx.doi.org/10.1021/es3050955

Verlicchi P. and Zambello E. (2014). How efficient are constructed wetlands in removing pharmaceuticals from untreated and treated urban wastewaters? A review. *Science of the Total Environment*, **470-471**, 1281–1306.

Verstraete W. and Vlaeminck S. E. (2011). ZeroWasteWater: short-cycling of wastewater resources for sustainable cities of the future. *International Journal of Sustainable Development & World Ecology*, **18**, 253–264. https://doi.org/10.1080/13504509.2011.570804.

Verstraete W., Van de Caveye P. and Diamantis V. (2009). Maximum use of resources present in domestic 'used water'. *Bioresource Technology*, **100**(23), 5537–5545.

Vinnera B. (2002) Possibilities for Sustainable Nutrient Recycling by Faecal Separation Combined with Urine Diversion *PhD thesis*, Agraria 353, Swedish University of Agricultural Sciences, Uppsala, Sweden. https://pub.epsilon.slu.se/332/1/Agraria_353.pdf

Vinnerås B., Höglund C., Jönsson H. and Stenström T. A. (1999). Characterisation of sludge in urine separating sewerage systems. In: B. Klöve, C. Etniner, P. Jenssen and T. Maehlum (eds) Proceedings of the 4th International Conference – Managing the Wastewater Resource Ecological Engineering for Wastewater Treatment. Norway, June 7–11, 1999.

Vinnerås B., Palmquist H., Balmér P., Weglin J., Jensen A., Andersson Å and Jönsson H. (2006). The characteristics of household wastewater and biodegradable waste – a proposal for new Swedish norms. *Urban Water*, **3**, 3–11.

Vinnerås B., Nordin A., Niwagaba C. and Nyberg K. (2008). Inactivation of bacteria and viruses in human urine depending on temperature and dilution rate. *Water Research*, **42**, 4067–4074.

Vogel C., Krüger O., Herzel H., Amidani L. and Adam C. (2016). Chemical state of mercury and selenium in sewage sludge ash based P-fertilizers. *Journal of Hazardous Materials*, **313**, 179–184.

Walker R. W., Wun C. K. and Litsky W. (1982). Coprostanol as an indicator of fecal pollution. *Critical Reviews in Environmental Control*, **12**, 91–112.

Wang L., Peng J., Wang B. and Yang L. (2006). Design and operation of an eco-system for municipal wastewater treatment and utilization. *Water Science and Technology*, **54**(11–12), 429–436, doi: 10.2166/wst.2006.923, PMID:17302348

Ward R. L., Bernstein D. I., Young E. C., Sherwood J. R., Knowlton D. R. and Schiff G. M. (1986). Human rotavirus studies in volunteers: determination of infectious dose and serological response to infection. *The Journal of Infectious Diseases*, **154**(5), 871–880.

Ward R. L., Knowlton D. R., Stober J., Jakubowki W., Mills T., Graham P. and Camman D. E. (1989). Effect of wastewater spray irrigation on rotavirus infection rates in an exposed population. *Water Research*, **23**(12), 1503–1509. https://doi.rg/10.1016/0043-1354(89)9015-2.

WCED (World Comission on Environment and Development) (1987). Our Common Future. Oxford University Press, Oxford, UK.

WE&RF (Water Environment Research Foundation) (2011). Nutrient Recovery State of the Knowledge. www.werf.org/c/2011Challenges/Nutrient_Recovery.aspx (accessed 18 June 2018).

WEF (Water and Environment Federation) (2010). Land Application and Composting of Biosolids: Q&A/Fact Sheet. https://www.wef.org/globalassets/assets-wef/3---resources/topics/a-n/biosolids/technical-resources/wef-land-app-fact-sheet---rev0510.pdf.

Weigand H., Bertau M., Hübner W., Bohndick F. and Bruckert A. (2013). RecoPhos: full-scale fertilizer production from sewage sludge ash. *Waste Management*, **33**(3), 540–544.

Welsh School of Architecture (2008). Case study: The Bromma biogas plant. Stockholm. http://www.cardiff.ac.uk/archi/programmes/cost8/case/watersewerage/bromma.html (accessed 12 August 2008).

WERF (Water Environmental Research. Foundation) (2011). Nutrient recovery: state of the knowledge as of December 2010. www.werf.org/c/2011Challenges/Nutrient_Recovery.aspx (accessed 6 November 2013).

Werner C., Fall P. A., Schlick J. and Mang H. P. (2003). Reasons for and principles of ecological sanitation. Proceedings of the 2nd international symposium on ecological sanitation, Lübeck, Germany, 7 to 11 April 2003.

Werner C., Panesar A., Rud S. B. and Olt C. U. (2009). Ecological sanitation: principles, technologies and project for examples sustainable wastewater and excreta management. *Desalination*, **248**, 392–401.

Westerman P. W., Zering K. D. and Rashash D. (2009). Struvite Crystallizer for Recovering Phosphorous from Lagoon and Digester Liquid. Unpublished article. NC State University collaborative Extension program, USA. http://www.bae.ncsu.edu/programs/extension/manure/lagoon/ag-724w_struvite_crystallizer.pdf (accessed 23 June 2018).

Wett B., Buchauer K. and Fimml C. (2007). Energy self-sufficiency as a feasible concept for wastewater treatment systems. In: IWA Leading Edge Tehnology Conference, Asia Water Singapore, pp. 21–24.

WHO (World Health Organization) (2006a). Guidelines for the safe use of wastewater, excreta and greywater, Volume I, Policy and regulatory aspects. http://apps.who.int/iris/bitstream/handle/10665/78265/9241546824_eng.pdf?sequence=1 (accessed 2 June 2018).

WHO (World Health Organization) (2006b). Guidelines for the Safe Reuse of Wastewater, Excreta and Greywater, Volume III, Wastewater and Excreta Use in Aquaculture. WHO, Geneva, Switzerland.

WHO (World Health Organization) (2016). Sanitation Safety Planning: Manual for Safe Use and Disposal of Wastewater, Greywater and Excreta. WHO Press, Geneva, Switzerland.

WHO (World Health Organization) (2017). Guidelines for drinking-water quality, 4th edition, incorporating the 1st addendum. http://www.who.int/water_sanitation_health/publications/drinking-water-quality-guidelines-4-including-1st-addendum/en/ (accessed 15 June 2018).

Wielemaker R. C., Weijma J. and Zeeman G. (2016). Harvest to harvest: recovering nutrients with new sanitation systems for reuse in urban agriculture. *Resources, Conservation and Recycling*. http://dx.doi.org/10.1016/j.resconrec.2016.09.015

Willetts J., Gero A., Susamto A. A., Sanjaya R., Trieu T. D., Murta J. and Carrard N. (2017). Sanitation value chains in low density settings in Indonesia and Vietnam: impetus for a rethink to achieve pro-poor outcomes. *Journal of Water, Sanitation and Hygiene for Development*, **7**(3), doi: 10.2166/washdev.2017.141

Wilsenach J. A., Schuurbiers C. A. H. and van Loosdrecht M. C. M. (2007). Phosphate and potassium recovery from source separated urine through struvite precipitation. *Water Research*, **41**(2), 458–466, doi: 10.1016/j.watres.2006.10.014, PMID:17126877

Winblad U. (1994) Small-scale systems for recycling of human excreta. Integrated Measures to Overcome Barriers to Minimizing Harmful Fluxes from Land to Water; Book Series: 3rd Stockholm Water Symposium, **3**, 225–231.

Winker M., Vinnerås B., Muskolus A., Arnold U. and Clemens J. (2009). Fertiliser products from new sanitation systems: their potential values and risks. *Bioresource Technology*, **100**, 4090–4096, doi: 10.1016/j.biortech.2009.03.024

WSP (Water and Sanitation Program) (2004). The Case for Marketing Sanitation. Field Note. http://www.wsp.org/sites/wsp.org/files/publications/af_marketing.pdf (accessed 13 June 2018).

WSP (Water and Sanitation Program) (2011a). Sanitation Markets at the Bottom of the Pyramid a Win-Win Scenario for Government, the Private Sector, and Communities: Domestic Private Sector Participation in Peru. Learning Note. https://www.wsp.org/sites/wsp.org/files/publications/WSP-Sanitation-Markets-Bottom-Pyramid-Peru.pdf (accessed 12 June 2018).

WSP (Water and Sanitation Program) (2011b). Sanitation Marketing at Scale: Experiences from Rural Benin. Sanitation and Hygiene Series. https://www.wsp.org/sites/wsp.org/files/publications/wsp-benin-sanitation-marketing.pdf (accessed 12 June 2018).

Xavier L. D., Cammarota M. C., Yokoyama L. and Volschan I. (2014). Study of the recovery of phosphorus from struvite recipitation in supernatant line from anaerobic digesters of sludge. *Water Science and Technology*, **69** (7), 1546–1551. https://doi.org/10.2166/wst.2014.033.

Yan J., Wang R. and Wang M. (1998). The fundamental principles and ecotechniques of wastewater aquaculture. *Ecological Engineering*, **10**, 191–208, PII S0925-8574(98)00019-6.

Yan J., Zhang Y., Ding S., Xu Y., Li W. (1993a). The study on the ecological engineering for treating and purifying the wastewater from township and village enterprise. In: EPAC (Environmental Protection Agency of China) (ed.) Technology for Wastewater Control and Treatment with Utilization Municipal Wastewater. Science Press, Beijing, China, pp. 979–988 (in Chinese).

Yan J., Zhang Y. and Wu X. (1993b). Advance of ecological engineering in China. *Ecological Engineering*, **2**, 193–216.

Yates M. V. and Gerba C. P. (1998). Microbial considerations in wastewater reclamation and reuse. In: Wastewater Reclamation and Reuse, T. Asano (ed.), Technomic Publishing Company, Inc., Lancaster, USA, pp. 437–488.

Ye Y., Ngo H. H., Guo W., Liu Y., Li J., Li Y., Zhang X. and Jia H. (2017). Insight into chemical phosphate recovery from municipal wastewater. *Science of the Total Environment*, **576**, 159–171, http://dx.doi.org/10.1016/j.scitotenv.2016.10.078

Yuan Z., Shi J., Wu H. and Bi J. (2011). Understanding the anthropogenic phosphorus pathway with substance flow analysis at the city level. *Journal of Environmental Management*, **92**, 2021–2028.

Zamalloa C., Vulsteke E., Albrecht J. and Verstraete W. (2011). The techno-economic potential of renewable energy through the anaerobic digestion of microalgae. *Bioresource Technology*, **102**(2), 1149–1158.

Zeeman G. (2012). New Sanitation: Bridging Cities and Agriculture. Wageningen University, Wageningen, Netherlands. http://edepot.wur.nl/246120 (accessed 23 June 2018).

Zeeman G., Kujawa K., Meulman B. and Kwant F. (2007). Full Scale Demonstration of Vacuum Collection, Transport & Treatment of Blackwater. Poster Presentation, Advanced Sanitation Conference, Aachen, Germany, March 2007.

Zhang J. F., She Q. H., Chang W. C., Tang C. Y. and Webster R. D. (2014). Mining nutrients (N, K, P) from urban source-separated urine by forward osmosis dewatering. *Advances in Environmental Science and Technology*, **48**, 3386–3394.

Zhou C., Liu J., Wang R., Yang W. and Jin J. (2010). Ecological-economic assessment of ecological sanitation development in the cities of Chinese Loess Plateau. *Ecological Complexity*, **7**, 162–169.

Zomerplaag J. and Mooijman A. (2005). Child-Friendly Hygiene and Sanitation Facilities in Schools: Indispensable to Effective Hygiene Education. IRC International Water and Sanitation Centre & UNICEF, Delft, Netherlands.

Zurbrügg C. and Tilley E. (2007). Evaluation of Existing Low Cost Conventional as Well as Innovative Sanitation System and Technologies – NETSSAF Deliverable D22&23. Swiss Federal Institute of Aquatic Science and Technology (EAWAG), Dübendorf, Switzerland. https://www.sswm.info/sites/default/files/reference_attachments/NETSSAF_Evaluation%20of%20existing%20low-cost%20conventional%20as%20well%20as%20innovative%20sanitation%20system%20and%20technologies.pdf (accessed 15 March 2018).

Chapter 7

Sanitation 4.0

'The test of our progress is not whether we add more to the abundance of those who have much; it is whether we provide enough for those who have too little'

Franklin D. Roosevelt, 32nd president of US

7.1 INTRODUCTION

Sanitation 4.0 is the strategic and practical application of ReGenSan (i.e. the fourth paradigm of sanitation), which acknowledges the dynamic complexities of sanitation systems through the integration of all subsystems (SES, TeS, and ReS), dimensions and components (see Chapter 3). It seeks to align with the targets of the SDG 6 and recognizes that household sanitation infrastructure (e.g. collection) is just one element in a full sanitation service delivery chain from collection to treatment and recovery of SDPs with minimal disposal of residues (Sparkman & Sturzenegger, 2018). This programme will be designed to provide sani-solutions for the vulnerable population, particularly those with the greatest burden, in order to meet the SDG targets of 'adequate and equitable' sanitation and hygiene for all by 2030. In this case, 'adequate' implies a system of 'safely managed' by-products of human digestion right through the sanitation service chain (whether with safe containment and disposal in situ or safe transport and treatment/reuse off-premises), which could be critical for public health and environmental protection (Prüss *et al.*, 2002; UN-Water, 2016). This, consequently, doubles the MDGs' service shortfall of 2.3 billion people (without improved sanitation facilities) to the SDG's current estimated 4.5 billion population without 'safely managed' sanitation facilities. In short, about three-fifths of the global population (much more than before the MDGs) do not have access to 'safely managed' sanitation (Mason *et al.*, 2017). This is probably because the 'safely managed' indicator demands (for the first time at global level) to know what happens to human excreta after its discharge into the toilet or latrine (WHO/UNICEF, 2017a) as against merely counting existing facilities. These new insights have given birth to an increased urgency towards improved service, expanded access, recovered resources (of value) and synergy with other sectors such as agriculture, horticulture, aquaculture etc. Furthermore, traditional

© IWA Publishing 2019. Regenerative Sanitation: A New Paradigm for Sanitation 4.0
Editor(s): Thammarat Koottatep, Peter Emmanuel Cookey and Chongrak Polprasert
doi: 10.2166/9781780409689_0283

dichotomies of small and large, modern and conventional, sewer and non-sewer, centralised and non-centralised technologies need to shift to more appropriate, holistic, integrated and hybridized options with transdisciplinary perspectives (Brands, 2014; Reymond *et al.*, 2016; van Vliet *et al.*, 2011).

Over the past four decades, sanitation has progressed along the lines of three major paradigms, which this book presented in Chapter 1 (sanitation 1.0 [see 1.4.1], sanitation 2.0 [see 1.4.2] and sanitation 3.0 [see 1.4.3]), but they were mainly technocentric (Black & Fawcett, 2008) (Figure 7.1). It climbed up the development agenda after the UN General Assembly in November 1980 proclaimed 1981–1990 as the International Drinking Water Supply and Sanitation Decade (IDWSSD) and all countries adopted the declared target of achieving 100% sanitation coverage by 1990. The launch of the Decade programme gave sanitation a publicity boost and led to concerted efforts to speed up the progress. The 1980s experienced real progress in technology development for improving sanitation services for low-income communities (WELL, 1998). But, at the end of the decade, sanitation challenges were still huge and continued to escalate as more than 3 billion people remained without access or service.

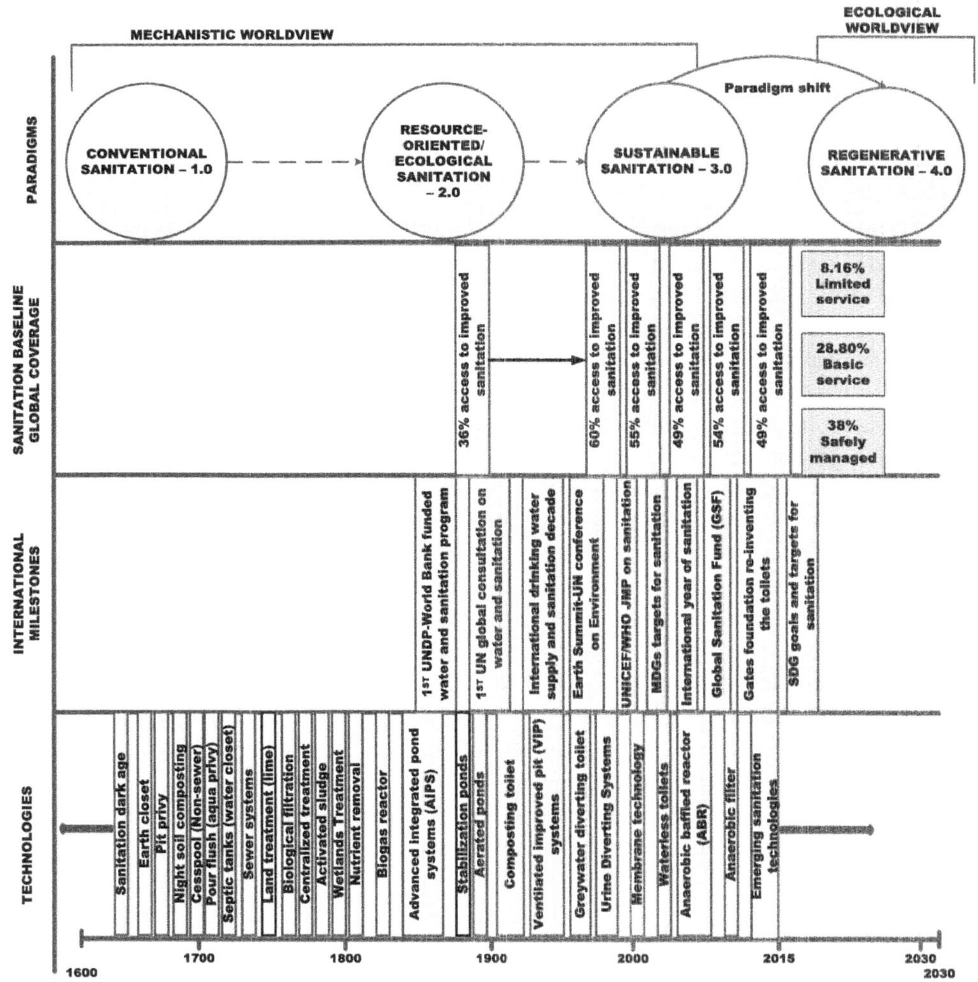

Figure 7.1 Sanitation historical timelines.

The MDGs came in 1990 with global targets that sought to reduce by half 'the proportion of the population without sustainable access to basic sanitation between 1990 and 2015'. Access to basic sanitation was measured with the indicator of 'proportion of population using an improved sanitation facility', which meant that such facilities should hygienically separate human excreta from human contact (UNDG, 2003). This target for 'basic sanitation' was further established at the World Summit on Sustainable Development (WSSD) and subsequently incorporated into the MDGs. The WSSD defined the term in broader perspectives than the 'improved sanitation' initially used by the MDG-JMP (Lenton *et al.*, 2005). According to the WSSD, 'basic sanitation' includes:

(I) Developing and implementing efficient household sanitation systems;
(II) Improving sanitation in public institutions, especially schools;
(III) Promoting education and outreach focused on children as agents of behaviour change;
(IV) Promoting affordable and socially and culturally acceptable technologies and practices;
(V) Developing innovative financing and partnership mechanisms;
(VI) Integration of sanitation into water resources management strategies in a manner that does not negatively affect the environment (i.e. protection of water resources from biological or faecal contamination).

The WSSD was more impact-oriented (with particular focus on communities with very low coverage and levels of sanitation services) and pursued the creation of processes and approaches that improved sanitation services and expanded coverage rather than on just constructing a certain number of toilets (Lenton *et al.*, 2005). The MDGs era peaked in 2008 when the UN General Assembly declared that year as the International Year of Sanitation (IYS), geared towards averting a global crisis. The message was clear: sanitation is vital for health; enhances social development; supports economic investment; improves the environment; and best of all, it is achievable (UN-Water, 2008). But, after fifteen years of the MDGs, there are still about 2.3 billion people or more without basic services (or using unimproved facilities) and roughly 892 million others still practising open defecation (UNICEF/WHO, 2015). The era of the SDGs has come in with higher expectations that go beyond 'halving' to 'universal access to safely managed' sanitation driven with immediate urgency at the global level (WHO/UNICEF, 2017a).

However, 'basic sanitation' or 'safely managed sanitation' will only be achieved through a combination of psycho-social, cultural and ecological factors as well as techno-solutions (Reymond *et al.*, 2016; Rosenqvist *et al.*, 2016). For instance, sanitation systems can experience social failure from non-use or misuse of available technology due to sociocultural or socio-economic deficiencies related to lack of demand, acceptance, knowledge of use, unwillingness or inability to pay as well as psycho-social and physical limitations; all of which could lead to system breakdown and environmental/hygiene risks (Starkl *et al.*, 2013). In addition, resource recovery from sanitation matter could also be affected by differences in social groups' values and attitudes. This calls for a critical and persistent drive towards aligning existing proven and emerging approaches/technologies, especially in the areas of management and governance, technologies, social marketing and financial mechanisms (among others) that will effectively guide practitioners, donors, policymakers and academics (World Water Council, 2012). This is where Sanitation 4.0 comes in to integrate all previous and emerging paradigms, concepts and approaches. It captures the core essence of sanitation as 'solutions' rather than as an 'arbiter of problems' that must be eradicated from other areas (water, health, hygiene and environment). This application platform sees sanitation as a major entity that could contribute to other sectors and should not be merely attached to seemingly more important concerns. There are approximately 7 billion humans on earth at the current time and about 97% of them will exercise the unavoidable physiological activity of excreting faeces and urine (regardless of age, status, race, gender, geographical region etc) every single day. This is a tremendous amount of material that could be viewed as 'dangerous waste' or 'valuable resource'. Sanitation 4.0 considers excreta a valuable resource and works from this structural premise, and ReGenSan will provide Sanitation 4.0 with emerging and proven strategies to focus on providing sani-solutions for where the needs are greatest, particularly the vulnerable in society (elders, people with disabilities etc), BoPs and Hard-to-Reach-Area (HtRAs).

The Sanitation 4.0 programme will aspire to use the ReGenSan framework to advocate for total rejuvenation and a comprehensive overhaul of all relevant and related architecture/infrastructure (soft and hard) so as to ensure retrofitting of dysfunctional, failed and abandoned facilities and devices, and of the rehabilitation of aged sewered systems (Lopes *et al.*, 2012). It will serve as a platform that actively seeks innovative and appropriate sani-solutions, better synergy and cooperation between actors (formal and informal) as well as ownership by users. Unsafe discharge of faecal matter into the environment will be discouraged and reduced through resource capturing for reuse and proper treatment before disposal. The preconception is that sanitation will deliver maximal benefits to society only when there exists an inter-synergistic integration of the subsystems (SES, TeS and ReS) to enable appropriate linkages between 'technological design' and 'delivery platform' (e.g. Sanitation 4.0) in order to achieve optimal and sustained sani-solutions within the sani-shads. In essence, sanitation is directly connected to natural systems, agriculture, horticulture, aquaculture, constructions and other industries through cyclical interactions that create and support a sanitation economy with blurred boundaries between different enterprises and actors in the IFSVC (see Section 6.3).

The Sanitation 4.0 delivery platform will use a set of resources (i.e. KaS, financial resources and social networks and functions, culture and traditions, etc.) to encourage the adoption and utilization of sanitation infrastructure guided by ReGenSan subsystems, dimensions and components (Bill & Melinda Gates Foundation, 2006; Ramani *et al.*, 2012; Sethi & King, 1998). It will also use market-based approaches to tackle the challenges of improving coverage and service by integrating innovative technological designs with sociotechnical and contextual factors to provide novel solutions (Bocken *et al.*, 2014). Creative and innovative solutions and strategies that guarantee provision that are safe, functional, serviceable, reliable and facilitate optimized resource capturing will be particularly empathized. The key leverage points for this delivery platform will be to seek to understand users' circumstances better, clarify the nature of their demands, require service and product performance quality and certify that proposed interventions carefully match the contexts of targeted sani-sheds (Bill & Melinda Gates Foundation, 2006). To this effect, some key strategic considerations for this proposed Sanitation 4.0 are highlighted in the next sections (see Box 7.1 for summary).

BOX 7.1 SANITATION 4.0 KEY CONSIDERATIONS FOR EXISTING SANITATION CHALLENGES

(I) Special considerations for providing ReGenSan solutions to the population at the base-of-the-pyramid (BoP),
(II) Special considerations for integrating ReGenSan into urban regeneration
(III) Special considerations for integrating ReGenSan into rural development
(IV) Special considerations to integrate ReGenSan into provision of public sanitation services
(V) Special considerations for contextual factors and place in delivering ReGenSan solutions
(VI) Special considerations for governance and standardization
(VII) Special considerations for data collection, analysis and management
(VIII) Special considerations for skills development

7.2 KEY STRATEGIC APPROACHES FOR SANITATION 4.0

There are eight special considerations for the strategic approaches for Sanitation 4.0, as described below:

7.2.1 Special considerations for the population at the base-of-the-pyramid (BoP)

The main strategy for Sanitation 4.0 involves special considerations for the four billion of the population at the base-of-the-pyramid (BoP). Focus will be on providing creative and innovative services and products that are appropriate and affordable and could support livelihood for the BoPs without losing sight of quality, functionality and aesthetics for those at the top-of-the-pyramid (ToP). This is fundamental because most BoPs (commonly found in Africa, Asia, eastern Europe, Latin America and the Caribbean) are exposed to inadequate, unimproved, limited and at best basic sanitation infrastructure (WHO/UNICEF, 2017a). They earn less than USD 2 a day, which adds up to an annual income of about USD 3,000 for local purchasing power and sums up to approximately USD 5 trillion globally (i.e. one-third of the world's economy) (Hammond *et al.*, 2007). The private sector recognizes this untapped market segment, especially in the sanitation sector (Pedi *et al.*, 2013), and so an in-depth and clear insight into this group of consumers and their spending patterns could provide appropriate innovative and focused market-based solutions that could attract investors and entrepreneurs (see Section 5.4.1.3). For example, BoPs globally spend about USD 2,895 billion on food (i.e. about half of their household budget) annually and putting sufficient food on the table is a constant struggle for many of these homes (Hammond *et al.*, 2007); therefore, sani-solutions that support agriculture (even at subsistence level) could improve food supply and generate employment and extra income for BoPs as well as enhance marketability. Sanitation technological designs that enhance the production of SDNs for limited energy supply could reduce food and energy budgets and at the same time boost income and livelihood (see Section 6.4.1.3).

A prime example is the domestic private sector participation in biogas development facilities in the Peru (Prahalad & Hart, 2000) model, which relies on key alliances and interactions among public, private and community actors (WSP, 2011). The local and global purchasing power of the BoPs consumer market constitute one-third of the world's economy (Hammond *et al.*, 2007) and they are increasingly recognized by the private sector as a major untapped market segment, especially in the sanitation industry (Pedi & Davies, 2013). More understanding of this segment would ensure better targeted market-based solutions for them. An understanding of the BoPs' spending patterns could also provide an idea of future innovative sani-solutions investment by entrepreneurs (see Section 5.4.1.3). Therefore, market-based innovative sani-solutions that support agriculture as an essential part of the food supply chain to generate employment and income for the BoPs would greatly impact their lives.

Therefore, leveraging sanitation technological designs that enhance production of SDNs and SDFs as a better substitute for inorganic fertilizers for farmers and provides sanitation derived energy (SDEs) can help reduce their food and energy budgets and at the same time boost their income generation and livelihood (see Section 6.4.1.3). A case example is biogas facilities in the Kibera slum, Kenya, and its rich benefits, which are economical, viable and environment-friendly (Figure 7.2). Biogas production and usage could actually improve sanitation, provide low-cost green energy and SDFs for agricultural purposes and also result in less contamination for water resources and increased business entrepreneurship; all of which are vital to sustainability.

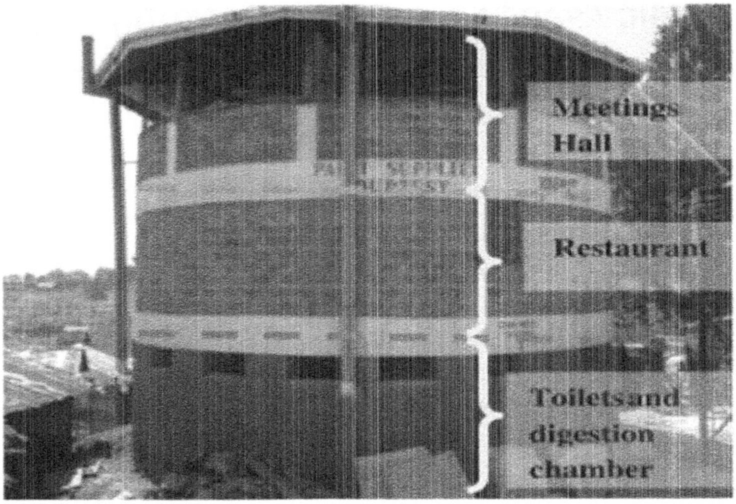

Figure 7.2 Biogas toilet in Kibera slum, Kenya (Reprinted from Katukiza, A.Y., Ronteltap, M., Niwagaba, C.B., Foppen, J.W.A., Kansiime, F. and Lens, P.N.L. (2012), Sustainable sanitation technology options for urban slums, *Biotechnology Advances*, 30, 964–978, with permission from Elsevier.)

To this effect, Sanitation 4.0 should pursue radical innovative solutions targeted at the BoPs in partnership with the private sector (mainly those in the sanitation supply and value chains) so as to improve services and products in sani-sheds (Prahalad & Hart, 2000), which relies on key alliances and interactions among public, private and community actors (WSP, 2011).

The priority of this strategy is to improve sanitation and living conditions of residents in such a way that supports business-oriented solutions to create incentives for investments in sanitation (Katukiza *et al.*, 2012; Yarime & Mutisya, 2011). Sanitation 4.0 demands 'safely managed' facilities that could substantially contribute (directly and indirectly) to the provisions of products and services for BoPs to trade on (see Section 6.4.1.1). The ReGenSan TeS provides a three-pronged dimensional spread of EIDT (see Section 5.4.1.1), ReDT (see Section 5.4.1.2) and NoDT (see 5.4.1.3), which could be adopted to contend with the complex sanitation challenges of BoPs (see Section 5.4.1). These techno-sani-solutions should be functional, serviceable, reliable, acceptable and affordable with innovative quality services and products that are appropriate across the three major segments of BoPs (i.e. extreme poor, subsistence and low income) as described below:

(I) The extreme poor BoPs constitute about 1 billion people residing at the lowest part of the pyramid and are mostly found in rural areas. They lack basic services, earn about USD 1 a day and are often exposed to open defecation in most cases. This group deserves special innovative sani-solutions that could benefit from subsidy programmes of government as key actors with support from NGOs/CBOs and other donors. Several countries' field experiences (supported by evidence-based research) found out that a well-designed, affordable and sustainable subsidy programme undertaken by country governments could help the poor gain access to improved sanitation (WSP, 2012). For instance, the Central Rural Sanitation Programme (CRSP) of the Government of India (GoI) in 1986 subsidized latrine constructions at 80% for households below poverty line (BPL) that desired to build latrines. Also, the Total Sanitation Campaign

(TSC) of 1999–2012 in India used subsidies targeting poor households with health education for hygiene behaviour change to drive demand for latrine coverage in rural communities (O'Reilly *et al.*, 2017). Therefore, Sanitation 4.0 advocates for redesign of sani-solutions to align and support the extreme poor within the BoPs (Pedi *et al.*, 2013; Rangan *et al.*, 2011). A critical feature of sanitation technological solutions at this level is the integration of RRR to provide inputs like SDFs and/or SDNs to support rural agriculture and other related community livelihood activities.

(II) The subsistence BoPs constitute roughly 1.6 billion people who live on USD 1 to 3 a day and typically have some income even though not steady. They, in most cases, use limited, inadequate and/or dysfunctional sanitation facilities and often need improved sanitation, food security, healthcare and education. Governments, private sectors, NGOs/CBOs, donors, communities and others should join forces to provide innovative 'safely managed' sanitation facilities with resource recovery elements as this could support food security and livelihood activities for these BoPs. In addition, institutionalized financial credit schemes and market-based approaches could also produce sanitation entrepreneurs from the BoPs (Pedi *et al.*, 2013; Rangan *et al.*, 2011; WSP, 2011).

(III) The low-income BoPs, roughly 1.4 billion people, who live on USD 3 to 5 a day (Rangan *et al.*, 2011) may require full market-based approaches and mostly use basic sanitation infrastructure. The key role of government will be to facilitate and regulate other players, strongly backed by the public sector and supported by the private sector (as investors, suppliers, contractors or consultants), NGOs/CBOs (for policy dialogue and advocacy), academics and researchers (for innovative solutions and comprehension), donors (for financial support and intervention) and the users/customers (for buy-in and ownership). In this case, innovative safely managed sanitation facilities with resource recovery options that support food security and livelihood activities of BoPs with full-cost recovery could be a sine qua non for success.

This Sanitation 4.0 special consideration for BoPs seeks to link infrastructure upgrading programmes with sanitation markets that will also contribute to improved policies aimed at achieving real impacts in the poor. This will require a shift in outlook from 'beneficiary' to 'consumer' in order to harness viable local entrepreneurs' participation, shared investments (between public-private-community partnerships), community resources, networks and natural systems as well as long-term return on investment (WSP, 2011).

7.2.2 Special considerations for integrating ReGenSan into urban regeneration

Urban sanitation facilities in most developing countries are characterized by unimproved services and limited access to 'safely managed' infrastructure. These facilities are usually poorly operated and maintained, which causes very high abandonment rates due to failure and non-standard procedures for proper faecal sludge management, thus increasing risks to public health and environmental quality (Cookey, 2013, Cookey *et al.*, 2016a, b, c; CSIR, 2007; Kaminsky & Javernick, 2015; Rodgers *et al.*, 2007; WSP, 2007) (see Section 5.2.4). Meanwhile, urban sanitation in developed countries is challenged with ageing sewered infrastructure dating back more than 100 years (when brick sewers were common) and now deterioration in performance occurs as blockages and leakages lead to infiltration and exfiltration, resulting in poor service delivery) (Ahmadi *et al.*, 2015; Andersson *et al.*, 2016; Ashley &

Hopkinson, 2002; Brown & Caldwell, 1984; Corcoran *et al.*, 2010; Diagne, 2013; El-Sheikh, 2011; Fenner & Saward, 2004; Kemp, 2000; Miszta-Kruk, 2016; Reynolds & Barrett, 2003; Qasem *et al.* 2009; USEPA, 1991) (see Section 5.2.5). It is estimated that globally 1.5 billion urban sewered facilities receive inadequate treatment and produce effluents of questionable standards (Baum *et al.*, 2013; Mara, 2003); these deplorable conditions impact urban sanitation management negatively, as is evident in major infrastructural deterioration, dysfunctionalities and failures that disrupt the urban environments' support systems (e.g. urban food production, green space and ecosystem services and functions) (Corcoran *et al.*, 2010) (See Section 5.4.1.2).

Giving special considerations to sanitation and urban regeneration (UR) could complement the focus on BoPs by ensuring complete coverage of all those residing in urban centres. In this regard, Sanitation 4.0 aims for functional and progressive approaches that support food security, urban greening, zero waste emissions and environmental protection, while also ensuring improved services and expanded access at 'scale'. Primarily, this strategy will push for continuous improvement of existing fit urban sanitation infrastructure, retrofitting of dysfunctional facilities and emerging/innovative solutions that incorporate RRR into technologies and infrastructure into the urban contexts of specific SES. Consequently, the platform should:

- (I) Ensure infrastructural standardizations, especially with non-sewered sanitation systems in cities of developing countries;
- (II) Ensure that urban sanitation infrastructure delivers products and services, particularly in harnessing resources from sanitation systems to support urban cities' renewable energy, local food supply and value chain and green landscapes;
- (III) Ensure that urban sanitation solutions are designed to address cities' population at the base-of-the-pyramid (BoP); and
- (IV) Ensure that urban sanitation infrastructure aligns with urban landscape to support biodiversity by mimicking nature's modes, processes and ecosystem and adapting them to solve sanitation challenges.

Providing sanitation services for the fast-growing (about 4 billion) urban population (i.e. 54% of the global population) is one of the world's most urgent challenges (Andersson *et al.*, 2016). An estimated 756 million urban dwellers lack access to 'safely managed' sanitation facilities (Figure 7.3) (Galli *et al.*, 2014; UNICEF/WHO, 2015; van Welie & Romijn, 2018), probably because UR debates often centred on the astonishingly dynamic infrastructural and technical complexities.

The fact that there are also people in this mix who engage in more toilet and hygiene-related activities than any other known human-biophysical activities in their lifetime is downplayed. And yet this also impacts urban life and society even as the core sanitation requirements for the sustainability of urban cities are not treated (Anastasiadis & Metaxas, 2013, Czischke *et al.*, 2015; Jewitt, 2011; Lehmann, 2010a; Parkinson *et al.*, 2009). Most of the time people will avoid the subject of faeces and urine and tuck them away under less unsavoury concepts such as wastewater and solid waste management while the real concerns overrun our metropolitan areas, which then hinders the SDGs targets for inclusive, safe and resilient cities hinged on a resource-based paradigm (RBP) (Bai *et al.*, 2016; Schmitt *et al.*, 2017; UNDP, 2016).

Sanitation 4.0 draws from the RBP and considers sanitation processes as a production factory that could provide high-quality bioenergy resources, safe fertilizers, soil conditioners, construction and pharmaceutical materials that will enhance ecosystem sustainability through urban greenery (horticulture), tree planting, nature preservation, increased biodiversity, climate change mitigation, water recycling for urban agriculture (crop and animal), aquaculture and so on (Esrey, 2001; Winker *et al.*, 2009; Zaręba *et al.*,

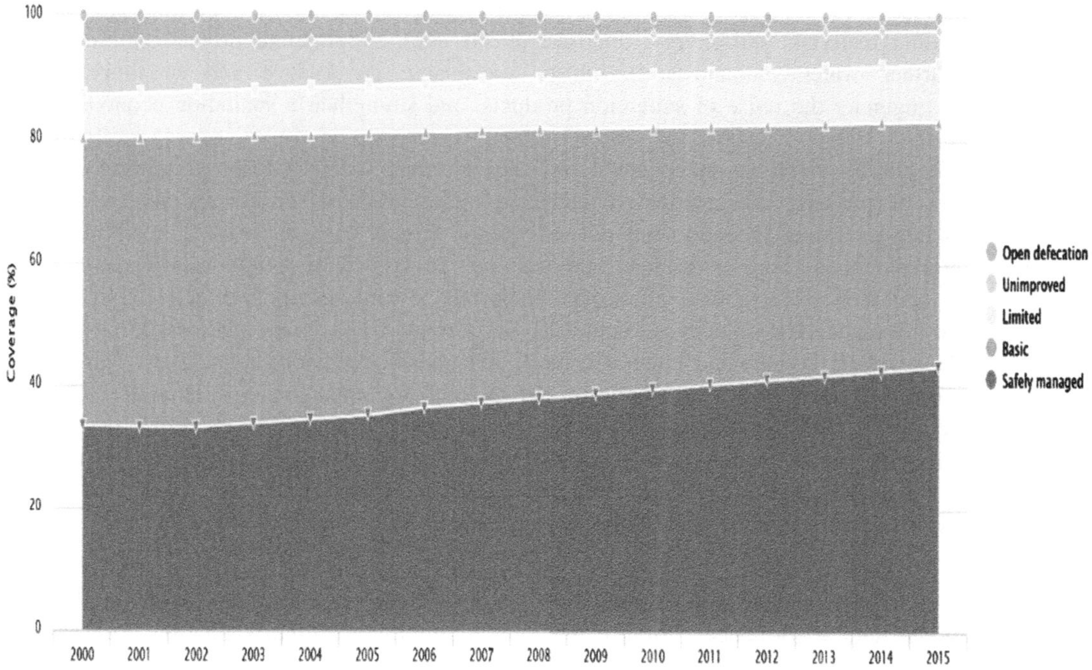

Figure 7.3 Global trend urban sanitation services level 2000–2015 (generated from online WASH database – WHO/UNICEF JMP, 2017b).

2016). The intent is to contribute to the overarching objectives of UR (Beatley, 2010, 2012a, b, 2005; Lehmann, 2010a; Piana, 2010), mainly by:

(I) Building livable cities, land use and urban reforms, innovative housing strategies, sustainable transportation, urban ecology and greenery, recycling and upcycling, energy conservation and renewable energy, sustainable building practices and sustainable economy/governance (Beatley, 2005, 2012a/b, 2015) that could benefit immensely from a synergy with Sanitation 4.0.; and

(II) Building livable cities around a circular economy that pursues reduced waste, resource recycling, natural systems adaptation, biophilic designs and plans, alternative energy and emission controls (Andersson *et al.*, 2016; Beatley, 2005, 2010, 2012a, b, 2015; Beatley & Newman, 2013, EMF, 2017, 2015; Fullerton, 2015; Kvarnström *et al.*, 2011; Lehmann, 2010a, b, c; Potter *et al.*, 2011 will require cost-effective ways to recover, transform and reuse sanitation resources. Any disregard (deliberately or not) of core sanitation elements in urban planning and design could end up with sewer blockages and leakages, blackwater and sludge overflow into surface and ground water, foul odours and the ugly sight of exposed excreta, health costs and mortality and the loss of revenue from SDPs.

UR could also benefit from incorporating Sanitation 4.0 into its programming in order to increase access to safely managed sanitation in the world's cities (to protect public health and environmental quality, reduce attendant costs, etc), improve sanitation service delivery (e.g. emptying, collection, storage, transportation, treatment/disposal), upgrade designs and development of sanitation devices and facilities (i.e. fabrication, construction, installation and connections of toilets, sewers, septic tanks, washing, hygiene devices etc), provide organic resources (for horticulture, urban greenery/landscapes, food – both crop and animal – production, building materials and energy, pharmaceuticals etc.), support and protect natural systems (ground and surface water, air and soil), ensure sustainable governance and economy through instruments that engender the reuse of sanitation products, and strengthen a sanitation economy within the city.

ReGenSan should be given an independent and active status within UR programming as urban sanitation cannot be properly and adequately managed under urban water management and merely recycling greywater for urban greenery will not deliver the targets of SDG 6 by 2030. It, therefore, behooves urban planners and built environment professionals to work with sanitation professionals so as to embed core sanitation expectations into urban blueprints toward inspiring effective and practical solutions that do not negate UR objectives. Sanitation 4.0 presents a systematic platform to tackle urban sanitation challenges from holistic and integrated multi and transdisciplinary perspectives. This synergy could drive innovations, through system thinking and research for management, planning, technology and governance in UR and sanitation, that support ecosystem sustainability. It could also provide solutions for sanitation management in urban renewal projects, supply resources for greenery projects, support livelihood and upgrade projects and strengthen regenerative agriculture/aquaculture to enhance and sustain food production and security. This special consideration could deliver the following and more to UR:

(I) Improve service and access expansion to 'safely managed' infrastructure that will deliver functional and performance currently inherent in 'improved' urban sanitation infrastructures (sewered and non-sewered);
(II) Retrofit and upgrade unimproved, failed, dysfunctional and abandoned non-sewered sanitation infrastructure in developing countries and rehabilitate poorly performing ageing sewered systems in developed countries;
(III) Redesign existing urban sanitation infrastructure to be more regenerative by adopting technologies and processes that are systemic and contextual, which could integrate eco-efficiency with ecological-based approaches, recover and reuse value-added resources and mimic natural systems to deliver improved functional and safely managed systems within urban sani-sheds; and
(IV) Support local physical and socio-economic, cultural, spatial and decision-making structures of the 'place' to support the adoption of appropriate and best applicable technological solutions because the dissonance between users, service providers, institutions, ecosystems and infrastructure could hinder progress in urban sanitation pursuits (Marshall & Farahbakhsh, 2013).

7.2.3 Special considerations for integrating REGENSAN into rural development

Sanitation 4.0 will work towards integrating ReGenSan into rural development (RD) as a synergistic strategy to improve the livelihood of rural populations through innovative sanitation initiatives while at the same time providing much-needed coverage for expansion and service improvement to the rural poor. RD has not given much attention to sanitation in the past, but has mostly focused on (Ashley & Carney, 1999; Ellis & Biggs, 2001; Stockbridge & Dorward, 2018):

(I) Agriculture and natural resources – crops, livestock, fishing, forestry;
(II) Non-farm sector-services to agriculture (including input supply, marketing, transport, finances, agricultural processing), rural manufacturing, mining and other rural services;

(III) Rural infrastructure – roads, transport, energy, water;
(IV) Education; and
(V) Health

The critical contributions of sanitation to rural residents and areas are often overlooked in the intervention agenda (e.g. providing expanded access and improved service as well as the socioeconomic and ecological benefits). This could be related to the fact that most RD experts often assume that considerations of water issues already cover sanitation, but often water does overshadow sanitation needs. Perhaps this is one explanation for the increased progress in rural water supply compared to the very slow rate of rural sanitation. Rural water supply has made evident improvement since the MDGs; however, rural sanitation has remained a huge and unabating challenge, and the absolute poor in low-income countries are mostly off track of the sanitation targets (SDGs 6) (WHO/UNICEF, 2017a). Statistics (Figure 7.4) confirm that about 50% of people living in rural areas lack access to improved sanitation facilities, compared to 18% of urban residents (SDKP, 2017).

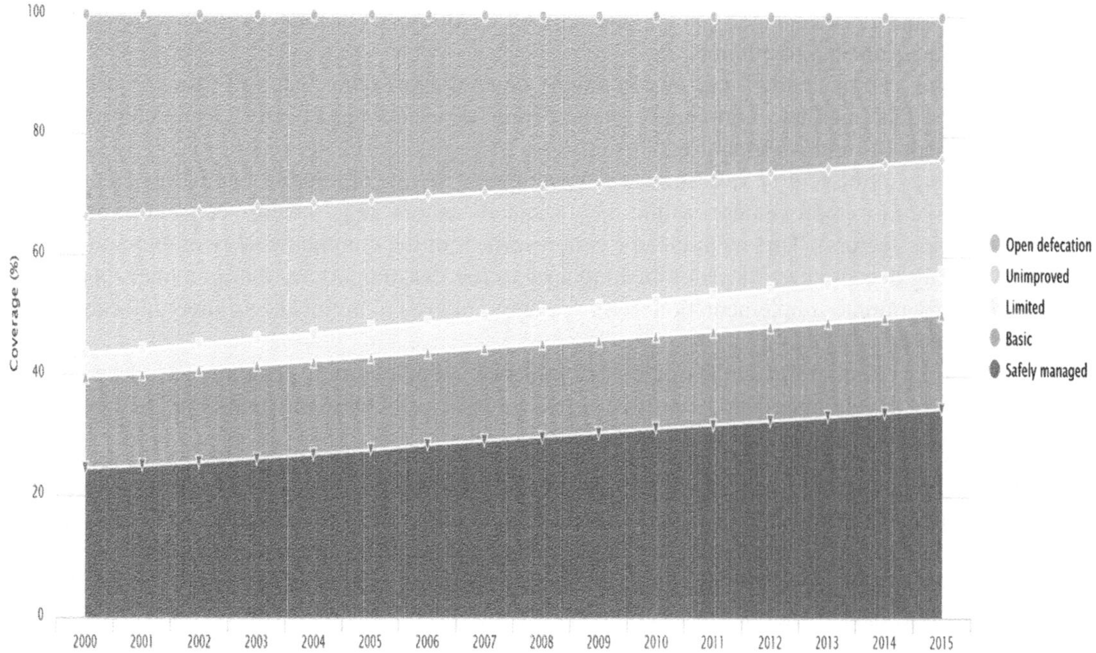

Figure 7.4 Global trend in rural sanitation services level 2000–2015 (generated from online WASH database – WHO/UNICEF JMP, 2017c).

To this effect, it becomes expedient that RD approaches consider a significant upscale (that could be simultaneously applied across contexts) for community mobilization that encourages rural agriculture (RA) and related households (and entire communities) to consider the resource potentials of urine and excreta for agriculture (crop and animal) and aquaculture (Lenton *et al.*, 2005) as well as other livelihood opportunities. In addition, rural sanitation and hygiene behaviour change approaches (e.g. community-led total sanitation [CLTS], school-led total sanitation [SLTS] etc.) that advocate for negative images of communities that still practise open defecation need to balance their messages with more positive narratives. They should focus more on the need for proper and adequate containment as well as management through the SSC and IFSVC in a bid to show the reuse potential of sanitation resources and the accruing benefits to RA and other ventures, which could also serve as positive incentives for communal change. In essence, integrating ReGenSan into the broader agenda of RA and RD could provide greater contributions to rural regeneration due to its socioeconomic gains through market-based and financial mechanisms, especially when the supply and value chains for sanitation products and services are strengthened. Then, rural sanitation systems could be designed to provide ways to recover and utilize transformed SDFs and/or SDNs and even SDEs to support RA activities, energy supply and rural economies. This special consideration also complements the one for the BoPs as it gives attention to the 1 billion population at the lowest level who are probably exposed to open defecation.

A synergy between Sanitation 4.0 and RD could result in the adoption of proven approaches that contribute to the overall goals of sustainable rural livelihoods improvement (Jacob, 2018; WELL, 1998) by:

(I) Adopting a target-oriented approach such as providing integrated solutions to the rural poor, especially small-scale farmers, as principal clients and beneficiaries, and those who live in remote and disadvantaged geographic regions by designing sanitation to provide the needed nutrient inputs to agriculture;

(II) Adopting people-centred and/or community-centred approaches that will focus on establishing systems that facilitate community participation to encourage effective sani-solutions within sani-sheds of rural communities;

(III) Adopting multisectoral infrastructure development by incorporating sanitation technological infrastructure improvement within key societal sectors (e.g. health, education, water and economic sectors). This would lead to improvement of the communities' overall quality of life;

(IV) Adopting decentralized decision-making approaches that support bottom-up models of planning and programme implementation with greater emphasis on local resource allocations and utilization;

(V) Adopting intersectoral collaborations to enhance social-cultural, economic, environment and political development and ensure better interactions between participating actors and stakeholders.

Sanitation 4.0 emphasizes the nexus approach to sustainable RD, which seeks to create synergistic linkages between development factors such as energy, agriculture, aquaculture, food security, livelihood supports and sanitation (SDKP, 2017). Therefore, integrating sanitation into RD remains an important approach in increasing access and service for the rural poor because of its focus on rural and remote areas of a country (province, state, territory, district or other geographic locations) (Jacob, 2018; UNDP, 2006, 2014). It also supports the process of socio-economic change in rural areas based on mobilization of local, natural and human resources (Cohen, 1980; Fasbender, 1982; Jacob, 2018; Leupolt, 1977) by linking the output of sanitation to products like SDFs, SDNs and SDEs to aid agricultural production and other livelihood activities, which could in turn improve the overall quality of rural life (Campbell *et al.*, 2015; Langergrabera & Muellegger, 2005; Larsen *et al.*, 2009; Murray & Buckley, 2009, Park & Chertow, 2014).

7.2.4 Special considerations to integrate REGENSAN into provision of public sanitation services

Sanitation 4.0 seeks to integrate ReGenSan into the provision of public sanitation services (PSS), especially in densely peopled places such as public toilets, government buildings, markets, shopping malls, airports, bus/train stations and transport facilities, schools, cinemas/theatres, stadia, event centres, hotels and tourism facilities, parks, playgrounds, recreation or community centres, hospitals and healthcare facilities, prisons and correctional facilities etc. However, most PSS have remained inadequate, particularly in developing countries with low sanitation access and services (Anthony & Dufresne, 2007; Greed, 1996). In many situations, these facilities pose a lot of health and safety problems, particularly for women, children, the elderly and the disabled. PSS limitations significantly restrict people's mobility and their ability to take part in communal and public life, especially the elderly and disabled (Afacan & Gurel, 2015).

Therefore, improved public sanitation infrastructure is seen as a core component of urban, small-town and rural environmental designs that add to the quality and viability of these environments (Afacan & Gurel, 2015; Greed, 2004). The Sanitation 4.0 platform will seek to ensure healthy participation in communities' social activities, increase student attendance rates in schools, reduce opportunistic open defecation and provide opportunities for PSS facilities to serve as demonstration points for adopting ReGenSan concept into the wider society as well as achieve sustainable, healthy and inclusive living environments (Bichard *et al.*, 2004). The convenient location, maintenance and management of these infrastructures could promote the use of public space, public institutions and transport and generally contribute to active living and positive public health outcomes (Government of South Australia, 2018). In addition, the harvesting of resources from such facilities could attract big investors due to the volume potential. Other specific benefits to society include:

(I) Contributing to making living areas and communities, as well as open spaces, parks, playgrounds, recreation or community centres, markets, shopping malls, airports, bus/train stations and transport facilities among others more accessible, inclusive and convenient for all members of the society (Afacan & Gurel, 2015; Greed, 2004);

(II) Ensuring that healthcare facilities provide people-centred extra services that are one of the key elements of quality within the context of the rapidly evolving landscape of universal health coverage that ensures the safety of patients and workers (UHC) (WHO, 2018); and

(III) Ensuring that schools are good learning environments that could influence school attendance, especially for girls, and consequently affect educational outcomes (Dreibelbis *et al.*, 2013; Freeman *et al.*, 2012; Garn *et al.*, 2013; Morgan *et al.*, 2017).

The ReGenSan application platform will encourage the adoption of the 'streets-for-life' principles, which is a concept for achieving inclusive design goals at neighbourhood scale and enabling the outdoor environment to be as friendly as possible to provide PSS (Burton & Mitchell, 2006). These principles include: convenient location; legibility in use with minimum confusion and disorientation; ease and comfort in access, use and operations; comfortable path of travel with environmental features; safety in use; hygiene in use. And they ensure that the facilities incorporate the SSC with special focus on resource utilization (Afacan & Gurel, 2015; Burton & Mitchell, 2006). On this ground, PSS infrastructure should be designed to integrate resource recovery, harvesting/transformation facilities and, perhaps, reuse within scale to support landscaping, greening, vegetable/flower gardens, groundwater recharge, biogas production etc. PSS facilities should also be designed with special consideration given to women, children, disabled people and the elderly. It is essential that these facilities are available for everyone because it could remove currently grievous barriers to wider participation in public and social life of communities (Afacan & Gurel, 2015; Knight & Bichard, 2011). The provision of public sanitation facilities affects how people use land, participate in social life and attend schools. Many people, such as those who are elderly and disabled, seldon travel and/or go outside and avoid long journeys due to the lack of and/or inaccessible designs of public sanitation infrastructure (Greed, 1996). Therefore, Sanitation 4.0 advocates for special attention to PSS in urban, rural, HtRA, highways and other key locations through public, private and community partnerships.

7.2.5 Special considerations for contextual factors and place in the delivery of ReGenSan solutions

As different people and locations have challenges that are exacerbated by their unique peculiarities, Sanitation 4.0 advocates that special considerations be given to context and place in the conceptualization and design of sani-solutions. Previously, interventions paid little attention to core contextual and place issues, as most programmes are built on the assumptions of research and technological experts without any concrete consideration for such issues as: culture, traditions, soil type, density, users, preference, ecological and geographical characteristics etc. Often failure of sanitation systems could be traced to contextual (e.g. rusted sewer lines due to saltwater intrusion) and place (e.g. users preferring toilets some distance from living area) features of specific areas that set them apart. These identity-features go beyond sociocultural and socio-economic factors to include the geographical, geological, ecological and ecosystem structures that affect coverage and access to sanitation services. They represent the observable characteristics of the PSC that influence the shape, design, maintenance and durable and functional quality of sanitation provisions (Novotný *et al.*, 2018). This could be exemplified in coastal and waterfront communities where geographical and environmental conditions limit options for workable sani-solutions and even affect functionality and durability of sanitation facilities, which makes it difficult to provide individual sanitation systems, especially in communities built above inundated land or above surface water (Navarro, 1994). These communities may be within or by sea coasts, estuaries, mangrove swamps, lake shores, riverbeds and, in most cases, extend right above the surface waters. They may also be built on low-lying areas such as swampy sites, wetlands, marshland and other flood-prone areas that are often vulnerable to such natural calamities as flooding, bank erosion and siltation (Ahmed & Hassan, 2012; Navarro, 1994); and most of the inhabitants could be at the BoP.

Studies have shown that contextual and place factors, as they relate to the natural environment, greatly influence the type, quality and functionality of sanitation infrastructure (Ahmed & Hassan, 2012; Dias, 2015; Kellert *et al.*, 2008; Mang & Reed, 2012a, b; Navarro, 1994; van Vliet *et al.*, 2011). This has been succinctly captured in the 'hard-to-reach areas' (HtRAs) concept that identifies certain areas with poor sanitation coverage based on adverse geological, geographical, ecological and environmental conditions and/or frequent occurrence of natural disasters that could hinder adequate sanitation services, especially coastal, wetlands, mountains, arid and low-lying lands communities, among others (Ahmed & Hassan, 2012). HtRAs, like most coastal communities globally, are daily confronted with hard choices as they grapple with inadequate and inappropriate sanitation services and almost no basic social amenities, largely due to the contexts of their geographical locations. The end result is usually indiscriminate disposal of human waste and pollution of water sources and, subsequently, episodes of faecal-oral infection transmission.

This aptly describes the daily experiences of most communities in Nigeria's Niger Delta, located on sea coasts, estuaries and mangrove swamps, where improved sanitation facilities are mostly absent. The overhung latrine (called Jetty latrine/toilets) is the major available sanitation infrastructure and these superstructures are built over the rivers, marine and swamps environment of these coastal communities. The greatest hazard with these latrines is direct defecation into the water body (in fact even privately collected faecal matter ends up directly in the environment), which then pollutes the already threatened surface water resources (Cookey *et al.*, 2008). Other HtRAs are arid and semi-arid lands (ASAL), where the fragile nature of the environment requires particular considerations in designing appropriate sani-solutions and hygiene promotion because most ASAL households do not have access to safely managed sanitation (WELL, 1998).

For Sanitation 4.0 to develop approaches for HtRAs, they will need to be classified based on their countries and contextual peculiarities rather than generic descriptions. In some countries, such as Bangladesh, remote geographical locations and populations residing out of the range of development activities will be the focus and most of these people are BoPs at the second and third levels. In this regard, based on six indicators (groundwater table, drinking water coverage, sanitation coverage, hotspots, poverty and child mortality) about 1,114 unions (the lowest administrative tier representing 10 to 15 villages with around 5,000 households) were considered HtRAs in Bangladesh in 2012 (Ahmed & Hassan, 2012). In India, remote areas with absolute distance, inaccessibility, poor roads and infrequent transportation (disconnecting them from urban centres), erratic electricity and poor healthcare and education services were used to identify HtRAs (Cook, 2013; Jakimow, 2012; Mitra *et al.*, 2008, O'Reilly *et al.*, 2017). The ReGenSan platform will draw from these examples to develop specific criteria for determining HtRAs and their peculiar characteristics in order to design and deliver appropirate sanitation infrastructure for them regardless of their rural and/or undeveloped status.

This special consideration is based on the ReGenSan principle of PaS (see Section 2.4.7) that runs through the three subsystems (SES, TeS and ReS), their dimensions and components, which specifically provide sani-solutions within sani-sheds (Section 4.2.1). It requires that sani-solutions be tailored to the unique characteristics of locations and their contexts by comprehending a wide range of social, cultural, institutional, geographical, geological, scientific, engineering, ecological and economic factors so as to determine solutions and opportunities that are indigenous to specific sani-sheds (see Sections 2.4.7, 4.2.1.1). This will ensure appropriate technology and adequate infrastructure that will match the capacity of the locality and capabilities of the people. Therefore, in-depth and detailed studies of these issues (that give clear understanding) are essential to determine what particular combination of infrastructure, technology and management will work in specific contexts and place before any interventions are proffered. It will be critical to understand these complex systems of sanitation intervention as they may be simple in eventual design, but they are sensitive to contextual conditions and dynamic interactions between the psycho-social cultural-ecological modes as well as the bio-geo-chemical, geographical and other parameters of natural, political and economic environments that operate at different scales in specific sani-sheds (Novotný *et al.*, 2018; Shiell *et al.*, 2008).

7.2.6 Special considerations for sanitation governance

Fit-for-purpose governance is fundamental to successful and long-term sanitation infrastructure provisions, but governance is often corralled by overlapping and complex processes and relationship (Mitchell & Ross, 2016; Ostrom, 2005; Ross *et al.*, 2014) (see Section 2.4.3). Sanitation coverage and service quality is hindered by political neglect, poor funding, inadequate standards, sectoral and ministerial fragmentation; which further increases the complexities in governance and the risks of administrative and institutional overlaps and gaps (UNDP-SIWI Water Governance Facility, 2016). This is compounded by complications from managing the SSC, RRR and other psycho-social issues related to sanitation (unlike in the use of virgin materials such as groundwater for agricultural or water needs), even more so when perceptions of public versus private interests are involved (Mitchell & Ross, 2016; Ross *et al.*, 2014). Sanitation is generally perceived as a household responsibility and not a public concern even by those in the sector (mainly non-sewered systems) and so there is almost no oversight and harmonization of all stakeholders involved in the sector (Galli *et al.*, 2014;

Welle *et al.*, 2008). This could restrict monitoring and regulatory programmes as they are dependent on available governance instruments (i.e. policies and legislations).

The government of Indonesia attempted to combat this hindrance by introducing the 'local scale wastewater services' (LoSWaS) (i.e. sani-sheds) clause as a key element of its commitment to provide 100% sanitation services access to its citizens (Mitchell & Ross, 2016). This LoSWaS requires that local scale infrastructure be used to collect and treat wastewater from 50–200 households locally and/or close to the point of production, using broad and varied combinations of SMEs from local government systems, community-based organizations and/or local businesses (Eales *et al.*, 2013; Mitchell *et al.*, 2015; Mitchell & Ross, 2016). This innovative approach will still need to be properly embedded in the country's governance structure in synergy with other established mechanisms. Therefore, this special consideration will work to ensure that sanitation receives its due place in national and local governance systems and budgets in an equal and prioritized manner.

Nevertheless, for long-term operations, sanitation systems in sani-sheds need effective governance and stanadardization (Ross *et al.*, 2014) and this entails ongoing dialogue between all stakeholders to clarify expectations, problems and urgencies and then develop common and shared understanding, perspectives and values into established written criteria (such as policies, legislations, regulations, standards etc.); and how to ensure that these criteria are met (Ekane *et al.*, 2013). This is sanitation governance (SaG) and ReGenSan refers to it as the responsibility and obligations of governments to provide safely managed and standard sanitation (which includes capturing viable resources that could be used in other sectors and for livelihood support) for all its citizens, particularly the BoPs, without endangering human health and environment, and implementing it through functional institutions, efficient and recognized management entities and sustainable financing mechanisms that guarantee continuous improvement, skill development usage and ownership in the sani-sheds.

Policies are paramount to creating the right background for planning and supporting actions and decisions towards closing the access and service gaps in sanitation and meeting the SDG 6 (Ekane *et al.*, 2016; Ostrom, 2005), but several countries still have policies (where they exist) that are fraught with gaps and fragmentation. The Water and Engineering Development Centre (WEDC) of Loughborough University, UK, in 2005 conducted an assessment study of nine countries (Bangladesh, Cambodia, Ghana, Indonesia, Mozambique, Nepal, Nigeria, South Africa and Uganda) and found that (WEDC, 2005):

(I) None of these countries specified quantified targets or identified resources for sanitation in their budgets;
(II) The countries' contextual issues and situations were not adequately captured in some policies, nor were specific needs of target groups (e.g. rural and urban poor) with desperate conditions much met;
(III) Minimum levels of service were not specified in all the policies, and most policies allowed for subsidy of capital costs (only some policies included the costs of meeting sanitation needs);
(IV) There was no specific responsible ministry for policy formulation and even though most policies identified lead agencies that will coordinate policy implementation, only South Africa outlined clear roles and responsibilities for line ministries;
(V) Health and environmental considerations were key elements in all policies, but the magnitude of sanitation-related environmental problems was not specified; and
(VI) Most policies stressed the need for both technical and social aspects of sanitation provisions, although some emphasized the latter (WEDC, 2005).

Current trends indicate that not much has changed since 2005; meanwhile these gaps constitute expansive SaG and standardization challenges across the globe (primarily, developing countries) such as:

(I) No policy and adequate institutional frameworks (Ekane *et al.*, 2016; Elledge *et al.*, 2002);
(II) Poor implementation of policy, where it exists (Ekane *et al.*, 2016; Seppälä, 2002);
(III) Fragmentation and gaps in policy and institutional frameworks (Ekane *et al.*, 2016; Morella *et al.*, 2008);
(IV) Poor prioritization of sanitation and hygiene (Ekane *et al.*, 2014/2016; ECA, 2012);
(V) Implementation failures characterized by overemphasis on numbers of toilets rather than functionality, serviceability and reliability (Ekane *et al.*, 2016; Kvarnström *et al.*, 2011);
(VI) Poor coordination of actors and activities (ECA, 2012; Ekane *et al.*, 2014/2016; UN-Water, 2012);
(VII) Overreliance on technological solutions with less emphasis on other subsystems and components of sanitation such as psycho-social-cultural-ecological and resource systems (Curtis *et al.*, 2011; Ekane *et al.*, 2016; Morella *et al.*, 2008); and
(VIII) Inadequate financial and technical capacities to drive meaningful change (Ekane *et al.*, 2014/2016).
(IX) Inadequate standards that support technological innovations, scalability and replicability (NOWRA, 2007; Starkl *et al.*, 2015).

As Plummer and Slaymaker (2007) have argued, governance issues are responsible for the poor progress in sanitation services of most countries and so this special consideration of the Sanitation 4.0 application platform is critical to the promotion of good governance as a prerequisite for successful access and services delivery for everyone (Mjoli, 2015). It seeks to improve, strengthen and standardize infrastructure and management solutions through governance instruments at different levels of government (i.e. constitutions, laws and by-laws, policies, regulations, standards, norms etc) that determine technology and service delivery options and criteria as well as expanding and clarifying the occupational structure in the sanitation sector (Mitchell & Ross, 2016; WSP, 2003). It could be described as the rules, roles and relations that make sanitation systems work (or not) and at what cost, for whom (rural, urban, BoPs, ToPs, women, children, elderly, disabled etc) and how they differ based on context and place. This platform will pursue appropriate governance structures that guarantee sustainability, functionality and reliability of selected technologies and systems' parts in given sani-sheds (UNDP-SIWI Water Governance Facility, 2016). It will aim to work with governments to support and ensure key functions of facilitation, investment support, regulations, standards and partnership-building to enhance institutional, organizational and infrastructural capacities toward functional, serviceable, durable and reliable services and products (Bill & Melinda Gates Foundation, 2006).

The platform advocates for a SaG characterized by the principles of equity, efficiency, participation, decentralization, integration, transparency and accountability (Pahl-Wostl *et al.*, 2011; WaterAid, 2011), that includes critical components of good sanitation policies such as: defining institutional roles and responsibilities, legal frameworks, targeting population and resources, levels of service, technology-options (EIDT, ReDT and NoDT), RRR, place, context and SMEs structures, standardization, health considerations, environmental and financial considerations, etc. (Elledge *et al.*, 2002; Ekane *et al.*, 2016; Evans, 2003; Mitchell & Ross, 2016). Subsequently, SaG should be able to deliver day-to-day operational responsibilities and support institutional arrangements that will ensure functionality and continuous improvement of existing improved sanitation facilities as well as monitor and record the technical status of SMEs; ensure retrofitting of dysfunctional and failed systems; and rehabilitate and repair aged and damaged infrastructure. Governance instruments should also give opportunities for the development of new and innovative technologies, especially for new development areas (Mitchell & Ross, 2016).

7.2.7 Special considerations for data collection, analysis and management

Sanitation 4.0 recognizes the central role of adequate and reliable data Brewerton and Millward (2001) in the implementation path towards 2030 and beyond. Data, indeed, is the lifeblood of design and decision-making for evidence-based sani-solutions and underpins governance elements of accountability and transparency. Special considerations (7.2.1–7.2.6 and 7.2.8) will be highly dependent on the quality, diversity and volume of data available for different levels and activities of sanitation management in order to monitor progress and be able to hold service providers and development partners accountable (Andres *et al.*, 2018). A strategy to improve data management in the sanitation sector is crucial because many countries still lack the financial, institutional and human resources necessary to acquire and analyze data for improved service delivery and governance (in spite of the current emphasis on data usage in the sector). Reliable, consistent and disaggregated data are significant in stimulating political commitment, inform policymakers and decision-makers, and trigger well-placed investments (Andres *et al.*, 2018; Bartram *et al.*, 2014; Lenton *et al.*, 2005). But evidence indicates that the sanitation sector still lacks data on a diverse range of topics that form the underlying reality of coverage (e.g. qualitative aspects, equitability of access, etc.) and the real costs of interventions and their benefits (Bartram *et al.*, 2014; Lenton *et al.*, 2005).

There are also no systematic methods for analyzing contexts and determining in advance technologies and approaches fit for PaS in the sani-sheds (Bill & Melinda Gates Foundation, 2006; Levy & Ellis, 2006), and, although the compliance of key sanitation technologies (such as latrines) has often been measured successfully using surveys and field observations, there are still substantial limitations (Andres *et al.*, 2018; Wood *et al.*, 2008). The most up-to-date and widely used data on sanitation are the WHO and UNICEF Joint Monitoring Programme (JMP) and the Global Assessment and Analysis of Sanitation and Water (GLAAS) implemented by WHO and Sanitation and Water for ALL (SWA) under the aegis of UN-Water. While the JMP, used to monitor the progress of the UN system's global targets, informs national and global policymakers on sectoral status, GLAAS, on the other hand, also analyzes progress and obstacles in the sector (Bartram *et al.*, 2014; Lenton *et al.*, 2005). Nonetheless, the JMP and GLAAS are still limited and biased in the face of current applications and potential future monitoring needs such as:

(I) Large discrepancies in facility classifications, definitions and services across countries complicate monitoring and introduce errors through gaps or biases in sample frames of nationally representative surveys and the census from which they are derived, particularly in countries where certain populations – such as those living in marginal or informal settlements – are excluded from or under-represented in censuses (Bartram *et al.*, 2014);

(II) Errors introduced by limitations of survey instruments (Mortel, 2008);

(III) Lower precision (but not lower accuracy) intrinsic in the survey approach and errors introduced by rules concerning extrapolation (Wolf *et al.*, 2013);

(IV) Bias introduced through calculation and application of a single rate of shared sanitation use for each country across all time periods instead of periodically updating these figures to reflect changes (Bartram *et al.*, 2014);

(V) Bias introduced by reliance on household-level data for calculating rural/urban coverage rates; analysis is further complicated because there is no common definition of rural and urban, and the difficulty in capturing intra-urban disparities in most cases (Christenson *et al.*, 2014);

(VI) Implicit generalizations in facility classifications lead to some households being categorized as unserved despite receiving adequate services (Heijnen *et al.*, 2014);

(VII) Implicit assumptions that members of the same household use a single sanitation infrastructure, especially when both adults and children may use different sanitation facilities when at home and when away from the home (especially at work and school); and the varied concepts of households across cultures (Nnaji *et al.*, 2013).

These could lead to serious gaps in the accuracy of the data gained from these monitoring records; so some studies have suggested the use of several geospatial methods to capture big data that could prove useful in providing more detailed information about sanitation globally while also assisting decision-making at all levels (Bill & Melinda Gates Foundation, 2006). Big datasets are so complex that traditional data-processing applications, software and approaches can hardly handle them (AGUASAN, 2018; Kohavi *et al.*, 2002; Park, 2017; Sumser, 2017; Wikipedia, 2018). The use of information and communication technology (ICT), like mobile phones, has been proposed as a tool that could strengthen surveys and field observations in data collection and real-time reporting. A proof of concept was piloted to test the viability and efficacy of mobile technology for monitoring rural sanitation in two blocks (i.e. sub-districts) of the Rajgir block in the Nalanda district of Bihar, India. About 23,000 households were covered for about three months and smartphones were shown to be able to provide quick and credible information outcomes in near-real time, as well as incorporating features such as geo-tagging and photographs to support credibility and relevance and present user-friendly data maps that enable a quick grasp of the immediate status. This could be used to address the gaps in monitoring the usage of sanitation facilities at large-scale and periodic intervals (Kumar & Singh, 2012 – Figure 7.5).

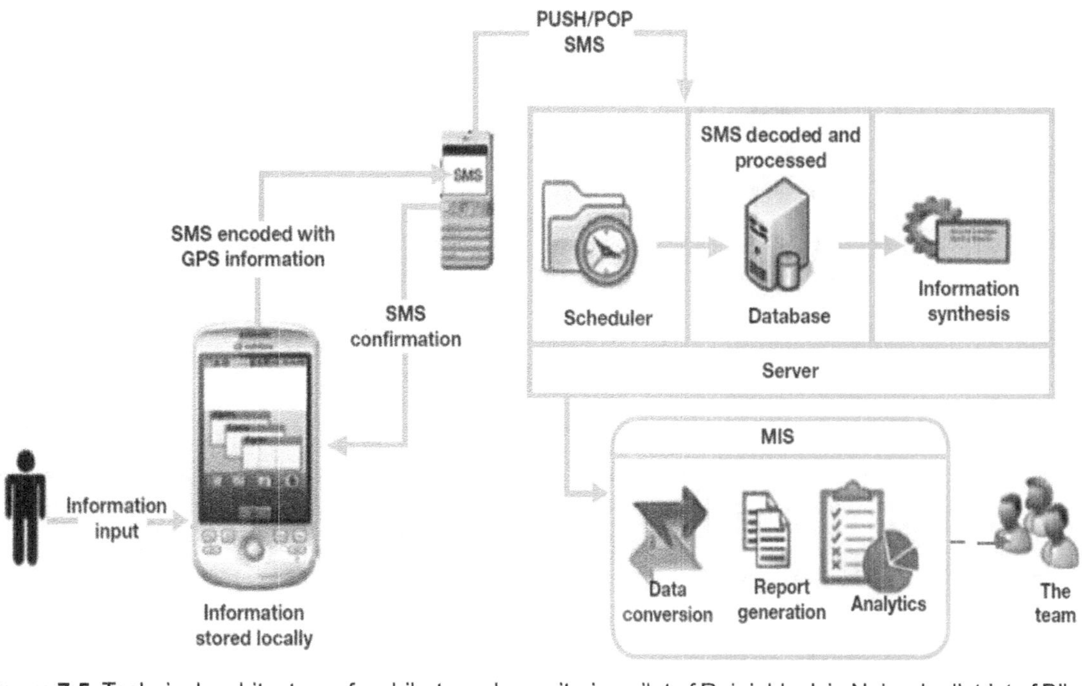

Figure 7.5 Technical architecture of mobile-to-web monitoring pilot of Rajgir block in Nalanda district of Bihar, India (*Source*: Kumar & Singh, 2012).

To this effect, innovative and integrated applications of digital and ICT data acquisition and analysis approaches (e.g. AM/FM radio stations, mobile phones, radio frequency identification technology (RFID), global positioning systems (GPS), remote sensing monitoring via satellite assets and distributed in situ sensors, geographic information systems (GIS), passive latrine use monitor (PLUM), closed-circuit television (CCTV), sewer inspection survey devices and other web-based applications among others) are potentially powerful tools (AGUASAN, 2018, Andres *et al.*, 2018; Delea *et al.*, 2017; Ndaw, 2015 – Box 7.2/Figure 7.6).

BOX 7.2 RURAL SANITATION MOBILE PHONE TEXT-MONITORING SHORT MESSAGE SERVICE, EAST JAVA, INDONESIA (NDAW, 2015)

A pilot study to monitor rural sanitation in two districts of East Java, Indonesia, was carried out by the Ministry of Health and supported by the World Bank's Water and Sanitation Program (WSP) using a mobile phone text-monitoring short message service (SMS) system. Information from the community's progress towards becoming open-defecation-free (ODF) and changes in households' access to improved sanitation was collected and sent via SMS and then stored in a district-level database. Based on the success of the pilot, in 2012 the mobile monitoring system was rolled out to all 119 districts and cities in five of the country's provinces, covering approximately 36 million households and 123 million individuals (Ndaw, 2015).

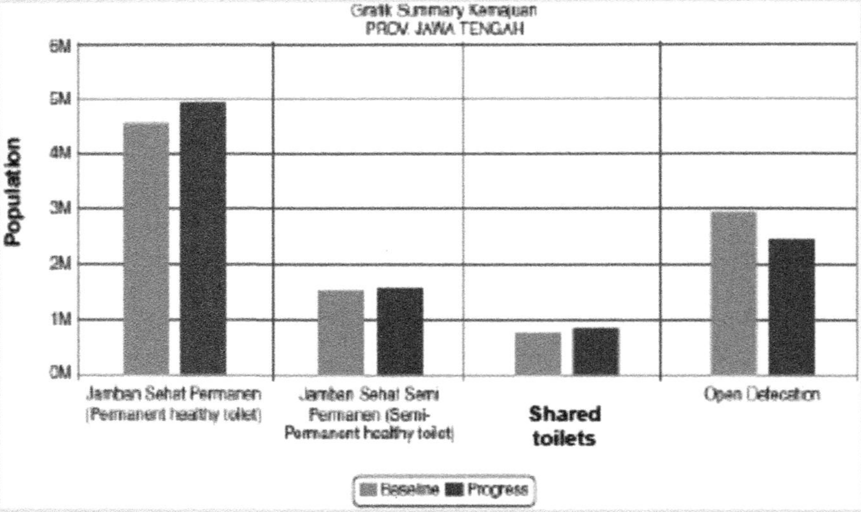

Figure 7.6 Summary of rural sanitation monitoring progress using mobile phone text-monitoring short message service for the Province of Central Java by type of Access (WSP 2014).

BOX 7.3 RURAL SANITATION MOBILE MONITORING SYSTEM – INDONESIA

These approaches and tools could leverage the utilization of actionable information that enables the development of NoDT (see Sections 3.3.2.3, 5.4.1.3), supports EIDT (see Sections 3.3.2.1, 5.3.1.1) and operationalizes ReDT (Sections 3.3.2.2, 5.4.1.2) in sani-sheds as well as assisting in data capturing, monitoring and management to:

(I) Strengthen monitoring and inventorying levels of infrastructure; create baseline information for benchmarking; identify areas where interventions/resources are required; and ensure sound management and decision-support systems;
(II) Improve and sustain planning and programming initiatives to achieve operational and service delivery efficiencies;
(III) Allow for more efficient data transfer and exchange of sanitation-related information between various sanitation infrastructure owners and operators that could strengthen consumer voice, and enhance partnership and governance in sani-sheds.

Sanitation 4.0 encourages special considerations for CiM of existing facilities and retrofitting of dysfunctional and aged infrastructure, which also supports innovative solutions in the sector. For example, big data could assist user behaviour analytics so as to deal with the discovery, interpretation and communication of meaningful patterns in captured data through statistics, computer programming and operations research. This will help to determine, predict and improve performance in regenerating sanitation technology and natural environments for better asset management (Kohavi *et al.*, 2002; Park, 2017; Sumser, 2017). It could also help to detect facility defects and failures early and accurately, which could in turn assist public health/environmental protection and assure infrastructure quality and reliability benefits for significant competitive value.

7.2.8 Special considerations for skills development

Special considerations are a critical strategy for growing and unlocking the employment/entrepreneurial investment potential of the sanitation economy and also exploring the systemic interactions and cultural assimilations necessary for acceptable, affordable and sustained sanitation services and product delivery (McGrath & Powell, 2016; Quendler *et al.*, 2013). The sector is often manned by unskilled/semi-skilled workers and external professionals from other fields whose major focus is not sanitation. This has affected the progress of sanitation management research and governance. Sanitation workers and professionals need to be equipped with specific appropriate knowledge and skills (KaS) so as to strengthen the sector to effectively deliver on the SDG 6 targets. They need to comprehend the core values of sanitation and commit to these principles and then acquire the precise competences, skills, knowledge and capacities necessary for quality sanitation service delivery and product handling according to the contexts and place of sani-sheds (Grün *et al.*, 2009). It is imperative to build a stream of adequately and appropriately trained human resources, from high-level experts to mid-/basic-level operators as well as governments, communities and other key and ancillary stakeholders. However, education and training for sanitation technicians and professionals and basic workers have not been easy, and it will need significant and innovative restructuring to keep pace with the demands of the sector (Heidebrecht & Mar, 2012).

Sanitation 4.0 demands deliberate and extensive efforts towards a comprehensive and systemic skills development blueprint for the sanitation sector that sufficiently captures formal and informal workers, contextual and place factors as well as the transdisciplinary structure of the sector's requirements so that an empowered and equipped workforce can emerge with indispensable capabilities for the challenges of sani-sheds (UNDP, 2009a, b). Notably, education and training programme designs will have to take account of the nuances of difference between rural and urban sanitation, contextual and location structures, existing skill gaps and occupational classifications and then determine the type of KaS relevant for the sector.

There is no holistic perspective or clear definition for skills in the sanitation sector (Nath *et al.*, 2017) and the skill acquisition process is considerably flawed by a lack of generic and sectoral competencies, unclear and inadequate training goals and targets, poor course contents, irregular course durations and improper standards and techniques. Skills acquired in one context can be difficult to apply in other contexts (and sometimes within the same sector), while several training courses are mostly inappropriate in terms of present and future sectoral needs (WRC, 2002). This could be traced to the fact that training institutions are ill-equipped, provide limited hands-on training, use outdated curricula, deliver inappropriate trainings that do not match local needs and are hardly ever standardized (Cavill *et al.*, 2011; IWA, 2014a, b, 2015; Crocker *et al.*, 2016; Nath *et al.*, 2017).

Furthermore, the updates to related knowledge and technology (from progressive and innovative research outputs) have generated new processes, approaches and procedures (for example, computer applications and the corresponding hard/software, databases, analyses and simulations as well as GPS systems and GIS application are now daily tools for operating personnel [Heidebrecht & Mar, 2012]). All of these affect graduates of universities and other training institutions, as they are not duly equipped for the labour market (IWA, 2014a, b, 2015; Nath *et al.*, 2017); this is compounded by the barriers to employee transition from academia to the work environment and a want of mentors to train novices – there are no opportunities for them to acquire core and practical sanitation-related experiences.

The application of ReGenSan is highly dependent on core sanitation skills that are drawn from multiple disciplines, especially as the occupational circle of the sector has expanded. Generally, skills gaps are often responsible for inadequate and poor service delivery, improper management and ownership structures and poor performance monitoring. Therefore, Sanitation 4.0 will adopt a combination of strategic approaches at secondary, vocational, tertiary and professional levels to cultivate knowledge and expertise for tackling current challenges through synergistic linkages and cooperation between all key players and stakeholders involved in skill and knowledge transfer (Heidebrecht & Mar, 2012). Learning content and skills relevant to all operations should be synchronized so that training is pertinent and focused on concrete performance and result-oriented activities (GWP, 2009) by designing adequate and fit-for-purpose training tools and innovative teaching/learning approaches that enhance workers' competences and strengthen institutional capacities to deliver effective and efficient services, safe and quality products and guaranteed OHS and environmental quality.

This application platform will also target intensive learning of professional competences to move sanitation workers across the stages of expertise (from novice to expert) in the skill acquisition process. To this effect, performance and self-managed competences that could function in open, complex and dynamic situations will form the bedrock of service and product delivery (Erpenbeck & Rosenstiel, 2004; IWA, 2014a, b, 2015). Four core competency areas will be explored (infrastructural/facility design and construction, maintenance management, user interface operations and cross-sectional networks) and further categorized into specific skill areas with a focus on sanitation project circle activities such as planning, designing, construction, installation, monitoring and evaluation, management, research, training and implementation. (Bartram, 2005; Gotoh, 1999; Kurz, 1999; Kurz *et al.*, 2004; WRC, 2002).

7.3 REGENSAN APPLICATION PLATFORM

The purpose of the Sanitation 4.0 delivery platform is to present a template for the application of ReGenSan concepts and solutions towards the SDG 6 targets of expanding access, improving services and ensuring safely managed facilities. It could then serve as a practical guide for global, national and local implementation of collective and specific actions at all levels. The application process could include defining key goals and objectives; exploring contextual issues of psycho-socio-ecophilia; assessing governance instruments; assessing technological systems; evaluating resource systems; assessing cross-cutting issues; identifying priority solutions and implementing action plans; and monitoring and evaluation. A detailed status assessment report to guide ReGenSan intervention programmes (ASHWAS, 2011; Bright-Davies *et al.*, 2016; GHK Research & Training, 2000; ISF-UTS and SNV, 2016; IWA, 2014b; Lüthi *et al.*, 2011a/b; Martínez-Jurado & Moyano-Fuentes, 2014; Parkinson *et al.*, 2014) (Figure 7.7) will provide an effective backdrop for sanitation management plans at global, national, regional and local levels. Details of the platform items are described below.

7.3.1 Contextual analysis of PSEP

This process starts at the 'place' where scalable solutions are produced within sani-sheds, which could be 'community-scale', 'neighbourhood-scale' or 'household-scale'. The contextual analysis could provide linkages between sanitation solutions and end users (Lüthi *et al.*, 2011b; Shove, 2003) for active and broad engagement and involvement of relevant stakeholders in decision-making (Cookey *et al.*, 2016c; Luyet *et al.*, 2012; Mang *et al.*, 2016; Rowe & Frewer, 2000; Webbler *et al.*, 2003). PSEP factors include a wide range of cultural and religious belief issues, societal values, health and hygiene practices and behaviours as well as user preferences and are linked to population and demographics (number of people per household, population density, type of housing, gender-related issues). They also include some influencing factors for investment in sanitation such as personal preferences, affordability, types and quality of systems, willingness-to-pay, livelihood activities and types of materials to be recovered from sanitation systems. Other parameters are water-hygiene-related, local climates, temperature and precipitation, biological, geophysical and hydrological parameters. Descriptions of project goals, objectives, scopes, purpose and scale should also be included.

7.3.2 Governance assessment

Determine and assure the suitability and adequacy of enabling environments for effective and efficient sanitation service delivery. Several governance issues influence delivery of sanitation services including the level of government support and political will, institutional frameworks (i.e. policy, laws, regulations, guidelines and standards), enforcement and compliance mechanisms, organizational competence, clear demarcation of roles, responsibilities of actors (formal and informal) and effective and efficient planning and management systems (especially issues of human resources, knowledge, skills and capacity) as well as financing and full cost recovery (Fonseca *et al.*, 2011; Fonseca & Verhoeven, 2013; Joshi *et al.*, 2015; Pahl-Wostl, 2009, 2010; Ross *et al.*, 2014).

7.3.3 Technological system assessment

Identify and categorize safely managed, basic and limited facilities as well as failing and dysfunctional facilities. The analysis will include existing sanitation systems, practices and technologies' performance as well as operations and maintenance. The effectiveness and efficiency of current faecal sludge management systems will be assessed and determined. The level of service delivery, which is basically the benefit that users derive from current sanitation systems, will also be determined. Such issues as: access to facilities, convenience of use and privacy, comfort, compatibility with the 'system of the place' and hygiene practices as well as downstream impacts on the environment, public health and ecological systems (IWA, 2014a, b) are critical: specifically, technical norms and standards that influence designs, operations and maintenance (whether official standards or those accepted in practice over time though not formally recognized) (Lüthi *et al.*, 2011a/b) will be explored. System integrity, quality control and assurance will also be assessed, while infrastructures and facility mappings in sani-sheds could be used

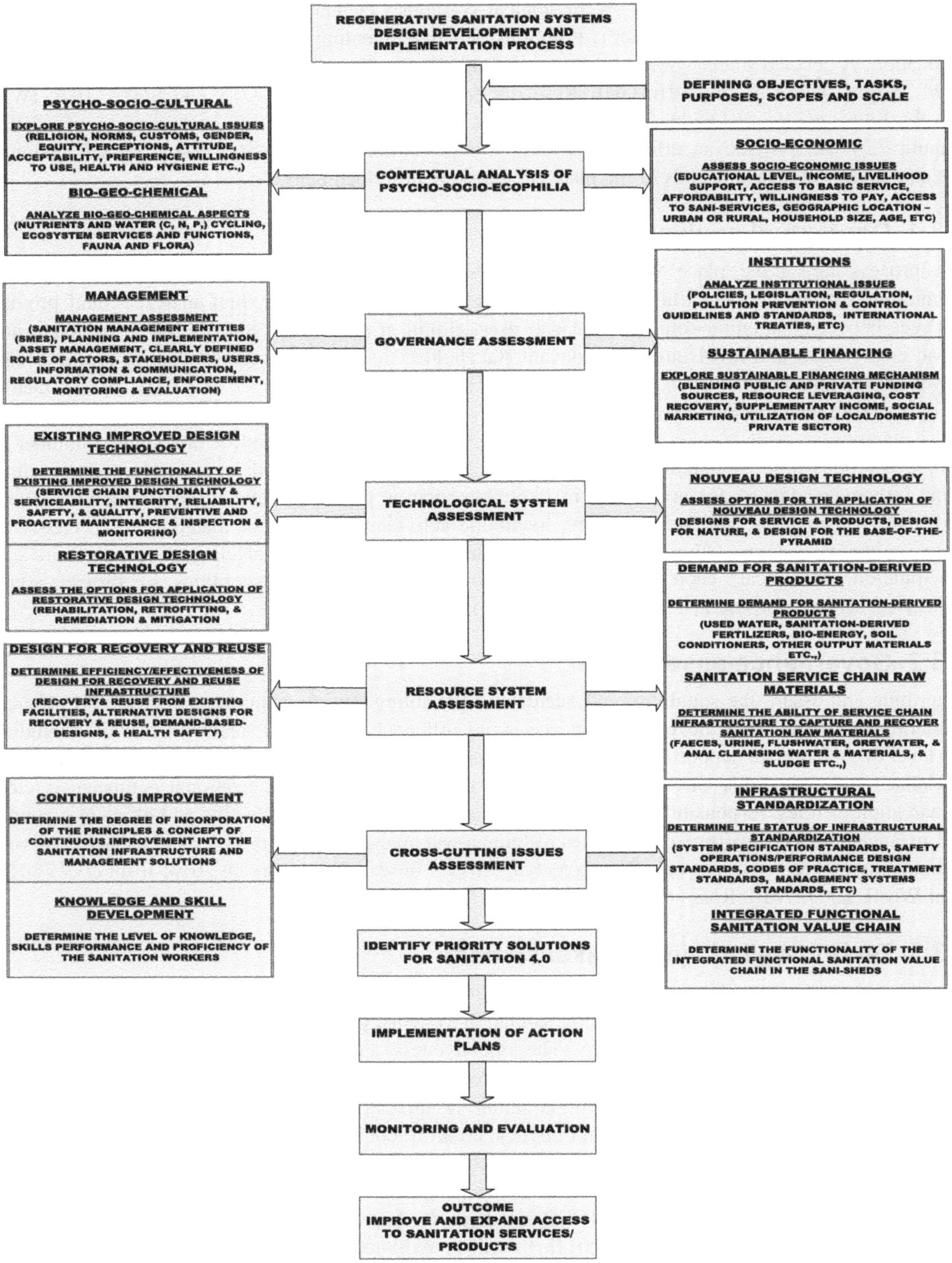

Figure 7.7 Integrated and multidimensional ReGenSan application platform.

to identify system types, components and other spatial information for a clearer picture. The feedback from this assessment could determine areas that need new designs and technologies, restorative and retrofitting solutions as well as continuous improvement programmes for existing improved sanitation facilities.

7.3.4 Resource system assessment

Explore current sanitation systems' capacity for resource recovery and reuse. In addition, the economic value of these resources and their utilization in the 'place' will be assessed in order to support fit technology designs (Murray & Buckley, 2009). The assessment shall focus on determining services and products that could be derived from sanitation systems based on contextual characteristics and livelihood support activities of the population in specific sani-sheds, demand for these services and products and business models for delivering these products and services, and the appropriate technological designs (EMF, 2012, 2013).

7.3.5 Cross-cutting issues assessment

Identify other actions and activities needed to achieve successful implementation and sustainability of ReGenSan solutions based on their integrating nature. For instance, knowledge and skill gaps assessments will help provide solutions for capacity development programmes, while infrastructure standardization assessment will enhance facilities' continuous improvement and value chain exploration will provide solutions for sustainable sanitation business models.

7.3.6 Identify priority solutions for Sanitation 4.0

The results, findings and conclusions from all previous activities (Sections 7.3.1–7.3.7) are put together to provide specific solutions, which are then prioritized for specific 'sani-sheds'. These priorities should be drawn from the eight special considerations highlighted in Box 7.1 above and could be selected through a workshop or a series of focus group discussions, research and media surveys. The aim is to ensure that problems considered are critical for access expansion and service improvement in sani-sheds, and for user preferences and solutions they are willing to support. This could result in well costed, time-guided activities and programmes, effective management and target achievements. Some up-front priority actions (global and national) could include:

(I) Develop framework for incorporating contextual factors and place into design and management of sanitation infrastructure;
(II) Develop a strategic comprehensive framework for data collection, analysis, processes and management;
(III) Develop framework for incorporating governance and standardization into design and management of sanitation infrastructure;
(IV) Develop a comprehensive and strategic framework to determine skill and knowledge gap, appropriate solutions and effective delivery options;
(V) Develop a strategic framework for management/technology solutions to incorporate ReGenSan into public service facilities
(VI) Develop a strategic framework and design appropriate solutions for the population at the base-of-the-pyramid (BoP);
(VII) Develop a strategic framework for management/technology solution to incorporate ReGenSan into rural development;
(VIII) Develop a strategic framework for management/technology solutions to incorporate ReGenSan into urban regeneration;

(IX) Develop a strategic priority areas framework to guide donors and intervention organizations' programmes.

7.3.7 Implementation of action plans

Implementation of the resulting action plans is the final stage of the planning process, which should generally lead to launching the ReGenSan project. However, after preparation of the final draft of the ReGenSan plan, there is the need for further consultations to ensure that every stakeholder (formal and informal) is represented in the final product before moving to solution designs and development. Operations and maintenance (O/M) should be incorporated into all plans from the inception as well as knowledge and skills, promotion, advocacy and awareness programmes for all categories of stakeholders (IWA, 2014a, b).

7.3.8 Monitoring and evaluation

Establishing monitoring and evaluation processes ensures that inputs and activities lead to their intended results and outcomes. Adjustments should be made where necessary while the progress of established goals is ascertained (UNICEF, 2014) and the logic of programmes and projects as well as the resulting activities and their implementation are regularly validated to make adjustments as needed (UNDP, 2009b).

Sanitation 4.0 will accommodate other existing or new frameworks for different subsystems, dimensions and components in order to achieve the aim of the whole, i.e., access expansion and service improvement.

7.4 CONCLUSION

The ReGenSan concept and framework presented in this book is innovative on four levels:

(I) It offers a comprehensive, holistic, integrated and systemic outlook for sanitation management rather than a singular focus;
(II) It embraces the dynamic complexities and contextual peculiarities of 'place and scale' in delivering sanitation solutions;
(III) It introduces the concept of sani-sheds; and
(IV) It is also the first time the regenerative concept has been applied to the field of sanitation research.

ReGenSan does not seek to replace previous frameworks or paradigms, but integrates all aspects of sanitation through holistic interactions with related issues (e.g. water, health, hygiene, gender etc.) to create a comprehensive framework of the best approaches and solutions for particular contexts. The need for a comprehensive, integrated and systemic framework becomes apparent and of utmost necessity after the dismal results of the MDGs' sanitation targets. With the recent JMP outlook, it is quite obvious that if the SDG 6 is to perform better by 2030, it cannot be business as usual. Sanitation management needs its own focused solutions provision and should also be thoroughly investigated for efficacy, effectiveness and acceptance; only then can the much-needed and desired access expansion and service improvement be realized.

The ReGenSan framework is unique in that it is applicable to both developed and developing countries, and to all stages, processes and cycles of delivering sanitation solutions. The framework has the potential to critically evaluate, analyze and provide credible, adequate and appropriate sanitation solutions by adopting diagnostic, prescriptive and normative analytical approaches. The diagnostic provides detailed existing status reports for interventions that are geared towards either restorative, existing or nouveau solutions. The prescriptive perspectives are built upon the diagnostic and play strategic roles in determining the appropriateness and adequacy of applied interventions, and at the same time promote the elements of continuous improvement and knowledge/skills, especially with existing improved-safely-managed (EIDT), retrofitting of unimproved-dysfunctional facilities (ReDT) and nouveau (NoDT) sanitation systems, as well as strategically incorporating resource recovery and reuse (RRR) into current conventional infrastructures (sewers and non-sewers). On the other hand, the normative is associated with the governance and PSEP perspectives that provide the optimum enabling and operating environment for the uptake of scalable solutions. These perspectives provide a better understanding of the interconnections of knowledge among the sanitation subsystems, dimensions and components, with overarching objectives to:

(I) Generate interdisciplinary system understanding for service improvement and access expansion;
(II) Develop and implement sani-shed-specific methods, models and technological solutions as well as evolve context-dependent governance structures and strategies for place- and scale-based sanitation service delivery;
(III) Integrate science-policy-technology-practice research into the sanitation space through a cyclical learning process;
(IV) Emphasize and support the existence and growth of the sanitation economy.

The Sanitation 4.0 application platform will be able to show how SES, TeS and ReS subsystems are inter-synergistically linked in a systemic whole to deliver comprehensive integrated sanitation solutions. The intention is to improve management, planning and technological innovations so as to help guide the activities of practitioners, policymakers and academics working in the sanitation space. For instance, the challenge of expanding access to sanitation in developing countries and addressing the impacts of conventional centralized systems in developed countries could be strategically addressed through the window of retrofitting and rehabilitation rather than building facilities, and could then provide more opportunities for well-thought-out novel designs for unserved and underserved areas. In addition, a comprehensive integration of all components of the sanitation system could strengthen the capacity of sustained service delivery and also improve participatory processes that enhance ownership and investment of would-be users. The diversity that often exists in social and ethnic groups of any locality greatly impacts design, access, acceptability, affordability, construction and use of sanitation systems; therefore, this framework could strengthen understanding of the unique dynamics-potential of a location to influence types and processes of technologies and determine systems' capacity to improve own performance.

This proposed ReGenSan framework could give rise to improvement and new concepts in sanitation system-thinking, management, planning and technology innovations without compromising the sustainability of the ecosystem. It will work with other pre-existing frameworks and perspectives to provide functional and progressive approaches towards improving and expanding access to sanitation services at 'scale'. There will be comprehensive, explorative, experimental and descriptive assessments using qualitative and quantitative datasets to gain better understanding of sanitation subsystems, dimensions and components. Sanitation 4.0 could serve as a guide for funding/donor and intervention agencies by assisting them to identify where and when funds and support for more effective and practical outcomes are urgent. It could also contribute to strengthening global planning and implementation towards expanding access and improving service delivery of sanitation solutions. At the same time, it could serve as a pathway or portal for innovative designs and development of sanitation technologies and social solutions, especially when there is integrated focus on the key special strategic considerations.

The major constraint, however, will be the seeming disregard of conceptual frameworks in sanitation research. Nonetheless, this trend is changing in the recent upsurge of frameworks, perspective and paradigm researches in the sanitation space. The perception of sanitation only through the focus of toilets as isolated structures for defecation, with products and by-products only considered as waste and nuisances that contaminate, pollute and degrade the ecological system, is changing. Sanitation should rather be seen as a catalyst for systemic community regeneration and transformation.

7.5 EXERCISES

(I) Critically describe Sanitation 4.0 and its significance as a global platform towards expanding access and improving services.
(II) Critically discuss the milestones of sanitation paradigms and technologies presented in Figure 7.1.
(III) Critically discuss the impacts and differences between the MDGs' target to 'reduce by half' and the SDGs 'universal access' on sanitation and relate them to the goals of the sanitation decade.
(IV) Critically examine the differences in 'improved sanitation' and 'basic sanitation' and how they affect perceptions, analysis, management and results.
(V) Critically and briefly examine the core objectives and purpose of Sanitation 4.0 and the key special strategic considerations used to improve sanitation delivery to BoPs.
(VI) Critically examine some governance challenges that Sanitation 4.0 could face in your own country.
(VII) Critically examine the Sanitation 4.0 application process in line with the special considerations and ReGenSan Principles.
(VIII) How does the Sanitation 4.0 application process relate to the ReGenSan subsystems and their dimensions and components?
(IX) What is the place and impact of ReGenSan in urban regeneration?
(X) Explore the possible integration of ReGenSan into rural development in your own country.
(XI) Briefly and critically describe the type of public sanitation services that could serve different public and open space for different kinds of people (e.g. elderly, women/girls, children, people with disabilities, religions and cultures, etc) in different contexts and places.
(XII) What are HtRAs and how can they be effectively provided with safely managed sanitation with resource recovery elements?
(XIII) What are the benefits of reliable and accurate data for ReGenSan's Sanitation 4.0 platform?

7.6 RECOMMENDATIONS FOR FUTURE RESEARCH

(I) How can sanitation solutions at household level connect to agriculture and what are the likely impacts?
(II) Explore integrated sanitation infrastructure/design options that work with agriculture, aquaculture and horticulture, that could serve at household scale to sani-sheds levels while remaining affordable, reliable, safe and functional.
(III) Explore the significance and limitations of Sanitation 4.0 as a global platform towards expanding access and improving service.
(IV) What kind of indicators are appropriate to classify HtRAs for sanitation?
(V) Explore options for dealing with place and contextual issues (such as coastal, arid, mountainous, temperate, tropical, cultural religious, status etc) in delivering sanitation solutions.
(VI) How can sanitation be used to leverage the livelihood of BoP in a given sani-shed and at different scales?

(VII) How effectively could the special considerations of Sanitation 4.0 support the 2030 SDG 6 targets?
(VIII) Explore sanitation technological options and solutions for the three different segments of the BoPs.
(IX) Determine indicators for measuring parameters of safely managed resource recovery sanitation facilities such as functionality, serviceability, reliability, acceptability and affordability.
(X) Explore the ICT option for data gathering and management in the Sanitation 4.0 platform.
(XI) Determine the key and relevant knowledge and skills that sanitation professionals and workers (at formal and informal levels) require to fulfil ReGenSan requirements on the Sanitation 4.0 application platform.

REFERENCES

Afacan Y. and Gurel M. O. (2015). Public toilets: an exploratory study on the demands, needs, and expectations in Turkey. *Environment and Planning B: Planning and Design*, **42**, 242–262, doi: 10.1068/b130020p, http://yoksis.bilkent.edu.tr/pdf/files/7424.pdf (accessed 14 August 2018).

AGUASAN (2018) Leveraging the Data Revolution: Informed Decision-Making for Better Water and Sanitation Management. AGUASAN Works, 25–29 June 2018, Spiez, Switzerland. http://www.aguasan.ch/ws2018/Announcement-AGUASAN-Workshop-2018.pdf (accessed 23 July 2018).

Ahmadi M., Cherqui F., Aubin J.B. and Le Gauffre P. (2015) Sewer asset management: impact of sample size and its characteristics on the calibration outcomes of a decision-making multivariate model. *Urban Water Journal*, **13**(1), 41–56, doi:10.1080/1573062X.2015.1011668

Ahmed R. and Hassan S. (2012). Hard-to-Reach Areas: Providing Water Supply and Sanitation Services to All. Supporting Poor-Inclusive Water Supply and Sanitation Sector Reforms. Water and Sanitation Program: Guidance Note. http://documents.worldbank.org/curated/en/374301468208781989/pdf/795500REPLACEM00India0Guidance0Note.pdf (accessed 7 August 2018).

Anastasiadis P. and Metaxas G. (2013). Formulating the principles of an eco-city. *World Transactions on Engineering and Technology Education*, **11**(4), 394–399, http://www.wiete.com.au/journals/WTE&TE/Pages/Vol.11,%20No.4%20(2013)/07-Metaxas-G.pdf (accessed 29 June 2018).

Andersson K., Dickin S. and Rosemarin A. (2016). Towards 'sustainable' sanitation: challenges and opportunities in Urban Areas. *Sustainability*, **8**, 1289, doi: 10.3390/su8121289

Andres L., Boateng K., Borja-Vega C. and Thomas E. (2018). A Review of In-Situ and Remote Sensing Technologies to Monitor Water and Sanitation Interventions. *Water*, **10**, 756, doi: 10.3390/w10060756

Anthony K. H. and Dufresne M. (2007). Potty parity in perspective: gender and family issues in planning and designing public restrooms. *Journal of Planning Literature*, **21**, 267–294.

Ashley C. and Carney D. (1999). Sustainable Livelihoods: Lessons from Early Experience. Department for International Development (DFID), London, UK, pp. 55pp.

Ashley R. and Hopkinson P. (2002). Sewer systems and performance indicators—into the 21st century. *Urban Water*, **4**, 123–135, https://doi.org/10.1016/S1462-0758(02)00010-9

ASHWAS (2011). ASHWAS Process Handbook: Planning and Execution Guide for Participatory Surveys on Household Water and Sanitation. Arghyam, Bengaluru 560 008, Karnataka, India.

Bai X., Surveyer A., Elmqvist T., Gatzweiler F. W., Guneralp B., Parnell S., Prieur-Richard A., Shrivastava P., Siri J. G., Stafford-Smith M., Toussaint J. and Webb R. (2016). Defining and advancing a systems approach for sustainable cities. *Environmental Sustainability*, **23**, 69–78, http://dx.doi.org/10.1016/j.cosust.2016.11.010

Bartram D. (2005). The great eight competencies: a criterion-centric approach to validation. *Journal of Applied Psychology*, **90**, (6), 1185–1203, doi: 10.1037/0021-9010.90.6.1185, https://www.researchgate.net/publication/7453309_The_Great_Eight_Competencies_A_Criterion-Centric_Approach_to_Validation (accessed 4 August 2018).

Bartram J., Brocklehurst C., Fisher M. B., Luyendijk R., Hossain R., Wardlaw T. and Gordon B. (2014). Global monitoring of water supply and sanitation: history, methods and future challenges. *International Journal of Environmental Research and Public Health*, **11**(8), 8137–8165, http://doi.org/10.3390/ijerph110808137 (accessed 27 July 2018).

Baum R., Luh J. and Bartram J. (2013). Sanitation: a global estimate of sewerage connections without treatment and the resulting impact on MDG progress. *Advances in Environmental Science and Technology*, **47**(4), 1994–2000. doi:10.1021/es304284f

Beatley T. (2005). Native to Nowhere: Sustaining Home and Community in a Global Age, 2nd edn. Island Press, Washington D.C., USA, pp. 408.

Beatley T. (2010). Biophilic Cities: Integrating Nature into Urban Design and Planning. Island Press, Washington D.C., USA, 191pp. https://link.springer.com/book/10.5822%2F978-1-59726-986-5 (accessed 12 July 2018).

Beatley T. (2012a). Green Urbanism: Learning from European Cities, First printing edn. Island Press, Washington D.C., USA, pp. 512.

Beatley T. (2012b). Green Cities of Europe: Global Lessons on Green Urbanism. Island Press, Washington D.C., USA, 248 pp. https://doi.org/10.1111/juaf.12053

Beatley T. (2015). Planning for Sustainability in European Cities: A Review of Practice in Leading Cities. In: The City Reader, LeGates R. T. and Frederic (eds). https://www.taylorfrancis.com/books/e/9781317672173/chapters/10.4324%2F9781315770369-45 (accessed 23 July 2017).

Beatley T. and Newman P. (2013). Biophilic cities are sustainable, resilient cities. *Sustainability Journal*, **5**, 3328–3345, (MDPI).

Bichard J., Hanson J. and Greed C. (2004). Inclusive Design of Away from Home (Public) Toilets in City Centres: Pilot Study. Clerkenwell, London VivaCity 2020 EPSRC Research Report, University College, London, UK.

Bill & Melinda Gates Foundation (2006). Landscaping and Review of Approaches and Technologies for Water, Sanitation and Hygiene Opportunities for Action. Main Report Aguaconsult, Cranfield University, and IRC International Water and Sanitation Center. https://www.issuelab.org/resources/5953/5953.pdf (accessed 26 July 2018).

Black M. and Fawcett B. (2008). The Last Taboo: Opening the Door on the Global Sanitation Crisis. Earthscan Publications Ltd, London, UK.

Bocken N. M. P., Short S. W., Rana P. and Evans S. (2014). A literature and practice review to develop sustainable business model archetypes. *Journal of Cleaner Production*, **65**, 42–56.

Brands E. (2014). Prospects and challenges for sustainable sanitation in developed nations: a critical review. *Environmental Reviews*, **22**, 1–18, doi: 10.1139/er-2013-0082

Brewerton P. and Millward L. (2001). Organisational research methods. Sage, London, UK.

Bright-Davies L., Schmidt A., Duma L. and Mbuduka F. (2016). 5 step planning guide which solutions go where? Case Study: Dar es Salaam, Tanzania Solutions for residential & commercial areas. Produced by BORDA.

Brown and Caldwell (1984). Utility infrastructure rehabilitation. NTIS No. PB86-N14642, U.S. Department of Housing and Urban Development, Office of Policy Development and Research, Building Technology Division.

Burton E. and Mitchell L. (2006). Inclusive Urban Design: Streets for Life. Elsevier, Oxford, UK.

Campbell O. M. R., Benova L., Gon G., Afsana K. and Cumming O. (2015). Getting the basic rights – the role of water, sanitation and hygiene in maternal and reproductive health: a conceptual framework. *Tropical Medicine and International Health*, **20**(3), 252–267, doi: 10.1111/tmi.12439

Cavill S., Parkinson J. and de Vette K. (2011). Meeting the Water and Sanitation Millennium Development Goals: a Study of Human Resource Development Requirements in Five Countries. Synthesis report. London, UK: Department for International Development (DFID) [UK] and the International Water Association (IWA). http://www.iwa-network.org/downloads/1422745887-an-avoidable-crisis-wash-gaps.pdf (accessed 23 July 2018).

Christenson E., Bain R., Wright J., Aondoakaa S., Hossain R. and Bartram J. (2014). Examining the influence of urban definition when assessing relative safety of drinking-water in Nigeria. *Science of The Total Environment*, **490**, 301–312.

Cohen J. M. (1980). Integrated rural development: clearing out the underbrush. *Sociologia Ruralis*, **20**, 3, 195–212.

Cook P. (2013). Rural electrification and rural development. In: Rural Electrification through Decentralised off-Grid Systems in Developing Countries, S. Bhattacharyya (ed.), Springer-Verlag, London, pp. 13–38.

Cookey P. E. (2013). Public Health Risk Assessment of Decentralized Wastewater Treatment Systems in Port Harcourt City, Nigeria. Master's thesis, Asian Institute of Technology, Bangkok and Science at the UNESCO-IHE, Delft, Netherlands.

Cookey P. E., Agnes K. B., Richard A. M., Weridongha W. and Chukwuemeka J. P. (2008). Assessment of water supply and sanitation of some coastal communities of the Niger Delta. In: Access to Sanitation and Safe Water – Global Partnerships and Local Actions, H. Jones (ed.). In Proceedings of the 33rd WEDC International Conference, Accra, Ghana, 7–11 April 2008, pp. 383–388. https://dspace.lboro.ac.uk/dspace-jspui/bitstream/2134/28978/2/Cookey_P.pdf (accessed 7 August 2018).

Cookey P. E., Koottatep T., van der Steen P. and Lens P. N. L. (2016a). Public health risk assessment tool: strategy to improve public policy framework for onsite wastewater treatment systems (OWTS). *Journal of Water, Sanitation and Hygiene for Development*, **06.1**, 74–88, doi: 10.2166/washdev.2016.081

Cookey P. E., Darnsawasdi R. and Ratanachai C. (2016b). A Conceptual Framework for Assessment of Governance Performance of Lake Basins: Towards Transformation to Adaptive and Integrative Governance. *Hydrology*, **3** (1), 12, doi: 10.3390/hydrology3010012

Cookey P. E., Darnsawasdi R. and Ratanachai C. (2016c). Local people's perceptions of Lake Basin water governance performance in Thailand. *Ocean & Coastal Management*, **120**, 11–28.

Corcoran E., Nellemann C., Baker E., Bos R., Osborn D. and Savelli H. (eds) (2010). Sick Water? The central role of wastewater management in sustainable development. A Rapid Response Assessment. United Nations Environment Programme, UN-HABITAT, GRID-Arendal. https://gridarendal-website-live.s3.amazonaws.com/production/documents/:s_document/208/original/SickWater_screen.pdf?1486721310 (accessed 23 March 2018).

Crocker J., Shields K. F., Venkataramanan V., Saywell D. and Bartram J. (2016). Building capacity for water, sanitation, and hygiene programming: Training evaluation theory applied to CLTS management training in Kenya. *Social Science & Medicine*, **166**, 66–76, http://dx.doi.org/10.1016/j.socscimed.2016.08.008

CSIR (Council for Scientific and Industrial Research) (2007). Spot check assessments of MIG water and sanitation projects. http://www.dwaf.gov.za/Masibambane/documents/monitoring-evaluation/MIGSpotcheckApr07part1.pdf (accessed 4 June 2013).

Curtis V., Schmidt W., Luby S., Florez R., Touré O. and Biran A. (2011). Hygiene: new hopes, new horizons. *The Lancet Infectious Diseases*, **11**(4), 312–321.

Czischke D., Moloney C. and Turcu C. (2015). State of the Art on Sustainable regeneration in urban areas, URBACT II capitalization. URBACT, Saint Denis, France, http://urbact.eu/sites/default/files/soa_04-final-03.pdf (accessed 29 June 2018).

Delea M. G., Nagel C. L., Thomas E. A., Halder A. K., Amin N., Shoab A. K., Freeman M. C., Unicomb L. and Clasen T. F. (2017). Comparison of respondent-reported and sensor-recorded latrine utilization measures in rural Bangladesh: A cross-sectional study. *Transactions of the Royal Society of Tropical Medicine and Hygiene*, **111**, 308–315.

Diagne N. A. (2013). Evaluation of Sewer Leakage into the Stormwater Drainage System in Singapore. Master's thesis, Massachusetts Institute of Technology, Cambridge, USA. http://hdl.handle.net/1721.1/8281.

Dias B. D. (2015). Beyond sustainability – biophilic and regenerative design in architecture. *European Scientific Journal*. http://eujournal.org/index.php/esj/article/view/5385 (accessed 12 October 2017).

Dreibelbis R., Greene L. E., Freeman M. C., Saboori S., Chase R. P. and Rheingans R. (2013). Water, sanitation, and primary schoolattendance: a multi-level assessment of determinants of household-reportedabsence in Kenya. *International Journal of Educational Development*, **33**, 457–465.

Eales K., Siregar R., Febriani E. and Blackett I. (2013) *Review of Community Managed Decentralized Wastewater Treatment Systems in Indonesia*, Final Report. World Bank Water and Sanitation Program, Washington D.C., USA, http://documents.worldbank.org/curated/en/181941468260140100/Review-of-community-managed-decentralized-wastewater-treatment-systems-in-Indonesia (accessed 23 July 2018).

ECA (European Court of Auditors) (2012) *European Union Development Assistance for Drinking Water Supply and Basic Sanitation in Sub-Saharan C ountries*, Special Report no. 13, doi:10.2865/10664 http://www.

europarl.europa.eu/meetdocs/2009_2014/documents/droi/dv/1405_specialreport_/1405_specialreport_en.pdf (accessed 23 July 2018).

Ekane N., Kjellén M., Nykvist B., Noel S. and Fogde M. (2013). Sanitation Governance viewed through different lenses. *Sustainable Sanitation Practice*, **16**(7/2013), 4–9, https://www.sei.org/publications/sanitation-governance-viewed-through-different-lenses/ (accessed 5 August 2018).

Ekane N., Nykvist B., Kjellén M., Noel S. and Weitz N. (2014). Multi-level sanitation governance: an approach to understanding and overcoming the challenges in the sanitation sector in sub-Saharan Africa. *Waterlines*, **33**, (3), 242–256.

Ekane N., Weitz N., Nykvist B., Nordqvist P. and Noel S. (2016) Comparative assessment of sanitation and hygiene policies and institutional frameworks in Rwanda, Uganda and Tanzania. Stockholm Environment Institute, Sweden (accessed 5 August 2018). https://www.sei.org/mediamanager/documents/Publications/SEI-2016-WorkingPaper-Ekane-Sanitation.pdf (accessed 5 August 2018).

Elledge M. F., Rosensweig F., Warner D. B., Austin J. H. and Perez E. A. (2002). Guidelines for the Assessment of National Sanitation Policies. Strategic Report 2. Prepared for the Office of Health, Infectious Diseases and Nutrition, Bureau for Global Health, U.S. Agency for International Development, under EHP Project 26568/CESH.SANITATION.POLTOOL.

Ellen MacArthur Foundation (2012) Towards the Circular Economy: Opportunities for the consumer goods sector, vol. **1**. https://www.ellenmacarthurfoundation.org/assets/downloads/publications/Ellen-MacArthur-Foundation-Towards-the-Circular-Economy-vol.1.pdf (accessed 7 September 2017).

Ellen MacArthur Foundation (2013) Towards the Circular Economy: Opportunities for the consumer goods sector, vol. **2**. https://www.ellenmacarthurfoundation.org/assets/downloads/publications/TCE_Report-2013.pdf (accessed 7 September 2017).

Ellis F. and Biggs S. (2001). Evolving themes in rural development 1950s – 2000s. *Development Policy Review*, **19**(4), 437–448.

El-Sheikh M. A. (2011). Optimization and upgrading wastewater treatment plants. *Journal of Engineering Sciences*, Assiut University, **39**(4), 697–713, http://www.aun.edu.eg/journal_files/82_J_8366.pdf (accessed 30 April 2017).

EMF (2012) Towards the Circular Economy: Opportunities for the Consumer Goods Sector, vol. 1. Ellen MacArthur Foundation. https://www.ellenmacarthurfoundation.org/assets/downloads/publications/Ellen-MacArthur-Foundation-Towards-the-Circular-Economy-vol.1.pdf (accessed 7 September 2017).

EMF (2013) Towards the Circular Economy: Opportunities for the Consumer Goods Sector, vol. 2. Ellen MacArthur Foundation. https://www.ellenmacarthurfoundation.org/assets/downloads/publications/TCE_Report-2013.pdf (accessed 7 September 2017).

EMF (2015). Towards a Circular Economy: Business Rationale for an Accelerated Transition. Ellen MacArthur Foundation.

EMF (2017). Cities in the circular economy: an initial exploration. https://www.ellenmacarthurfoundation.org/assets/downloads/publications/Cities-in-the-CE_An-Initial-Exploration.pdf (accessed 5 August 2018).

Erpenbeck J. and Rosenstiel L. (2004). Handbuch Kompetenzmessung (Handbook of Competence Measurement). Schäffer-Poeschel Verlag, Stuttgart, Germany, http://www.ciando.com/img/books/extract/3791035126_lp.pdf (accessed 5 August 2018).

Esrey S. A. (2001). Towards a recycling society ecological sanitation – closing the loop to food security. *Water Science and Technology*, **43**(4), 177–87.

Evans B. E. (2003). Sanitation and Hygiene Promotion: Programming Guidance. Water Supply & Sanitation Collaborative Council, WHO.

Fasbender K. (1982). Strategy of integrated development: some remarks on the development of rural areas in developing countries. *Intereconomics*, **17**(6), 291–295.

Fenner R. A. and Saward G. (2004). Towards Assessing Sewer Performance and Serviceability Using Knowledge Based Systems. Ninth International Conference on Urban Drainage (9ICUD). https://ascelibrary.org/doi/pdf/10.1061/40644%282002%2982 (accessed 6 April 2018).

Fonseca C. and Verhoeven J. (2013). Monitoring the Finance needed for Sustainable Service Delivery. IRC Symposium 2013, Addis Ababa, Ethiopia. http://www.irc.nl/page/77982 (accessed 23 July 2018).

Fonseca C., Franceys R., Batchelor C., Mcintyre P., Klutse A., Komives K. and Snehalatha M. (2011). Briefing Note 1a Life-Cycle Costs Approach Costing Sustainable Services. IRC International Water and Sanitation Centre, The Hague, http://www.washcost.info/page/1557 (accessed 5 August 2018).

Freeman M. C., Greene L. E., Dreibelbis R., Saboori S., Muga R., Brumback B. and Rheingans R. (2012). Assessing the impact ofa school-based water treatment, hygiene and sanitation programme on pupilabsence in Nyanza Province Kenya: a cluster-randomized trial. *Tropical Medicine & International Health*, **17**(3), 380–391. https://doi.org/10.1111/j.1365-3156.2011.02927.x

Fullerton J. (2015). Regenerative Capitalism: how universal principles and patterns will shape our new economy. http://capitalinstitute.org/wp-content/uploads/2015/04/2015-Regenerative-Capitalism-4-20-15-final.pdf (accessed 12 June 2018).

Galli G., Nothomb C. and Baetings E. (2014). Towards Systemic Change in Urban Sanitation. IRC Working Paper. IRC, The Hague, https://www.ircwash.org/sites/default/files/201411_wp_towardssyschangeinurbansan_web.pdf (accessed 5 August 2018).

Garn J. V., Greene L. E., Dreibelbis R., Saboori S., Rheingans R. D. and Freeman M. C. (2013). A cluster-randomized trial assessing the impact of school water, sanitation, and hygiene improvements on pupil enrollment and gender parity in enrollment. *Journal of Water, Sanitation and Hygiene for Development*, **3**, http://dx.doi.org/10.2166/washdev.013.217

GHK Research and Training (2000). Strategic Planning for Municipal Sanitation: A guide. Published in association with Water, Engineering and Development Centre (WEDC) and Water and Sanitation Program for South Asia (WSP-SA), London. https://www.ircwash.org/sites/default/files/302.5-00ST-16187.pdf (accessed 13 May 2017).

Gotoh A. (1999). The evaluation of competencies predictors across nine studies in five countries. Unpublished Master's thesis, Goldsmith College, London.

Government of South Australia (2018). Code of Practice for the Provision of Facilities for Sanitation and Personal Hygiene. Department for Health and Ageing, Government of South Australia. https://www.sahealth.sa.gov.au/wps/wcm/connect/df72a18047b216389e61de7c69742d6b/17178.2+Sharps+Disposal+Code+A4+Booklet-ONLINE.pdf?MOD=AJPERES&CACHEID=ROOTWORKSPACE-df72a18047b216389e61de7c69742d6b-m6sJRSt (accessed 14 August 2018).

Greed C. H. (1996). Planning for women and other disenabled groups, with reference to the provision of public toilets in Britain. *Environment and Planning A*, **28**, 573–588.

Greed C. H. (2004). Public toilets: the need for compulsory provision. *Proceedings of the Institution of Civil Engineers–Municipal Engineers*, **157**, 77–85.

Grün G., Tritscher-Archan S. and Weiß S. (2009). Guidelines for the Description of Learning Outcomes. http://ibw4.mservices.at/zoom/pdf/wp2/Guidelines_EN_final_2.pdf (accessed 5 July 2018).

GWP (German Water Partnership) (2009). Skill Development in the Water Sector – Guidelines. German Water Partnership e. V. Reinhardtstr. 32-10117 Berlin, Germany. https://www.germanwaterpartnership.de/fileadmin/pdfs/gwp_materialien/gwp-broschuere_skill_development_in_the_water_sector.pdf (accessed 3 August 2018).

Hammond A. L., Kramer W. J., Katz R. S., Tran J. T. and Walker C. (2007). The Next 4 Billion: Market size and business strategy at the base of the pyramid. https://www.ifc.org/wps/wcm/connect/3c2787004cc75e6094d7b59ec86113d5/Pub_009_The%2BNext%2B4%2BBillion.pdf?MOD=AJPERES (accessed 23 July 2018).

Heidebrecht R. and Mar N. Y. (2012). Skills Challenges in the Water and Wastewater Industry. UNESCO-UNEVOC International Centre for Technical and Vocational Education and Training, Bonn, Germany, https://unevoc.unesco.org/fileadmin/user_upload/docs/WaterBooklet.pdf (accessed 3 August 2018).

Heijnen M., Cumming O., Peletz R., Chan G. K.-S., Brown J., Baker K. and Clasen T. (2014). Shared sanitation vs. individual household latrines: a systematic review of health outcomes. *PLoS One*, **9**, doi: 10.1371/journal.pone.0093300

ISF-UTS and SNV (2016). Are We Doing the Right Thing? Critical Questioning for City Sanitation Planning. Prepared by Institute for Sustainable Futures, University of Technology Sydney and SNV Netherlands Development

Organization. https://www.uts.edu.au/sites/default/files/Arewedoingtherightthing-criticalquestioningforcitysanitationplanning.pdf (accessed 2 May 2017).

IWA (International Water Association) (2014a). AN AVOIDABLE CRISIS: wash human resource capacity gaps in 15 developing economies. http://www.iwa-network.org/downloads/1422745887-an-avoidable-crisis-wash-gaps.pdf (accessed 3 August 2018).

IWA (2014b). Sanitation 21 – a planning framework for improving city-wide sanitation services. Publication supported by Sandec-EAWAG and Switzerland, GIZ, Germany. http://www.iwa-network.org/wp-content/uploads/2016/03/IWA-Sanitation-21_22_09_14-LR.pdf (accessed 23 June 2018).

IWA (International Water Association) (2015). Mapping human resource capacity gaps in the water supply and sanitation sector. Country briefing note Tanzania. http://www.iwa-network.org/wp-content/uploads/2015/12/1422744593-Briefing-Note-Tanzania-LoRes.pdf (accessed 3 August 2018).

Jacob W. J. (2018). Integrated rural development from a historical and global perspective, Asian Education and Development Studies. https://doi.org/10.1108/AEDS-02-2018-0022

Jakimow T. (2012). Peddlers of information: Unintended consequences of information-centered development for north Indian non-government organizations. *VOLUNTAS: International Journal of Voluntary and Nonprofit Organizations*, **23**, 1014–1035.

Jewitt S. (2011). Geographies of shit: Spatial and temporal variations in attitudes towards human waste. *Progress in Human Geography*, **35**(5), 608–626.

Joshi D. K., Hughes B. B. and Sisk T. D. (2015). Improving Governance for the Post-2015 Sustainable Development Goals: Scenario Forecasting the Next 50 years. *World Development*, **70**, 286–302, 2015 0305-750X, http://dx.doi.org/10.1016/j.worlddev.2015.01.013

Kaminsky J. and Javernick-Will A. (2015). Theorizing the Internal Social Sustainability of Sanitation Organizations. *Journal of Water, Sanitation and Hygiene for Development*, **141**(2), 04014071, doi: 10.1061/(ASCE)CO.1943-7862.0000933

Katukiza A. Y., Ronteltap M., Niwagaba C. B., Foppen J. W. A., Kansiime F. and Lens P. N. L. (2012). Sustainable sanitation technology options for urban slums. *Biotechnology Advances*, **30**, 964–978.

Kellert S. R., Heerwagen J. and Mador M. (2008). Biophilic Design: The Theory, Science and Practice of Bringing Buildings to Life. John Wiley & Son Inc., New York, USA.

Kemp G. (2000). Sewerage Assets, Their Condition and Relationship to Performance. Proceedings of IWA/IAHR/CIWEM Joint 1-day Conference on Performance Indicators for Urban Drainage, University of Hertfordshire, 6 December 2000.

Knight G. and Bichard J. (2011). Publicly Accessible Toilets: An Inclusive Design Guide. Helen Hamlyn Centre for Design, London, UK.

Kohavi R., Othleder N. J. R. and Simoudis E. (2002). Evolving data mining into solutions for insights. *Communications of the ACM*, **45**(8), 45–48, http://delivery.acm.org/10.1145/550000/545177/p45-kohavi.pdf?ip=203.159.30.134&id=545177&acc=ACTIVE20SERVICE&key=E2F11C531D2FCAD32E13E298C97051D1C62E4D4702B0C3E38B352E4D4702B0C3E38B35&__acm__=1532335755_ac79ce13dd19485615990e7c12e7f908 (accessed 23 July 2018).

Kumar C. A. and Singh U. (2012). You Manage What You Measure: Using Mobile Phones to Strengthen Outcome Monitoring in Rural Sanitation. Scaling Up Rural Sanitation, Water and Sanitation Program. http://documents.worldbank.org/curated/en/492591468333026927/pdf/NonAsciiFileName0.pdf (accessed 7 August 2018).

Kurz R. (1999). Automated Prediction of Managerial Competencies from Personality and Ability Variables. Proceedings of the BPS Test User Conference, British Psychological Society, Leicester, UK, pp. 96–101.

Kurz R., Bartram D. and Baron H. (2004). Assessing Potential and Performance at Work: The Great Eight Competencies. Proceedings of the British Psychological Society Occupational Conference, British Psychological Society, Leicester, UK, pp. 91–95.

Kvarnström E., McConville J., Bracken P., Johansson M. and Fogde M. (2011). The Sanitation Ladder – a Need for a Revamp? *Journal of Water Sanitation and Hygiene for Development*, **1**(1), 312, https://doi.org/10.2166/washdev.2011.014 (accessed 5 July 2018).

Langergrabera G. and Muelleggera E. (2005). Ecological Sanitation – a way to solve global sanitation problems? *Environment International*, **31**(2005), 433–444.

Larsen T. A., Alder A. C., Eggen R. I., Maurer A. and Lienert J. (2009). Source separation: Will we see a paradigm shift in wastewater handling? *Environmental Science & Technology*, **43**(16), 6121–6125, doi: 10.1021/es803001r

Lehmann S. (2010a). Green Urbanism: Formulating a Series of Holistic Principles. S.A.P.I.EN.S, 3.2. http://sapiens.revues.org/1057 (accessed 5 July 2018).

Lehman S. (2010b). The principles of Green Urbanism: transforming the city for sustainability. Earthscan, London, UK, pp. 911 pp.

Lehman S. (2010c). The principles of Green Urbanism: the post-industrial city. Earthscan, London, UK, pp. 911.

Lenton R. L., Wright A. M. and Lewis K. (2005). Health, Dignity and Development: What Will It Take? UN Milliennium Project, Task Force on Water and Sanitation. http://www.unoosa.org/documents/pdf/psa/activities/2005/graz/watercomplete-lowres.pdf (accessed 27 July 2018).

Leupolt M. (1977). Integrated rural development: key elements of an integrated rural development strategy. *Sociologia Ruralis*, **17**(1/2), 7–28.

Levy Y. and Ellis T. J. (2006). A systems approach to conduct an effective literature review in support of information systems research. *Informing Science and Technology* , **9**, 181–212.

Lopes A. M., Fam D. and Williams J. (2012). Designing sustainable sanitation: Involving design in innovative, transdisciplinary research. *Design Studies*, **33**, 298–317, doi: 10.1016/j.destud.2011.08.005

Lüthi C., Morel A., Tilley E. and Ulrich L. (2011a). Community-Led Urban Environmental Sanitation Planning (CLUES); Swiss Federal Institute of Aquatic Science and Technology (EAWAG) Dübendorf: Switzerland. http://www.susana.org/_resources/documents/default/2-1300-cluesguid.pdf (accessed 14 May 2017).

Lüthi C., Panesar A., Schütze T., Norström A., McConville J., Parkinson J., Saywell D. and Ingle R. (2011b). Sustainable Sanitation in Cities: A Framework for Action . Papiroz Publishing House, Netherlands

Luyet V., Schlaepfer R., Parlange M. B. and Buttler A. (2012). A framework to implement Stakeholder participation in environmental projects. *Journal of Environmental Management*, **111**, 213–219.

Mang P. and Reed B. (2012a). Designing from place: a regenerative framework and methodology. *Building Research and Information*, **40**(1), 23–38.

Mang P. and Reed B. (2012b). Regenerative development and design. Regenesis Group and Story of Place, Encyclo. *Sustainability Science & Technology*, doi: 10.1007/978-1-4614-5828-9_303, https://www.researchgate.net/publication/273379786_Regenerative_Development_and_Design (accessed 5 August 2018).

Mang P., Haggard B. and Regenesis (2016). Regenerative Development and Design: A Framework for Evolving Sustainability. John Wiley & Sons, Inc., New York, USA.

Mara D. (2003). Water, sanitation and hygiene for the health of developing nations. *American Journal of Public Health*, **117**, 452–456.

Marshall R. E. and Farahbakhsh K. (2013). Systems approaches to integrated solid waste management in developing countries. *Waste Management*, **33**, 988–1003, http://dx.doi.org/10.1016/j.wasman.2012.12.023

Martínez-Jurado P. J. and Moyano-Fuentes J. (2014). Lean Management, Supply Chain Management and Sustainability: A Literature Review. *Journal of Cleaner Production*, **85**, 134–150.

Mason N., Le Sève M.D. and Calow R. (2017) Future flows Global trends to watch on water and sanitation. ODI Working Paper 520. Overseas Development Institute, London. https://www.odi.org/sites/odi.org.uk/files/resource-documents/11721.pdf (accessed 30 July 2018).

McGrath S. and Powell L. (2016). Skills for sustainable development: Transforming vocational education and training beyond 2015. *International Journal of Educational Development*, **50**, 12–19, http://dx.doi.org/10.1016/j.ijedudev.2016.05.006

Miszta-Kruk K. (2016). Reliability and failure rate analysis of pressure, vacuum and gravity sewer systems based on operating data. *Engineering Failure Analysis*, **61**, 37–45, http://dx.doi.org/10.1016/j.engfailanal.2015.07.034

Mitchell C. and Ross K. (2016). Governance of local scale sanitation: How to design governance for lasting service? Guidance Material: Introduction. Prepared by the Institute for Sustainable Futures, University of Technology Sydney, as part of the Australian Development Research Award Scheme (ADRAS) Project: Effective governance for the successful long-term operation of local scale wastewater systems.

Mitchell C., Ross K., Abeysuriya K., Puspowardoyo P. and Wedahuditama F. (2015). Effective governance for the successful long-term operation of community scale air limbah systems: Mid-term observations report. Prepared by the Institute for Sustainable Futures, University of Technology Sydney as part of the Australian Scheme (ADRAS) Project: Effective governance for the successful long-term operation of local scale wastewater systems. https://www.academia.edu/30675424/Effective_Governance_for_The_Successful_Long_Term_Operation_of_Local_Scale_Wastewater_Systems_Review_of_Regulatory_Framework_for_Local_Scale_Air_Limbah_ (accessed 5 June 2018).

Mitra S., Dangwal R. and Thadani L. (2008). Effects of remoteness on the quality of education: A case study from north Indian schools. *Australasian Journal of Educational Technology*, **24**(2), 168–180.

Mjoli N. P. (2015) A Framework for Sanitation Governance in South African Municipalities. Report to the Water Research Commission by Hlathi Development Services. WRC Report No. TT 629/15. http://www.wrc.org.za/Knowledge%20Hub%20Documents/Research%20Reports/TT%20629-15.pdf (accessed 6 August 2018).

Morella E., Foster V. and Ghosh Banerjee S. (2008). Climbing the Ladder: The State of Sanitation in Sub-Saharan Africa. Background Paper 13 (Phase 1). Africa Infrastructure Country Diagnostic. Ministry of Health/WSP (2004). Strengthening budget mechanisms for sanitation in Uganda. http://www.wsp.org/sites/wsp.org/files/publications/af_strengtheningbudget_execsum.pdf

Morgan C., Bowling M., Bartram J. and Kayser G. L. (2017). Water, sanitation, and hygiene in schools: Status and implications of low coverage in Ethiopia, Kenya, Mozambique, Rwanda, Uganda, and Zambia. *International Journal of Hygiene and Environmental Health*, **220**, 950–959, http://dx.doi.org/10.1016/j.ijheh.2017.03.015 (accessed 14 August 2018).

Mortel T. F. V. D. (2008). Faking it: Social desirability response bias in self – report research. *The Australian Journal of Advanced Nursing* , **25**, 40–48.

Murray A. and Buckley C. (2009). Designing Reuse-Oriented Sanitation Infrastructure: The Design for Service Planning Approach. In: Wastewater Irrigation and Health Assessing and Mitigating Risk in Low-income Countries, C.A. S. Drechsel, L. Raschid-Sally, M. Redwood and A. Bahri (eds), Earthscan, London, UK.

Nath V., Saxena S., Shiva R. and Prakash M. (2017) Skill gap in sanitation sector: challenges & opportunities. This document is a product of the Insights Series; developed by IRC, India Sanitation Coalition and TARU Leading Edge. https://www.ircwash.org/sites/default/files/insights3_skilling_in_sanitation_ver_final_2.0.pdf (accessed 3 August 2018).

Navarro R. G. (1994). Improving Sanitation in Coastal Communities with Special Reference to Puerto Princesa, Palawan Province, Philippines. PhD thesis, McGill University, Montreal, Canada. http://www.bvsde.paho.org/texcom/cd050999/navarro.pdf

Ndaw M. F. (2015). Unlocking the Potential of Information Communications Technology to Improve Water and Sanitation Services: Summary of Findings and Recommendations. http://www.worldbank.org/en/topic/water/publication/unlocking-the-potential-of-information-communications-technology-to-improve-water-and-sanitation-services (accessed 24 July 2018).

Nnaji C. C., Eluwa C. and Nwoji C. (2013). Dynamics of domestic water supply and consumption in a semi-urban Nigerian city. *Habitat International*, **40**, 27–135.

Novotný J., Hasman J. and Lepič M. (2018). Contextual factors and motivations affecting rural community sanitation in low- and middle-income countries: A systematic review. *International Journal of Hygiene and Environmental Health*, **221**, 121–133, https://doi.org/10.1016/j.ijheh.2017.10.018

NOWRA (National Onsite Wastewater Recycling Association) (2007). Model Code Framework for the Decentralized Wastewater Infrastructure. Volume I: Workbook for Writing the Code; Volume II: Code Design Philosophy and Guidance. http://082511a.membershipsoftware.org/files/About%20NOWRA/Model%20Code/VOLUMEone-WorkbookfortheCode.pdf (accessed 16 March 2018).

O'Reilly K., Dhanju R. and Goel A. (2017). Exploring 'The Remote' and 'The Rural': Open Defecation and Latrine Use in Uttarakhand, India. *World Development*, **93**, 193–205, http://dx.doi.org/10.1016/j.worlddev.2016.12.022

Ostrom E. (2005). Understanding Institutional Diversity. Princeton University Press, Princeton, USA.

Pahl-Wostl C. (2009). A conceptual framework for analysing adaptive capacity and multi-level learning processes in resource governance regimes. *Global Environmental Change*, **19**(3), 354–365, doi: 10.1016/j.gloenvcha.2009.06.001

Pahl-Wostl C., Holtz G., Kastens B. and Knieper C. (2010). Analyzing complex water governance regimes: the Management and Transition Framework. *Environmental Science & Policy*, **13**, 571–581.

Pahl-Wostl C., Jeffrey P., Isendahl N. and Brugnach M. (2011). Maturing the New Water Management Paradigm: Progressing from Aspiration to Practice. *Water Resources Management*, **25**, 837–856, doi: 10.1007/s11269-010-9729-2

Park D. (2017) Analysis vs. Analytics: Past vs. Future. *EE Times*. https://www.eetimes.com/author.asp?section_id=36&doc_id=1332172 (accessed 23 July 2018).

Park J. Y. and Chertow M. R. (2014). Establishing and testing the 'reuse potential' indicator for managing wastes as resources. *Journal of Environmental Management*, **137**, 45–53, http://dx.doi.org/10.1016/j.jenvman.2013.11.053

Parkinson J.N., Goldenfum J.A. and Tucci C. E. M. (eds). (2009). Integrated Urban Water Management: Humid Tropics. Urban Water Series 6-UNESCO-IHP. United Nations Educational Scientific Organization (UNESCO), Paris, France and Taylor and Francis, Leiden, Netherlands. ISBN 978-92-3-104065-8. ISSN 1749-0790.

Parkinson J., Lüthi C., Walther D. (GIZ) (2014) Sanitation21 – A Planning Framework for Improving City-wide Sanitation Services. IWA, EAWAG-Sandec, GIZ. http://www.iwa-network.org/filemanager-uploads/IWA-Sanitation-21_22_09_14-LR.pdf (accessed 2 May 2017).

Pedi D. and Davies W. (2013). Transforming Markets, Increasing Access: Early Lessons on Base-of-the-Pyramid Market Development in Sanitation. SMARTLESSONS – OCTOBER 2013. https://openknowledge.worldbank.org/bitstream/handle/10986/18691/843480BRI0Box3000Selling0Sanitation.pdf?sequence=1&isAllowed=y (accessed 25 July 2018).

Piana V. (2010). Urban Regeneration. Economics Web Institute. http://www.economicswebinstitute.org/glossary/urbanregeneration.htm (accessed 5 May 2018).

Plummer J. and Slaymaker T. (2007) Rethinking governance in water services. Working Paper No. 284, ODI, London.

Potter A., Klutse A., Snehalatha M., Batchelor C., Uandela A., Naafs A., Fonseca C. and Moriarty P. (2011) Assessing sanitation service levels. IRC International Water and Sanitation Centre, Working Paper 3, 2nd edn. https://www.ircwash.org/sites/default/files/Potter-2011-Assessing.pdf (accessed 18 May 2017).

Prahalad C. K. and Hart S. L. (2000). Raising the Bottom of the Pyramid: Strategies for Sustainable Growth. http://www.bus.tu.ac.th/usr/wai/xm622/conclude%20monsanto/Bottom%2520of%2520the%2520pyramid.pdf (accessed 25 July 2018).

Prüss A., Kay D., Fewtrell L. and Bartram J. (2002). Estimating the burden of disease from water, sanitation, and hygiene at a global level. *Environmental Health Perspectives*, **110**, 537–542.

Qasem A., Zayed T. and Chen Z. (2009). A condition rating system for wastewater treatment plants infrastructures. *World Academy of Science, Engineering and Technology International Journal of Environmental and Ecological Engineering*, **3**, 4, https://waset.org/publications/15241/a-condition-rating-system-for-wastewater-treatment-plants-infrastructures (accessed 23 March 2018).

Quendler E., Van der Luit J., Monteteone M., Aguado P., Pfeiffenscheider M., Wagner K., Valente F. and Cunha-Queda C. (2013). Employers' needs on Competences, Knowledge and Skills for Sustainable Development as a Reference Framework for Higher Education in Life Sciences. *Procedia – Social and Behavioral Sciences*, **106**, 1063–1085, https://doi.org/10.1016/j.sbspro.2013.12.121

Ramani S. V., SadreGhazi S. and Duysters G. (2012). On the diffusion of toilets as bottom of the pyramid innovation: Lessons from sanitation entrepreneurs. *Technological Forecasting & Social Change*, **79**, 676–687, doi: 10.1016/j.techfore.2011.06.007

Rangan V.K., Chu M. and Petkoski D. (2011) The globe: segmenting the base of the Pyramid. *Harvard Business Review*, **June**, Issue. Harvard Business School Publishing. https://hbr.org/2011/06/the-globe-segmenting-the-base-of-the-pyramid (accessed 5 August 2018).

Reymond P., Renggli S. and Lüthi C. (2016) Towards sustainable sanitation in an urbanising world, sustainable urbanization. In: I. Mustafa Ergen, Dr. (ed.) InTech, doi:10.5772/63726. https://www.intechopen.com/books/sustainable-urbanization/towards-sustainable-sanitation-in-an-urbanising-world (accessed 5 June 2018).

Reynolds J. H. and Barrett M. H. (2003). A review of the effects of sewer leakage on groundwater quality. *Water and Environment Journal*, **17**(11), 34–39, https://doi.org/10.1111/j.1747-6593.2003.tb00428.x

Rodgers A. F., Ajono L. A., Gyapong J. O., Hagan M. and Emerson P. M. (2007). Characteristics of latrine promotion participants and non-participants; inspection of latrines; and perceptions of household latrines in northern Ghana. *Tropical Medicine & International Health*, **12**(6), 772–782.

Rosenqvist T., Mitchell C. and Willetts J. (2016). A short history of how we think and talk about sanitation services and why it matters. *Journal of Water, Sanitation and Hygiene for Development*, **06**(2), doi: 10.2166/ washdev. 2016.118

Ross K., Abeysuriya K., Mikhailovich N. and Mitchell C. (2014) Governance for Decentralized Sanitation: Global Practice Scan. A Working Document, Prepared by the Institute for Sustainable Futures, UTS as part of the Australian Development Research Award Scheme (ADRAS).

Rowe G. and Frewer L. J. (2000). Public participation methods: a framework for evaluation. *Science Technology & Human Values*, **25**(1), 3–29.

Schmitt R. J. P., Morgenroth E. and Larsen T. A. (2017). Robust planning of sanitation services in urban informal settlements: An analytical framework. *Water Research*, **110**, 297–312, http://dx.doi.org/10.1016/j.watres.2016.12.007

SDKP (Sustainable development knowledge platform) (2017). CSD Multi-Year Programme of Work 2004/2005 to 2016/2017. https://sustainabledevelopment.un.org/index.php?menu=1483 (accessed 11 August 2018).

Seppälä O. (2002). Effective water and sanitation policy reform implementation: need for systemic approach and stakeholder participation. *Water Policy*, **4**(4), 367–388.

Sethi V. and King W. R. (1998). Organizational transformation through business process reengineering. Prentice Hall, Upper Saddle River, USA.

Shiell A., Hawe P. and Gold L. (2008) Complex interventions or complex systems? Implications for health economic evaluation. *BMJ*, **336**, 1281–1283. http://dx.doi.org/10.1136/bmj.39569.510521.AD

Shove E. (2003). Comfort, Cleanliness and Convenience: The Social Organization of Normality. Berg, Oxford, UK.

Sparkman D. and Sturzenegger G. (2018) *The Need for a Sanitation Revolution in LAC: Conclusions from World Water Week*. Infrastructure and Energy Sector, Water and Sanitation Division. Technical Note N° IDB-TN-1394. Inter-American Development Bank. https://publications.iadb.org/bitstream/handle/11319/8891/The-Need-for-a-Sanitation-Revolution-in-LAC-Conclusions-from-World-Water-Week.pdf?sequence=1&isAllowed=y (accessed 30 July 2018).

Starkl M., Brunner N. and Stenstrom T. (2013). Why do water and sanitation systems for the poor still fail? Policy analysis in economically advanced developing countries. *Advances in Environmental Science and Technology*, **47**(12), 6102–6110. doi: 10.1021/es3048416

Starkl M., Brunner N., Feil M. and Hauser A. (2015). Ensuring sustainability of non-networked sanitation technologies: an approach to standardization. *Advances in Environmental Science and Technology*, **49**, 6411–6418. doi:10.1021/acs.est.5b00887.

Stockbridge M. and Dorward A. (2018). Rural Development. P530: Module Introduction of Centre for Development, Environment and Policy. SOAS University of London, https://www.soas.ac.uk/cedep-demos/000_P530_RD_K3736-Demo/module/pdfs/p530_module_introduction.pdf (accessed 12 August 2018).

Sumser J. (2017) People analytics. *HR Examiner*. https://www.hrexaminer.com/people-analytics-mit-july-24-2017/ (accessed 23 July 2018).

UN-Water (2008). Tackling a Global Crisis: International Year of Sanitation 2008. https://esa.un.org/iys/docs/IYS_flagship_web_small.pdf (accessed 8 August 2018).

UN-Water (2012). UN-Water Global Annual Assessment of Sanitation and Drinking-Water (GLAAS) 2012 Report: The Challenge of Extending and Sustaining Services. UN-Water and World HealthOrganization (WHO), Geneva, Switzerland.

UN-Water (2016). Integrated Monitoring Guide for SDG 6 Targets and global indicators. http://webcache.googleusercontent.com/search?q=cache:j3BHSIpS7fIJ:www.unwater.org/app/uploads/2017/03/SDG-6-targets-and-global-indicators_2016-07-19.pdf+&cd=6&hl=en&ct=clnk&gl=th&client=firefox-b (accessed 8 August 2018).

UNDG (United Nations Development Group) (2003). Indicators for Monitoring the Millennium Development Goals. ST/ESA/STAT/SER.F/95. United Nations, New York.

UNDP (2006). Human Development Report 2006: Beyond Scarcity: Power, Poverty and the Global Water Crisis. Palgrave Macmillan, Basingstoke, New York, http://www.undp.org/content/dam/undp/library/corporate/HDR/200620Global20HDR/HDR-2006-Beyond%20scarcity-Power-poverty-and-the-global-water-crisis.pdf (accessed 5 August 2018).

UNDP (2009a). Capacity Development: A UNDP Primer. United Nations Development Programme, New York, USA, http://www.undp.org/content/dam/aplaws/publication/en/publications/capacity-development/capacity-development-a-undp-primer/CDG_PrimerReport_final_web.pdf (accessed 3 August 2018).

UNDP (2009b). Handbook on Planning, Monitoring and Evaluating for Development Results. United Nations Development Programme, One United Nations Plaza, New York, USA, http://web.undp.org/evaluation/evaluations/handbook/english/documents/pme-handbook.pdf (accessed 14 May 2017).

UNDP (2014). Human Development Report 2014: Sustaining Human Progress – Reducing Vulnerability and Building Resilience. United Nations Development Programme, New York.

UNDP (2016). UNDP Support to the Implementation of the 2030 Agenda. http://www.undp.org/content/dam/undp/library/SDGs/SDG%20Implementation%20and%20UNDP_Policy_and_Programme_Brief.pdf (accessed 20 June 2018).

UNDP-SIWI Water Governance Facility (2016). WGF Issue sheet: Sanitation Governance. SIWI, Stockholm, http://www.watergovernance.org/resources/no5-sanitation-governance/ (accessed 5 August 2018).

UNICEF (2014). Sanitation Monitoring Toolkit. http://www.communityledtotalsanitation.org/sites/communityledtotalsanitation.org/files/UNICEF_SanitationMonitoring_Toolkit.pdf (accessed 14 May 2017).

UNICEF/WHO (2015) United Nations Children's Fund/World Health Organization. Progress on Sanitation and Drinking Water: 2015 Update and MDG Assessment, UNICEF/WHO, New York, USA. http://mdgs.un.org/unsd/mdg/Resources/Attach/Indicators/HandbookEnglish.pdf (accessed 8 August 2018).

USEPA (U.S. Environmental Protection Agency) (1991). Sewer System Infrastructure Analysis and Rehabilitation. EPA 625/6-91/030. U.S. Environmental Protection Agency, Office of Research and Development, Cincinnati, USA.

Van Vliet B. J. M., Spaargaren G. and Oosterveer P. (2011). Sanitation under challenge: contributions from the social sciences. *Water Policy*, **13**, 797–809.

Van Welie M. J. and Romijn H. A. (2018). NGOs fostering transitions towards sustainable urban sanitation in low income countries: Insights from Transition Management and Development Studies. *Environmental Science & Policy*, **84**, 250–260, https://doi.org/10.1016/j.envsci.2017.08.011

WaterAid (2011) A Study on working with Parliament towards improving WASH governance in Uganda. www.wateraid.org https://washmatters.wateraid.org/sites/g/files/jkxoof256/files/LICSU%20Uganda%20case%20study_1.pdf (accessed 6 August 2018).

Webbler T., Tuler S., Shockey I., Stern P. and Beattie R. (2003). Participation by local government officials in watershed management planning. *Journal Society & Natural Resources*, **16**, 105–121.

WEDC (2005) Comparing National Sanitation Policy Content: An Initial Review of Nine Country Profiles. Briefing Note: Review. WEDC, London, UK.

WELL (1998). Guidance Manual on Water Supply and Sanitation Programmes. Department for International Development, London, UK, https://www.kpesic.com/sites/default/files/guia20manual20de%20sanemamiento.pdf (accessed 8 August 2018).

Welle K., Nicol A. and van Steenbergen F. (2008). Why is Harmonisation and Alignment Difficult for Donors? Lessons from the Water Sector. ODI – Overseas Development Institute, London, (ODI Project Briefing 6) [pdf], metameta.nl/wp-content/uploads/2012/07/ha-in-water-sector.pdf.

WHO (2017) Water and Sanitation for Health Facility Improvement Tool (WASH FIT). Licence: CC BY-NC-SA 3.0 IGO, World Health Organization, Geneva, Switzerland. https://reliefweb.int/sites/reliefweb.int/files/resources/9789241511698-eng.pdf (accessed 14 August 2018).

WHO (2018). World health Statistics: Monitoring Health for the SDGs, Sustainable Development Goals. World Health Organization, Geneva, Switzerland. ISBN 978-92-4-156558-5.

WHO/UNICEF (2017a) Progress on Drinking Water, Sanitation and Hygiene: 2017 Update and SDG Baselines. Licence: CC BY-NC-SA 3.0 IGO. World Health Organization (WHO) and the United Nations Children's Fund (UNICEF), Geneva, Switzerland.

WHO/UNICEF (2017b) washdata_Trends-in urban sanitation service levels (2000–2015) _20180629142634 https://washdata.org/data#!/dashboard/1324 (accessed/modified new grid 29 June 2018).

WHO/UNICEF (2017c) washdata_Trends-in rural sanitation service levels (2000–2015) https://washdata.org/data#!/dashboard/new (accessed 12 August 2018).

Wikipedia Contributors (2018). Big Data. In: Wikipedia, The Free Encyclopedia. https://en.wikipedia.org/w/index.php?title=Bigdata&oldid=875634829 (accessed 23 July 2018).

Winker M., Vinnerås B., Muskolus A., Arnold U. and Clemens J. (2009). Fertiliser products from new sanitation systems: Their potential values and risks. *Bioresource Technology*, **100**, 4090–4096, doi: 10.1016/j.biortech.2009.03.024

Wolf J., Bonjour S. and Prüss-Ustün A. (2013). An exploration of multilevel modeling for estimating access to drinking-water and sanitation. *Journal of Water and Health*, **11**, 64–77.

Wood L., Egger M., Gluud L. L., Schulz K. F., Jüni P., Altman D. G., Gluud C., Martin R. M., Wood A. J. G. and Sterne J. A. C. (2008). Empirical evidence of bias in treatment effect estimates in controlled trials with different interventions and outcomes: Meta-Epidemiological study. *British Medical Journal*, **336**, 601–605.

World Water Council (WWC) (2012). Priority Action 1.2: Improve access to integrated sanitation services for all. 6th World Water Forum, Coordinated by International Water Association. http://www.worldwaterforum6.org/commissions/thematic/priorities-for-action-and-conditions-for-success/priority-for-action-12/ (accessed 15 June 2017).

WRC (Water Research Commission) (2002) *Development of generic and sectoral competencies in the water supply and sanitation training sector Report to the Water Research Commission by National Community Water and Sanitation Training Institute (NCWSTI)*, WRC Report No. 1020/1/02. http://www.wrc.org.za/Knowledge%20Hub%20Documents/Research%20Reports/1020-1-02.pdf (accessed 4 August 2018).

WSP (Water and Sanitation Program) (2007). Community-led total sanitation in rural areas. World Bank, New Delhi, India.

WSP (Water and Sanitation Program) (2003) Governance and Financing of Water Supply and Sanitation in Ethiopia, Kenya and South Africa: A Cross Country Synthesis. Sector Finance Working Papers: No. 5. http://documents.worldbank.org/curated/en/744961468036303198/pdf/463620WSP0Box31UBLIC10af1governance.pdf (accessed 6 August 2018).

WSP (Water and Sanitation Program) (2011). Sanitation Markets at the Bottom of the Pyramid A Win-Win Scenario for Government, the Private Sector, and Communities. Learning Note: Domestic Private Sector Participation in Peru. https://www.wsp.org/sites/wsp.org/files/publications/WSP-Sanitation-Markets-Bottom-Pyramid-Peru.pdf (accessed 25 July 2018).

WSP (Water and Sanitation Program) (2012) What Does It Take to Scale Up Rural Sanitation? Water and Sanitation Programme: Working Paper – Scaling Up Rural Sanitation. https://www.wsp.org/sites/wsp.org/files/publications/WSP-What-does-it-take-to-scale-up-rural-sanitation.pdf (accessed 29 July 2018).

WSP (Water and Sanitation Program) (2014). Scaling Up Indonesia's Rural Sanitation Mobile Monitoring System Nationally. Water and Sanitation Program: Scaling Up Rural Sanitation Learning Note, December 2014. https://www.wsp.org/sites/wsp.org/files/publications/WSP-Indonesia-Mobile-Monitoring.pdf (accessed 24 July 2018).

Yarime M. and Mutisya E. (2011). Investing at the Bottom of the Pyramid for Sustainable Development. https://ourworld.unu.edu/en/investing-at-the-bottom-of-the-pyramid-for-sustainable-development (accessed 25 July 2018).

Zaręba A., Krzemińska A. and Widawski K. (2016). Green urbanism for the greener future of metropolitan areas. *IOP Conference Series: Earth and Environmental Science*, **44**(5), 20–62, doi: 10.1088/1755-1315/44/5/052062, http://iopscience.iop.org/article/10.1088/1755-1315/44/5/052062/pdf (accessed 29 June 2018).

Index

E

F

Y

Z

IWA Publishing's authorised EU representative for General Product Safety Regulations is Diane D'Arras, 15 rue Duret, 75116 Paris, France, e-mail: safety@iwap.co.uk.

Printed and bound by CPI Group (UK) Ltd, Croydon, CR0 4YY
12/05/2026
02108870-0013